Communications and Control Engineering

Communications and Control Engineering is a high-level academic monograph series publishing research in control and systems theory, control engineering and communications. It has worldwide distribution to engineers, researchers, educators (several of the titles in this series find use as advanced textbooks although that is not their primary purpose), and libraries.

The series reflects the major technological and mathematical advances that have a great impact in the fields of communication and control. The range of areas to which control and systems theory is applied is broadening rapidly with particular growth being noticeable in the fields of finance and biologically-inspired control. Books in this series generally pull together many related research threads in more mature areas of the subject than the highly-specialised volumes of *Lecture Notes in Control and Information Sciences*. This series's mathematical and control-theoretic emphasis is complemented by *Advances in Industrial Control* which provides a much more applied, engineering-oriented outlook.

Indexed by SCOPUS and Engineering Index.

Publishing Ethics: Researchers should conduct their research from research proposal to publication in line with best practices and codes of conduct of relevant professional bodies and/or national and international regulatory bodies. For more details on individual ethics matters please see:
https://www.springer.com/gp/authors-editors/journal-author/journal-author-helpdesk/publishing-ethics/14214

More information about this series at http://www.springer.com/series/61

Christoforos N. Hadjicostis

Estimation and Inference in Discrete Event Systems

A Model-Based Approach with Finite Automata

Springer

Christoforos N. Hadjicostis
Department of Electrical
and Computer Engineering
University of Cyprus
Nicosia, Cyprus

ISSN 0178-5354 ISSN 2197-7119 (electronic)
Communications and Control Engineering
ISBN 978-3-030-30820-9 ISBN 978-3-030-30821-6 (eBook)
https://doi.org/10.1007/978-3-030-30821-6

This Springer imprint is published by the registered company Springer Nature Switzerland AG
The registered company address is: Gewerbestrasse 11, 6330 Cham, Switzerland

To my family.

Preface

This book addresses the growing need for systematic state estimation and event inference techniques, as well as efficient verification of related properties, in discrete event systems (DES). Such systems have become prevalent in the past three decades, primarily due to the proliferation of digital technologies, interconnectivity, and sensor technology. These developments have led to the emergence of many highly complex systems, in which a significant portion of the activity is determined by rules designed by human engineers. Examples include automated manufacturing systems, communication networks and transmission protocols, traffic control and transportation systems, autonomous vehicles, digital controllers for automotive and printing devices, and others. The dynamics of such systems are driven by possibly asynchronous occurrences of discrete events, some of which are controlled (e.g., conscious decisions by the system controller or the system user) and some of which are uncontrolled (e.g., unexpected disturbances by the environment or unauthorized users). As a result, when DES are coupled with dynamic equations, they can capture, completely or at some level of abstraction, many human-designed technological systems including several emerging cyber-physical systems.

The above applications span a number of disciplines (e.g., systems and control, communications, computer science and engineering, manufacturing engineering, to name a few), but a common denominator in all of the above examples is that an accurate model of the underlying system (at least for part of its activity) is available to the engineers that built it. The model-based techniques for state estimation and event inference in DES that are described in this book have been brewing over the past three decades. Pieces of them can be found in research papers and existing DES books, but a single reference that includes such discussions was largely absent.

The book is suitable for providing a basic introduction to DES to undergraduate and graduate student audiences, as well as practitioners, who are interested in learning more about these systems, and related techniques for state estimation, event inference, and the verification of properties of interest, such as detectability, diagnosability, or opacity. The book could certainly be used as a textbook for a first-year graduate course in (state estimation and event inference in) DES. Parts of the book could also be used for a senior-level undergraduate course (for example, by excluding the later material on decentralized and distributed estimation/ inference).

One of the objectives in writing this book was to provide a comprehensive, yet relatively concise, treatment of the subjects of state estimation and event inference. This is one of the reasons for focusing the discussions on systems that are modeled as deterministic or nondeterministic finite automata, which allows the reader who is interested in these topics (e.g., a graduate student or researcher) to quickly obtain sufficient background to pursue these issues further, perhaps in other types of DES (e.g., Petri nets) where similar techniques can be used. Therefore, despite the fact that the primary purpose of writing the book was to provide a self-sufficient exposition of the material, discussions, and pointers to related work in other types of DES are provided at the end of each chapter for the reader's benefit. The intention was not to provide a comprehensive list of the relevant literature, as this would be a formidable task given the large and continuously growing literature in the field. Clearly, the references provided at the end of each chapter are rather focused and by no means complete, and we apologize in advance for any omissions.

Contents of the Book and Organization of the Material

The book focuses on estimation and inference techniques for DES that are modeled as deterministic or nondeterministic finite automata. The organization of the material in the book follows a linear structure, starting from some motivating discussions, moving on to online state estimation, then exploring offline verification of several properties of interest, and finally concluding with more advanced topics on decentralized/distributed state estimation.

- Chapter 1 serves as a motivational chapter that situates the discussions in the book within the more general context of DES research (in state estimation and event inference, but also in other relevant DES research topics). In particular, this chapter provides pointers and identifies related literature that the interested reader can further investigate.
- Chapters 2 and 3 introduce some necessary notation for capturing finite automata models in terms of their input/output behavior and their state transition functionality. Particular attention is paid to the observation models in these automata (e.g., the presence of silent or unobservable transitions).
- In Chap. 4, we develop state estimation techniques, focusing on three different types of state estimation tasks: current-state estimation, delayed-state estimation (or smoothing), and initial-state estimation. This chapter essentially develops recursive (online) state estimation algorithms for current-, delayed-, or initial-state estimation. In all cases, a distinction is made between finite automata without silent transitions and finite automata with silent transitions.

- Chapter 5 is an introduction to the topic of verification of observation-related properties. It focuses on the analysis of so-called state isolation properties, which capture our ability to determine, following any feasible sequence of observations, that the state (current, delayed, or initial) of the given finite automaton falls within a given set of states of interest.
- In the next three chapters, we discuss three important system properties and ways to verify them. More specifically, Chap. 6 discusses several variants of detectability, i.e., our ability to be in position to (almost always) determine exactly the state of the system; Chap. 7 discusses diagnosability, i.e., our ability to eventually infer that a certain type of event has occurred or not; and Chap. 8 discusses opacity, i.e., our inability to conclude, based on the observations we make about a given system, that the system has necessarily executed secret behavior.
- Extensions to decentralized and distributed observation settings are addressed in Chaps. 9 and 10, respectively. In these settings, there are multiple observation points, which may communicate with a coordinator (decentralized setting) or among themselves (distributed setting). The two chapters discuss and analyze the implementation and verification of several protocols that can be used for decentralized/distributed state estimation and event inference.

There are many ways in which readers can study the material, depending on their interests and their familiarity with DES modeled as finite automata. Below we discuss some of these possible ways.

1. One should certainly obtain an understanding of the notation by reading (or at least skimming through) Chaps. 2 and 3. The amount of time a reader spends on this material depends on the familiarity of the reader with these concepts.
2. The next two chapters, namely Chaps. 4 and 5, comprise the core of the book: the former chapter discusses online (recursive) state estimation and event inference, whereas the latter chapter discusses offline verification of a class of properties that relate to state estimation.
3. The next three chapters could be read rather independently, depending on the interests of a reader. For example, a reader who is interested in opacity could jump directly to Chap. 8 (skipping Chaps. 6 and 7, which focus on detectability and fault diagnosis, respectively). Similarly, a reader who is interested in fault diagnosis can focus on Chap. 7, and skip the other two chapters.
4. The last two chapters focus on decentralized/distributed state estimation and event reference. We expect that the reader will first study Chap. 9 and then Chap. 10, though this is not strictly necessary. A reader can read these two chapters without explicitly studying detectability, diagnosis, or opacity in Chaps. 6–8 (though some parts of the discussions in Chaps. 9 and 10 discuss decentralized/distributed verification of these properties).

The hope is that the readers will find the contents of the book interesting and inspiring for further work in the exciting area of DES. Clearly, in a project of this size, there is always room for improvement and comments/suggestions are most welcome. Finally, I would like to take this opportunity to express my gratitude to numerous persons and organizations, whose help and assistance made the completion of this book possible. I am grateful to all of them, particularly to Oliver Jackson of Springer, for his encouragement, patience, support, and enthusiasm throughout the later stages of the writing of this book.

Nicosia, Cyprus
July 2019

Christoforos N. Hadjicostis

Acknowledgements

This book has grown out of research work of more than two decades, starting from the Massachusetts Institute of Technology and continuing at the University of Illinois at Urbana-Champaign and the University of Cyprus. There are many colleagues, students, and friends who have been extremely generous with their advice and help during these years, and to whom I am indebted, since the writing of this book would not have been made possible without them.

In particular, I would like to express my most sincere thanks to George Verghese for directing me toward DES research and for being an inspirational role model throughout my research career. I would also like to thank Jan van Schuppen and Carla Seatzu, for their enthusiasm for this book and for reading earlier versions of many chapters, as well as Christos Cassandras and Stephane Lafortune for offering encouragement to pursue this writing project. In addition, I am grateful to my colleagues at the University of Illinois and the University of Cyprus, several of whom played a key supportive role in my efforts while preparing the book. Finally, I would like to take this opportunity to acknowledge my many DES colleagues, collaborators, and friends, who offered insightful suggestions for my DES research and/or the writing of the book. I am grateful to all of them.

There are also several graduate students who worked with me on DES research topics (both at the University of Illinois and the University of Cyprus) as well as many students who took my course on *State Estimation and Event Inference in DES* at the University of Cyprus. All of them have helped (directly or indirectly) in the writing of this book; they are too many to explicitly mention, but their efforts are gratefully acknowledged. I would like to extend special thanks to Christoforos Keroglou, Apostolos Rikos, and Martha Christou, who read versions of the book and helped with the preparation of examples and diagrams. Their generosity in devoting their time to the development of material in this book is much appreciated.

I would also like to take this opportunity to thank the University of Illinois and the University of Cyprus for providing stimulating research environments for my professional development. The University of Cagliari and Xidian University also helped greatly by hosting me at various points while writing this book. Funding from many other organizations and agencies was used to drive some of the results reported in this book: special thanks go to the U.S. National Science Foundation and the Air Office for Scientific Research, the European Commission, and the Cyprus Research Promotion Foundation.

In producing this book, I was fortunate to receive assistance, suggestions, and support from several people at Springer. I feel extremely fortunate to have worked with such a highly professional group of editors. In particular, I would like to express my sincere thanks to Oliver Jackson for his encouragement, patience, support, and enthusiasm throughout the later stages of the writing of this book.

Contents

Abbreviations and Notation

Commonly Used Abbreviations

CSO	Current-State Opacity
DEDS	Discrete Event Dynamic System
DES	Discrete Event System
DFA	Deterministic Finite Automaton
DFS	Depth-First Search
DiSIR	Distributed Set Intersection Refinement
DSO	Delayed-State Opacity
FA	Finite Automaton
FSM	Finite-State Machine
ISO	Initial-State Opacity
LDFA	Labeled Deterministic Finite Automaton
LFA	Labeled Finite Automaton
LNFA	Labeled Nondeterministic Finite Automaton
NFA	Nondeterministic Finite Automaton
PO	Partially Ordered
QFA	Moore Machine
UIO	Unique Input–Output (sequence)
Weak CSO	Weak Current-State Opacity
Weak ISO	Weak Initial-State Opacity

Commonly Used Symbols

$\circ$	Composition operator for state mappings and relations
δ	Transition function for an FA (DFA or NFA)
$\Delta_{\sigma^{(i)}}$	Transition matrix corresponding to observation $\sigma^{(i)}$

$\delta_{seq}(q_k, \sigma_k^{k+m})$ or $\delta(q_k, \sigma_k^{k+m})$	Function that provides final state of DFA (or final states of NFA), starting from state q_k at time epoch k and sequentially applying the sequence of inputs $\sigma[k], \sigma[k+1], \ldots, \sigma[k+m]$
$AC(G)$	Accessible part of finite automaton G
$\mathcal{DFA}$	Deterministic finite automaton
$\mathcal{DFA}_m$	Marked deterministic finite automaton
$\mathcal{FA}$	Finite automaton
$\mathcal{FA}_m$	Marked finite automaton
G_{obs}	Current-state estimator (observer) for a given DFA or NFA G
G_{Dobs}	D-delayed-state estimator for a given DFA or NFA G
G_{Iobs}	Initial-state estimator for a given DFA or NFA G
$L(G)$ or $\mathcal{L}(G)$	Language of FA G
$L_m(G)$ or $\mathcal{L}_m(G)$	Marked language of FA G
M	State mapping
M_y or $M_{y_k^{k+m}}$	State mapping induced by observation y or observation sequence y_k^{k+m}
$M_{y_0^m}^{(m+2)}$	State trajectory induced by observation sequence y_0^m
$\mathcal{NFA}$	Nondeterministic finite automaton
$\mathcal{NFA}_m$	Marked nondeterministic finite automaton
$\mathcal{NFA}_1 \times \mathcal{NFA}_2$	Product of $\mathcal{NFA}_1$ and $\mathcal{NFA}_2$
$\mathcal{NFA}_1 \|\| \mathcal{NFA}_2$	Parallel composition of $\mathcal{NFA}_1$ and $\mathcal{NFA}_2$
P	Natural projection (with respect to a set of observable events Σ_{obs})
Q_F	Set of states reachable after a fault event in set F
$\hat{q}_0(y_0^k)$	Initial-state estimate, following sequence of observations $y[0], y[1], \ldots, y[k]$
$\hat{q}_{y[i]}(y_0^k)$	Delayed-state estimate, following sequence of observations $y[0], y[1], \ldots, y[k]$, at the point immediately after the observation of $y[i]$
$\hat{q}_{y[k]}(y_0^k)$	Current-state estimate, following sequence of observations $y[0], y[1], \ldots, y[k]$
$s \cdot t$ or st	Concatenation of string s with string t
sync	Synchronization event
$UR(q)$ or $UR(S)$	Unobservable reach from state q or from set of states S

Chapter 1
Introduction to Estimation and Inference in Discrete Event Systems

1.1 Introduction and Motivation

Discrete event (dynamic) systems (DES or DEDS) form a continuously growing class of automation systems that has become popular in the past three decades due to the proliferation of digital computing technologies, network interconnectivity, and improved/affordable sensing capability, as well as the emergence of cyber-physical systems and Internet of Things applications. The DES class includes a variety of systems, the common denominator of which is the presence of *events* that occur at discrete points in time and cause abrupt state changes in the system. Another typical feature of DES is that they are mostly technological (human-made), which implies that an underlying model is available for them, at least in a nominal form.

Examples of DES include computer and communication networks, autonomous vehicles, traffic and automated transportation systems, manufacturing systems, monitoring and control mechanisms (in aircrafts, automobiles, buildings, or printing devices), and others. Depending on the underlying system, a discrete event could be the command issued by a computer user, the change of a traffic light from red to green, the occurrence of a paper jam in the printer, the speed of a car reaching a certain level, the altitude of an airplane reaching a certain value, and so forth.

It should be clear from the above examples that the state of a DES could be discrete, continuous, or hybrid, whereas events can have a variety of characteristics, some of which are listed below.

- *Exogenous versus endogenous events.* Exogenous events are events caused by the environment (e.g., disturbance) or a user (e.g., a user command) or other entities that interact with the given system. On the other hand, endogenous events are events that are generated by the system itself (e.g., due to one of its state variables reaching a certain value).
- *Controllable versus uncontrollable events.* Uncontrollable events are events that cannot be prevented from occurring (e.g., disturbances that are external to the system). Controllable events, on the other hand, are events over which a

C. N. Hadjicostis, *Estimation and Inference in Discrete Event Systems*,
Communications and Control Engineering,
https://doi.org/10.1007/978-3-030-30821-6_1

monitor/control mechanism has some partial control. For instance, in the supervisory control framework, the assumption is that controllable events can be disabled (forbidden from occurring) by the monitor/control unit.

- *Unobservable versus observable events.* Unobservable events are events whose occurrence does not get recorded directly, i.e., there is no sensor that responds to their occurrence. Even though unobservable events are not directly observed, one might be able to infer their occurrence by analyzing other measurements (or sequences of other measurements) collected from the system. On the opposite side, observable events are events whose occurrence gets recorded, i.e., there is a sensor that responds to their occurrence. In certain cases, the sensor response to an observable event may be shared by more than one event, implying that, despite the sensor reading, there is some ambiguity about which event has occurred; further analysis of measurements collected from the system, along with knowledge of the system model, might be required to resolve this ambiguity.

The modeling capabilities of DES are vast and can be used to capture many of the complex systems that have emerged due to the convergence of computing, communication, and sensing technologies. The class of DES encompasses systems with discrete, continuous, and hybrid dynamics, and their analysis borrows tools from a variety of disciplines, such as systems and control, electrical and computer engineering, computer science, operations research, manufacturing engineering, and others. Popular DES models include automata and Petri nets (both untimed in their basic form), statecharts, and process algebras, as well as timed automata, timed Petri nets, max-plus algebra, Markov chains, stochastic (timed) models, and others.

While the relevance and modeling capability of DES are unquestionable, one of the major challenges in their analysis is algorithmic complexity, which can quickly become overwhelming. For example, verifying that the design of a central locking mechanism in an automobile indeed functions as intended, without reaching any deadlock or undesirable states (e.g., locking the passengers outside the automobile), appears to be a straightforward task but its complexity, even in simple cases, can be quite formidable. Similarly, analyzing a collection of routing protocols to select the one that has the best performance could be a challenging task.

1.2 State Estimation and Event Inference

This book addresses the growing need to systematically perform state estimation and event inference in DES. Determining the state of the system (or at least determining a small subset of states that includes the true state of the system) can be key to achieving a variety of performance and control objectives (such as avoiding entrance of the system to undesirable states or avoiding the execution of illegal sequences of events). Similarly, the ability to infer the occurrence of events that might not be directly observable is important for diagnosing faults or special events that need to be handled in specific ways.

State estimation and event inference tasks in DES can be broken in two main categories. In both cases, the underlying DES is assumed to be completely (or at least partially) known, i.e., a (partial) DES model for the system is available.

Online Estimation/Inference. The first category comprises *online* tasks, which address the following basic problem: given a (streaming) sequence of observations that are generated due to some unknown (or partially unknown) underlying activity in the given DES, determine the possible states that the system might be in (state estimation) or whether certain special events, such as fault events, may have occurred in the recent past (event inference). One could also enrich online estimation/inference tasks with additional challenges (e.g., determining past states, determining when or how long in the past the special events occurred). Another key challenge in online estimation/inference is the ability to perform the task recursively, i.e., the ability to update the state estimate or inference decision when a new observation becomes available.

Offline Estimation/Inference. The second category comprises *offline* tasks, which essentially amount to verification of various system properties of interest. For instance, a property of interest might be the ability to determine the occurrence of a fault event, at least after a few subsequent events take place (following the occurrence of the fault). A fault that allows, under all possible activity that may occur in the system, the external observer to determine its occurrence is called *diagnosable* and the system is said to have the property of *diagnosability* (with respect to this fault). Determining whether or not diagnosability holds for a given DES is an offline (verification) task as it does not depend on the specific activity that might take place in the system. There are also many other properties of interest that depend on observations that one makes as a result of activity in a given system, and we describe some of them later in this section. Different properties may be verified using different constructions and verification algorithms, some of which may be simpler than others.

Not surprisingly, a major challenge in state estimation and event inference (and the verification of related system properties) is complexity; this challenge if further accentuated by the continuously increasing size and diversity of emerging DES. For this reason, this book focuses on a popular DES class, namely finite automata (more generally, finite automata with outputs), and systematically builds online (recursive) state estimation and event inference mechanisms, as well as offline verification algorithms. The focus on automata is natural, due to their dominant role in DES literature, and results in a relatively concise treatment of the topic, while also enabling extensions to richer automata models (e.g., timed finite automata) as well as other DES models. For instance, extensions to *bounded* Petri nets are relatively straightforward if one is willing to construct the reachability graph of the given Petri net (which in many ways will resemble a finite automaton); the challenge in the case of Petri nets is to try to take advantage of the structure of the given Petri net to reduce the complexity of online/offline state estimation and event inference tasks.

The book discusses system properties that depend on observations that are generated by underlying activity in a given DES. The three main such properties discussed in the book are the following:

- *Detectability*. Detectability comes in many variations but its basic version is concerned with the ability of the external observer to almost always determine the state of the system exactly. In other words, under all possible activity in the system, the corresponding sequence of observations (generated by the underlying activity in the system) allows the observer to almost always determine the state of the system exactly.
- *Diagnosability*. As discussed earlier, diagnosability is concerned with the ability of the external observer to determine the occurrence of a fault event, at least after a certain (finite) number of events takes place, regardless of the subsequent activity in the system. In other words, for any underlying system activity that involves a fault, the external observer is able to conclude, based on the sequence of observations that are recorded (both before and after the occurrence of the fault), that the fault has definitely occurred. Fault diagnosis and diagnosability also come in many variations, depending on the number of faults, number of fault classes, need to determine combinations of faults or combinations of faults from different classes, and so forth.
- *Opacity*. Unlike the two properties described above, opacity is concerned with hiding information from the external observer (rather than revealing information to it). It also comes in many variations but its most basic form assumes that there is a certain subset of states that is secret, and one is concerned whether the external observer is able to conclude, under certain circumstances, that the state of the system definitely belongs in this secret set. In other words, to verify opacity with respect to a certain set of secret states, one needs to ensure, that under all possible underlying activity in the given system, the corresponding sequence of observations (generated by the underlying activity in the system) does not allow the external observer to determine that the state of the system definitely lies within the set of secret states.

1.3 Examples of Applications

It should be clear from the discussion in the previous section that DES are prevalent in many technological applications. In this section, we provide examples of applications to motivate our discussion on challenges pertaining to models that can be captured by (possibly interacting) finite automata under partial observation. We choose simple examples so that we can discuss them without necessarily having established all necessary notation and machinery. Note that the concepts of finite automata, product and parallel compositions of finite automata, and marked states mentioned below are all discussed at length in Chap. 3. One should also keep in mind that, compared to the examples in this section, the number and size of the interacting components in

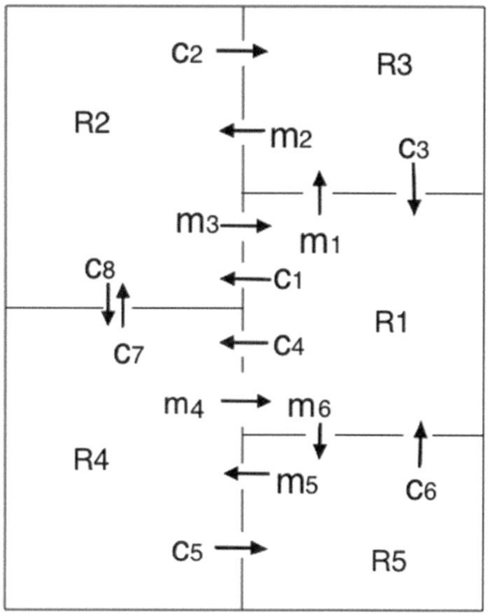

Fig. 1.1 Maze for cat and mouse problem

the types of systems mentioned in the previous section would be higher by orders of magnitude.

Example 1.1 (Cat and Mouse Problem) This example is a variation of the classical cat and mouse problem from Ramadge and Wonham (1989), which was introduced in the context of supervisory control. In Ramadge and Wonham (1989), the problem was modeled using two interacting finite automata, one capturing the movement of the mouse, and one capturing the movement of the cat. Later on the problem was also modeled using Petri nets.

Consider the maze of five rooms, referred to as $R1$, $R2$, $R3$, $R4$, and $R5$, in Fig. 1.1. A cat and a mouse circulate in this maze with the cat moving from room to room through unidirectional doors $c_1, c_2, \ldots, c_8$, and the mouse moving from room to room through unidirectional doors $m_1, m_2, \ldots, m_6$. Given the initial locations of the two animals, the goal is to control the various doors, so as to ensure that the two animals are never in the same room together.

We can use a finite automaton (called generator in Ramadge and Wonham 1989) to capture the movement of each animal. The generator for the cat is a finite automaton with five states $\{1, 2, 3, 4, 5\}$ (each representing the corresponding room in the maze), with the state of the generator capturing the location of the cat. Events in the cat generator include $c_1, c_2, \ldots, c_8$, and their occurrence captures the movement of the cat from one room to another. For instance, from room 1 (cat generator state 1), the cat can move to state 2 (via event c_1) or to state 4 (via event c_4). Similarly, we can have a generator for the mouse movement (with the difference being that the events are $m_1, m_2, \ldots, m_6$).

We can combine the two generators to model the simultaneous movement of the two animals. Formally, this is done by taking the parallel product of the two finite automata as described in Chap. 3, but for the purposes of this example, it suffices to think of the state of the combined system as a pair of integers (M, C) where $M \in \{1, 2, \ldots, 5\}$ denotes the location of the mouse and $C \in \{1, 2, \ldots, 5\}$ denotes the location of the cat. For instance, $(2, 1) \xrightarrow{c_1} (2, 2)$ indicates that initially the mouse was in room 2 and the cat was in room 1, and then event c_1 took place leading us to state (2, 2), i.e., both the mouse and the cat are in room 2, which is a problem. More generally, in terms of the resulting product automaton, *forbidden* (undesirable) states are states in the set $\{(1, 1), (2, 2), \ldots, (5, 5)\}$ (in which both animals are located in the same room).

Given an instance of the problem where we are given the maze, descriptions of which events are observable/unobservable, and which events are controllable/uncontrollable, the objective is to devise a strategy to determine how to control the various events so as to ensure that the two animals are never in the same room together. In the presence of unobservable transitions, such strategy will presumably need to determine (or at least obtain a reasonable estimate of) the locations of the two animals. The task can become challenging due to a variety of reasons, some of which are listed below.

- Desire to allow maximum freedom to the animals (maximal permissiveness).
- Certain doors could be uncontrollable (controllability limitations).
- Knowledge about the position of the various animals may be partial (observability limitations).
- Erroneous operation, communication delays, or faulty signals (fault diagnosis).

Despite its simplicity, the above example gives a sense of the power of finite automata models (e.g., in terms of modeling multiple interacting agents) and the challenges associated with them. □

Example 1.2 (Privacy Analysis and Enforcement) In this example, we illustrate some of the issues regarding privacy and security in DES that are modeled as interacting finite automata. The system in Fig. 1.2 can be viewed as an interconnection of

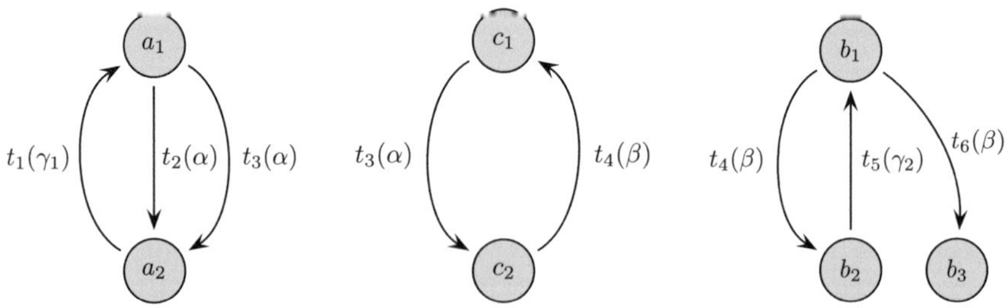

Fig. 1.2 Interacting finite automata: subsystem A (left), subsystem C (middle), and subsystem B (right)

two finite-state transition subsystems (e.g., automata or ordinary Petri nets) A and B, together with another finite-state transition subsystem C. Each transition subsystem (or module) A, B or C has (i) a finite set of states (denoted in the figure by circles), (ii) a finite set of transitions (each of which is denoted by a directed arc, that is associated with a pair of states, one starting state and one ending state, and with a—possibly nonunique—label), and (iii) an initial state.

More specifically, for the subsystems in Fig. 1.2 we have the following:

- The initial states of the subsystems are assumed to be state a_1 for subsystem A, state b_1 for subsystem B, and state c_1 for subsystem C; we denote this concisely by the triplet (a_1, c_1, b_1).
- Activity in the system of Fig. 1.2 is captured by transitions $t_1, t_2, \ldots, t_6$, some of which are *private* to a subsystem and some of which are *shared* between subsystems. For instance, transition t_1 is private to subsystem A and can take place when subsystem A is in state a_2; if/when t_1 occurs, it takes subsystem A to state a_1. Transition t_4 is shared between subsystems B and C, and can only take place when subsystem B is in state b_1 *and* subsystem C is in state c_2; if/when t_4 occurs, it takes subsystem C to state c_1 *and* subsystem B to state b_2.
- Each transition is associated with a label that gets emitted (and can be observed) when the transition occurs. For instance, in case transition t_1 occurs, it emits label γ_1 (which happens to be unique to this transition). On the other hand, when t_4 occurs, it emits label β (which happens to be shared with transition t_6). If a transition is associated with a unique label, then its occurrence can be immediately inferred based on the observed label; however, if the label is shared among multiple transitions, then it might not be possible to immediately infer which exact transition has occurred.

What couples together transition systems A and B is the fact that, apart from their private transitions, namely transitions t_1 and t_2 for subsystem A, and transitions t_5 and t_6 for subsystem B, they also have transitions that affect their common interface C, namely transitions t_3 and t_4. For example, when the overall state is (a_1, c_1, b_1), if private transition t_2 takes place in subsystem A, the new state of subsystem A will be a_2, whereas the states of the other subsystems will not be affected; we can denote this concisely with $(a_1, c_1, b_1) \xrightarrow{t_2} (a_2, c_1, b_1)$. Similarly, if shared transition t_3 takes place in subsystems A and C, the new state of subsystem A will be a_2 and the new state of subsystem C will be c_2, leading to the overall state (a_2, c_2, b_1); we can denote this concisely with $(a_1, c_1, b_1) \xrightarrow{t_3} (a_2, c_2, b_1)$.

Consider now the labels associated with each transition in Fig. 1.2. These labels represent the observations that are made available to an external observer about particular transitions (activity in the system). For instance, if label β is observed, then the observer can only infer that either transition t_4 or t_6 has occurred (because label β is shared by these two transitions). Depending, however, on the observer's knowledge of the system state, it may be able to resolve this ambiguity: specifically, both transitions require that subsystem B is in state b_1, however, transition t_4 also requires the interface C to be in state c_2.

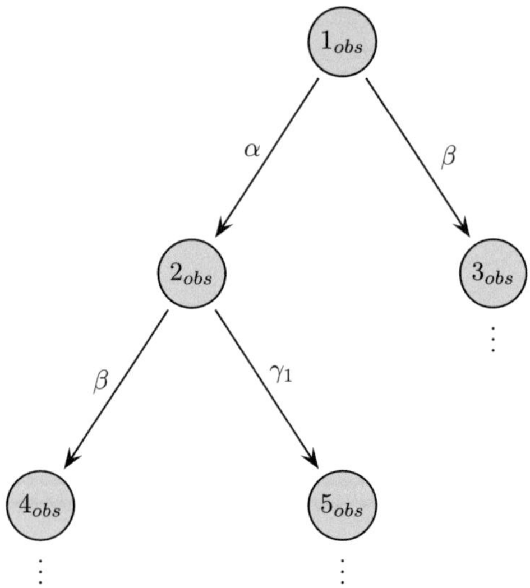

Observer State	Corresponding Set of States
1_{obs}	$\{(a_1, c_1, b_1)\}$
2_{obs}	$\{(a_2, c_1, b_1), (a_2, c_2, b_1)\}$
3_{obs}	$\{(a_1, c_1, b_3)\}$
4_{obs}	$\{(a_2, c_1, b_2), (a_2, c_1, b_3), (a_2, c_2, b_3)\}$
5_{obs}	$\{(a_1, c_1, b_1), (a_1, c_2, b_1)\}$
⋮	⋮

Fig. 1.3 Observer (partially shown) for the interacting finite automata in Fig. 1.2

Consider an external observer that is able to observe the labels α, β, γ_1, and γ_2 (generated by underlying activity in the system) and uses the sequence of observations (and knowledge of the system model) to make inferences about the system. We can codify the knowledge of this observer regarding the possible states of the system using the so-called *observer* construction, part of which is shown in Fig. 1.3. In this construction, each state (circle) is associated with a subset of the possible states of the overall system in Fig. 1.2. For instance, the initial state 1_{obs} is associated with the set of states $\{(a_1, c_1, b_1)\}$ (because that is the initial state of the system in Fig. 1.2). From state 1_{obs}, only two observations are possible:
(i) If α is observed, the observer moves to state 2_{obs} associated with $\{(a_2, c_1, b_1), (a_2, c_2, b_1)\}$ because these are the two possible states of the overall system (depending on whether transition t_2 or transition t_3 occurred).
(ii) If β is observed, the observer moves to state 3_{obs} associated with $\{(a_1, c_1, b_3)\}$ because the only possibility is for transition t_6 to have occurred (transition t_4 is not possible).

We can continue in this fashion, each time considering all possible observations to complete the observer construction. For example, if we consider state 2_{obs} (associated with states in the set $\{(a_2, c_1, b_1), (a_2, c_2, b_1)\}$), we realize that the only possible observations are β or γ_1, leading, respectively, to states 4_{obs} and 5_{obs}. For instance, β can be generated via the following activity:

$$(a_2, c_2, b_1) \xrightarrow{t_4} (a_2, c_1, b_2)$$
$$(a_2, c_1, b_1) \xrightarrow{t_6} (a_2, c_1, b_3)$$
$$(a_2, c_2, b_1) \xrightarrow{t_6} (a_2, c_2, b_3).$$

Thus, the state set associated with 4_{obs} is the set $\{(a_2, c_1, b_2), (a_2, c_1, b_3), (a_2, c_2, b_3)\}$. Similarly, state 5_{obs} can be found to be associated with the set $\{(a_1, c_1, b_1), (a_1, c_2, b_1)\}$. This process can continue until no new observer states (sets of possible states of the system in Fig. 1.2) can be obtained anymore; it is a finite process and provides a summary of the information available to the external observer under all possible scenarios. Observer constructions are discussed in detail in Chaps. 4 and 5.

We now illustrate some of the security/privacy concerns that arise in the context of DES (some of which may be answered using an observer construction). Suppose that the user of subsystem B does not want the world (external observer) to know with certainty whether the subsystem is in state b_2 (but does not care if the observer knows with certainty that it is in state b_1 or b_3). If the external observer does not receive information from subsystems A and B (i.e., observations α and γ_1, as well as the possible states of A and B are *not* available to the observer), then this desirable property holds: if an observer sees β, then this observer is unsure whether subsystem B is in state b_2 or b_3; when the observer sees γ_2, then it knows the system is back in state b_1 (and also that it was in state b_2 before, but this is not a privacy concern for subsystem B).

Even though there are no privacy concerns when subsystem B operates in isolation, if the observer has additional information (e.g., partial knowledge of the structure of the interconnected system and/or ability to observe α and γ_1), then the privacy of subsystem B may be violated. In terms of the terminology that will be introduced in Chap. 8, the above privacy concerns require that the overall system is current-state opaque with respect to the set of secret states $S = \{(a, c, b_2) \mid a \in \{a_1, a_2\}, c \in \{c_1, c_2\}\}$. Whether current-state opacity holds or not, and how it can be enforced in case of a violation, are questions that can be answered using constructions like the observer in Fig. 1.3. Chapters 6 and 8 systematically explore ways to check and verify related properties concerning detectability and opacity. In particular, state-based notions of opacity, including current-state opacity, are discussed in Chap. 8. □

1.4 Book Coverage

Apart from this introductory/motivational chapter, the book is organized as follows:

- Chapters 2 and 3 provide some basic notation and detailed description of the various finite automata models that are of interest in this book. Different observation models are also introduced and discussed in several examples.
- Chapter 4 develops recursive (online) state estimation techniques, focusing on three different types of state estimation tasks: current-state estimation, delayed-state estimation (or smoothing), and initial-state estimation. In all cases, a distinction is made between finite automata without or with silent transitions. Attention is paid to the complexity of the recursive state estimation task and on techniques that can be used to reduce this complexity.
- Chapter 5 introduces and analyzes the verification of system properties that pertain to the perceived states of an external observer with access to output activity of a given (known) finite automaton. More specifically, the chapter discusses system properties that relate to state isolation, i.e., the ability of the external observer to isolate the (estimated) state of the given finite automaton within a certain subset of states (or within certain subsets of states). Prime examples of such state isolation properties are the properties of detectability, diagnosability, and opacity, which are discussed in the subsequent three chapters.
- Chapter 6 discusses the property of detectability and its verification. The chapter also describes several variants of detectability (e.g., strong periodic detectability, which requires that the external observer is periodically, but perhaps not all the time, able to determine exactly the state of the system).
- Chapter 7 discusses fault detection and fault diagnosis under various possible scenarios, including the cases of multiple faults and multiple fault classes. The chapter also describes at length diagnosability, a system property that asks whether the external observer will always be able to detect/identify the fault event(s), and the verification of diagnosability using state estimator or verifier constructions. Finally, the chapter establishes connections between variants of diagnosability and state isolation.
- Chapter 8 discusses opacity and its many variations, including language-based opacity and state-based opacity.
- Extensions to decentralized and distributed observation settings are addressed in Chaps. 9 and 10, respectively. In these settings, there are multiple observation points, which may communicate with a coordinator (decentralized setting) or among themselves (distributed setting). The two chapters discuss and analyze the implementation (online, at run time) and the verification (offline, during the design of the system) of several protocols that can be used for decentralized/distributed state estimation and event inference.

1.5 Comments and Further Reading

As mentioned earlier, the book aims to provide a comprehensive study of state estimation and event inference techniques in (deterministic and nondeterministic) finite automata. There exist many outstanding textbooks and review articles that treat related subjects; below, we provide a list of some important references that the interested reader can pursue. We try to be concise, so the list of references is by no means exhaustive and we apologize for any omissions. We break the references into the following related categories: (i) discrete event systems, (ii) finite automata, (iii) Petri nets, (iv) state estimation and diagnosis.

Discrete Event Systems. A very comprehensive study of discrete event systems and related techniques (ranging from state estimation and fault diagnosis to supervisory control and discrete event system simulation) can be found in Cassandras and Lafortune (2007). This book covers not only automata, but also Petri nets, timed models, hybrid models, stochastic timed automata, Markov chains, queueing systems, and others. Several of these topics are also covered in the collection of chapters in Seatzu et al. (2013), which includes discussions on finite automata (analysis, diagnosis, supervisory control, and decentralized control), Petri nets (structural analysis, diagnosis, and supervisory control), as well as some aspects of timed and stochastic models. Some important topics within DES, mostly in the context of finite automata, include the following:

- Supervisory control methodologies are discussed extensively in the context of finite automata in the recent book (Wonham and Cai 2019) and also in Kumar and Garg (2012); one could also look for inspiration at the classical papers (Ramadge and Wonham 1989, 1987).
- The topic of fault diagnosis is discussed at length in the review article (Zaytoon and Lafortune 2013); one could also refer to the classical papers on fault diagnosis in discrete event systems (Sampath et al. 1995, 1998) as well as Lafortune et al. (2018) which offers historical perspectives on the evolution of the topic of fault diagnosis within the DES community. Fault-tolerant control in discrete event systems is discussed in a series of separate chapters in Blanke et al. (2006). A major breakthrough in diagnosability was established by Jiang et al. (2001), Yoo and Lafortune (2002), which show that many variations of diagnosability in finite automata can be verified with complexity that is polynomial in the size of the given finite automaton.
- Discussions on perturbation analysis can be found in Cassandras and Lafortune (2007), Ho and Cao (2012).

Finite Automata. The reader who is interested in learning more about finite automata, their languages, and/or related analysis methodologies, can use one of many existing excellent treatments, such as Hopcroft et al. (2006), Lawson (2003). An algebraic treatment of finite automata can be found in Arbib (1968), whereas a recent treatment of finite-state machines can be found in Kohavi and Jha (2009). Probabilistic finite automata are discussed in detail in Paz (2014).

Petri nets. The other popular class of DES models, namely Petri nets and related analysis, are covered in a range of review articles and books. For example, Murata (1989) is an excellent review article for Petri net properties and analysis techniques; it has withstood the test of time and serves as a good summary of Petri net knowledge up to the time the article was written. More recent books on Petri nets include (Reisig 2012; Zhou 2012; David and Alla 2010). Some control aspects for Petri nets can be found in Zhou (2012), Moody and Antsaklis (1998), whereas resource allocation and deadlock avoidance problems using Petri nets are discussed in Reveliotis et al. (2017), Park and Reveliotis (2001), Li and Zhou (2009).
State Estimation and Event Inference. The literature on state estimation and event inference for finite automata consists primarily of research articles, which is one of the reasons for the writing of this book. We provide references to these research articles at the end of individual chapters of this book and we do not repeat them here. Instead we make brief remarks about state estimation and event inference in Petri nets, a topic that is also mostly covered in research articles.

Clearly, if one is interested in bounded Petri nets (in which the set of reachable markings is effectively finite), one could utilize the approaches for finite automata described in this book by first constructing the reachability graph of the given Petri net, and subsequently using this reachability graph as the underlying (finite state) model. A potential problem with such an approach is that the reachability graph could have a size that is exponential in the size of the given Petri net. In order to avoid this problem, one might want to take advantage of the Petri net structure while performing estimation/inference. Systematic online state estimation and event inference techniques for Petri nets have gained attention recently, starting with work in Giua (1997), Giua and Seatzu (2002), extending to labeled Petri nets (without or with silent transitions) (Giua et al. 2005, 2007), and evolving to event inference (specifically, fault diagnosis) techniques that take advantage of the structure of the given Petri net via basis markings and minimal explanations (Cabasino et al. 2010, 2013). In certain cases, these ideas have been extended to the verification of properties of interest (offline estimation and inference), as done, for example, for the case of diagnosability in labeled Petri nets in (Cabasino et al. 2012, 2014).

There are also other works that relate to online state estimation and event inference in Petri nets. For example, Declerck and Bonhomme (2014), Basile et al. (2015) perform state estimation using linear programming formulations. Other contributions address the problem of estimating firing sequences (related to event inference) using algebraic properties of Petri net models (Lefebvre and El Moudni 2001; Lefebvre 2008), whereas the work in Li and Hadjicostis (2011) uses a cost criterion to estimate a least cost matching firing sequence.

References about detectability, fault diagnosis, and opacity using Petri net models can also be found in the concluding sections of Chaps. 6, 7, and 8 respectively.

References

Arbib MA (ed) (1968) Algebraic theory of machines, languages, and semigroups. Academic Press, New York

Basile F, Cabasino MP, Seatzu C (2015) State estimation and fault diagnosis of labeled time Petri net systems with unobservable transitions. IEEE Trans Autom Control 60(4):997–1009

Blanke M, Kinnaert M, Lunze J, Staroswiecki M, Schröder J (2006) Diagnosis and fault-tolerant control. Springer, Berlin

Cabasino MP, Giua A, Seatzu C (2010) Fault detection for discrete event systems using Petri nets with unobservable transitions. Automatica 46(9):1531–1539

Cabasino MP, Giua A, Lafortune S, Seatzu C (2012) A new approach for diagnosability analysis of Petri nets using verifier nets. IEEE Trans Autom Control 57(12):3104–3117

Cabasino MP, Giua A, Seatzu C (2013) Diagnosis using labeled Petri nets with silent or undistinguishable fault events. IEEE Trans Syst Man Cybern Syst 43(2):345–355

Cabasino MP, Giua A, Seatzu C (2014) Diagnosability of discrete-event systems using labeled Petri nets. IEEE Trans Autom Sci Eng 11(1):144–153

Cassandras CG, Lafortune S (2007) Introduction to discrete event systems. Springer, Berlin

David R, Alla H (2010) Discrete, continuous, and hybrid petri nets. Springer, Berlin

Declerck P, Bonhomme P (2014) State estimation of timed labeled Petri nets with unobservable transitions. IEEE Trans Autom Sci Eng 11(1):103–110

Giua A (1997) Petri net state estimators based on event observation. In: Proceedings of 36th IEEE conference on decision and control (CDC), vol 4, pp 4086–4091

Giua A, Seatzu C (2002) Observability of place/transition nets. IEEE Trans Autom Control 47(9):335–440

Giua A, Corona D, Seatzu C (2005) State estimation of λ-free labeled Petri nets with contact-free nondeterministic transitions. Discret Event Dyn Syst 15(1):85–108

Giua A, Seatzu C, Corona D (2007) Marking estimation of Petri nets with silent transitions. IEEE Trans Autom Control 52(9):1695–1699

Ho YCL, Cao XR (2012) Perturbation analysis of discrete event dynamic systems, vol 145. Springer Science & Business Media, Berlin

Hopcroft JE, Motwani R, Ullman JD (2006) Automata theory, languages, and computation. Pearson Education

Jiang S, Huang Z, Chandra V, Kumar R (2001) A polynomial algorithm for testing diagnosability of discrete-event systems. IEEE Trans Autom Control 46(8):1318–1321

Kohavi Z, Jha NK (2009) Switching and finite automata theory. Cambridge University Press, Cambridge

Kumar R, Garg VK (2012) Modeling and control of logical discrete event systems, vol 300. Springer Science & Business Media, Berlin

Lafortune S, Lin F, Hadjicostis CN (2018) On the history of diagnosability and opacity in discrete event systems. Annu Rev Control 45:257–266

Lawson MV (2003) Finite automata. Chapman and Hall/CRC

Lefebvre D (2008) Firing sequences estimation in vector space over Z_3 for ordinary Petri nets. IEEE Trans Syst Man Cybern Part A Syst Hum 38(6):1325–1336

Lefebvre D, El Moudni A (2001) Firing and enabling sequences estimation for timed Petri nets. IEEE Trans Syst Man Cybern Part A Syst Hum 31(3):153–162

Li L, Hadjicostis CN (2011) Least-cost transition firing sequence estimation in labeled Petri nets with unobservable transitions. IEEE Trans Autom Sci Eng 8(2):394–403

Li Z, Zhou M (2009) Deadlock resolution in automated manufacturing systems: a novel petri net approach. Springer Science & Business Media, Berlin

Moody JO, Antsaklis PJ (1998) Supervisory control of discrete event systems using Petri nets. Springer Science & Business Media, Berlin

Murata T (1989) Petri nets: properties, analysis and applications. Proc IEEE 77(4):541–580

Park J, Reveliotis SA (2001) Deadlock avoidance in sequential resource allocation systems with multiple resource acquisitions and flexible routings. IEEE Trans Autom Control 46(10):1572–1583

Paz A (2014) Introduction to probabilistic automata. Academic

Ramadge PJ, Wonham WM (1987) Supervisory control of a class of discrete event processes. SIAM J Control Optim 25(1):206–230

Ramadge PJ, Wonham WM (1989) The control of discrete event systems. Proc IEEE 77(1):81–97

Reisig W (2012) Petri nets: an introduction. Springer Science & Business Media, Berlin

Reveliotis S, et al (2017) Logical control of complex resource allocation systems. Found Trends® Syst Control 4(1–2):1–223

Sampath M, Sengupta R, Lafortune S, Sinnamohideen K, Teneketzis D (1995) Diagnosability of discrete-event systems. IEEE Trans Autom Control 40(9):1555–1575

Sampath M, Lafortune S, Teneketzis D (1998) Active diagnosis of discrete-event systems. IEEE Trans Autom Control 43(7):908–929

Seatzu C, Silva M, van Schuppen J (eds) (2013) Control of discrete-event systems. Lecture notes in control and information sciences, Springer, Berlin

Wonham WM, Cai K (2019) Supervisory control of discrete-event systems. Springer International Publishing, Berlin

Yoo TS, Lafortune S (2002) Polynomial-time verification of diagnosability of partially observed discrete-event systems. IEEE Trans Autom Control 47(9):1491–1495

Zaytoon J, Lafortune S (2013) Overview of fault diagnosis methods for discrete event systems. Annu Rev Control 37(2):308–320

Zhou M (2012) Petri nets in flexible and agile automation. Springer Science & Business Media, Berlin

Chapter 2
Preliminaries and Notation

2.1 Set Theory

A *set* A is an unordered collection of objects, which are called *elements* or *members* of A. We write $a \in A$ ($a \notin A$) if element a is (not) contained in set A. Two sets A and B are *equal*, denoted $A = B$, if and only if (iff) they have the same elements. The *empty set* contains no elements and is denoted by $\emptyset$ (or by $\{\ \}$).

A set A is called a *subset* of set B, denoted as $A \subseteq B$, if each element of A is also contained in B. Set A is called a *strict subset* of set B, denoted as $A \subset B$, if A is a subset of B and also $A \neq B$. The empty set is a subset of any set A (i.e., $\emptyset \subseteq A$).

The *power set* of a set A is the set of all subsets of A and is denoted by 2^A, i.e.,

$$2^A = \{B \mid B \subseteq A\} \; .$$

Note that the elements of 2^A are sets of elements of A (and not elements of A).

The *cardinality* of a set A is the number of elements contained in A and is denoted by $|A|$. A set is called *singleton* set if it has cardinality $|A| = 1$. For a finite set A (i.e., $|A| < \infty$), we have $|2^A| = 2^{|A|}$.

The *union* of two sets A and B is the set of elements that contains elements that are either in A or in B, or in both; it is denoted by

$$A \cup B = \{x \mid x \in A \text{ or } x \in B\} \; .$$

The *intersection* of two sets A and B is the set of elements that contains elements that are both in A and in B; it is denoted by

$$A \cap B = \{x \mid x \in A \text{ and } x \in B\} \; .$$

Two sets A and B are said to be *disjoint* if $A \cap B = \emptyset$. Clearly, we have $(A \cap B) \subseteq (A \cup B)$.

C. N. Hadjicostis, *Estimation and Inference in Discrete Event Systems*,
Communications and Control Engineering,
https://doi.org/10.1007/978-3-030-30821-6_2

The *difference* of two sets A and B, denoted by $A \backslash B$ (also by $A - B$), is defined as

$$A \backslash B = \{x \mid x \in A \text{ and } x \notin B\} ,$$

i.e., it is the set that contains elements in A that are not in B. In general, $A \backslash B \neq B \backslash A$.

If we use U to denote the *universal set*, i.e., the set that contains all known elements (so that $A \subseteq U$ for any set A), we can define the *complement* of a set A, denoted by $\overline{A}$ (or A^c), as $\overline{A} = U \backslash A$. In other words, the complement of A is the set that contains all elements that are not in A.

It is not hard to argue that, for any sets A and B, we have $A \backslash B = A \cap \overline{B}$. Other important properties of set operations (see, for example, Rosen 2011) are the following:

1. $A \cup \emptyset = A$, $A \cap U = A$ (identity).
2. $A \cap \emptyset = \emptyset$, $A \cup U = U$ (domination).
3. $A \cup A = A$, $A \cap A = A$ (idempotent).
4. $\overline{(\overline{A})} = A$ (complementation).
5. $A \cup B = B \cup A$, $A \cap B = B \cap A$ (commutative).
6. $(A \cup B) \cup C = A \cup (B \cup C)$, $(A \cap B) \cap C = A \cap (B \cap C)$ (associative).
7. $A \cap (B \cup C) = (A \cap B) \cup (A \cap C)$, $A \cup (B \cap C) = (A \cup B) \cap (A \cup C)$ (distributive).
8. $\overline{A \cup B} = \overline{A} \cap \overline{B}$, $\overline{A \cap B} = \overline{A} \cup \overline{B}$ (De Morgan's).

Since the union and intersection operations are associative, we write

$$\cup_{i=1}^{N} A_i = A_1 \cup A_2 \cup \cdots \cup A_N$$

and

$$\cap_{i=1}^{N} A_i = A_1 \cap A_2 \cap \cdots \cap A_N ,$$

because the order of performing the union/intersection operations does not matter. A collection of pairwise disjoint sets $A_1, A_2, \ldots, A_N$ (i.e., $A_i \cap A_j = \emptyset$ for $i \neq j \in \{1, 2, \ldots, N\}$) is said to form a *partition* of S if $S = \cup_{i=1}^{N} A_i$. Sometimes, the union of two (or more) disjoint sets A and B is also written as $A \dot{\cup} B$ where the "·" serves to indicate that the two sets are disjoint.

Example 2.1 Consider the universe $U = \{1, 2, 3, \ldots, 8, 9\}$ (consisting of all positive integers from 1 to 9) and the following sets that are defined in this universe:

$$\begin{aligned} A &= \{1, 2, 3\} , \\ B &= \{1, 2, 3, 4, 5\} , \\ C &= \{3, 4, 7, 8, 9\} , \\ D &= \{4, 5\} . \end{aligned}$$

We observe that $A \subset B$ (strict subset); however, neither A is a subset of C nor C is a subset of A. Also note that A and D are disjoint ($A \cap D = \emptyset$) and form a partition of B, i.e., $B = A \cup D$.

The cardinality of set A is $|A| = 3$. The power set 2^A is given by

$$2^A = \{\emptyset, \{1\}, \{2\}, \{3\}, \{1, 2\}, \{1, 3\}, \{2, 3\}, \{1, 2, 3\}\}$$

and has cardinality $|2^A| = 8$ (which indeed matches $2^{|A|} = 2^3$).

When performing set operations on sets A and C, we obtain

$$\begin{aligned} A \cup C &= \{1, 2, 3, 4, 7, 8, 9\} , \\ A \cap C &= \{3\} , \\ A \backslash C &= \{1, 2\} , \\ C \backslash A &= \{4, 7, 8, 9\} , \\ \overline{A} &= \{4, 5, 6, 7, 8, 9\} . \end{aligned}$$

Note that $(A \cap C) \subseteq (A \cup C)$ (as expected) and $A \backslash C \neq C \backslash A$. The set $A \cap C$ is singleton as it contains a single element. □

2.2 Relations

The *Cartesian product* $A \times B$ of two sets A and B is the set that contains ordered pairs of the form (a, b) where $a \in A$ and $b \in B$, i.e.,

$$A \times B = \{(a, b) \mid a \in A \text{ and } b \in B\} .$$

Note that in general $A \times B \neq B \times A$ (though, for A and B of finite cardinality, we have $|A \times B| = |A| \times |B| = |B \times A|$). A subset R of $A \times B$ is called a *binary relation* from A to B. In particular, when $(a, b) \in R$ we say that *a is related to b* and denote it as aRb.

A relation R *on the set* A is a relation of the form $R \subseteq A \times A$. Some important properties of a relation R on the set A are the following:

- *Reflexivity*: R is reflexive if $(a, a) \in R$ (or aRa) for every element $a \in A$.
- *Symmetry*: R is symmetric if $(a, b) \in R$ (or aRb) whenever $(b, a) \in R$ (or bRa).
- *Transitivity*: R is transitive if whenever (a, b), $(b, c) \in R$, then $(a, c) \in R$.

When all of the above properties hold, the relation R on the set A is called an *equivalence relation*. More specifically, a relation R on the set A is an equivalence relation if it is reflexive, symmetric, and transitive. It can be shown that an equivalence relation R on the set A induces a partition of A into pairwise disjoint subsets P_1, P_2, ..., (such that $P_i \cap P_j = \emptyset$ if $i \neq j$ and $\cup_i P_i = A$) with the property that aRb iff the elements a and b belong in the same partition (i.e., we can find a unique partition P_i such that $a, b \in P_i$).

Remark 2.1 Relations can also be defined with respect to more than two sets. For instance, given sets $A_1, A_2, \ldots, A_N$, we can define an N-ary relation R as a subset of $A_1 \times A_2 \times \cdots \times A_N$ (i.e., $R \subseteq A_1 \times A_2 \times \cdots \times A_N$). □

Since relations are subsets of sets, we can compose new relations using set operations. For example, given two relations $R_1, R_2 \subseteq A \times B$, we can define the relations $R_1 \cup R_2$, $R_1 \cap R_2$, $R_1 \backslash R_2$, and so forth. One can also compose relations as follows: given relation $R \subseteq A \times B$ and $S \subseteq B \times C$, the composite relation $S \circ R$ is defined as

$$S \circ R = \{(a, c) \in A \times C \mid \exists b \text{ such that } (a, b) \in R \text{ and } (b, c) \in S\}\ .$$

Note that $S \circ R$ is a subset of $A \times C$.

Example 2.2 Consider the set $A = \{1, 2, 3\}$ and the following relations defined on it:

$$\begin{aligned} R_1 &= \{(1, 1), (1, 2), (2, 1)\}\ , \\ R_2 &= \{(1, 1), (2, 2), (2, 3), (3, 1), (3, 3)\}\ , \\ R_3 &= \{(1, 1), (1, 2), (2, 1), (2, 2), (3, 3)\}\ . \end{aligned}$$

It is not hard to verify that the above relations have the following properties:

1. R_1 is not reflexive (e.g., $(2, 2) \notin R_1$); it is symmetric; it is not transitive (since $(2, 1)$ and $(1, 2)$ are both in R_1, we need $(2, 2)$ to also be in R_1 for the relation to be transitive).
2. R_2 is reflexive, not symmetric (e.g., $(3, 2) \notin R_2$ whereas $(2, 3) \in R_2$), and not transitive (since $(2, 3)$ and $(3, 1)$ are both in R_2, we need $(2, 1)$ to also be in R_2 for the relation to be transitive).
3. R_3 is reflexive, symmetric, and transitive. This implies that R_3 is an equivalence relation, which means that A can be partitioned into disjoint sets, with the property that aRb iff the elements a and b belong in the same partition. In this particular case, we have two partitions, namely $P_1 = \{1, 2\}$ and $P_2 = \{3\}$.

   We illustrate the composition of relations with two examples below:

$$\begin{aligned} R_1 \circ R_1 &= \{(1, 1), (1, 2), (2, 1), (2, 2)\}, \\ R_3 \circ R_2 &= \{(1, 1), (1, 2), (2, 1), (2, 2), (2, 3), (3, 1), (3, 2), (3, 3)\}. \end{aligned}$$ □

2.3 Alphabets, Strings, and Languages

Consider a non-empty finite set of symbols (*alphabet*) $\Sigma = \{\sigma^{(1)}, \sigma^{(2)}, \ldots, \sigma^{(K)}\}$. We often refer to the symbols in the set Σ as *letters* (even though there is no restriction to use letters from a specific alphabet). We can use letters in the alphabet to compose *words* or *strings*; more specifically, given a positive integer n, a word of length n is a sequence of n symbols $\sigma^{(i_1)}, \sigma^{(i_2)}, \ldots, \sigma^{(i_n)}$ (where $i_1, i_2, \ldots, i_n \in \{1, 2, \ldots, K\}$, with repetitions allowed). We denote a string as

$$s = \sigma^{(i_1)}\sigma^{(i_2)} \dots \sigma^{(i_n)} .$$

We use the symbol ϵ (or sometimes λ) to denote a special string, called the *empty string*, that contains no symbols.

Note that two strings $s = \sigma^{(i_1)}\sigma^{(i_2)} \dots \sigma^{(i_n)}$ and $t = \sigma^{(j_1)}\sigma^{(j_2)} \dots \sigma^{(j_{n'})}$ are equal if they have the same length (i.e., $n = n'$) and the exact same elements ($\sigma^{(i_k)} = \sigma^{(j_k)}$ for $k = 1, 2, \dots, n$).

Given two strings $s = \sigma^{(i_1)}\sigma^{(i_2)} \dots \sigma^{(i_n)}$ and $t = \sigma^{(j_1)}\sigma^{(j_2)} \dots \sigma^{(j_{n'})}$ of the lengths n and n' respectively, the *concatenation* of s and t is the string

$$s \cdot t = \sigma^{(i_1)}\sigma^{(i_2)} \dots \sigma^{(i_n)}\sigma^{(j_1)}\sigma^{(j_2)} \dots \sigma^{(j_{n'})} ,$$

i.e., the sequence of symbols in s followed by the sequence of symbols in t. This is also written as st. It is easy to verify that the concatenation of strings (operation $\cdot$) is an associative operation (but not commutative in general). For any string s, we define $s \cdot \epsilon = \epsilon \cdot s = s$.

The length ℓ of a string s is formally defined recursively as

$$\ell(s) = \begin{cases} 1 + \ell(s'), & \text{when } s = \sigma s', \text{ for } \sigma \in \Sigma, s' \in \Sigma^* , \\ 0, & \text{when } s = \epsilon. \end{cases}$$

It can be shown that $\ell(st) = \ell(s) + \ell(t)$ (Rosen 2011). Note that the length of string s is also denoted by $|s|$.

Given an alphabet Σ, we denote the set of all strings of length k with Σ^k, i.e., for $k = 0, 1, 2, \dots$

$$\Sigma^k = \{s \mid s \text{ is a word over } \Sigma \text{ and } \ell(s) = k\} .$$

In particular, $\Sigma^0 = \{\epsilon\}$ and $\Sigma^1 = \Sigma$. The set of all finite length strings that can be generated by alphabet Σ is denoted by Σ^* and includes the empty string ϵ. In other words,

$$\Sigma^* = \Sigma^0 \cup \Sigma^1 \cup \Sigma^2 \cup \cdots$$

We also define Σ^+ as

$$\Sigma^+ = \Sigma^1 \cup \Sigma^2 \cup \Sigma^3 \cup \cdots$$

The superscript $*$ is also used to concisely describe sets of strings. For example, ab^*c denotes the set of strings that start with a, followed by an arbitrary number of zero of more b's, followed by c, i.e.,

$$ab^*c = \{ac, abc, abbc, abbbc, \dots\} .$$

For any integer $n \geq 0$, we use b^n to denote the concatenation of symbol b n times. For example, ab^nc denote the string that starts with a, is followed by exactly n b's, and ends with c. Using this notation, we can write

$$ab^*c = \cup_{i=0}^{\infty} \{ab^ic\} .$$

Finally, we use the notation + to denote unions of different types of strings. For example, $a(b+c)^{10}d$ denotes the set of strings that start with a, followed by exactly 10 symbols from the set $\{b, c\}$, followed by d. This notation can be combined to create more complex sets of strings, e.g.,

$$a(b+cd)^*e = \{ae, abe, acde, abcde, acdbe, abbe, acdcde, \ldots\}$$

(i.e., the set of strings that start with a and end with e, and in between they have any number of zero or more instances of b or cd).

A *prefix* of a string $s \in \Sigma^*$ is a string t for which there exists a string t' such that $tt' = s$. The *prefix closure* of string s is the set of all of its prefixes (including the empty string ϵ and string s itself) and is denoted by the set $\overline{s} = \{t \in \Sigma^* \mid \exists t' \in \Sigma^* \text{ such that } tt' = s\}$. The *post-string* of string s after $t \in \overline{s}$ is denoted by s/t and is defined as the string t' such that $tt' = s$. In other words, if $s = tt'$ for $s, t, t' \in \Sigma^*$, then t is a prefix of s (i.e., $t \in \overline{s}$) and $s/t = t'$.

Following the above convention, for a string s, $s \in \Sigma^*$, we can similarly define the string t' for which there exists a string t such that $tt' = s$ as a *suffix* of string s. The *suffix closure* of string s is the set of all of its suffixes (including the empty string ϵ and string s itself) and is denoted by the set $\underline{s} = \{t' \in \Sigma^* \mid \exists t \in \Sigma^* \text{ such that } tt' = s\}$. The *pre-string* of string s before $t' \in \underline{s}$ is defined as the string t such that $tt' = s$.

Example 2.3 Given the alphabet $\Sigma = \{a, b, c\}$, we have the following sets:

$$\begin{aligned}
\Sigma^1 &= \{a, b, c\} , \\
\Sigma^2 &= \{aa, ab, ac, ba, bb, bc, ca, cb, cc\} , \\
\Sigma^3 &= \{aaa, aab, aac, aba, abb, abc, aca, acb, acc, \ldots., ccc\} ,
\end{aligned}$$

and so forth. The string $s = abbca$ has as set of prefixes

$$\overline{s} = \{\epsilon, a, ab, abb, abbc, abbca\}$$

and set of suffixes

$$\underline{s} = \{\epsilon, a, ca, bca, bbca, abbca\} .$$

If we take any prefix t in $\overline{s}$, we can define the post-string s/t of string s, e.g., for $t = ab$, we have $s/t = bca$. □

Given an alphabet Σ, a language $\mathcal{L}$ is a subset of Σ^*, i.e., a subset (of possibly infinite cardinality) of finite length strings composed of symbols in the alphabet. We can define the union and intersection of languages in terms of the sets they represent, i.e., for two languages $\mathcal{L}_1$ and $\mathcal{L}_2$ we have

$$\begin{aligned}
\mathcal{L}_1 \cup \mathcal{L}_2 &= \{s \mid s \in \mathcal{L}_1 \text{ or } s \in \mathcal{L}_2\}, \\
\mathcal{L}_1 \cap \mathcal{L}_2 &= \{s \mid s \in \mathcal{L}_1 \text{ and } s \in \mathcal{L}_2\}.
\end{aligned}$$

Furthermore, the concatenation of two languages is taken to be the set of strings that is generated by concatenating any string from the first language with any string in the second language

$$\mathcal{L}_1 \cdot \mathcal{L}_2 = \mathcal{L}_1\mathcal{L}_2 = \{s_1 s_2 \mid s_1 \in \mathcal{L}_1,\ s_2 \in \mathcal{L}_2\} .$$

Finally, the prefix closure of a language $\mathcal{L}$ is denoted by $\overline{\mathcal{L}}$ and is defined to be the union of all sets of strings that are prefixes of strings in the language

$$\begin{aligned}\overline{\mathcal{L}} &= \{t \in \Sigma^* \mid \exists s \in \mathcal{L} \text{ such that } t \in \overline{s}\} \\ &= \cup_{s \in \mathcal{L}} \overline{s} .\end{aligned}$$

A language $\mathcal{L}$ is said to be prefix closed if $\mathcal{L} = \overline{\mathcal{L}}$.

In Chap. 3, we look at *regular languages*, i.e., languages that can be generated by finite automata models.

Example 2.4 Consider the following languages over the alphabet $\Sigma = \{a, b, c\}$:

$$\begin{aligned}\mathcal{L}_1 &= \{\epsilon, a, b, ab, bc, abc, bca\} , \\ \mathcal{L}_2 &= \{\epsilon, a, c, bb, bc\} .\end{aligned}$$

It can be verified that $\mathcal{L}_1$ is a prefix closed language whereas $\mathcal{L}_2$ is not a prefix closed language. The prefix closure of $\mathcal{L}_2$ is given by

$$\overline{\mathcal{L}_2} = \{\epsilon, a, b, c, bb, bc\} .$$

The concatenation of $\mathcal{L}_1$ and $\mathcal{L}_2$ is given by

$$\mathcal{L}_1 \cdot \mathcal{L}_2 = \{\epsilon, a, b, ab, bc, abc, bca, aa, ba, aba, abca, bcaa, \ldots, bcabc\}$$

and is not prefix closed. □

Given an alphabet Σ and a string $s \in \Sigma^*$, we will frequently view s as a sequence of events and be interested in how this sequence of events is perceived by external entities (observer). A very commonly studied setting is one where only a subalphabet Σ_{obs} of events ($\Sigma_{obs} \subseteq \Sigma$) can be observed by some external entity, whereas the remaining events, $\Sigma_{uo} := \Sigma \backslash \Sigma_{obs}$ cannot be observed. In this case, the *natural projection* $P_{\Sigma_{obs}}$ is a convenient way of mapping s to the sequence of observations ω ($\omega \in \Sigma_{obs}^*$) that it generates. Specifically, the natural projection is a mapping

$$P_{\Sigma_{obs}} : \Sigma^* \to \Sigma_{obs}^* ,$$

which can be defined recursively as follows: $P_{\Sigma_{obs}}(\epsilon) = \epsilon$ and for $s \in \Sigma^*$, $\sigma \in \Sigma$, we have

$$P_{\Sigma_{obs}}(s\sigma) = P_{\Sigma_{obs}}(s) \cdot P_{\Sigma_{obs}}(\sigma) ,$$

where $P_{\Sigma_{obs}}(\sigma) = \sigma$ if $\sigma \in \Sigma_{obs}$ and $P_{\Sigma_{obs}}(\sigma) = \epsilon$ if $\sigma \in \Sigma_{uo}$.

The *inverse projection* $P^{-1}_{\Sigma_{obs}}$ of a sequence of observations

$$\omega = \sigma_o[1]\sigma_o[2], \ldots, \sigma_o[k] ,$$

(where $\sigma_o[i] \in \Sigma_{obs}$ for $i = 1, 2, \ldots, k$) is defined as the set of sequences of events in Σ^* that project, under the natural projection mapping $P_{\Sigma_{obs}}$, to the output sequence ω. This set can be described concisely as

$$\begin{aligned} P^{-1}_{\Sigma_{obs}} &= \{s \in \Sigma^* \mid P_{\Sigma_{obs}}(s) = \omega\} \\ &= \Sigma^*_{uo}\ \sigma_o[1]\ \Sigma^*_{uo}\ \sigma_o[2]\ \Sigma^*_{uo}\ \ldots\ \Sigma^*_{uo}\ \sigma_o[k]\ \Sigma^*_{uo} \ . \end{aligned}$$

Note that $P^{-1}_{\Sigma_{obs}}$ is a mapping from Σ^*_{obs} to 2^{Σ^*} and we define $P^{-1}_{\Sigma_{obs}}(\epsilon) = \Sigma^*_{uo}$.

Example 2.5 Consider the alphabet $\Sigma = \{a, b, c, d\}$ where the set of observable events is given by $\Sigma_{obs} = \{a, b\}$ (so that $\Sigma_{uo} = \{c, d\}$). Then, we have

$$\begin{aligned} P_{\Sigma_{obs}}(abcd) &= ab \ , \\ P_{\Sigma_{obs}}(cadbcd) &= ab \ , \\ P_{\Sigma_{obs}}(abcbd) &= abb \ . \end{aligned}$$

Moreover, we have

$$\begin{aligned} P^{-1}_{\Sigma_{obs}}(ab) &= \Sigma^*_{uo} a \Sigma^*_{uo} b \Sigma^*_{uo} \\ &= (c+d)^* a (c+d)^* b (c+d)^* \ , \\ P^{-1}_{\Sigma_{obs}}(abb) &= \Sigma^*_{uo} a \Sigma^*_{uo} b \Sigma^*_{uo} b \Sigma^*_{uo} \\ &= (c+d)^* a (c+d)^* b (c+d)^* b (c+d)^* \ . \end{aligned}$$

As expected $P^{-1}_{\Sigma_{obs}}(ab)$ includes $abcd$ and $cadbcd$, whereas $P^{-1}_{\Sigma_{obs}}(abb)$ includes $abcbd$. □

2.4 Miscellaneous Notation

In this section, we describe some of the notation that we will be using frequently in the book.

- $P \equiv Q$ means that the two quantities, P and Q, are equivalent.
- $P := Q$ means the quantity on the left (P) is defined to be the same as the quantity on the right (Q).
- $P =: Q$ means the quantity on the right (Q) is defined to be the same as the quantity on the left (P).
- $\sigma[k]$ is used to denote the symbol (event) that occurs at the time epoch indexed by k. Moreover, $\sigma_m^{m'}$ (for $m' \geq m$) indicates the sequence of one or more consecutive events, starting at time epoch m and ending at time epoch m' (inclusive), i.e.,

$$\sigma_m^{m'} := \sigma[m], \sigma[m+1], \ldots, \sigma[m'] .$$

In particular, $\sigma_m^m = \sigma[m]$. For $m' < m$ we define $\sigma_m^{m'} = \epsilon$ (the empty sequence). Note that if one ignores the indices for the time epochs, $\sigma_m^{m'}$ can be viewed as a string of length $m + 1$.

- $A = [a_{ij}]$ or $A = [a(i, j)]$ denotes the matrix A with entry a_{ij} or $a(i, j)$ at its ith row and jth column position.
- The number of r permutations of a set of n distinct objects (i.e., the number of ways of choosing, in a specific order, r elements out of the n elements, with no repetition allowed) is given by

$$P(n,r) = n(n-1)(n-2)\ldots(n-r+1) = \frac{n!}{(n-r)!}$$

- The number of r combinations of a set of n distinct objects (i.e., the number of ways of choosing, in no specific order, r elements out of the n elements, with no repetition allowed) is given by

$$C(n,r) = \frac{P(n,r)}{r!} = \frac{n!}{(n-r)!r!}$$

We also write $C(n,r) = \binom{n}{r}$.

- A deterministic finite automaton (DFA) will be denoted as a four-tuple $\mathcal{DFA} = (Q, \Sigma, \delta, q[0])$, whereas a nondeterministic finite automaton (NFA) will be denoted as a four-tuple $\mathcal{NFA} = (Q, \Sigma, \delta, Q_0)$. Both definitions are elaborated upon in Chap. 3.

2.5 Comments and Further Reading

Several books provide interesting discussions on set theory and relations, including Rosen (2011). Extensive discussions on alphabets, words, and languages can be found in Hopcroft (2008), Cassandras and Lafortune (2007), Eilenberg (1974).

References

Cassandras CG, Lafortune S (2007) Introduction to discrete event systems. Springer, Berlin

Eilenberg S (1974) Automata, languages, and machines. Academic Press, Orlando

Hopcroft JE (2008) Introduction to automata theory, languages, and computation. Pearson Education, India

Rosen KH (2011) Discrete mathematics and its applications. McGraw-Hill, New York

Chapter 3
Finite Automata Models

3.1 Introduction and Motivation

This chapter provides an introduction to the theory of finite automata, including deterministic and nondeterministic models. We discuss various aspects of automata, such as state transition functions, valid sequences of inputs, languages, and relevant observation models. Essentially, this chapter establishes the notation that will be necessary for our developments in the remainder of this book on state estimation and related applications, and property verification. As discussed in Chap. 1, finite automata models are critical in many applications of discrete event systems.

Our discussion in this chapter assumes basic familiarity with sets and strings. The reader who would like to refresh these concepts can do so by referring to Chap. 2.

3.2 Finite Automata and Languages

3.2.1 Finite Automata

3.2.1.1 Deterministic and Nondeterministic Finite Automata

A deterministic finite automaton (DFA) is a dynamic system $\mathcal{DFA}$ with a finite number of states and inputs, denoted by the sets $Q = \{q^{(1)}, q^{(2)}, \ldots, q^{(N)}\}$ and $\Sigma = \{\sigma^{(1)}, \sigma^{(2)}, \ldots, \sigma^{(K)}\}$ respectively. The state of $\mathcal{DFA}$ at time epoch $k+1$ is denoted by $q[k+1]$ ($q[k+1] \in Q$) and is uniquely determined by its state $q[k]$ at the previous time epoch k ($q[k] \in Q$) and the input $\sigma[k]$ applied to the system at time epoch k ($\sigma[k] \in \Sigma$). This is captured via the (possibly partially defined) next-state transition function $\delta : Q \times \Sigma \to Q$:

$$q[k+1] = \delta(q[k], \sigma[k]) \ . \tag{3.1}$$

C. N. Hadjicostis, *Estimation and Inference in Discrete Event Systems*, Communications and Control Engineering,
https://doi.org/10.1007/978-3-030-30821-6_3

The next state of the DFA is considered undefined for pairs of the form $(q, \sigma), q \in Q$, $\sigma \in \Sigma$, for which δ is not specified. The automaton is assumed to start from a known initial state $q[0]$ at time epoch zero.

Remark 3.1 Note that the term "time epoch k" is used to indicate that the time instances at which the state of the automaton changes are not necessarily regular; nevertheless, instead of "time epoch k," we will also frequently use the term "time step k," which is more appropriate for cases where the intervals at which the state of the automaton is updated are regular (e.g., in the implementation of a finite automaton as a clocked digital system). Note that the function δ is assumed to be time-invariant, i.e., it does not change over time. □

Definition 3.1 (*Deterministic Finite Automaton (DFA)*) A deterministic finite automaton (DFA) is a four-tuple $\mathcal{DFA} = (Q, \Sigma, \delta, q[0])$ where $Q = \{q^{(1)}, q^{(2)}, \ldots, q^{(N)}\}$ is a finite set of states, $\Sigma = \{\sigma^{(1)}, \sigma^{(2)}, \ldots, \sigma^{(K)}\}$ is a finite set of inputs, $\delta : Q \times \Sigma \to Q$ is the (possibly partially defined) next-state transition function, and $q[0] \in Q$ is the initial state.

Note that the set Σ should not necessarily be viewed as a set of exogenous inputs. In many cases, Σ also captures endogenous events that are possible at different states of the system. For this reason, we will interchangeably refer to elements of Σ as either *inputs* or *events*.

In a deterministic automaton $\mathcal{DFA} = (Q, \Sigma, \delta, q[0])$, the state $q[0]$ ($q[0] \in Q$) at initialization (time epoch 0) is unique, so that its state at any given time is unique and known as long as the initial state and the applied sequence of inputs is known. In other words, $q[0]$ together with $\sigma[0], \sigma[1], \ldots, \sigma[m]$ uniquely specify $q[1], q[2], \ldots, q[m+1]$. [Note that the state will be taken to be undefined if the applied sequence of inputs leads—at any point—to a state from which the next-state transition function based on the subsequent input is undefined.] The trajectory followed by the system starting from state $q[0]$ under the input sequence $\sigma[0], \sigma[1], \ldots, \sigma[m]$ will be denoted by

$$q[0] \xrightarrow{\sigma[0]} q[1] \xrightarrow{\sigma[1]} q[2] \xrightarrow{\sigma[2]} \cdots \xrightarrow{\sigma[m]} q[m+1] ,$$

and implies that $\delta(q[k], \sigma[k]) = q[k+1]$ (and is, of course, assumed to be defined) for $k = 0, 1, 2, \ldots, m$. We refer to the $(m+2)$-tuple $(q[0], q[1], \ldots, q[m+1])$ as a state $(m+2)$-trajectory (which is an element of the set $Q \times Q \times \cdots \times Q$, where the product is taken $m+2$ times).

Remark 3.2 As we will see later, in nondeterministic automata a trajectory due to a sequence of inputs is not necessarily unique due to uncertainty in the initial state as well as uncertainty in the state transition mechanism; thus, the above notation will have to be generalized. □

We will use the notation $\sigma_m^{m'}$ (for $m' \geq m$) to indicate the sequence of one or more events

$$\sigma_m^{m'} := \sigma[m], \sigma[m+1], \ldots, \sigma[m'] .$$

For convenience, we will also use the notation σ_m^{m-1} to capture the empty sequence. Similarly, we will use the notation $q_m^{m'}$ to denote a trajectory of states starting from $q[m]$ and ending at $q[m']$, i.e.,

$$q_m^{m'} := q[m], q[m+1], \ldots, q[m'] .$$

When the index of the time epoch is not important in our discussions, we will often denote $\sigma_m^{m'}$ as a generic string $s \in \Sigma^*$ (sequence of events from Σ) of the form

$$s = \sigma[m], \sigma[m+1], \ldots, \sigma[m'] ,$$

where $\sigma[m+i] \in \Sigma$ for $i = 0, 1, 2, \ldots, m - m'$.

The next-state transition function δ can be extended to a sequence of $m+1$ inputs (events) $\sigma_k^{k+m} = \sigma[k], \sigma[k+1], \ldots, \sigma[k+m]$ as

$$\delta^{m+1}(q[k], \sigma_k^{k+m}) = \delta(\delta(\delta(\ldots\delta(\delta(q[k], \sigma[k]), \sigma[k+1]), \ldots), \sigma[k+m-1]), \sigma[k+m]), \tag{3.2}$$

where $\delta^{m+1}(q[k], \sigma_k^{k+m})$ is taken to be undefined if $\delta(q[k+i], \sigma[k+i])$ is undefined for some $i \in \{0, 1, \ldots, m\}$ (where $q[k+i] = \delta^i(q[k], \sigma_k^{k+i-1})$). With a slight abuse of notation, we will usually drop the superscript $^{m+1}$ so that $\delta^{m+1}(q[k], \sigma_k^{k+m})$ for a sequence of $m+1$ events σ_k^{k+m} will be represented by

$$\delta(q[k], \sigma_k^{k+m}) := \delta^{m+1}(q[k], \sigma_k^{k+m}) \ , \ m = 0, 1, 2, 3, \ldots \tag{3.3}$$

Note that one can also define δ recursively as

$$\delta(q[k], \sigma_k^{k+i}) \equiv \delta(\delta(q[k], \sigma_k^{k+i-1}), \sigma[k+i])) \ , \ i = 0, 1, 2, 3, \ldots ,$$

with $\delta(q[k], \sigma_k^{k-1}) = q[k]$ for all $q[k] \in Q$.

Remark 3.3 The fact that the next-state transition function may be partially defined can be a notational nuisance in some cases, but this can be easily circumvented by introducing an extra state to our automaton and by using it to capture all inconsistent transitions. If this extra state is denoted by q_{inc}, then the modified DFA is given by $\mathcal{DFA}' = (Q', \Sigma, \delta', q[0])$ where $Q' = Q \cup \{q_{inc}\}$ and $\delta'(q, \sigma)$ for $q \in Q'$ and $\sigma \in \Sigma$ is defined as

$$\delta'(q, \sigma) = \begin{cases} \delta(q, \sigma), & \text{for } q \in Q, \sigma \in \Sigma \text{ such that } \delta(q, \sigma) \text{ is defined,} \\ q_{inc}, & \text{otherwise.} \end{cases}$$

This ensures that the next-state transition function is always defined. □

In some cases, the initial state of a DFA might not be precisely known (due to a variety of reasons) but the automaton might still be considered "deterministic" in the sense that its next-state transition function δ specifies a unique next state given a known current state and an input. For this reason, we will feel free to also talk about a DFA of the form (Q, Σ, δ, Q_0) where Q_0, $Q_0 \subseteq Q$, is the set of possible initial states.

Definition 3.2 (*Reachable States of DFA*) Given $\mathcal{DFA} = (Q, \Sigma, \delta, Q_0)$, we say that state $q \in Q$ is *reachable* if there exists a sequence of events $s \in \Sigma^*$ and an initial state $q_0 \in Q_0$ such that $q = \delta(q_0, s)$. The set of reachable states is denoted by

$$Q_R = \{q \in Q \mid \exists s \in \Sigma^*, \exists q_0 \in Q, \text{ such that } q = \delta(q_0, s)\} .$$

Typically, we assume that all states in Q are reachable (i.e., $Q_R = Q$); if an automaton contains unreachable states, we can define the accessible part of the automaton to be an identical automaton that excludes the states that are not reachable.

Definition 3.3 (*Accessible Part of DFA*) Given $\mathcal{DFA} = (Q, \Sigma, \delta, Q_0)$, its accessible part is denoted by $AC(Q, \Sigma, \delta, Q_0) = (Q_R, \Sigma, \delta_R, Q_0)$, where Q_R is as defined above and δ_R is defined as

$$\delta_R(q_r, \sigma) = \delta(q_r, \sigma)$$

for $q_r \in Q_R$ and $\sigma \in \Sigma$ (note that $\delta_R(q_r, \sigma)$ remains undefined if $\delta(q_r, \sigma)$ is undefined).

A finite automaton can optionally include a set of marked states $Q_m \subseteq Q$. In such case, we will say the automaton is in a marked state at time epoch k if $q[k] \in Q_m$; equivalently, we say that a sequence of inputs σ_0^k leads to a marked state from initial state $q[0]$ if $\delta(q[0], \sigma_0^k) \in Q_m$.

Definition 3.4 (*Marked Deterministic Finite Automaton*) A marked deterministic finite automaton (marked DFA) is a five-tuple $\mathcal{DFA}_m = (Q, \Sigma, \delta, q[0], Q_m)$ where $(Q, \Sigma, \delta, q[0])$ is a DFA, and $Q_m \subseteq Q$ is the set of marked states.

Example 3.1 On the left of Fig. 3.1, we see a DFA $(Q, \Sigma, \delta, q[0])$ where $Q = \{q^{(1)}, q^{(2)}, q^{(3)}, q^{(4)}\}$ is the set of states, $\Sigma = \{\alpha, \beta, \gamma\}$ is the set of inputs, and $q[0] = q^{(1)}$ is the initial state. The next-state transition function δ is defined by the arrows in the figure: for instance, from state $q^{(1)}$ with input α we transition to $q^{(2)}$, from state $q^{(3)}$ with β we transition to $q^{(2)}$, and so forth. Note that the next-state transition function δ is partially defined (e.g., from state $q^{(3)}$ there is no transition with input γ).

On the right of Fig. 3.1, we see a marked DFA that is identical to the one on the left and has as set of marked states $Q_m = \{q^{(2)}\}$, which is indicated by the double white circle in the diagram on the right of Fig. 3.1. There are various sequences

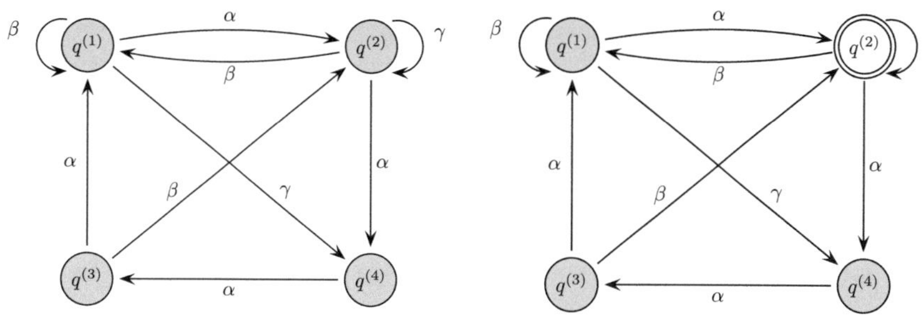

Fig. 3.1 Deterministic finite automaton (left) and marked deterministic finite automaton (right) discussed in Example 3.1

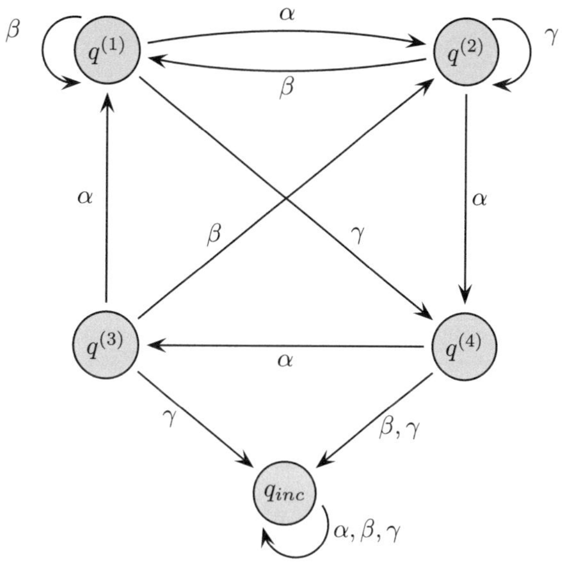

Fig. 3.2 Deterministic finite automaton with absorbing state q_{inc} and fully defined next-state transition function

of inputs that, starting from initial state $q[0] = q^{(1)}$, take us to the marked state $q^{(2)}$: for example, the sequences $\sigma_0^2 = \gamma\alpha\beta$ (i.e., $\sigma[0] = \gamma$, $\sigma[1] = \alpha$, $\sigma[2] = \beta$), or $\sigma_0^3 = \gamma\alpha\alpha\alpha$, and others. In terms of the notation introduced in this section, we have $\delta(q^{(1)}, \gamma\alpha\beta) = q^{(2)}$ and $\delta(q^{(1)}, \gamma\alpha\alpha\alpha) = q^{(2)}$.

In Fig. 3.2 we show an automaton $(Q_c, \Sigma, \delta_c, q[0])$ that is essentially identical to the finite automaton on the left of Fig. 3.1, but includes an additional state, namely q_{inc}, to capture all inconsistent transitions. As a result, $Q_c = Q \cup \{q_{inc}\}$, and for all $q \in Q_c, \sigma \in \Sigma$, we have

$$\delta_c(q, \sigma) = \begin{cases} \delta(q, \sigma) \ , & \text{when } \delta(q, \sigma) \text{ is defined,} \\ q_{inc} \ , & \text{when } \delta(q, \sigma) \text{ is not defined.} \end{cases}$$

This is essentially the finite automaton discussed in Remark 3.3. □

A nondeterministic finite automaton (NFA) is a generalization of a finite automaton that does not require its next-state transition function to map to a unique next state from a given current state and a given input. More specifically, the possibly partially defined next-state transition function is now defined as $\delta : Q \times \Sigma \rightarrow 2^Q$, where 2^Q denotes the set of all subsets of the set of states Q. This is motivated by several applications where the state of the system under a particular input is not uniquely defined due to unknown external or endogenous inputs (in fact, as we will see when we talk about observation models later in this chapter and in Chap. 4, non-determinism arises quite naturally in situations where one tries to model the uncertainty in the system state following a sequence of partial observations). One convenient way of thinking about an NFA is to think of the state $q[k]$ of the NFA as a subset of the set of states Q (and not an element of Q). Given that at time epoch k, state $q \in Q$ is one of the possible states of the NFA (i.e., given that $q \in q[k]$) and given that input $\sigma[k] \in \Sigma$ is applied, the states captured by $\delta(q, \sigma[k])$ are possible states at time epoch $k + 1$. Moreover, the next state $q[k + 1]$ at time epoch $k + 1$ is given by

$$q[k+1] = \cup_{q \in q[k]} \delta(q, \sigma[k]) \ .$$

The fact that the next-state transition function of an NFA may only be partially defined is not really an issue. If $\delta(q, \sigma)$ is undefined for a certain state $q \in Q$ and input $\sigma \in \Sigma$, this is equivalent to $\delta(q, \sigma) = \emptyset$ (where $\emptyset$ represents the empty set).

A marked nondeterministic finite automaton (marked NFA) is an NFA for which a subset of states $Q_m \subseteq Q$ is designated to be the set of marked states; extending the case of marked DFA, we will say the NFA reaches a marked state at time epoch $k + 1$ if $q[k + 1] \cap Q_m \neq \emptyset$.

Definition 3.5 (*Nondeterministic Finite Automaton (NFA)*) A nondeterministic finite automaton (NFA) is a four-tuple $\mathcal{NFA} = (Q, \Sigma, \delta, Q_0)$ where $Q = \{q^{(1)}, q^{(2)}, \ldots, q^{(N)}\}$ is a finite set of states, $\Sigma = \{\sigma^{(1)}, \sigma^{(2)}, \ldots, \sigma^{(K)}\}$ is a finite set of inputs, $\delta : Q \times \Sigma \rightarrow 2^Q$ is the next-state transition function, and $Q_0 \subseteq Q$ is the set of initial states.

Definition 3.6 (*Marked Nondeterministic Finite Automaton*) A marked nondeterministic finite automaton (marked NFA) is a five-tuple $\mathcal{NFA}_m = (Q, \Sigma, \delta, Q_0, Q_m)$ where (Q, Σ, δ, Q_0) is an NFA, and $Q_m \subseteq Q$ is the set of marked states.

The set of possible states of a given NFA at initialization is given by the set Q_0 and is denoted by $q[0] = Q_0$ (note that $q[\cdot]$ denotes a subset of Q, whereas it denoted an element of Q in the case of a DFA). When input $\sigma[0]$ is applied to the system, the set of possible states at time epoch 1 is given by $q[1] = \cup_{q \in q[0]} \delta(q, \sigma[0])$. Continuing in this fashion, we can iteratively define

$$q[k+1] = \cup_{q \in q[k]} \delta(q, \sigma[k]) \ .$$

In fact, with a slight abuse of notation, we will write

$$q[k+1] = \delta(q[k], \sigma[k]) := \cup_{q \in q[k]} \delta(q, \sigma[k]) \; .$$

Using the above (abuse of) notation and given the set of possible states $q[k]$ at time epoch k and the input sequence $\sigma_k^{k+m} = \sigma[k], \sigma[k+1], \ldots, \sigma[k+m]$ of $m+1$ inputs, the next-state transition function can be expressed as

$$\delta^{m+1}(q[k], \sigma_k^{k+m}) = \delta(\delta(\delta(\ldots\delta(\delta(q[k], \sigma[k]), \sigma[k+1]), \ldots), \sigma[k+m-1]), \sigma[k+m]) \; . \tag{3.4}$$

As in the case of DFA, we will typically drop the superscript $^{m+1}$ so that $\delta^{m+1}(q[k], \sigma_k^{k+m})$ for a sequence of $m+1$ inputs σ_k^{k+m} will be represented as

$$\delta(q[k], \sigma_k^{k+m}) := \delta^{m+1}(q[k], \sigma_k^{k+m}) \; , \; m = 0, 1, 2, 3, \ldots$$

Also note that we can define δ recursively as

$$\delta(q[k], \sigma_k^{k+i}) = \delta(\delta(q[k], \sigma_k^{k+i-1}), \sigma[k+i])) \; , \; i = 1, 2, \ldots, m,$$

with $\delta(q[k], \sigma_k^{k-1}) = q[k]$.

As in the case of DFA, we can also define the set of reachable states and the accessible part of a given NFA.

Definition 3.7 (*Reachable States of NFA*) Given $\mathcal{NFA} = (Q, \Sigma, \delta, Q_0)$, we say that state $q \in Q$ is *reachable* if there exists a sequence of events $s \in \Sigma^*$ such that $q \in \delta(Q_0, s)$. The set of reachable states is denoted by

$$Q_R = \{q \in Q \mid \exists s \in \Sigma^*, \text{ such that } q \in \delta(Q_0, s)\} \; .$$

Typically, we assume that all states in Q are reachable (i.e., $Q_R = Q$); if an NFA contains unreachable states, we can define the accessible part of the automaton to be an identical NFA that excludes the states that are not reachable.

Definition 3.8 (*Accessible Part of NFA*) Given $\mathcal{NFA} = (Q, \Sigma, \delta, Q_0)$, its accessible part is denoted by $AC(Q, \Sigma, \delta, Q_0) =: (Q_R, \Sigma, \delta_R, Q_0)$, where Q_R is as defined above and δ_R is defined as

$$\delta_R(q_r, \sigma) = \delta(q_r, \sigma)$$

for $q_r \in Q_R$ and $\sigma \in \Sigma$.

Example 3.2 On the left of Fig. 3.3, we see an NFA $\mathcal{NFA} = (Q, \Sigma, \delta, Q_0)$ where $Q = \{q^{(1)}, q^{(2)}, q^{(3)}, q^{(4)}\}$ is the set of states, $\Sigma = \{\alpha, \beta, \gamma\}$ is the set of inputs, and $Q_0 = \{q^{(1)}, q^{(2)}\}$ is the set of initial states. The next-state transition function δ is

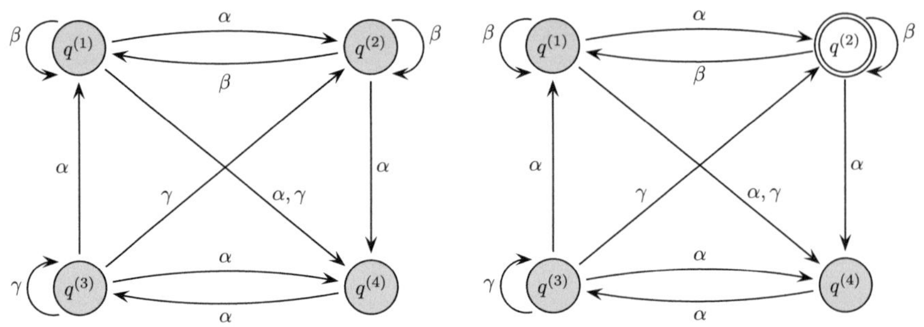

Fig. 3.3 Nondeterministic finite automaton (left) and marked nondeterministic finite automaton (right) discussed in Example 3.2

defined by the arrows in the figure: for instance, from state $q^{(1)}$ with input α we transition to the set of states $\{q^{(2)}, q^{(4)}\}$, from state $q^{(3)}$ with input γ we transition to the set of states $\{q^{(2)}, q^{(3)}\}$, and so forth. Note that the next-state transition function δ is partially defined (e.g., from state $q^{(3)}$ there is no transition with input β), but one can think of such transitions as transitions to the empty set $\emptyset$ (equivalently, to the inconsistent state q_{inc}). From the initial set of states Q_0, if we apply the sequence of inputs $\sigma_0^2 = \alpha\alpha\gamma$ (i.e., $\sigma[0] = \alpha$, $\sigma[1] = \alpha$, $\sigma[2] = \gamma$), we have:

$$\begin{aligned} q[0] &= Q = \{q^{(1)}, q^{(2)}\} \\ q[1] &= \delta(q[0], \alpha) = \{q^{(2)}, q^{(4)}\} \\ q[2] &= \delta(q[1], \alpha) = \{q^{(3)}, q^{(4)}\} \\ q[3] &= \delta(q[2], \gamma) = \{q^{(2)}, q^{(3)}\} \; . \end{aligned}$$

In terms of the notation introduced before the example, we have $\delta(q[0], \alpha\alpha\gamma) = \{q^{(2)}, q^{(3)}\}$.

On the right of Fig. 3.3, we see a marked NFA, denoted by $\mathcal{NFA}_m = (Q, \Sigma, \delta, Q_0, Q_m)$, that is identical to the one on the left of the figure and has as set of marked states the set $Q_m = \{q^{(2)}\}$ (again, we indicate the marked state with the double white circle in the diagram). There are various sequences of inputs which, starting from an initial state in the set $Q_0 = \{q^{(1)}, q^{(2)}\}$ take us to the marked state $q^{(2)}$: for example, the sequences $\sigma_0^2 = \gamma\alpha\gamma$ (i.e., $\sigma[0] = \gamma$, $\sigma[1] = \alpha$, $\sigma[2] = \gamma$), or $\sigma_0^3 = \gamma\alpha\alpha\alpha$, and others. Note that these sequences may also take us to other states: for example, $\gamma\alpha\gamma$ also takes us to state $q^{(3)}$ (from state $q^{(1)}$). □

Remark 3.4 One might be tempted to capture the trajectories followed by a given NFA, starting from some state $q[0] = Q_0$ and under the input sequence $\sigma[0]$, $\sigma[1]$, ..., $\sigma[m]$, as

$$q[0] \stackrel{\sigma[0]}{\longrightarrow} q[1] \stackrel{\sigma[1]}{\longrightarrow} q[2] \stackrel{\sigma[2]}{\longrightarrow} \cdots \stackrel{\sigma[m]}{\longrightarrow} q[m+1] \; ,$$

where each $q[\cdot]$ in the above notation represents the set of possible states at the corresponding time epoch. The above notation suggests that the set of possible trajectories is captured by the set of state $(m + 2)$-trajectories

$$\{(q_0, q_1, q_2, \ldots, q_{m+1}) \mid q_i \in q[i] \text{ for } i = 0, 1, \ldots, m+1\}\ .$$

However, this is not correct because not all such trajectories are necessarily possible in the system (refer to the example below). In order to represent the possible trajectories of states in response to the input sequence $\sigma_0^m \equiv \sigma[0], \sigma[1], \ldots, \sigma[m]$, we need to ensure that state trajectories are compatible with the input sequence. In fact, the state $(m+2)$-trajectory that is compatible with the sequence of $m+1$ inputs σ_0^m is called the state $m+2$-trajectory induced by σ_0^m and is denoted by

$$M^{(m+2)}(\sigma_0^m) = \{(q_0, q_1, \ldots, q_m, q_{m+1}) \mid q_0 \in Q_0, q_i \in Q, q_{i+1} \in \delta(q_i, \sigma[i]) \text{ for } 0 \leq i \leq m\}\ .$$

More details about state trajectories and their graphical representations (via trellis diagrams) are provided in Chap. 4. □

Example 3.3 We now revisit the NFA on the left of Fig. 3.3. In Example 3.2, we saw that the sequence of inputs $\alpha\alpha\gamma$ results in

$$\begin{aligned} q[0] &= \{q^{(1)}, q^{(2)}\} \\ q[1] &= \{q^{(2)}, q^{(4)}\} \\ q[2] &= \{q^{(3)}, q^{(4)}\} \\ q[3] &= \{q^{(2)}, q^{(3)}\}\ . \end{aligned}$$

As mentioned in Remark 3.4, not all state trajectories in the set

$$\{(q_0, q_1, q_2, q_3) \mid q_i \in q[i] \text{ for } i = 0, 1, 2, 3\}$$

are possible. For example, $(q^{(2)}, q^{(2)}, q^{(3)}, q^{(2)})$ is not possible. In fact, out of the 16 trajectories implied by the above formulation, only the following four are possible:

$$\begin{aligned} &(q^{(1)}, q^{(4)}, q^{(3)}, q^{(2)}) \\ &(q^{(1)}, q^{(4)}, q^{(3)}, q^{(3)}) \\ &(q^{(2)}, q^{(4)}, q^{(3)}, q^{(2)}) \\ &(q^{(2)}, q^{(4)}, q^{(3)}, q^{(3)})\ . \end{aligned}$$

The other sequences are not possible. □

Remark 3.5 We will frequently deal with finite automata that take the form (Q, Σ, δ, Q_0) where the next-state transition function $\delta(q, \sigma)$, for all $q \in Q$ and for all $\sigma \in \Sigma$, is either empty or a singleton set. Thus, apart from the initial state (which could be stated in Q_0), such automata behave essentially as deterministic automata. We will refer to them as deterministic finite automata (DFA) with uncertainty in the initial state. □

3.2.1.2 Determinizing a Nondeterministic Finite Automaton

Given $\mathcal{NFA} = (Q, \Sigma, \delta, Q_0)$ (where $\delta : Q \times \Sigma \rightarrow 2^Q$), we can construct an equivalent DFA $(2^Q, \Sigma, \delta_D, Q_0)$ (where 2^Q is the set of all subsets of the set Q) by defining δ_D in the following way: for all $S \in 2^Q$ (i.e., $S \subseteq Q$) and all $\sigma \in \Sigma$, let

$$\delta_D(S, \sigma) = \cup_{q \in S} \delta(q, \sigma) .$$

This leads to an automaton that has at most 2^N states where N is the number of states in the nondeterministic automaton ($N = |Q|$). Since states that are not reachable from state Q_0 are not of interest, we usually assume that this deterministic automaton consists only of states that are accessible (via some sequence of inputs) from the initial state Q_0. This is denoted by

$$\mathcal{DFA}_D = AC(2^Q, \Sigma, \delta_D, Q_0) .$$

The deterministic automaton $\mathcal{DFA}_D$ is equivalent to the $\mathcal{NFA}$ we started off with in the sense that, following a sequence of inputs, there is a one-to-one mapping between the set of states represented by each state of $\mathcal{DFA}_D$ and the set of states in which the $\mathcal{NFA}$ resides. This is illustrated in the following example:

Example 3.4 Consider the NFA, denoted by $\mathcal{NFA}$, on the left of Fig. 3.3 with initial set of states $Q_0 = \{q^{(1)}\}$. In Fig. 3.4 we see the finite automaton $\mathcal{DFA}_D$ that results from the determinization procedure discussed above. To avoid confusion we refer to the states of $\mathcal{NFA}$ as "NFA states" and to the states of $\mathcal{DFA}_D$ as the "DFA states." The DFA states are subsets of the set of NFA states $Q = \{q^{(1)}, q^{(2)}, q^{(3)}, q^{(4)}\}$, which we now denote as $Q = \{1, 2, 3, 4\}$ for simplicity.

To better understand the construction of $\mathcal{DFA}_D$, we start from its initial state which is given by $Q_0 = \{1\}$. From this initial DFA state $\{1\}$, we have the following transitions: (i) with input α, we transition to $\{2, 4\}$; with input β we remain at $\{1\}$; with input γ, we transition to $\{4\}$. In each case, the set of NFA states we transition to is the set of NFA states reached from initial NFA state 1 when applying the corresponding input. From these new DFA states, we can define transitions in a similar fashion: for example, from DFA state $\{2, 4\}$ we transition with input β to DFA state $\{1, 2\}$ (because this is the set of NFA states we transition to from an NFA state in the set $\{2, 4\}$ with input β), and so forth.

The finite automaton $\mathcal{DFA}_D$ in Fig. 3.4 can be used to obtain the set of possible states that the NFA resides in following some sequence of inputs. For example, following the sequence of inputs $\sigma_0^3 = \alpha\alpha\alpha\alpha$ (i.e., $\sigma[0] = \alpha$, $\sigma[1] = \alpha$, $\sigma[2] = \alpha$, $\sigma[3] = \alpha$), the automaton $\mathcal{DFA}_D$ reaches state $\{1, 2, 3, 4\}$; this means that the set of possible states in the NFA on the left of Fig. 3.3 (with initial state $Q_0 = \{1\}$) is the set $\{1, 2, 3, 4\}$. In the same way, $\mathcal{DFA}_D$ can be used to obtain the set of NFA states following any input sequence; however, it cannot easily provide sequences of NFA states that are compatible with a particular sequence of inputs (see Remark 3.4 and Example 3.3).

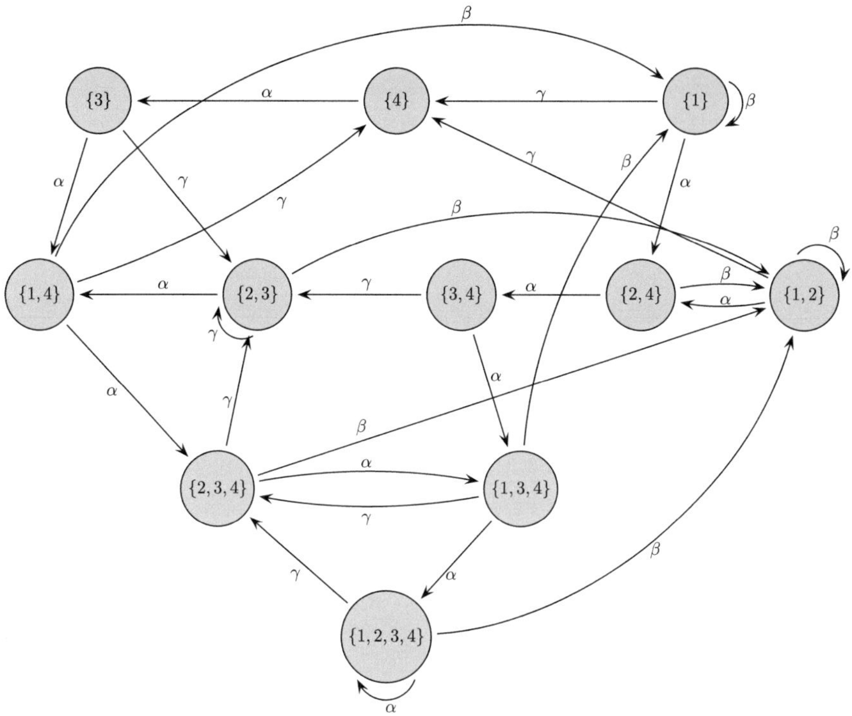

Fig. 3.4 Finite automaton $\mathcal{DFA}_D$ resulting from the determinization of nondeterministic finite automaton $\mathcal{NFA}$ on the left of Fig. 3.3

Note that $\mathcal{DFA}_D$ has 11 states out of a possible $2^4 = 16$ states. In reality, there is implicitly one additional state, which has not been drawn (following trend in such constructions). More specifically, the DFA state { }, which corresponds to the empty set of NFA states, is the state we reach from any DFA state at which there is an undefined transition: for instance, from DFA state {3, 4} with input β we could have drawn a transition to DFA state { }; similarly, from DFA state {4} with input β or input γ we could have drawn a transition to DFA state { }, and so forth. Note that once we transition to DFA state { } we stay there under any input (i.e., it is an absorbing state). □

3.2.1.3 Transition Matrix Notation

Given $\mathcal{DFA} = (Q, \Sigma, \delta, q[0])$ where $Q = \{q^{(1)}, q^{(2)}, \ldots, q^{(N)}\}$ is the finite set of states and $\Sigma = \{\sigma^{(1)}, \sigma^{(2)}, \ldots, \sigma^{(K)}\}$ is the finite set of inputs, we can use an $N \times 1$ binary indicator vector $\mathbf{q}[k]$ to represent its state at time epoch k: more specifically, if the automaton is in state $q^{(j)}$ at time epoch k (i.e., $q[k] = q^{(j)}$), then $\mathbf{q}[k]$ is a vector of zeros except a single nonzero entry with value "1" at the jth position. With this notation in hand, we have

$$\mathbf{q}[k+1] = \mathbf{\Delta}_{\sigma[k]}\mathbf{q}[k] ,$$

where $\mathbf{\Delta}_{\sigma^{(i)}}$ for $i = \{1, 2, \ldots, K\}$ are K binary transition matrices (of size $N \times N$), each associated with one of the K inputs in Σ. Specifically, each column of the matrix $\mathbf{\Delta}_{\sigma^{(i)}}$ has at most one nonzero entry with value "1" (so that $\Delta_{\sigma^{(i)}}$ has at most N nonzero entries, all with value "1"). Moreover, a nonzero entry at the j'th-jth position of $\mathbf{\Delta}_{\sigma^{(i)}}$ denotes a transition from state $q^{(j)}$ to state $q^{(j')}$ under input $\sigma^{(i)}$. Clearly, the constraint that each column of $\Delta_{\sigma^{(i)}}$ has at most one nonzero entry simply reflects the requirement that, in a DFA, there can be at most one transition out of a particular state under the given input. If column j of matrix $\mathbf{\Delta}_{\sigma^{(i)}}$ is zero, then input $\sigma^{(i)}$ is undefined at state $q^{(j)}$.

The above notation can be easily extended to the case of a sequence of observations $\sigma_k^{k+m} = \sigma[k], \sigma[k+1], \ldots, \sigma[k+m]$ as

$$\mathbf{q}[k+m+1] = \mathbf{\Delta}_{\sigma[k+m]}\Delta_{\sigma[k+m-1]} \cdots \mathbf{\Delta}_{\sigma[k]}\mathbf{q}[k] ,$$

with $\mathbf{q}[k+m+1]$ being the all zero vector if the sequence of inputs σ_k^{k+m} is undefined from the state represented by the indicator vector $\mathbf{q}[k]$.

For an $\mathcal{NFA} = (Q, \Sigma, \delta, Q_0)$, where $Q = \{q^{(1)}, q^{(2)}, \ldots, q^{(N)}\}$ is the finite set of states and $\Sigma = \{\sigma^{(1)}, \sigma^{(2)}, \ldots, \sigma^{(K)}\}$ is the finite set of inputs, the transition matrix notation extends in a similar way, but has three major differences:

- Transition matrices $\mathbf{\Delta}_{\sigma^{(i)}}$, $i = \{1, 2, \ldots, K\}$, are still binary with entries that are "0" or "1", but are not required to have at most one nonzero entry in each column. This reflects the fact that, in an NFA, a transition under input $\sigma^{(i)}$ from a particular state does not necessarily lead to a unique state.
- For the same reason, vector $\mathbf{q}[k]$ is not required to have a single nonzero entry. For example, if $q[k] = \{q^{(1)}, q^{(2)}\}$, then $\mathbf{q}[k]$ will be a vector with the first two entries "1" and the remaining entries zero. This also applies to the initial vector $\mathbf{q}[0]$.
- Also note that unless matrix–vector and matrix–matrix multiplication are taken over the ring $(\{0, 1\}, \max, \times)$, the vector $\mathbf{q}[k]$ is not necessarily binary (its entries take nonnegative integer values—see the example below). Nevertheless, $\mathbf{q}[k]$ can still be viewed as an indicator vector because the following property can be verified easily: the jth entry of $\mathbf{q}[k]$ is nonzero iff (if and only if) state $q^{(j)} \in q[k]$ (i.e., state $q^{(j)}$ is a possible state in the given NFA at time epoch k).

Example 3.5 Consider again the DFA on the left of Fig. 3.1. It has four states and three inputs, therefore we can define the following three 4×4 transition matrices:

$$\mathbf{\Delta}_\alpha = \begin{bmatrix} 0&0&1&0 \\ 1&0&0&0 \\ 0&0&0&1 \\ 0&1&0&0 \end{bmatrix}, \quad \mathbf{\Delta}_\beta = \begin{bmatrix} 1&1&0&0 \\ 0&0&1&0 \\ 0&0&0&0 \\ 0&0&0&0 \end{bmatrix}, \quad \mathbf{\Delta}_\gamma = \begin{bmatrix} 0&0&0&0 \\ 0&1&0&0 \\ 0&0&0&0 \\ 1&0&0&0 \end{bmatrix} .$$

For instance, the first column of matrix $\mathbf{\Delta}_\alpha$ indicates that from state 1 we transition to state 2 with input α. Similarly, the fourth column of matrix $\mathbf{\Delta}_\gamma$ indicates that there is no transition from state 4 under input γ. More generally, the (j', j) entry of matrix $\mathbf{\Delta}_\sigma$ indicates a transition from state j to state j' under input σ. As expected, each column has at most a single nonzero entry (with value 1) because we are dealing with a deterministic automaton with a possibly partially defined transition function.

If we multiply any sequence involving the above transition matrices, the resulting matrix will also be a 4×4 transition matrix. For example,

$$\mathbf{\Delta}_{\alpha\alpha\alpha\alpha} = \mathbf{\Delta}_\alpha \mathbf{\Delta}_\alpha \mathbf{\Delta}_\alpha \mathbf{\Delta}_\alpha = \begin{bmatrix} 1\,0\,0\,0 \\ 0\,1\,0\,0 \\ 0\,0\,1\,0 \\ 0\,0\,0\,1 \end{bmatrix} .$$

This implies that, following the sequence of inputs $\alpha\alpha\alpha\alpha$, we go from state 1 to state 1 (as indicated by the first column of matrix $\mathbf{\Delta}_{\alpha\alpha\alpha\alpha}$), from state 2 to state 2 (as indicated by the second column of matrix $\mathbf{\Delta}_{\alpha\alpha\alpha\alpha}$), from state 3 to state 3 (as indicated by the third column of matrix $\mathbf{\Delta}_{\alpha\alpha\alpha\alpha}$), and from state 4 to state 4 (as indicated by the fourth column of matrix $\Delta_{\alpha\alpha\alpha\alpha}$).

Let us now consider the NFA on the left of Fig. 3.3. It has four states and three inputs, therefore we can define the following three 4×4 transition matrices:

$$\mathbf{\Delta}'_\alpha = \begin{bmatrix} 0\,0\,1\,0 \\ 1\,0\,0\,0 \\ 0\,0\,0\,1 \\ 1\,1\,1\,0 \end{bmatrix} , \mathbf{\Delta}'_\beta = \begin{bmatrix} 1\,1\,0\,0 \\ 0\,1\,0\,0 \\ 0\,0\,0\,0 \\ 0\,0\,0\,0 \end{bmatrix} , \mathbf{\Delta}'_\gamma = \begin{bmatrix} 0\,0\,0\,0 \\ 0\,0\,1\,0 \\ 0\,0\,1\,0 \\ 1\,0\,0\,0 \end{bmatrix} .$$

For instance, the first column of matrix $\mathbf{\Delta}'_\alpha$ indicates that from state 1 we transition to states 2 and 4 with input α. Similarly, the fourth column of matrix $\mathbf{\Delta}'_\gamma$ indicates that there is no transition from state 4 under input γ. More generally, the (j', j) entry of matrix $\mathbf{\Delta}'_\sigma$ indicates a transition from state j to state j' under input σ. As expected, each column is binary and can have zero, one, two, three or four nonzero entries (with value 1) because we are dealing with an NFA.

If we multiply any sequence involving the above transition matrices over the ring $(\{0, 1\}, \max, \times)$, the resulting matrix will also be a 4×4 transition matrix of the same form. For example, following the sequence of inputs $\alpha\alpha\alpha\alpha$, we have

$$\mathbf{\Delta}'_{\alpha\alpha\alpha\alpha} \overset{(\{0,1\},\max,\times)}{=} \mathbf{\Delta}'_\alpha \mathbf{\Delta}'_\alpha \mathbf{\Delta}'_\alpha \mathbf{\Delta}'_\alpha \overset{(\{0,1\},\max,\times)}{=} \begin{bmatrix} 1\,0\,1\,1 \\ 1\,1\,1\,0 \\ 1\,1\,1\,1 \\ 1\,1\,1\,1 \end{bmatrix} .$$

This implies that, following the sequence of inputs $\alpha\alpha\alpha\alpha$, we go from state 1 to states 1, 2, 3, and 4 (as indicated by the first column of matrix $\mathbf{\Delta}'_{\alpha\alpha\alpha\alpha}$), from state 2 to states 2, 3, and 4 (as indicated by the second column of matrix $\mathbf{\Delta}'_{\alpha\alpha\alpha\alpha}$), from state

3 to all states (as indicated by the third column of matrix $\mathbf{\Delta}'_{\alpha\alpha\alpha\alpha}$), and from state 4 to states 1, 3, and 4 (as indicated by the fourth column of matrix $\mathbf{\Delta}'_{\alpha\alpha\alpha\alpha}$).

If we multiply any sequence involving the above transition matrices over the ring of integers, the resulting matrix will be a 4×4 integer matrix but will not necessarily have binary entries. For the case we saw above, we have

$$\mathbf{\Delta}''_{\alpha\alpha\alpha\alpha} = \mathbf{\Delta}'_{\alpha}\mathbf{\Delta}'_{\alpha}\mathbf{\Delta}'_{\alpha}\mathbf{\Delta}'_{\alpha} = \begin{bmatrix} 1 & 0 & 1 & 1 \\ 1 & 1 & 1 & 0 \\ 1 & 1 & 2 & 1 \\ 2 & 1 & 2 & 2 \end{bmatrix} .$$

Notice that the zero/nonzero structure of matrix $\mathbf{\Delta}''_{\alpha\alpha\alpha\alpha}$ is identical to the zero/nonzero structure of matrix $\mathbf{\Delta}'_{\alpha\alpha\alpha\alpha}$, and can be interpreted in exactly the same way (Rosen 2011): for example, following the sequence of inputs $\alpha\alpha\alpha\alpha$, we go from state 1 to states 1, 2, 3, and 4 (because all entries of the first column of matrix $\mathbf{\Delta}''_{\alpha\alpha\alpha\alpha}$ are nonzero), and so forth. In fact, matrix $\mathbf{\Delta}''_{\alpha\alpha\alpha\alpha}$ contains additional information not contained in $\mathbf{\Delta}'_{\alpha\alpha\alpha\alpha}$. More specifically, entry $\mathbf{\Delta}''_{\alpha\alpha\alpha\alpha}(4, 1) = 2$ indicates that with input sequence $\alpha\alpha\alpha\alpha$ one can go from state 1 to state 4 following *two* different state trajectories (namely, $1 \to 2 \to 4 \to 3 \to 4$ and $1 \to 4 \to 3 \to 1 \to 4$); similarly, entry $\mathbf{\Delta}''_{\alpha\alpha\alpha\alpha}(1, 1) = 1$ indicates that with input sequence $\alpha\alpha\alpha\alpha$ one can go from state 1 to state 1 following *one* state trajectory (namely, $1 \to 2 \to 4 \to 3 \to 1$); finally, entry $\mathbf{\Delta}''_{\alpha\alpha\alpha\alpha}(2, 4) = 0$ indicates that with input sequence $\alpha\alpha\alpha\alpha$ one cannot go from state 4 to state 2.

It can be shown by induction on the length of the input sequence (see, for example, Rosen 2011), that the above property of matrices of the form of $\mathbf{\Delta}''_{\sigma_0^m}$ (obtained by multiplication over the ring of integers) holds in general for all input sequences σ_0^m. □

Remark 3.6 As we will see in Chap. 4 in more detail, the transition matrix $\mathbf{\Delta}_{\sigma^{(i)}}$ introduced above is equivalent to the notion of a state mapping $M_{\sigma^{(i)}}$ under input $\sigma^{(i)}$. They are both different ways of capturing all pairs of states of the form $(q^{(j')}, q^{(j)})$ where $q^{(j')}$ is a state reachable from state $q^{(j)}$ under input $\sigma^{(i)}$. □

3.2.1.4 Compositions of Finite Automata

Automata can be composed in various ways to obtain more complex automata. Given a pair of automata, two common compositions are the product composition and the parallel composition, which are described next.

Definition 3.9 (*Product of Automata*) Given two deterministic (or, respectively, nondeterministic) finite automata $\mathcal{DFA}_1 = (Q_1, \Sigma_1, \delta_1, q_{01})$ and $\mathcal{DFA}_2=(Q_2, \Sigma_2, \delta_2, q_{02})$ (or, respectively, $\mathcal{NFA}_1 = (Q_1, \Sigma_1, \delta_1, Q_{01})$ and $\mathcal{NFA}_2 = (Q_2, \Sigma_2, \delta_2, Q_{02})$), their product is a deterministic (respectively, nondeterministic) finite automaton $\mathcal{DFA}_{1\times 2} \equiv \mathcal{DFA}_1 \times \mathcal{DFA}_2 = (Q, \Sigma, \delta, q_0)$ (or, respectively, $\mathcal{NFA}_{1\times 2} \equiv \mathcal{NFA}_1 \times \mathcal{NFA}_2 = (Q, \Sigma, \delta, Q_0)$) where

1. $Q = Q_1 \times Q_2$,
2. $\Sigma = \Sigma_1 \cap \Sigma_2$,
3. $\delta : Q \times \Sigma \to Q$ (respectively, $\delta : Q \times \Sigma \to 2^Q$) is defined for all $q_1 \in Q_1, q_2 \in Q_2$ as

$$\delta((q_1, q_2), \sigma) = \begin{cases} \delta_1(q_1, \sigma) \times \delta_2(q_2, \sigma), & \text{when } both\ \delta_1(q_1, \sigma) \text{ and } \\ & \delta_2(q_2, \sigma) \text{ are defined,} \\ \text{undefined,} & \text{otherwise,} \end{cases}$$

4. $q_0 = (q_{01}, q_{02})$ (respectively, $Q_0 = Q_{01} \times Q_{02}$).

Note that in the DFA case, $\delta_1(q_1, \sigma) \times \delta_2(q_2, \sigma)$ is simply the pair of states $(\delta_1(q_1, \sigma), \delta_2(q_2, \sigma))$, whereas in the NFA case, $\delta_1(q_1, \sigma) \times \delta_2(q_2, \sigma)$ is the set of all pairs with first element taken from the set $\delta_1(q_1, \sigma)$ and second element taken from the set $\delta_2(q_2, \sigma)$. Also, the initial set of states $Q_0 = Q_{01} \times Q_{02}$ in the nondeterministic case represents the set of all pairs of the form (q_1, q_2) with $q_1 \in Q_{01}, q_2 \in Q_{02}$. In the NFA case, if we follow the convention that δ being undefined for a particular pair of state and input is equivalent to δ mapping to the empty set for that particular pair of state and input, then we could simply remove the condition "*both* $\delta_1(q_1, \sigma)$ and $\delta_2(q_2, \sigma)$ are defined" in the above definition (also, under this interpretation, the second clause in the definition of δ is not needed as long as we keep in mind that δ is only defined for inputs in $\Sigma = \Sigma_1 \cap \Sigma_2$).

The parallel composition described below is identical to the product composition as far as common inputs are concerned, but also allows for each of the two automata to act separately on non-shared inputs.

Definition 3.10 (*Parallel Composition of Automata*) Given two deterministic (or, respectively, nondeterministic) finite automata $\mathcal{DFA}_1 = (Q_1, \Sigma_1, \delta_1, q_{01})$ and $\mathcal{DFA}_2 = (Q_2, \Sigma_2, \delta_2, q_{02})$ (or, respectively, $\mathcal{NFA}_1 = (Q_1, \Sigma_1, \delta_1, Q_{01})$ and $\mathcal{NFA}_2 = (Q_2, \Sigma_2, \delta_2, Q_{02})$), their parallel composition is a deterministic (respectively, nondeterministic) finite automaton $\mathcal{DFA}_{1||2} \equiv \mathcal{DFA}_1 || \mathcal{DFA}_2 = (Q, \Sigma, \delta, q_0)$ (or, respectively, $\mathcal{NFA}_{1||2} \equiv \mathcal{NFA}_1 || \mathcal{NFA}_2 = (Q, \Sigma, \delta, Q_0)$) where

1. $Q = Q_1 \times Q_2$,
2. $\Sigma = \Sigma_1 \cup \Sigma_2$,
3. $\delta : Q \times \Sigma \to Q$ (respectively, $\delta : Q \times \Sigma \to 2^Q$) is defined for all $q_1 \in Q_1, q_2 \in Q_2$ as

$$\delta((q_1, q_2), \sigma) = \begin{cases} \delta_1(q_1, \sigma) \times \delta_2(q_2, \sigma), & \text{for } \sigma \in \Sigma_1 \cap \Sigma_2, \text{ and } both \\ & \delta_1(q_1, \sigma) \text{ and } \delta_2(q_2, \sigma) \text{ defined,} \\ \delta_1(q_1, \sigma) \times q_2, & \text{for } \sigma \in \Sigma_1 \setminus \Sigma_2, \text{ and} \\ & \delta_1(q_1, \sigma) \text{ defined,} \\ q_1 \times \delta_2(q_2, \sigma), & \text{for } \sigma \in \Sigma_2 \setminus \Sigma_1, \text{ and} \\ & \delta_2(q_2, \sigma) \text{ defined,} \\ \text{undefined,} & \text{otherwise.} \end{cases}$$

4. $q_0 = (q_{01}, q_{02})$ (respectively, $Q_0 = Q_{01} \times Q_{02}$).

As mentioned earlier in this section, in the DFA case, $\delta_1(q_1, \sigma) \times \delta_2(q_2, \sigma)$ is simply the pair $(\delta_1(q_1, \sigma), \delta_2(q_2, \sigma))$, whereas in the NFA case, the following hold: (i) $\delta_1(q_1, \sigma) \times \delta_2(q_2, \sigma)$ is the set of all pairs with first element taken from the set $\delta_1(q_1, \sigma)$ and second element taken from the set $\delta_2(q_2, \sigma)$, (ii) $\delta_1(q_1, \sigma) \times q_2$ is the set of all pairs with first element taken from the set $\delta_1(q_1, \sigma)$ and second element q_2, and (iii) $q_1 \times \delta_2(q_2, \sigma)$ is the set of all pairs with first element q_1 and second element taken from the set $\delta_2(q_2, \sigma)$. Also, note that the initial set of states $Q_0 = Q_{01} \times Q_{02}$ in the nondeterministic case represents the set of all pairs of the form (q_1, q_2) with $q_1 \in Q_{01}, q_2 \in Q_{02}$. If we follow the convention that δ being undefined for a particular state and input is equivalent to δ mapping to the empty set, then we could simply remove the conditions "*both* $\delta_1(q_1, \sigma)$ and $\delta_2(q_2, \sigma)$ defined", "$\delta_1(q_1, \sigma)$ defined", "$\delta_2(q_2, \sigma)$ defined" in the above definition.

The above definitions for the product and the parallel composition can be easily extended to the case where the two deterministic or nondeterministic automata are marked, with sets of marked states given by Q_{m1} and Q_{m2} respectively. In both cases (product and parallel composition), the marked states of the resulting automaton are taken to be the set $Q_m = Q_{m1} \times Q_{m2}$.

Example 3.6 Consider the nondeterministic finite automata $\mathcal{NFA}_1$ (left) and $\mathcal{NFA}_2$ (right) shown in Fig. 3.5. For the first automaton, we have $\mathcal{NFA}_1 = (Q_1, \Sigma_1, \delta_1, Q_{01})$ where $Q_1 = \{1_1, 2_1\}$, $\Sigma_1 = \{a, b, c\}$, $Q_{01} = \{1_1\}$, and δ_1 as defined in the figure. For the second automaton, we have $\mathcal{NFA}_2 = (Q_2, \Sigma_2, \delta_2, Q_{02})$ where $Q_2 = \{1_2, 2_2, 3_2\}$, $\Sigma_2 = \{a, b, d\}$, $Q_{02} = \{1_2\}$, and δ_2 as defined in the figure (note that the second automaton is actually deterministic).

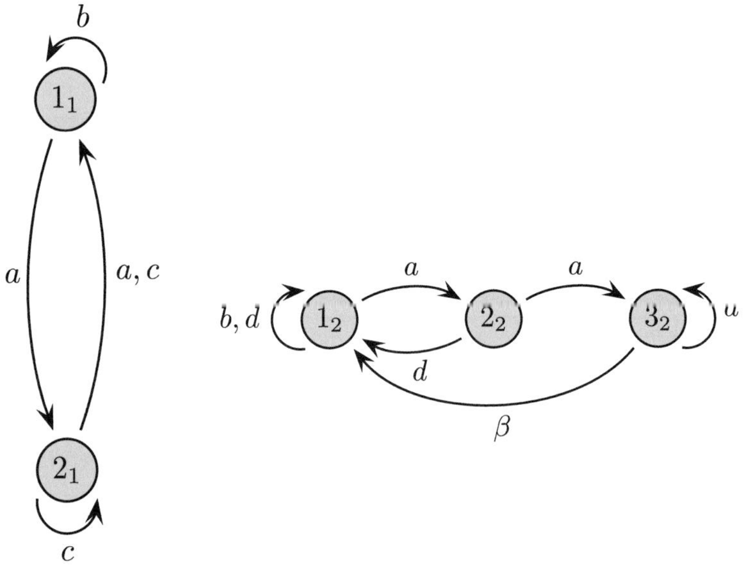

Fig. 3.5 Nondeterministic finite automata $\mathcal{NFA}_1$ (left) and $\mathcal{NFA}_2$ (right) used to demonstrate product and parallel compositions in Example 3.6

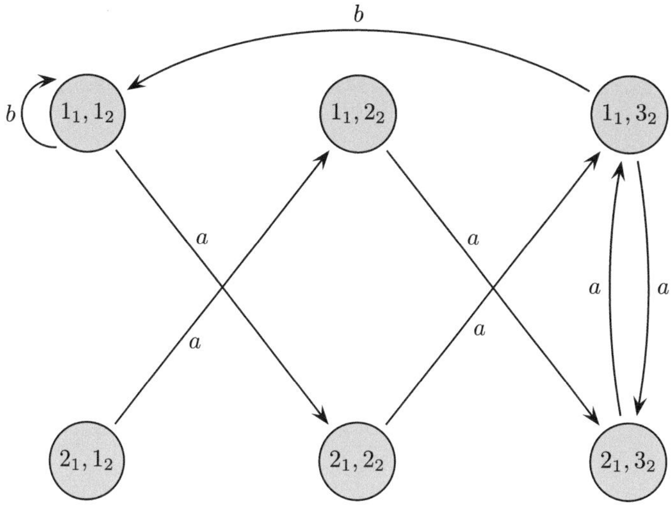

Fig. 3.6 Automaton $\mathcal{NFA}_{1\times 2}$ resulting from the product of the two automata in Fig. 3.5

The automaton $\mathcal{NFA}_{1\times 2} \equiv \mathcal{NFA}_1 \times \mathcal{NFA}_2 = (Q, \Sigma_\times, \delta_\times, Q_0)$ (resulting from taking the product of the above automata) is shown in Fig. 3.6. It has states

$$Q = \{(1_1, 1_2), (1_1, 2_2), (1_1, 3_2), (2_1, 1_2), (2_1, 2_2), (2_1, 3_2)\}$$

(involving all pairs of states from each of the two automata), inputs $\Sigma_\times = \Sigma_1 \cap \Sigma_2 = \{a, b\}$ (i.e., the common inputs for the two automata), and initial state $Q_0 = Q_{01} \times Q_{02} = \{(1_1, 1_2)\}$ (involving all possible pairs of initial states in each of the two automata). The next-state transition function $\delta_\times$ is as shown in the figure. For example, from state $(1_1, 2_2)$ with input a the product automaton transitions to state $(2_1, 3_2)$ (because automaton $\mathcal{NFA}_1$ from state 1_1 transitions to state 2_1 with input a, and automaton $\mathcal{NFA}_2$ from state 2_2 transitions to state 3_2 with input a). Note that from state $(1_1, 2_2)$ there is no other transition: inputs c and d are not common to the two automata, whereas input b is not allowed from state 2_2 of $\mathcal{NFA}_2$. It turns out that the product automaton is deterministic in this case, but, in general, this will not necessarily be true. Note that there are certain states in $\mathcal{NFA}_{1\times 2}$ (namely, $(2_1, 1_2)$ and $(1_1, 2_2)$) that are not reachable from the initial state $(1_1, 1_2)$ and can be dropped; however, if the initial states were different (e.g., if $Q_{01} = \{2_1\}$ and $Q_{02} = \{1_2\}$), these states would have to be retained.

The automaton $\mathcal{NFA}_{1||2} = \mathcal{NFA}_1 || \mathcal{NFA}_2 \equiv (Q, \Sigma_{||}, \delta_{||}, Q_0)$ (resulting from taking the parallel composition of the two automata in Fig. 3.5) is shown in Fig. 3.7. It has states

$$Q = \{(1_1, 1_2), (1_1, 2_2), (1_1, 3_2), (2_1, 1_2), (2_1, 2_2), (2_1, 3_2)\}$$

(involving all pairs of states from each of the two automata, as in the case of the product automaton), inputs $\Sigma_{||} = \{a, b, c, d\}$ (i.e., the union of the inputs of the two

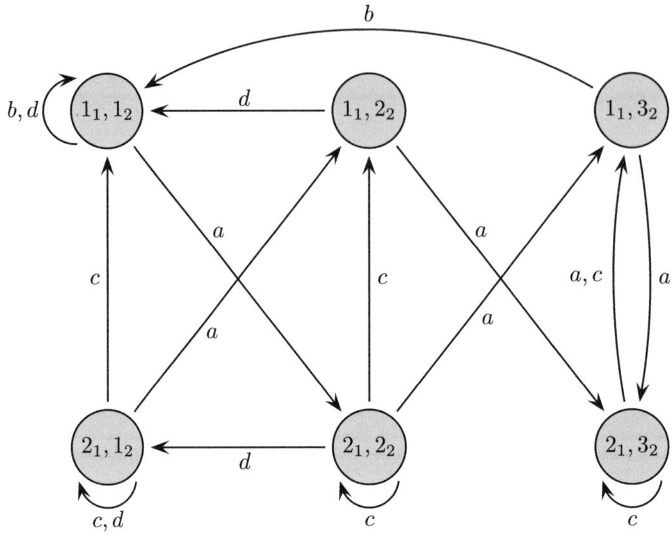

Fig. 3.7 Automaton $\mathcal{NFA}_{1||2}$ resulting from the parallel composition of the two automata in Fig. 3.5

automata), and initial state $Q_0 = \{(1_1, 1_2)\}$ (involving all possible pairs of initial states in each of the two automata, as in the case of the product automaton). The next-state transition function $\delta_{||}$ is as shown in Fig. 3.6. Consider, for example, state $(1_1, 2_2)$:

1. With input a, the parallel composition automaton transitions to state $(2_1, 3_2)$ (because, with input a, automaton $\mathcal{NFA}_1$ transitions to state 2_1 from state 1_1, and automaton $\mathcal{NFA}_2$ transitions to state 3_2 from state 2_2).
2. With input b, the parallel composition automaton does not have any transition out of state $(1_1, 2_2)$ because input b (which is common to both automata) is not allowed from state 2_2 of $\mathcal{NFA}_2$.
3. With input c (which is private to $\mathcal{NFA}_1$), the parallel composition automaton does not have any transition out of state $(1_1, 2_2)$ (because $\mathcal{NFA}_1$ does not have any transition with input c from state 1_1).
4. With input d (which is private to $\mathcal{NFA}_2$), the parallel composition automaton transitions to state $(1_1, 1_2)$ (because $\mathcal{NFA}_2$ from state 2_2 transitions to state 1_2 under input d); the component of the state corresponding to $\mathcal{NFA}_1$ does not change.

As expected, the first two cases (which correspond to inputs that are common to the two automata) are identical to what we had in the case of the product of the two automata which was discussed earlier in this example. What is added in the case of parallel composition, are the inputs that are private to the two automata. Note that the parallel composition automaton is not deterministic as it inherits the non-determinism that $\mathcal{NFA}_1$ has due to its private events (e.g., from state $(2_1, q_2)$ with input c, which is private to $\mathcal{NFA}_1$, the parallel composition goes to states $(2_1, q_2)$ and $(1_1, q_2)$, where $q_2 \in \{1_2, 2_2, 3_2\}$). [In this argument, it is important that c is a private event; inheritance of non-determinism in the case of common events is not automatic due to the fact that common events may be disallowed by the other component.] □

3.2.2 *Languages*

3.2.2.1 Strings and Languages

Note that one can also refer to Chap. 2 for a preliminary discussion on alphabets, words, and languages. Some of the definitions are repeated here for completeness.

Given an alphabet of symbols (or events or letters) $\Sigma = \{\sigma^{(1)}, \sigma^{(2)}, \ldots, \sigma^{(K)}\}$, we refer to the concatenation of $m+1$ such symbols as a string $s = \sigma_0\sigma_1 \ldots \sigma_m$ of length $m+1$ ($\sigma_i \in \Sigma$ for $i = 0, 1, \ldots, m$). The terminology is natural if one thinks of Σ as an alphabet so that sequences of symbols generate strings (or words) in this alphabet. Note that in terms of the notation used for automata in the previous section, the sequence of events $\sigma_k^{k+m} = \sigma[k], \sigma[k+1], \ldots, \sigma[k+m]$ is equivalent to the string $s = \sigma[k]\sigma[k+1]\ldots\sigma[k+m]$ together with the index k of the time epoch at which this sequence of events starts.

The set of all finite length strings that can be generated by alphabet Σ is denoted by Σ^* and includes the empty string ϵ (note that the empty string is sometimes also denoted by λ). The concatenation of two strings $s = \sigma_0\sigma_1 \ldots \sigma_m$, $\sigma_i \in \Sigma$ for $i = 0, 1, \ldots, m$, and $s' = \sigma'_0\sigma'_1 \ldots \sigma'_{m'}$, $\sigma'_{i'} \in \Sigma$ for $i' = 0, 1, \ldots, m'$ is denoted by $s \cdot s' := ss'$ and is given by

$$s \cdot s' \equiv ss' = \sigma_0\sigma_1 \ldots \sigma_m\sigma'_0\sigma'_1 \ldots \sigma'_{m'} \ .$$

Moreover, we define $s\epsilon = \epsilon s = s$ for all $s \in \Sigma^*$. The length ℓ of a string s can be defined recursively as follows: we take $\ell(s) = 0$ for $s = \epsilon$, whereas when $s = \sigma s'$ for $\sigma \in \Sigma$, $s' \in \Sigma^*$ (i.e., when $s \neq \epsilon$) we set

$$\ell(s) = 1 + \ell(s') \ .$$

One can easily show that $\ell(ss') = \ell(s) + \ell(s')$ (Rosen 2011).

A prefix of a string $s \in \Sigma^*$ is a string t for which there exists a string t' such that $tt' = s$. The prefix closure of string s is the set of all of its prefixes (including the empty string ϵ and string s itself) and is denoted by the set $\overline{s} = \{t \in \Sigma^* \mid \exists t' \in \Sigma^* \text{ such that } tt' = s\}$. The post-string of string s after $t \in \overline{s}$ is denoted by s/t and is defined as the string t' such that $tt' = s$. In other words, if $s = tt'$ for $s, t, t' \in \Sigma^*$, then t is a prefix of s (i.e., $t \in \overline{s}$) and $s/t = t'$.

Following the above convention, for a string s, $s \in \Sigma^*$, we can similarly define the string t' for which there exists a string t such that $tt' = s$ as a suffix of string s. The suffix closure of string s is the set of all of its suffixes (including the empty string ϵ and string s itself) and is denoted by the set $\underline{s} = \{t' \in \Sigma^* \mid \exists t \in \Sigma^* \text{ such that } tt' = s\}$. The pre-string of string s before $t' \in \underline{s}$ is defined as the string t such that $tt' = s$.

Given an alphabet Σ, a language $\mathcal{L}$ is a subset of Σ^*, i.e., a subset (of possibly infinite cardinality) of finite length strings composed of symbols in the alphabet. We can define the union and intersection of languages in terms of the sets they represent, i.e., for two languages $\mathcal{L}_1$ and $\mathcal{L}_2$ we have

$$\mathcal{L}_1 \cup \mathcal{L}_2 = \{s \mid s \in \mathcal{L}_1 \text{ or } s \in \mathcal{L}_2\},$$
$$\mathcal{L}_1 \cap \mathcal{L}_2 = \{s \mid s \in \mathcal{L}_1 \text{ and } s \in \mathcal{L}_2\}.$$

Furthermore, the concatenation of two languages is taken to be the set of strings that is generated by concatenating any string from the first language with any string in the second language

$$\mathcal{L}_1 \cdot \mathcal{L}_2 := \mathcal{L}_1\mathcal{L}_2 = \{s_1s_2 \mid s_1 \in \mathcal{L}_1,\ s_2 \in \mathcal{L}_2\}\ .$$

Finally, the prefix closure of a language $\mathcal{L}$ is denoted by $\overline{\mathcal{L}}$ and is defined to be the union of all sets of strings that are prefixes of strings in the language

$$\overline{\mathcal{L}} = \{t \in \Sigma^* \mid \exists s \in \mathcal{L} \text{ such that } t \in \overline{s}\}$$
$$= \cup_{s \in L} \overline{s}\ .$$

A language $\mathcal{L}$ is said to be prefix closed if $\mathcal{L} = \overline{\mathcal{L}}$.

3.2.2.2 Automata as Language Generators

Given $\mathcal{DFA} = (Q, \Sigma, \delta, q[0])$, the language generated by the automaton is the set of all sequences of events (starting at initialization and having arbitrary lengths) that are possible from the initial state $q[0] \in Q$ (i.e., they lead to a well-defined state of the finite automaton). For this terminology to make sense, the sequence of events $\sigma_0^m = \sigma[0], \sigma[1], \ldots, \sigma[m]$ needs to be viewed as a string of length $m + 1$, which (together with index 0 that indicates the time epoch at which this sequence of events is applied) provides completely equivalent information. If we let $s = \sigma_0\sigma_1 \ldots \sigma_m$ where $\sigma_0 = \sigma[0], \sigma_1 = \sigma[1], \ldots, \sigma_m = \sigma[m]$ are the corresponding events, we can define the next-state transition function for this string s starting from state $q[0]$ (or from any state for that matter) as

$$\delta(q[0], s) := \delta(q[0], \sigma_0^m)\ ,$$

where $\delta(q, \sigma_0^m)$ was defined in Eq. (3.3). For $s = \epsilon$, we define $\delta(q, \epsilon) = q$ for all $q \in Q$.

With this notation in hand, it is not very hard to define the language for each of the types of automata we have considered in Sect. 3.2.1. For $\mathcal{DFA} = (Q, \Sigma, \delta, q[0])$, a string $s \in \Sigma^*$ is in the language $\mathcal{L}(\mathcal{DFA})$ of the automaton if it leads to a well-defined state from state $q[0]$ (i.e., if $\delta(q[0], s)$ is defined). Similarly, for marked $\mathcal{DFA}_m = (Q, \Sigma, \delta, q[0], Q_m)$, a string $s \in \Sigma^*$ is in the language $\mathcal{L}(\mathcal{DFA}_m)$ of the automaton if it leads to a well-defined state from state $q[0]$ (i.e., if $\delta(q[0], s)$ is defined); in addition, string $s \in \Sigma^*$ is in the *marked* language $\mathcal{L}_m(\mathcal{DFA}_m)$ of the automaton if it leads to a marked state from state $q[0]$ (i.e., if $\delta(q[0], s) \in Q_m$).

Definition 3.11 (*Language of DFA*) Given $\mathcal{DFA} = (Q, \Sigma, \delta, q[0])$, the language $\mathcal{L}(\mathcal{DFA}) \subseteq \Sigma^*$ generated by the automaton is defined as

$$\mathcal{L}(\mathcal{DFA}) = \{s \in \Sigma^* \mid \delta(q[0], s) \text{ is defined}\} .$$

Definition 3.12 (*Languages of Marked DFA*) Given marked $\mathcal{DFA}_m = (Q, \Sigma, \delta, q[0], Q_m)$, the language $\mathcal{L}(\mathcal{DFA}_m) \subseteq \Sigma^*$ and the marked language $\mathcal{L}_m(\mathcal{DFA}_m) \subseteq \Sigma^*$ generated by the automaton are defined as

$$\begin{aligned}
\mathcal{L}(\mathcal{DFA}) &= \{s \in \Sigma^* \mid \delta(q[0], s) \text{ is defined}\} ,\\
\mathcal{L}_m(\mathcal{DFA}_m) &= \{s \in \Sigma^* \mid \delta(q[0], s) \text{ is defined and } \delta(q[0], s) \in Q_m\} .
\end{aligned}$$

For the NFA case, the definitions are similar.

Definition 3.13 (*Language of NFA*) Given $\mathcal{NFA} = (Q, \Sigma, \delta, Q_0)$, the language $\mathcal{L}(\mathcal{NFA}) \subseteq \Sigma^*$ generated by the automaton is defined as

$$\mathcal{L}(\mathcal{NFA}) = \{s \in \Sigma^* \mid \exists q_0 \in Q_0 \text{ such that } \delta(q_0, s) \text{ is defined}\} .$$

Definition 3.14 (*Languages of Marked NFA*) Given marked $\mathcal{NFA}_m = (Q, \Sigma, \delta, Q_0, Q_m)$, the language $\mathcal{L}(\mathcal{NFA}_m) \subseteq \Sigma^*$ and the marked language $\mathcal{L}_m(\mathcal{NFA}_m) \subseteq \Sigma^*$ generated by the automaton are defined as

$$\begin{aligned}
\mathcal{L}(\mathcal{NFA}_m) &= \{s \in \Sigma^* \mid \exists q_0 \in Q_0 \text{ such that } \delta(q_0, s) \text{ is defined}\} ,\\
\mathcal{L}_m(\mathcal{NFA}_m) &= \{s \in \Sigma^* \mid \exists q_0 \in Q_0 \text{ such that } \delta(q_0, s) \text{ is defined and}\\
&\qquad \delta(q_0, s) \cap Q_m \neq \emptyset\} .
\end{aligned}$$

We will also use the notation $L(G, q)$ and $L(G, Q')$ to denote respectively strings that can be generated by a DFA or an NFA $G = (Q, \Sigma, \delta, Q_0)$, starting from state q or starting from a state in the set Q', i.e.,

$$\begin{aligned}
L(G, q) &= \{s \in \Sigma^* \mid \delta(q, s) \text{ is defined}\} ,\\
L(G, Q') &= \{s \in \Sigma^* \mid \exists q' \in Q' \{\delta(q', s) \text{ is defined}\}\} .
\end{aligned}$$

Similar definitions could be reproduced for the marked language of a marked DFA or a marked NFA.

The language $\mathcal{L}(\mathcal{DFA})$ generated by a DFA and the language $\mathcal{L}(\mathcal{NFA})$ generated by an NFA capture all sequences of events that can be generated by the underlying automaton. These languages are prefix closed (i.e., $\mathcal{L}(\mathcal{DFA}) = \overline{\mathcal{L}}(\mathcal{DFA})$ and

$\mathcal{L}(\mathcal{NFA}) = \overline{\mathcal{L}}(\mathcal{NFA})$) and provide a behavioral description of allowable activity in the underlying automaton. Note that the languages $\mathcal{L}_m(\mathcal{DFA}_m)$ and $\mathcal{L}_m(\mathcal{NFA}_m)$ are *not* necessarily prefix closed. Languages generated by finite automata (deterministic or nondeterministic) are called *regular* languages.

One can also prove interesting properties of the languages generated by automata obtained via the product and parallel composition operations on two given (deterministic or nondeterministic) finite automata. For instance, one can prove the theorem below (a proof can be found in Cassandras and Lafortune 2007). One can also obtain extensions of such results to the case of marked languages and marked automata obtained via product and parallel composition operations (see, for example, Cassandras and Lafortune 2007).

Theorem 3.1 *Consider two DFA (or NFA) $\mathcal{DFA}_1 = (Q_1, \Sigma_1, \delta_1, q_{01})$ and $\mathcal{DFA}_2 = (Q_2, \Sigma_2, \delta_2, q_{02})$ (or $\mathcal{NFA}_1 = (Q_1, \Sigma_1, \delta_1, Q_{01})$ and $\mathcal{NFA}_2 = (Q_2, \Sigma_2, \delta_2, Q_{02})$), and the DFA (or NFA) $\mathcal{DFA}_{1\times 2} \equiv \mathcal{DFA}_1 \times \mathcal{DFA}_2$ (respectively, $\mathcal{NFA}_{1\times 2} \equiv \mathcal{NFA}_1 \times \mathcal{NFA}_2$) resulting from the product composition as defined in Definition 3.9. Then, the languages generated by these three automata satisfy $\mathcal{L}(\mathcal{DFA}_{1\times 2}) = \mathcal{L}(\mathcal{DFA}_1) \cap \mathcal{L}(\mathcal{DFA}_2)$ (respectively, $\mathcal{L}(\mathcal{NFA}_{1\times 2}) = \mathcal{L}(\mathcal{NFA}_1) \cap \mathcal{L}(\mathcal{NFA}_2)$).*

Example 3.7 Consider again $\mathcal{NFA}_1$ (left) and $\mathcal{NFA}_2$ (right) shown in Fig. 3.5, where $\mathcal{NFA}_1 = (Q_1, \Sigma_1, \delta_1, Q_{01})$ with $Q_1 = \{1_1, 2_1\}$, $\Sigma_1 = \{a, b, c\}$, $Q_{01} = \{1_1\}$, and δ_1 as defined in the figure; and $\mathcal{NFA}_2 = (Q_2, \Sigma_2, \delta_2, Q_{02})$ with $Q_2 = \{1_2, 2_2, 3_2\}$, $\Sigma_2 = \{a, b, d\}$, $Q_{02} = \{1_2\}$, and δ_2 as defined in the figure (recall that the second automaton is actually deterministic).

The language generated by $\mathcal{NFA}_1$ is given by

$$\mathcal{L}(\mathcal{NFA}_1) = \{\epsilon, a, b, aa, ac, ba, bb, aaa, aab, aca, acb, acc, baa, bac, bba, bbb, \ldots\} ,$$

where strings have been ordered according to their length (first criterion) and lexicographically (second criterion). Similarly, the language generated by $\mathcal{NFA}_2$ is given by

$$\mathcal{L}(\mathcal{NFA}_2) = \{\epsilon, a, b, d, aa, ad, ba, bb, bd, da, db, dd, aaa, aab, ada, adb, add, baa, bad, bba, bbb, bbd, bda, bdb, bdd, daa, dad, dba, dbb, dbd, dda, ddb, ddd, \ldots\} ,$$

where strings have again been ordered according to their length (first criterion) and lexicographically (second criterion). Note that the intersection of the two languages given above is

$$\mathcal{L}(\mathcal{NFA}_1) \cap \mathcal{L}(\mathcal{NFA}_2) = \{\epsilon, a, b, aa, ba, bb, aaa, aab, baa, bba, bbb, \ldots\} .$$

The above is actually identical to the language $\mathcal{L}(\mathcal{NFA}_{1\times 2})$ of the automaton $\mathcal{NFA}_{1\times 2}$ resulting from the product composition of automata $\mathcal{NFA}_1$ and $\mathcal{NFA}_2$, and shown in Fig. 3.6. □

3.3 Observation Models: Finite Automata with Outputs

The activity at time epoch k in a given automaton $\mathcal{DFA} = (Q, \Sigma, \delta, q[0])$ usually generates an observation that we will capture via an output $y[k]$. In general, this output depends on the state of the automaton $q[k]$ at time epoch k as well as the input $\sigma[k]$ applied at time epoch k. More specifically, we assume that there is a finite set of possible outputs $Y = \{y^{(1)}, y^{(2)}, \ldots, y^{(R)}\}$ and an output function $\lambda : Q \times \Sigma \to Y$ that specifies the output (including, in some cases, the absence of it) depending on the particular state of the automaton and the applied input at a given time epoch.

One important distinction we have to keep in mind is between automata that provide an output at each time epoch and automata which, depending on the specific state they are in and the input that is applied, might produce no output (at those particular time epochs). To model such cases, we will assume that at these instances the automaton produces the unobservable output which we will denote with the empty symbol ϵ. This distinction implies that we will have to treat two cases: finite automata models *without ϵ-observations* (i.e., without silent transitions) and finite automata models *with ϵ-observations* (i.e., with silent transitions). Essentially, the latter models allow silent state transitions to occur, which is an important characteristic to keep in mind when developing our state estimation and event inference procedures in later chapters. Both of these procedures have to be based on observable transitions, which implies that they will be driven by events in the output set Y. Among other things, this also implies that estimation and fault diagnosis procedures will be driven by sequences of events that occur at time epochs that do not necessarily coincide with the time epochs in the underlying system because some of the time epochs in the original system will be "erased" (more precisely, they will be unobservable) due to the absence of any outputs. We revisit these issues in Chaps. 4 and 5, where we develop state estimation and fault diagnosis procedures for finite automata in detail.

3.3.1 Finite Automata Without Silent Transitions

We start by considering observation models that are associated with DFA, including the special (and relatively well known) cases of Moore automata and labeled automata. We then discuss observation models for NFA.

3.3.1.1 Observation Models for DFA Without Silent Transitions

Definition 3.15 (*DFA with Outputs*) A DFA with outputs is a six-tuple $\mathcal{DFA} = (Q, \Sigma, Y, \delta, \lambda, q[0])$ where $(Q, \Sigma, \delta, q[0])$ is a DFA and $\lambda : Q \times \Sigma \to Y$ is the (possibly partially defined) output function for a finite set of outputs $Y = \{y^{(1)}, y^{(2)}, \ldots, y^{(R)}\}$.

The output function λ is assumed without loss of generality to be surjective. Given state $q[k]$ and a sequence of inputs σ_k^{k+m} that is applied starting at time epoch k, the sequence of states q_k^{k+m+1} and the sequence of outputs y_k^{k+m} that are generated can be obtained iteratively starting from state $q[k]$ as follows:

$$\begin{aligned}
y[k] &= \lambda(q[k], \sigma[k]) \\
q[k+1] &= \delta(q[k], \sigma[k]) \\
\\
y[k+1] &= \lambda(q[k+1], \sigma[k+1]) \\
q[k+2] &= \delta(q[k+1], \sigma[k+1]) \\
&\vdots \\
y[k+m-1] &= \lambda(q[k+m-1], \sigma[k+m-1]) \\
q[k+m] &= \delta(q[k+m-1], \sigma[k+m-1]) \\
\\
y[k+m] &= \lambda(q[k+m], \sigma[k+m]) \\
q[k+m+1] &= \delta(q[k+m], \sigma[k+m])
\end{aligned} \tag{3.5}$$

Given an initial state $q[k]$ and a valid sequence of inputs σ_k^{k+m} (i.e., a sequence of inputs that results in a well-defined sequence of states q_k^{k+m+1} as defined above), the corresponding sequence of outputs y_k^{k+m} is also well-defined if the output function λ is defined for all pairs of states and inputs for which δ is defined. This is a sufficient condition for y_k^{k+m} to be well-defined and we will assume that it holds unless we specify otherwise. [Note that this assumption is mostly technical because this condition becomes necessary when all states in Q are reachable from the initial state $q[0]$.]

We will use the notation

$$\begin{aligned}
\delta_{seq}(q[k], \sigma_k^{k+m}) &= q_k^{k+m+1} \\
\lambda_{seq}(q[k], \sigma_k^{k+m}) &= y_k^{k+m}
\end{aligned} \tag{3.6}$$

to denote the fact that the sequence of inputs σ_k^{k+m} applied at state $q[k]$ generates the sequence of states q_k^{k+m+1} and the sequence of outputs y_k^{k+m}, as defined in the iteration (3.5) above. In general, for λ_{seq} to be known, we need knowledge of both the state transition function δ and the output function λ.

Example 3.8 In Fig. 3.8 we see a DFA with outputs (and without silent transitions) $(Q, \Sigma, Y, \delta, \lambda, q[0])$ where $Q = \{q^{(1)}, q^{(2)}, q^{(3)}, q^{(4)}\}$ is the set of states, $\Sigma = \{\alpha, \beta, \gamma\}$ is the set of inputs, $Y = \{0, 1\}$ is the set of outputs, and $q[0] = q^{(1)}$ is the initial state. The next-state transition function δ and the output function λ are defined by the arrows and the labels in the figure: for instance, from state $q^{(1)}$ with input α we transition to $q^{(2)}$ and generate output 0 (this is captured by the label "$\alpha/0$" on that arrow); similarly, from state $q^{(3)}$ with input β we transition to $q^{(2)}$ and generate output 1 (this is captured by the label "$\beta/1$" on that arrow); and so forth.

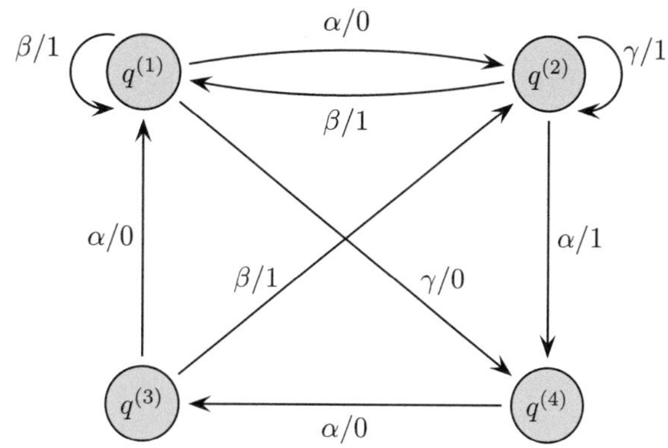

Fig. 3.8 Deterministic finite automaton with outputs discussed in Example 3.8

Notice that if one ignores the outputs and the output function, the finite automaton in Fig. 3.8 is identical to the one on the left of Fig. 3.1, which was discussed in Example 3.1. Thus, given a particular sequence of inputs, the two automata will follow the same sequence of states. The main difference is that the automaton in Fig. 3.8 also conveys information about the output sequence that is observed. For example, the sequence of inputs $\alpha\gamma\alpha$ will induce the state sequence $q^{(1)} \to q^{(2)} \to q^{(2)} \to q^{(4)}$ and the sequence of outputs 011 (i.e., $y[0] = 0$, $y[1] = 1$, and $y[2] = 1$). Knowing the output sequence does not necessarily allow us to determine the input or state sequence; for example, the input sequence $\alpha\beta\beta$ will induce the state sequence $q^{(1)} \to q^{(2)} \to q^{(1)} \to q^{(1)}$ and identical sequence of outputs 011. In terms of the notation introduced earlier, we have

$$\delta_{seq}(q^{(1)}, \alpha\gamma\alpha) = q^{(1)}\, q^{(2)}\, q^{(2)}\, q^{(4)} \;,$$
$$\lambda_{seq}(q^{(1)}, \alpha\gamma\alpha) = 011 \;,$$

and

$$\delta_{seq}(q^{(1)}, \alpha\beta\beta) = q^{(1)}\, q^{(2)}\, q^{(1)}\, q^{(1)} \;,$$
$$\lambda_{seq}(q^{(1)}, \alpha\beta\beta) = 011 \;.$$

As a final note, observe that the next-state transition function δ and the output function λ are both partially defined (e.g., from state $q^{(3)}$ there is no transition/output with input γ); nevertheless, the output function is defined whenever the transition function is defined, which we argued is a sufficient condition to ensure a well-defined output for each well-defined sequence of states. □

We now discuss two important special cases of observation functions.

(1) Output depends only on state: In this scenario, the output function at time epoch k depends only on the state $q[k]$ and it is not a function of the input applied at time epoch k. In other words, the output function λ is of the form $\lambda_Q : Q \to Y$. In this case, we have an automaton that is called a *Moore automaton* or (*Moore automaton*).

Definition 3.16 (*Deterministic Moore Automaton (QDFA)*) A deterministic Moore automaton is a six-tuple $\mathcal{QDFA} = (Q, \Sigma, Y, \delta, \lambda_Q, q[0])$ where $(Q, \Sigma, \delta, q[0])$

is a DFA and $\lambda_Q : Q \rightarrow Y$ is the output function for a finite set of outputs $Y = \{y^{(1)}, y^{(2)}, \ldots, y^{(R)}\}$.

Assuming that all states in Q are reachable from $q[0]$ (and nonterminating), we require the output function λ_Q of a Moore automaton to be defined for all possible states. The reason is that, if the output function is not defined from a reachable state q, then the output when receiving any valid input at that state will be undefined (unless state q is terminating state). We usually also assume without loss of generality that λ_Q is surjective.

(2) Output depends only on input: In this scenario, the output function at time epoch k only depends on the input $\sigma[k]$ applied at time epoch k and it is not a function of the state $q[k]$ at time epoch k. In other words, the output function λ is of the form $\lambda_\Sigma : \Sigma \rightarrow Y$. In this case, we have an automaton that is called a *labeled automaton*.

Definition 3.17 (*Labeled Deterministic Finite Automaton (LDFA)*) A labeled deterministic finite automaton (LDFA) is a six-tuple $\mathcal{LFA} = (Q, \Sigma, Y, \delta, \lambda_\Sigma, q[0])$ where $(Q, \Sigma, \delta, q[0])$ is a DFA and $\lambda_\Sigma : \Sigma \rightarrow Y$ is the output function for a finite set of outputs $Y = \{y^{(1)}, y^{(2)}, \ldots, y^{(R)}\}$.

Note that in the case of an LDFA, the function $\lambda_{seq}(q[k], \sigma_k^{k+m})$ is simply a function of the sequence of inputs σ_k^{k+m} (i.e., it is independent of $q[k]$ and can be defined without knowledge of the state transition function δ, as long as $\delta(q[k], \sigma_k^{k+m})$ is defined). In fact, with a slight abuse of notation, we will not even require that $\delta(q[k], \sigma_k^{k+m})$ be defined, and simply take $\lambda_{\Sigma,seq}(\sigma_k^{k+m})$ to be equal to y_k^{k+m} where

$$y[k+i] = \lambda_\Sigma(\sigma[k+i]) \tag{3.7}$$

for $i = 0, 1, \ldots, m$.

Example 3.9 On the top of Fig. 3.9, we see a deterministic Moore automaton $\mathcal{QDFA} = (Q, \Sigma, Y, \delta, \lambda_Q, q[0])$, where $Q = \{q^{(1)}, q^{(2)}, q^{(3)}, q^{(4)}\}$ is the set of states, $\Sigma = \{\alpha, \beta, \gamma\}$ is the set of inputs, $Y = \{0, 1\}$ is the set of outputs, and $q[0] = q^{(1)}$ is the initial state. What is important here is that the output function from each state is the same regardless of the input (i.e., for each state $q \in Q$, we have $\lambda(q, \sigma) = \lambda(q, \sigma')$ for all $\sigma, \sigma' \in \Sigma$—at least whenever λ is defined). In terms of the diagram on the top of Fig. 3.9, this translates to having all arrows out of each state be associated with the same output. For example, transitions out of state $q^{(1)}$ are all associated with output 0; transitions out of state $q^{(2)}$ are all associated with output 1; transitions out of state $q^{(3)}$ are all associated with output 1; and transitions out of state $q^{(4)}$ are all associated with output 0.

At the bottom of Fig. 3.9 we see an example of an LDFA, denoted by $\mathcal{LDFA} = (Q, \Sigma, Y, \delta, \lambda_\Sigma, q[0])$, where $Q = \{q^{(1)}, q^{(2)}, q^{(3)}, q^{(4)}\}$ is the set of states, $\Sigma = \{\alpha, \beta, \gamma\}$ is the set of inputs, $Y = \{0, 1\}$ is the set of outputs, and $q[0] = q^{(1)}$ is the initial state. What is important here is that the output function for each input is the same regardless of the state (i.e., for each input $\sigma \in \Sigma$, we have $\lambda(q, \sigma) = \lambda(q', \sigma)$ for all $q, q' \in Q$—at least whenever both of $\delta(q, \sigma)$ and $\delta(q', \sigma)$ are defined).

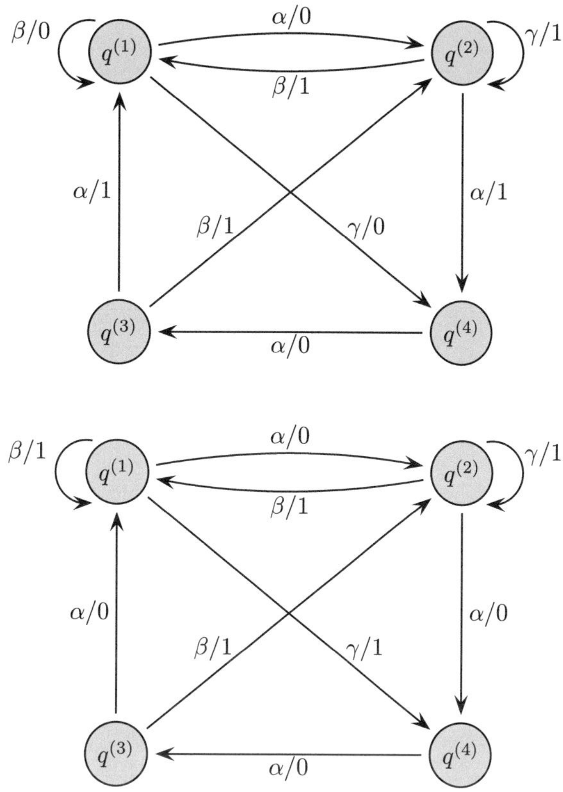

Fig. 3.9 Deterministic Moore automaton (top) and labeled deterministic finite automaton (bottom)

In terms of the diagram at the bottom of Fig. 3.9, this translates to having all arrows with the same input label be associated with the same output. For example, transitions with input label α are all associated with output 0; transitions with input label β are all associated with output 1; and transitions with input label γ are all associated with output 1. □

3.3.1.2 Observation Models for NFA Without Silent Transitions

The extension of observation models to NFA requires output functions λ of the form

$$\lambda : Q \times \Sigma \times Q \to Y \; ,$$

i.e., functions that take as an argument a triplet (q_i, σ, q_f) of an initial state $q_i \in Q$, an input $\sigma \in \Sigma$, and a final state $q_f \in Q$. This is necessary because from a state q_i there might be multiple possible transitions under input σ, each of which can, in general, generate a different output. Clearly, if λ is defined for all triplets (q_i, σ, q_f) such that $q_f \in \delta(q_i, \sigma)$, then the output is guaranteed to be defined for each possible state transition (so that, given an initial state and a sequence of inputs, an output sequence will be defined for each possible sequence of states). [Again, this is a sufficient condition that becomes necessary if all states are reachable from the initial set of states under some sequence of inputs.] An NFA with outputs is formally defined below.

Definition 3.18 (*NFA with Outputs*) An NFA with outputs is a six-tuple $\mathcal{NFA} = (Q, \Sigma, Y, \delta, \lambda, Q_0)$ where (Q, Σ, δ, Q_0) is an NFA and $\lambda : Q \times \Sigma \times Q \to Y$ is the (possibly partially defined) output function for a finite set of outputs $Y = \{y^{(1)}, y^{(2)}, \ldots, y^{(R)}\}$.

When trying to extend the definitions of functions δ_{seq} and λ_{seq} in Eq. (3.6) to the NFA case, the notation becomes significantly more complex because

(i) $\delta_{seq}(q[k], \sigma_k^{k+m})$ needs to return all possible sequences of states that are compatible with the current state(s) of the system (as captured by the set $q[k]$) and the applied input sequence σ_k^{k+m}; this should be contrasted with the case of a deterministic automaton where $\delta_{seq}(q[k], \sigma_k^{k+m})$ returns, if defined, a unique compatible sequence of states.
(ii) $\lambda_{seq}(q[k], \sigma_k^{k+m})$ needs to return all sequences of outputs that are generated by the above possible sequences of states; again, this should be contrasted with the case of a deterministic automaton where $\lambda_{seq}(q[k], \sigma_k^{k+m})$ returns, if defined, a unique compatible sequence of outputs (refer to Remark 3.4).

Also note that the same sequence of outputs could be generated by multiple state sequences; in fact, when considering compatible sequences of states (that are returned by δ_{seq}) and compatible sequences of outputs (that are returned by λ_{seq}) one should be aware that each compatible sequences of states can be matched with a single—but not multiple—compatible sequence of outputs. The definitions of δ_{seq} and λ_{seq} for the NFA case without silent transitions are provided below.

$$\begin{aligned}
\delta_{seq}(q[k], \sigma_k^{k+m}) &= \{(q_k, q_{k+1}, \ldots, q_{k+m}, q_{k+m+1}) \mid q_k \in q[k], q_{i+1} \in \delta(q_i, \sigma[i]) \\
&\qquad \text{for } k \le i \le k+m\}\ , \\
\lambda_{seq}(q[k], \sigma_k^{k+m}) &= \{(y_k, y_{k+1}, \ldots, y_{k+m}) \mid \\
&\qquad \exists (q_k, q_{k+1}, \ldots, q_{k+m}, q_{k+m+1}) \in \delta_{seq}(q[k], \sigma_k^{k+m}) \\
&\qquad \text{such that } y_i = \lambda(q_i, \sigma[i], q_{i+1}) \text{ for } k \le i \le k+m\}\ .
\end{aligned}$$

When the output function only depends on the state (Moore automaton) or the input (labeled automaton), the specification of λ is simplified significantly and we are naturally led to the notions of nondeterministic Moore automaton (without silent state visitations) and labeled nondeterministic finite automata (without silent inputs). **(1) Output depends only on state**: In this scenario, the output function at time epoch k depends only on the state $q[k]$ and it is not a function of the input applied at time epoch k. In other words, the output function λ is of the form $\lambda_Q : Q \to Y$. In this case, we have an automaton that is a nondeterministic Moore automaton without silent state visitations.

Definition 3.19 (*Nondeterministic Moore Automaton (QNFA)*) A nondeterministic Moore Automaton (QNFA) is a six-tuple $\mathcal{QNFA} = (Q, \Sigma, Y, \delta, \lambda_Q, Q_0)$ where (Q, Σ, δ, Q_0) is an NFA and $\lambda_Q : Q \to Y$ is the output function for a finite set of outputs $Y = \{y^{(1)}, y^{(2)}, \ldots, y^{(R)}\}$.

Assuming that all states in Q are reachable from states in Q_0 and nonterminating, we require the output function λ_Q of a Moore automaton to be defined for all possible states. We usually also assume without loss of generality that λ_Q is surjective. In this case, the functions δ_{seq} and λ_{seq} take the following form:

$$\begin{aligned}\delta_{seq}(q[k], \sigma_k^{k+m}) = \{&(q_k, q_{k+1}, \ldots, q_{k+m}, q_{k+m+1}) \mid q_k \in q[k], q_{i+1} \in \delta(q_i, \sigma[i]) \\ &\text{for } k \leq i \leq k+m\}\ , \\ \lambda_{seq}(q[k], \sigma_k^{k+m}) = \{&(y_k, y_{k+1}, \ldots, y_{k+m}) \mid \\ &\exists (q_k, q_{k+1}, \ldots, q_{k+m}, q_{k+m+1}) \in \delta_{seq}(q[k], \sigma_k^{k+m}) \\ &\text{such that } y_i = \lambda_Q(q_i) \text{ for } k \leq i \leq k+m\}\ .\end{aligned}$$

(2) Output depends only on input: In this scenario, the output function at time epoch k only depends on the input $\sigma[k]$ applied at time epoch k and it is not a function of the state $q[k]$ at time epoch k. In other words, the output function λ is of the form $\lambda_\Sigma : \Sigma \to Y$. In this case, we have an automaton that is called a labeled nondeterministic finite automaton (LNFA) without silent inputs.

Definition 3.20 (*Labeled NFA (LNFA)*) A labeled NFA (LNFA) is a six-tuple $\mathcal{LNFA} = (Q, \Sigma, Y, \delta, \lambda_\Sigma, Q_0)$ where (Q, Σ, δ, Q_0) is an NFA and $\lambda_\Sigma : \Sigma \to Y$ is the output function for a finite set of outputs $Y = \{y^{(1)}, y^{(2)}, \ldots, y^{(R)}\}$.

The output mapping λ_Σ is typically assumed (without loss of generality) to be surjective. As in the case of an LDFA, we can define the function $\lambda_{seq}(q[k], \sigma_k^{k+m})$ that generates the sequence of outputs y_k^{k+m} starting from state $q[k]$ with applied input sequence σ_k^{k+m} without knowledge of $q[k]$ and without even requiring $\delta(q[k], \sigma_k^{k+m})$ to be defined. Therefore, again with a slight abuse of notation, we can define $\lambda_{\Sigma,seq}(\sigma_k^{k+m})$ exactly as in Eq. (3.7).

Example 3.10 On the top of Fig. 3.10 we see an NFA with outputs (and without silent transitions) $(Q, \Sigma, Y, \delta, \lambda, Q_0)$ where $Q = \{q^{(1)}, q^{(2)}, q^{(3)}, q^{(4)}\}$ is the set of states, $\Sigma = \{\alpha, \beta, \gamma\}$ is the set of inputs, $Y = \{0, 1\}$ is the set of outputs, and $Q_0 = \{q^{(1)}, q^{(2)}\}$ is the set of initial states. The next-state transition function δ and the output function λ are defined by the arrows and the labels in the figure. Notice that if one ignores the outputs and the output function, the finite automaton in Fig. 3.10 is identical to the one on the left of Fig. 3.3, which was discussed in Example 3.2. Thus, given a particular sequence of inputs, the two automata will follow the same sequences of states. The main difference is that the automaton in Fig. 3.10 also conveys information about the output sequence that is observed. It is worth pointing out that this output sequence will in general be different for different sequences of states that are generated by the same sequence of inputs. For example, if input $\alpha\alpha\alpha$ is applied, it will generate the following state and corresponding output sequences:

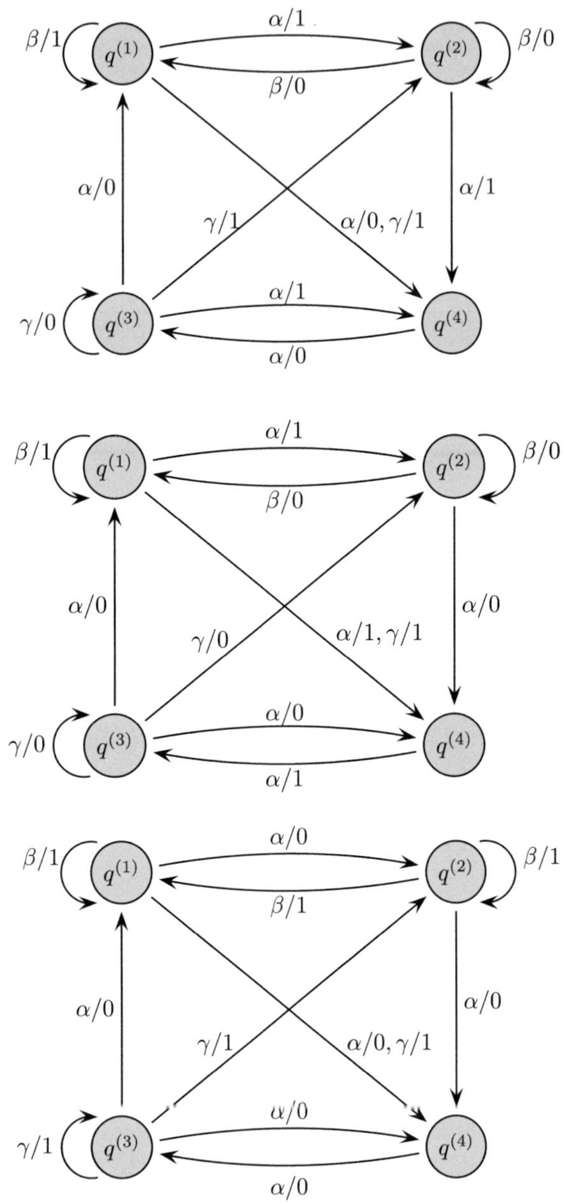

Fig. 3.10 Nondeterministic finite automaton with outputs (top), nondeterministic Moore automaton (middle), and labeled nondeterministic finite automaton (bottom)

State Sequence	Output Sequence
$q^{(1)}\ q^{(2)}\ q^{(4)}\ q^{(3)}$	110
$q^{(1)}\ q^{(4)}\ q^{(3)}\ q^{(4)}$	001
$q^{(1)}\ q^{(4)}\ q^{(3)}\ q^{(1)}$	000
$q^{(2)}\ q^{(4)}\ q^{(3)}\ q^{(1)}$	100
$q^{(2)}\ q^{(4)}\ q^{(3)}\ q^{(4)}$	101

In the middle of Fig. 3.10, we see an example of a QNFA. What is important here is that the output function from each state is the same regardless of the input (i.e., for each state $q \in Q$, we have $\lambda(q, \sigma) = \lambda(q, \sigma')$ for all $\sigma, \sigma' \in \Sigma$—at least whenever λ is defined). In terms of the diagram in the middle of Fig. 3.10, this translates to having all arrows out of each state be associated with the same output. For example, transitions out of state $q^{(1)}$ are all associated with output 1; transitions out of state $q^{(2)}$ are all associated with output 0; transitions out of state $q^{(3)}$ are all associated with output 0; and transitions out of state $q^{(4)}$ are all associated with output 1.

At the bottom of Fig. 3.10 we see an example of an LNFA. What is important here is that the output function for each input is the same regardless of the state (i.e., for each input $\sigma \in \Sigma$, we have $\lambda(q, \sigma) = \lambda(q', \sigma)$ for all $q, q' \in Q$—at least whenever λ is defined). In terms of the diagram at the bottom of Fig. 3.10, this translates to having all arrows with the same input label be associated with the same output. For example, transitions with input label α are all associated with output 0; transitions with input label β are all associated with output 1; and transitions with input label γ are all associated with output 1. Clearly, in the case of an LNFA, given a sequence of inputs, all state sequences that are generated are associated with a unique output sequence. For example, if input $\alpha\alpha\alpha$ is applied, it will generate the state sequences

$$
\begin{array}{l}
\text{State Sequence} \\
q^{(1)}\ q^{(2)}\ q^{(4)}\ q^{(3)} \\
q^{(1)}\ q^{(4)}\ q^{(3)}\ q^{(4)} \\
q^{(1)}\ q^{(4)}\ q^{(3)}\ q^{(1)} \\
q^{(2)}\ q^{(4)}\ q^{(3)}\ q^{(1)} \\
q^{(2)}\ q^{(4)}\ q^{(3)}\ q^{(4)}\ ,
\end{array}
$$

which are identical to the ones we had before (as expected since all three automata in Fig. 3.10 have identical next-state transition functions); the corresponding sequence of outputs for all state sequences will be 000. Note, however, that in the more general case of an NFA (top of Fig. 3.10) we had different output sequences for each of these state sequences. □

3.3.2 *Finite Automata with Silent Transitions*

3.3.2.1 Observation Models for DFA with Silent Transitions

When dealing with automata that do not necessarily generate an observation for each combination of state and input, we need ways to handle transitions that are not signified by the generation of an output. To model this, we assume that a special empty output, which we denote by ϵ, is generated in such cases. Thus, if we take the set of outputs to be $Y = \{y^{(1)}, y^{(2)}, \ldots, y^{(R)}\}$, the output function, in this case, will be of the form $\lambda : Q \times \Sigma \rightarrow Y \cup \{\epsilon\}$.

This viewpoint leads to natural extensions of the definitions for observation models for DFA without silent transitions, which were discussed in the previous section.

Definition 3.21 (*DFA with Outputs and Silent Transitions*) A DFA with outputs and silent transitions is a six-tuple $\mathcal{DFA} = (Q, \Sigma, Y \cup \{\epsilon\}, \delta, \lambda, q[0])$ where $(Q, \Sigma, \delta, q[0])$ is a DFA and $\lambda : Q \times \Sigma \to Y \cup \{\epsilon\}$ is the (possibly partially defined) output function for a finite set of outputs $Y = \{y^{(1)}, y^{(2)}, \ldots, y^{(R)}\}$ and ϵ being the empty output.

The output function λ is assumed without loss of generality to be surjective on the set of outputs Y (note that it is not explicitly required that a silent transition be present). Given state $q[k]$ and a sequence of inputs σ_k^{k+m} that is applied starting at time epoch k, the sequence of states that are visited and the sequence of outputs that are generated can be obtained iteratively as in the case of automata without silent transitions (refer to the iteration in Eq. (3.5)).

Given state $q[k]$ and a valid sequence of inputs σ_k^{k+m} (i.e., a sequence of inputs that results in a well-defined sequence of states q_k^{k+m+1}), the corresponding sequence of outputs y_k^{k+m} is well-defined if the output function λ is defined for all pairs of states and inputs for which δ is defined. As in the case of DFA without silent transitions, this is a sufficient condition for y_k^{k+m} to be well-defined and we will assume that it holds.

The major difference here is that the sequence of outputs that is observed is no longer necessarily of length $m + 1$; in fact, depending on how many silent transitions take place during the application of the input sequence σ_k^{k+m} starting from state $q[k]$, one might end up with sequences of outputs of length $m + 1$, m, ..., or even 0. Since we are studying models that are untimed and event driven, we will have no way of knowing when silent transitions take place, at least not based solely on the observation of outputs. Thus, any state estimation and diagnosis procedures we develop will have to rely on the sequence of observable outputs that are generated by the system and explicitly take into account silent transitions (based on knowledge of the system model). We will revisit this issue in the next chapter when we discuss state estimation procedures for finite automata.

To keep notation consistent, we will make use of the function $E : (Y \cup \{\epsilon\})^{m+1} \to \{\epsilon\} \cup Y \cup Y^2 \cdots \cup Y^{m+1}$ that takes a sequence of outputs (including empty outputs) and produces the compatible sequence of non-empty outputs (by simply removing the empty outputs). The function E can be defined as follows:

$$E((y_k, y_{k+1}, \ldots, y_{k+m})) = \begin{cases} (i)\ \epsilon, \\ \quad \text{if } y_k = y_{k+1} = \cdots = y_{k+m} = \epsilon\ , \\ (ii)\ (y_{j_\ell}, y_{j_{\ell+1}}, \ldots, y_{j_{\ell+m'}}), \\ \quad k \le j_\ell < j_{\ell+1} < \cdots < j_{\ell+m'} \le k+m, \\ \quad \text{if } y_j = \epsilon \text{ for } j \neq j_\ell, j_{\ell+1}, \ldots, j_{\ell+m'} \text{ and} \\ \quad y_j \neq \epsilon \text{ for } j = j_\ell, j_{\ell+1}, \ldots, j_{\ell+m'}\ . \end{cases} \tag{3.8}$$

We can use function E to express the sequence of outputs generated due to a sequence of inputs σ_k^{k+m} that is applied at state $q[k]$ as follows:

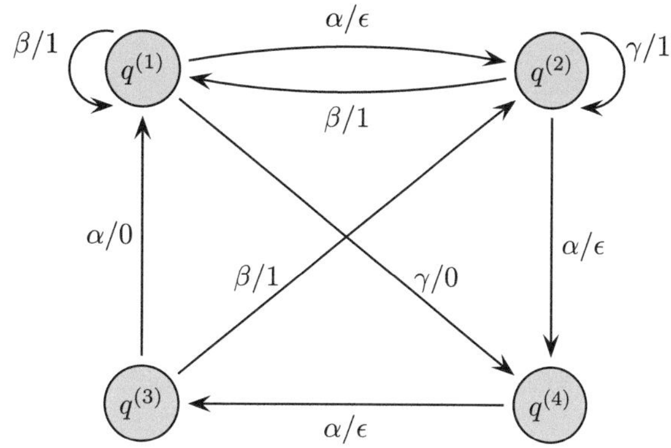

Fig. 3.11 Deterministic finite automaton with outputs and silent transitions discussed in Example 3.11

$$y_k^{k+m'} = E(\lambda_{seq}(q[k], \sigma_k^{k+m})) , \tag{3.9}$$

where λ_{seq} was defined in Eq. (3.6) and $-1 \leq m' \leq m$ (if $m' = -1$ then no outputs are produced or, equivalently, $E(\lambda_{seq}(q[k], \sigma_k^{k+m})) = \epsilon$). Note that Eq. (3.9) implies that $\delta(q[k], \sigma_k^{k+m})$ is defined and that for λ_{seq} to be known, we generally need knowledge of both the next-state transition function δ and the output function λ.

Example 3.11 In Fig. 3.11 we see a DFA with outputs and silent transitions $(Q, \Sigma, Y \cup \{\epsilon\}, \delta, \lambda, q[0])$, where $Q = \{q^{(1)}, q^{(2)}, q^{(3)}, q^{(4)}\}$ is the set of states, $\Sigma = \{\alpha, \beta, \gamma\}$ is the set of inputs, $Y = \{0, 1\}$ is the set of outputs, ϵ is the empty output, and $q[0] = q^{(1)}$ is the initial state. The next-state transition function δ and the output function λ are defined by the arrows and the labels in the figure: for instance, from state $q^{(1)}$ with input α we transition to state $q^{(2)}$ and generate no output (this is captured by the label "α/ϵ" on that arrow or sometimes simply by the label "α" on that arrow); similarly, from state $q^{(3)}$ with input β we transition to state $q^{(2)}$ and generate output 1 (this is captured by the label "$\beta/1$" on that arrow); and so forth.

Notice that if one ignores the outputs and the output function, the finite automaton in Fig. 3.11 is identical to the one on the left of Fig. 3.1, which was discussed in Example 3.1. Thus, given a particular sequence of inputs, the two automata will follow the same sequence of states. The main difference is that the automaton in Fig. 3.11 also conveys information about the output sequence that is observed. For example, the sequence of inputs $\alpha\gamma\alpha$ will induce the state sequence $q^{(1)} \to q^{(2)} \to q^{(2)} \to q^{(4)}$ and the sequence of outputs 1 (i.e., $y[0] = \epsilon$, $y[1] = 1$, and $y[2] = \epsilon$). Knowing the output sequence does not necessarily allow us to determine the input or state sequence; for example, the input sequence $\beta\alpha\alpha$ will induce the state sequence $q^{(1)} \to q^{(1)} \to q^{(2)} \to q^{(4)}$ and identical sequence of outputs 1 (i.e., $y[0] = 1$, $y[1] = \epsilon$, $y[2] = \epsilon$). In terms of the notation introduced earlier, we have

$$\begin{aligned} \delta_{seq}(q^{(1)}, \alpha\gamma\alpha) &= q^{(1)}\, q^{(2)}\, q^{(2)}\, q^{(4)} \\ \lambda_{seq}(q^{(1)}, \alpha\gamma\alpha) &= \epsilon\ 1\ \epsilon \\ E(\lambda_{seq}(q^{(1)}, \alpha\gamma\alpha)) &= 1 \end{aligned}$$

and

$$\delta_{seq}(q^{(1)}, \beta\alpha\alpha) = q^{(1)}\, q^{(1)}\, q^{(2)}\, q^{(4)}$$
$$\lambda_{seq}(q^{(1)}, \beta\alpha\alpha) = 1\;\epsilon\;\epsilon$$
$$E(\lambda_{seq}(q^{(1)}, \beta\alpha\alpha)) = 1\;.$$

As a final note, observe that the next-state transition function δ and the output function λ are both partially defined (e.g., from state $q^{(3)}$ there is no transition/output with input γ); nevertheless, the output function is defined whenever the transition function is defined. □

The two special cases of observation models we discussed for the case of automata without silent transitions extend naturally to the case of automata with silent transitions. When the output only depends on the state, we obtain Moore automata with silent state visitations, whereas when the output only depends on the inputs we obtain labeled automata with silent inputs.

Definition 3.22 (*Deterministic Moore Automaton with Silent State Visitations*) A deterministic Moore automaton (QDFA) with silent state visitations is a six-tuple $\mathcal{QFA} = (Q, \Sigma, Y \cup \{\epsilon\}, \delta, \lambda_Q, q[0])$ where $(Q, \Sigma, \delta, q[0])$ is a DFA and $\lambda_Q : Q \to Y \cup \{\epsilon\}$ is the output function for a finite set of outputs $Y = \{y^{(1)}, y^{(2)}, \ldots, y^{(R)}\}$ and ϵ being the empty output.

Again, if we assume that all states in Q are reachable from $q[0]$ (and nonterminating), we require the output function λ_Q of a Moore automaton to be defined for all possible states. We can also assume without loss of generality that λ_Q is surjective on the set of observable outputs Y.

Definition 3.23 (*Labeled Deterministic Finite Automaton (LDFA) with Silent Inputs*) A labeled deterministic finite automaton (LDFA) with silent transitions is a six-tuple $\mathcal{LDFA} = (Q, \Sigma, Y \cup \{\epsilon\}, \delta, \lambda_\Sigma, q[0])$ where $(Q, \Sigma, \delta, q[0])$ is a DFA and $\lambda_\Sigma : \Sigma \to Y \cup \{\epsilon\}$ is the output function for a finite set of outputs $Y = \{y^{(1)}, y^{(2)}, \ldots, y^{(R)}\}$ and ϵ being the empty output.

As in the case of an LDFA, the function $\lambda_{seq}(q[k], \sigma_k^{k+m})$ is simply a function of the sequence of inputs σ_k^{k+m} (i.e., it is independent of $q[k]$) and can be defined without knowledge of the state transition function δ. In fact, with a slight abuse of notation, we do not require the function $\delta(q[k], \sigma_k^{k+m})$ to be defined and simply take $\lambda_{\Sigma,seq}(\sigma_k^{k+m})$ to be equal to $E(y_k^{k+m})$ where E is the function in Eq. (3.8) that erases empty outputs and satisfies

$$y[k+i] = \lambda_\Sigma(\sigma[k+i]) \tag{3.10}$$

for $i = 0, 1, \ldots, m$; again, the major difference from the case of an LDFA without silent transitions is that in this case some of the outputs could correspond to the empty output ϵ.

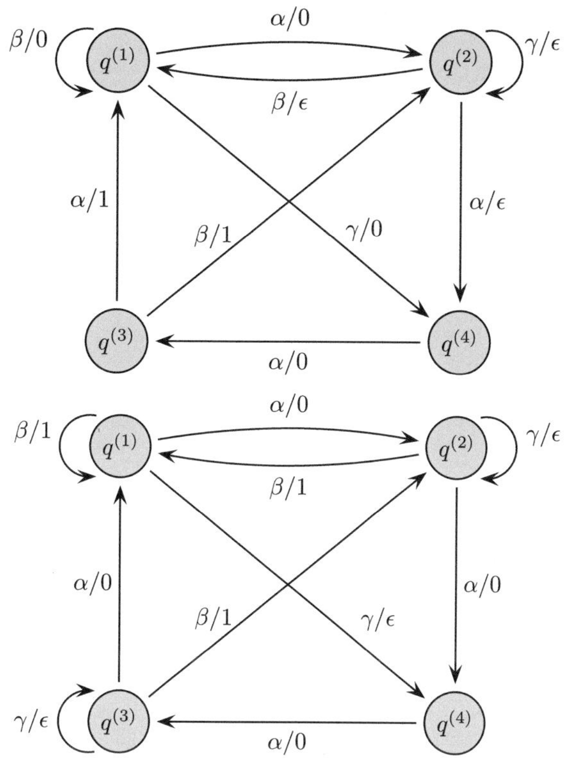

Fig. 3.12 Deterministic Moore automaton with silent state visitations (top) and labeled deterministic finite automaton with silent transitions (bottom)

Example 3.12 On the top of Fig. 3.12, we see an example of a QDFA with silent state visitations. What is important here is that the output function from each state is the same regardless of the input (i.e., for each state $q \in Q$, we have $\lambda(q, \sigma) = \lambda(q, \sigma')$ for all $\sigma, \sigma' \in \Sigma$—at least whenever λ is defined). In terms of the diagram on the top of Fig. 3.12, this translates to having all arrows out of each state be associated with the same output, including the empty output ϵ. For example, transitions out of state $q^{(1)}$ are all associated with output 0; transitions out of state $q^{(2)}$ are all associated with the empty output ϵ; transitions out of state $q^{(3)}$ are all associated with output 1; and transitions out of state $q^{(4)}$ are all associated with output 0. Note that visitations to state $q^{(2)}$ are silent in the sense that they generate no output: for instance, starting from initial state $q[0] = q^{(1)}$, the input sequence $\alpha\alpha\alpha$ generates the state sequence $q^{(1)} \rightarrow q^{(2)} \rightarrow q^{(4)} \rightarrow q^{(3)}$ and the sequence of outputs 00. Note that the same sequence of outputs 00, again starting from initial state $q[0] = q^{(1)}$, is generated from the input sequence $\beta\alpha\gamma$ which generates the state sequence $q^{(1)} \rightarrow q^{(1)} \rightarrow q^{(2)} \rightarrow q^{(2)}$.

At the bottom of Fig. 3.12 we see an example of an LDFA with silent inputs. What is important here is that the output function for each input is the same regardless of the state (i.e., for each input $\sigma \in \Sigma$, we have $\lambda(q, \sigma) = \lambda(q', \sigma)$ for all $q, q' \in Q$—at least whenever λ is defined). In terms of the diagram at the bottom of Fig. 3.12, this translates to having all arrows with the same input label be associated with the same output. For example, transitions with input label α are all associated with output 0; transitions with input label β are all associated with output 1; and transitions with input label γ are all associated with output ϵ. This means that any sequence of inputs, if defined, will generate a corresponding sequence of outputs: for example, if we assume that we are starting with state $q^{(1)}$, the sequence of inputs $\alpha\alpha\alpha$ generates the sequence of outputs 000 (it is well defined in this case with corresponding sequence of states $q^{(1)} \to q^{(2)} \to q^{(4)} \to q^{(3)}$); sequence of inputs $\gamma\beta\beta$ would have generated the sequence of outputs 11 (but it is not defined in this case); sequence of inputs $\gamma\alpha\beta$ generates the sequence of outputs 01 (it is well defined in this case with corresponding sequence of states $q^{(1)} \to q^{(4)} \to q^{(3)} \to q^{(2)}$). □

A special case of an LDFA is the case when a subset of the inputs Σ_{uo}, $\Sigma_{uo} \subseteq \Sigma$, map to the empty output ϵ, whereas the remaining inputs Σ_{obs}, $\Sigma_{obs} = \Sigma \setminus \Sigma_{uo}$, map to an output from the set Y that is dedicated to this type of transition (i.e., the restriction of mapping λ_Σ to the set Σ_{obs} is one to one). In such case, the alphabet of events Σ is essentially partitioned into two subsets, the set of observable events Σ_{obs} and the set of unobservable events Σ_{uo}, which are defined as

$$\begin{aligned} \Sigma_{obs} &= \{\sigma \in \Sigma \mid \lambda_\Sigma(\sigma) \in Y\} , \\ \Sigma_{uo} &= \{\sigma \in \Sigma \mid \lambda_\Sigma(\sigma) = \epsilon\} . \end{aligned}$$

Clearly, we have $\Sigma_{obs} \cap \Sigma_{uo} = \emptyset$ and $\Sigma_{obs} \dot{\cup} \Sigma_{uo} = \Sigma$; furthermore, since the restriction of λ_Σ to Σ_{obs} is one to one, we can take (without loss of generality) $Y = \Sigma_{obs}$ and define the output function as

$$\lambda_\Sigma(\sigma) = \begin{cases} \sigma, & \sigma \in \Sigma_{obs} , \\ \epsilon, & \sigma \in \Sigma_{uo} . \end{cases}$$

This mapping is called the natural projection with respect to Σ_{obs} and is denoted with $P_{\Sigma_{obs}}$ (the subscript is typically dropped if Σ_{obs} is obvious from the context of the discussion). To denote the sequence of outputs generated starting from state $q[k]$ due to a sequence of inputs σ_k^{k+m} (which we denoted earlier by $y_k^{k+m'} = E(\lambda_{seq}(q[k], \sigma_k^{k+m})) = E(\lambda_\Sigma(\sigma_k^{k+m})))$, we can again slightly abuse notation and set

$$y_k^{k+m'} = P_{\Sigma_{obs}}(\sigma_k^{k+m}) = P_{\Sigma_{obs}}(\sigma[k]), P_{\Sigma_{obs}}(\sigma[k+1]), \ldots, P_{\Sigma_{obs}}(\sigma[k+m]) \tag{3.11}$$

Note that the definition also appears in a recursive form as

$$y_k^{k+m'} = P_{\Sigma_{obs}}(\sigma_k^{k+m-1}) P_{\Sigma_{obs}}(\sigma[m]) ,$$

where $P_{\Sigma_{obs}}(\sigma) = \sigma$ if $\sigma \in \Sigma_{obs}$ and $P_{\Sigma_{obs}}(\sigma) = \epsilon$ if $\sigma \in \Sigma_{uo} \cup \{\epsilon\}$. [Recall that $\sigma_k^{k+m} = \sigma[k], \sigma[k+1], \ldots, \sigma[k+m]$ denotes the sequence of inputs applied at time epoch k, time epoch $k+1$, ..., up to and including time epoch $k+m$.]

Given $\mathcal{LDFA} = (Q, \Sigma, \Sigma_{obs} \cup \{\epsilon\}, \delta, P_{\Sigma_{obs}}, q[0])$ under a natural projection mapping $P_{\Sigma_{obs}}$ with respect to the set of observable events Σ_{obs}, we define the *inverse projection* $P^{-1}_{\Sigma_{obs}}$ as the set of sequences of inputs that generate output sequence $y_k^{k+m'}$ captured by

$$\Sigma_{uo}^* \; y[k] \; \Sigma_{uo}^* \; y[k+1] \; \Sigma_{uo}^* \; \ldots \; \Sigma_{uo}^* \; y[k+m'] \; \Sigma_{uo}^* \; ,$$

where $\Sigma_{uo} = \Sigma \setminus \Sigma_{obs}$ and $y[k+i] = \sigma[k+i]$ for $i = 0, 1, 2, \ldots, m'$. Note that it is possible that many of these sequences of observations cannot be generated by the system (i.e., there is no valid behavior in the system that will generate them).

3.3.2.2 Observation Models for NFA with Silent Transitions

Following the development for the case of NFA without silent transitions, the extension of observation models to NFA with silent transitions requires output functions λ of the form

$$\lambda : Q \times \Sigma \times Q \to Y \cup \{\epsilon\} \; ,$$

i.e., we allow a transition from a state q_i to a state q_f under input σ not to generate any output. As in the case of NFA with outputs (and without silent transitions), if λ is defined for all triplets (q_i, σ, q_f) such that $q_f \in \delta(q_i, \sigma)$, then the output is guaranteed to be defined for each possible state transition (so that, given an initial state and a sequence of inputs, a corresponding output sequence will be defined for each possible sequence of states). [As in the case of NFA with outputs and without silent transitions, this is a sufficient condition but it becomes necessary if all states are reachable from the initial set of states under some sequence of inputs.] Formally, an NFA with outputs and silent transitions is defined below.

Definition 3.24 (*NFA with Outputs and Silent Transitions*) An NFA with outputs and silent transitions is a six-tuple $\mathcal{NFA} = (Q, \Sigma, Y \cup \{\epsilon\}, \delta, \lambda, Q_0)$ where (Q, Σ, δ, Q_0) is an NFA and $\lambda : Q \times \Sigma \times Q \to Y \cup \{\epsilon\}$ is the (possibly partially defined) output function for a finite set of outputs $Y = \{y^{(1)}, y^{(2)}, \ldots, y^{(R)}\}$ and ϵ being the empty output.

The definition of δ_{seq} for the case of NFA with outputs and silent transitions is identical to the case of NFA with outputs and without silent transitions; however, the definition of λ_{seq} requires additional notation due to the possible presence of silent transitions. This notation is established below.

$$\begin{aligned}
\delta_{seq}(q[k], \sigma_k^{k+m}) &= \{(q_k, q_{k+1}, \ldots, q_{k+m}, q_{k+m+1}) \mid q_k \in q[k], \\
&\qquad q_{i+1} \in \delta(q_i, \sigma_i) \text{ for } k \leq i \leq k+m\} \; , \\
\lambda_{seq}(q[k], \sigma_k^{k+m}) &= \{(y_k, y_{k+1}, \ldots, y_{k+m}) \mid \\
&\qquad \exists (q_k, q_{k+1}, \ldots, q_{k+m}, q_{k+m+1}) \in \delta_{seq}(q[k], \sigma_k^{k+m}) \\
&\qquad \text{such that } y_i = \lambda(q_i, \sigma[i], q_{i+1}) \text{ for } k \leq i \leq k+m\} \; ,
\end{aligned}$$

$$E(\lambda_{seq}(q[k], \sigma_k^{k+m})) = \{E((y_k, y_{k+1}, \ldots, y_{k+m})) \mid \exists (q_k, q_{k+1}, \ldots, q_{k+m}, q_{k+m+1}) \in \delta_{seq}(q[k], \sigma_k^{k+m}) \text{ such that } y_i = \lambda(q_i, \sigma[i], q_{i+1}) \text{ for } k \le i \le k+m\} ,$$

where the E function was defined in Eq. (3.8).

When the output function depends only on the state (Moore automaton) or only on the input (labeled automaton), the specification of λ is simplified significantly and we are naturally led to the notions of nondeterministic Moore automaton (QNFA) with silent state visitations and labeled nondeterministic finite automaton (LNFA) with silent inputs.

(1) Output depends only on state: In this scenario, the output function at time epoch k depends only on the state $q[k]$ and it is not a function of the input applied at time epoch k. In other words, the output function λ is of the form $\lambda_Q : Q \to Y \cup \{\epsilon\}$. In this case, we have an automaton that is a nondeterministic Moore automaton (QNFA) with silent state visitations.

Definition 3.25 (*Nondeterministic Moore automaton (QNFA) with Silent State Visitations*) A nondeterministic Moore automaton (QNFA) with silent state visitations is a six-tuple $\mathcal{QFA} = (Q, \Sigma, Y \cup \{\epsilon\}, \delta, \lambda_Q, Q_0)$ where (Q, Σ, δ, Q_0) is an NFA and $\lambda_Q : Q \to Y \cup \{\epsilon\}$ is the output function for a finite set of outputs $Y = \{y^{(1)}, y^{(2)}, \ldots, y^{(R)}\}$ and ϵ being the empty output.

Assuming that all states in Q are reachable from states in Q_0 and nonterminating, we require the output function λ_Q of a Moore automaton to be defined for all possible states. We usually also assume without loss of generality that λ_Q is surjective on the set Y. In this case, the functions δ_{seq} and λ_{seq} take the following form:

$$\delta_{seq}(q[k], \sigma_k^{k+m}) = \{(q_k, q_{k+1}, \ldots, q_{k+m}, q_{k+m+1}) \mid q_k \in q[k], q_{i+1} \in \delta(q_i, \sigma[i]) \text{ for } k \le i \le k+m\} ,$$

$$\lambda_{seq}(q[k], \sigma_k^{k+m}) = \{(y_k, y_{k+1}, \ldots, y_{k+m}) \mid \exists (q_k, q_{k+1}, \ldots, q_{k+m}, q_{k+m+1}) \in \delta_{seq}(q[k], \sigma_k^{k+m}) \text{ such that } y_i = \lambda_Q(q_i) \text{ for } k \le i \le k+m\} ,$$

$$E(\lambda_{seq}(q[k], \sigma_k^{k+m})) = \{E((y_k, y_{k+1}, \ldots, y_{k+m})) \mid \exists (q_k, q_{k+1}, \ldots, q_{k+m}, q_{k+m+1}) \in \delta_{seq}(q[k], \sigma_k^{k+m}) \text{ such that } y_i = \lambda_Q(q_i) \text{ for } k \le i \le k+m\} ,$$

where the E function was defined in Eq. (3.8).

(2) Output depends only on input: In this scenario, the output function at time epoch k only depends on the input $\sigma[k]$ applied at time epoch k and it is not a function of the state $q[k]$ at time epoch k. In other words, the output function λ is of the form $\lambda_\Sigma : \Sigma \to Y \cup \{\epsilon\}$. In this case, we have an automaton that is called a labeled nondeterministic finite automaton (LNFA) with silent inputs.

Definition 3.26 (*Labeled NFA (LNFA) with Silent Inputs*) A labeled nondeterministic finite automaton (LNFA) with silent inputs is a six-tuple $\mathcal{LNFA} = (Q, \Sigma, Y \cup \{\epsilon\}, \delta, \lambda_\Sigma, Q_0)$ where (Q, Σ, δ, Q_0) is an NFA and $\lambda_\Sigma : \Sigma \to Y \cup \{\epsilon\}$ is the output function for a finite set of outputs $Y = \{y^{(1)}, y^{(2)}, \ldots, y^{(R)}\}$ and ϵ being the empty output.

The output function λ_Σ is usually assumed (without loss of generality) to be surjective on the set Y. Again, a special case of an LNFA with outputs and silent inputs is the case when the output function is defined as

$$\lambda_\Sigma(\sigma) = \begin{cases} \sigma, \ \sigma \in \Sigma_{obs} \ , \\ \epsilon, \ \sigma \in \Sigma_{uo} \end{cases}$$

for some partition of Σ into sets Σ_{obs} and Σ_{uo} (such that $\Sigma_{obs} \cap \Sigma_{uo} = \emptyset$ and $\Sigma_{obs} \cup \Sigma_{uo} = \Sigma$). As mentioned earlier, this leads to a natural projection mapping $P_{\Sigma_{obs}}$ for any given sequence of transitions σ_k^{k+m}, defined exactly as in Eq. (3.11).

Example 3.13 On the top of Fig. 3.13 we see an NFA with outputs and silent transitions $(Q, \Sigma, Y \cup \{\epsilon\}, \delta, \lambda, Q_0)$ where $Q = \{q^{(1)}, q^{(2)}, q^{(3)}, q^{(4)}\}$ is the set of states, $\Sigma = \{\alpha, \beta, \gamma\}$ is the set of inputs, $Y = \{0, 1\}$ is the set of outputs, ϵ is the empty output, and $Q_0 = \{q^{(1)}, q^{(2)}\}$ is the set of initial states. The next-state transition function δ and the output function λ are defined by the arrows and the labels in the figure: for instance, from state $q^{(1)}$ with input α we transition to state $q^{(2)}$ and generate output 1 (this is captured by the label "$\alpha/1$" on that arrow) and to state $q^{(4)}$ and generate output 1 (this is captured by the label "$\alpha/1$" on that arrow); similarly, from state $q^{(3)}$ with input γ we transition to state $q^{(3)}$ and generate output 0 (this is captured by the label "$\gamma/0$" on that arrow) and to state $q^{(2)}$ and generate output 1 (this is captured by the label "$\gamma/1$" on that arrow); and so forth.

Notice that if one ignores the outputs and the output function, the finite automaton on the top of Fig. 3.13 is identical to the one on the left of Fig. 3.3, which was discussed in Example 3.2. Thus, given a particular sequence of inputs, the two automata will follow the same sequences of states. The main difference is that the automaton in Fig. 3.13 also conveys information about the output sequence that is observed. It is worth pointing out that this output sequence will in general be different for different sequences of states (generated by the same sequence of inputs). For example, if input $\alpha\alpha\alpha$ is applied, it will generate the following state and corresponding output sequences:

State Sequence	Output Sequence
$q^{(1)}\ q^{(2)}\ q^{(4)}\ q^{(3)}$	101
$q^{(1)}\ q^{(4)}\ q^{(3)}\ q^{(1)}$	11
$q^{(1)}\ q^{(4)}\ q^{(3)}\ q^{(4)}$	11
$q^{(2)}\ q^{(4)}\ q^{(3)}\ q^{(1)}$	01
$q^{(2)}\ q^{(4)}\ q^{(3)}\ q^{(4)}$	01

In the middle of Fig. 3.13, we see an example of a nondeterministic Moore automaton (QNFA) with silent state visitations. What is important here is that the output

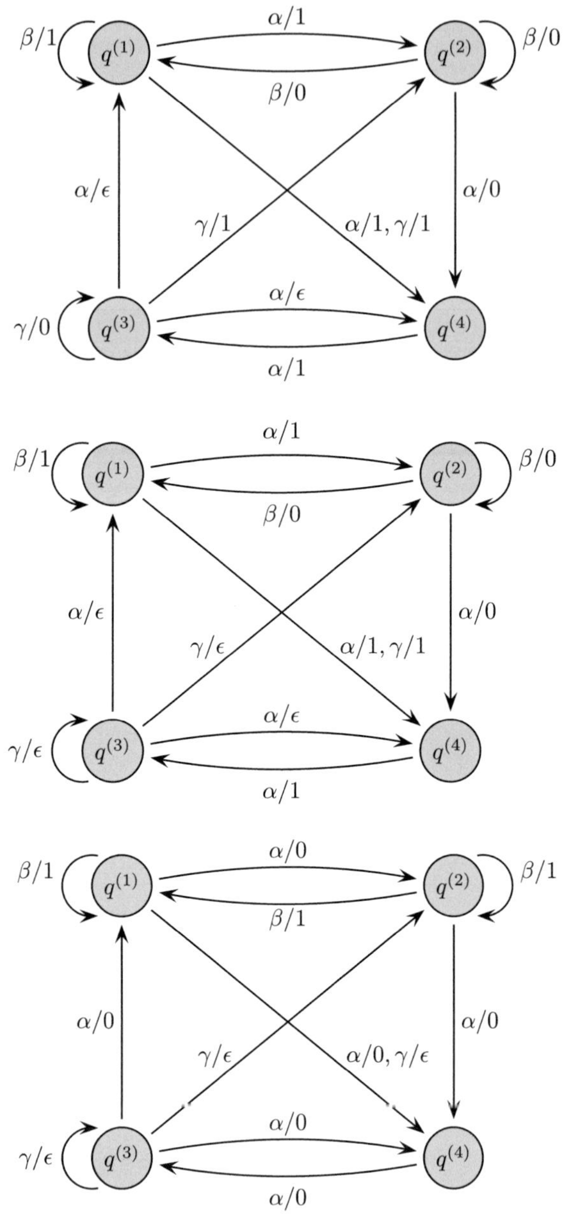

Fig. 3.13 Nondeterministic finite automaton with outputs and with silent transitions (top); nondeterministic Moore automaton with silent state visitations (middle); labeled nondeterministic finite automaton with silent inputs (bottom)

function from each state is the same regardless of the input (i.e., for each state $q \in Q$, we have $\lambda(q, \sigma) = \lambda(q, \sigma')$ for all $\sigma, \sigma' \in \Sigma$—at least whenever λ is defined). In terms of the diagram in the middle of Fig. 3.13, this translates to having all arrows out of each state be associated with the same output or with no output (ϵ). For example, transitions out of state $q^{(1)}$ are all associated with output 1; transitions out of state

$q^{(2)}$ are all associated with output 0; transitions out of state $q^{(3)}$ are all associated with output ϵ; and transitions out of state $q^{(4)}$ are all associated with output 1.

At the bottom of Fig. 3.13 we see an example of a labeled nondeterministic finite automaton (LNFA) with silent inputs. What is important here is that the output function for each input is the same regardless of the state (i.e., for each input $\sigma \in \Sigma$, we have $\lambda(q, \sigma) = \lambda(q', \sigma)$ for all $q, q' \in Q$—at least whenever λ is defined). In terms of the diagram at the bottom of Fig. 3.13, this translates to having all arrows with the same input label be associated with the same output. For example, transitions with input label α are all associated with output 0; transitions with input label β are all associated with output 1; and transitions with input label γ are all associated with no output. Clearly, in the case of an LNFA, given a sequence of inputs, all state sequences that are generated are associated with the same unique output sequence. For example, if input $\alpha\alpha\alpha$ is applied, it will generate the state sequences

$$\begin{array}{l}\text{State Sequence}\\ q^{(1)}\ q^{(2)}\ q^{(4)}\ q^{(3)}\\ q^{(1)}\ q^{(4)}\ q^{(3)}\ q^{(1)}\\ q^{(1)}\ q^{(4)}\ q^{(3)}\ q^{(4)}\\ q^{(2)}\ q^{(4)}\ q^{(3)}\ q^{(1)}\\ q^{(2)}\ q^{(4)}\ q^{(3)}\ q^{(4)}\ ,\end{array}$$

which are identical to the ones we had before (as expected since all three automata in Fig. 3.13 have identical next-state transition functions); in the case of the LNFA, the corresponding sequence of outputs for all of the above state sequences will be 000. □

3.3.3 *Unobservable Reach*

When attempting to perform state estimation and event inference in a deterministic or nondeterministic finite automaton with silent transitions, we often have to make decisions based on available (non-empty) observations. Therefore, we have to worry about silent transitions that can take place without generating any observations. Given a deterministic (respectively, nondeterministic) finite automaton with silent transitions $\mathcal{DFA} = (Q, \Sigma, Y \cup \{\epsilon\}, \delta, \lambda, q_0)$ (respectively, $\mathcal{NFA} = (Q, \Sigma, Y \cup \{\epsilon\}, \delta, \lambda, Q_0)$), the unobservable reach of a state $q \in Q$ is the set of states that can be reached from state q via sequences of zero, one or more silent transitions (i.e., sequences of zero, one or more inputs that do not generate any observable outputs). Note that the unobservable reach of state q necessarily includes state q.

Definition 3.27 (*Unobservable Reach*) Given a DFA with silent transitions $\mathcal{DFA} = (Q, \Sigma, Y \cup \{\epsilon\}, \delta, \lambda, q[0])$, the unobservable reach of a state $q \in Q$ is

$$UR(q) = \{q\} \cup \{q' \in Q \mid \exists m \geq 0, \sigma[k], \sigma[k+1], \ldots, \sigma[k+m] \in \Sigma \text{ such that } q' = \delta_{seq}(q, \sigma_k^{k+m}) \text{ and } \lambda_{seq}(q, \sigma_k^{k+m}) = \epsilon\} .$$

Given an NFA with silent transitions $\mathcal{NFA} = (Q, \Sigma, Y \cup \{\epsilon\}, \delta, \lambda, Q_0)$, the unobservable reach of state $q \in Q$ is

$$\begin{aligned} UR(q) = \{q\} \cup \{q' \in Q \mid \exists m \geq 0, \sigma[k], \sigma[k+1], \ldots, \sigma[k+m] \in \Sigma, \\ q_0 = q, q_1, q_2, \ldots, q_{m+1} = q' \in Q \text{ such that} \\ \text{(i) } q_{i+1} \in \delta(q_i, \sigma[k+i]) \text{ for } i = 0, 1, \ldots, m, \text{ and} \\ \text{(ii) } E(y_0^m) = \epsilon \text{ where } y_i = \lambda(q_i, \sigma[k+i], q_{i+1}) \text{ for } i = 0, 1, \ldots, m\} . \end{aligned}$$

The unobservable reach of a set of states $S \subseteq Q$ is defined as

$$UR(S) = \cup_{q \in S} UR(q) .$$

With this definition in hand, the set of states that a deterministic (respectively, nondeterministic) finite automaton with silent transitions $\mathcal{DFA} = (Q, \Sigma, Y \cup \{\epsilon\}, \delta, \lambda, q[0])$ (respectively, $\mathcal{NFA} = (Q, \Sigma, Y \cup \{\epsilon\}, \delta, \lambda, Q_0)$) could reside in once its starts operation and *before any observations become available* is the set captured by $UR(q[0])$ (respectively, $UR(Q_0)$). For this reason, we frequently assume that the set of possible initial states of a given DFA (or NFA) is such that it includes its unobservable reach, i.e., we take $Q_0 = UR(q[0])$ (or choose $Q_0 = UR(Q_0)$).

One can also define the concept of the unobservable reach with respect to a single output $y \in Y$. More specifically, we have the following definition:

Definition 3.28 (*Unobservable Reach with Respect to Single Output*) Given a DFA with silent transitions $\mathcal{DFA} = (Q, \Sigma, Y \cup \{\epsilon\}, \delta, \lambda, q[0])$, the unobservable reach of a state $q \in Q$ with respect to output $y \in Y$ is defined by

$$UR(q, y) = \{q' \in Q \mid \exists m \geq 0, \sigma[k], \sigma[k+1], \ldots, \sigma[k+m] \in \Sigma \text{ such that } q' = \delta(q, \sigma_k^{k+m}) \text{ and } E(\lambda_{seq}(q, \sigma_k^{k+m})) = y\} .$$

Given an NFA with silent transitions $\mathcal{NFA} = (Q, \Sigma, Y \cup \{\epsilon\}, \delta, \lambda, Q_0))$, the unobservable reach of state $q \in Q$ with respect to output $y \in Y$ is defined by

$$\begin{aligned} UR(q, y) = \{q' \in Q \mid \exists m \geq 0, \sigma[k], \sigma[k+1], \ldots, \sigma[k+m] \in \Sigma, \\ q_0 = q, q_1, q_2, \ldots, q_{m+1} = q' \in Q \text{ such that} \\ \text{(i) } q_{i+1} \in \delta(q_i, \sigma[i]) \text{ for } i = 0, 1, \ldots, m, \text{ and} \\ \text{(ii) } E(y_k^{k+m}) = y \text{ where } y_i = \lambda(q_i, \sigma[k+i], q_{i+1}) \text{ for} \\ i = 0, 1, \ldots, m\} . \end{aligned}$$

The unobservable reach of a set of states $S \subseteq Q$ with respect to output $y \in Y$ is defined as

$$UR(S, y) = \cup_{q \in S} UR(q, y) .$$

In both cases, the unobservable reach $UR(q, y)$ captures the set of states that can be reached from state q via sequences of inputs that generate observation y.

Example 3.14 Consider the LNFA with silent inputs shown at the bottom of Fig. 3.13. The unobservable reach of each state is given by

$$\begin{aligned}UR(q^{(1)}) &= \{q^{(1)}, q^{(4)}\}\\ UR(q^{(2)}) &= \{q^{(2)}\}\\ UR(q^{(3)}) &= \{q^{(2)}, q^{(3)}\}\\ UR(q^{(4)}) &= \{q^{(4)}\}\ .\end{aligned}$$

Similarly, in terms of the notation introduced in this section, we have

$$\begin{array}{ll}UR(q^{(1)}, 0) = \{q^{(2)}, q^{(3)}, q^{(4)}\} & UR(q^{(1)}, 1) = \{q^{(1)}, q^{(4)}\}\\ UR(q^{(2)}, 0) = \{q^{(4)}\} & UR(q^{(2)}, 1) = \{q^{(1)}, q^{(2)}, q^{(4)}\}\\ UR(q^{(3)}, 0) = \{q^{(1)}, q^{(4)}\} & UR(q^{(3)}, 1) = \{q^{(1)}, q^{(2)}\}\\ UR(q^{(4)}, 0) = \{q^{(2)}, q^{(3)}\} & UR(q^{(4)}, 1) = \{\}\end{array}$$

Note that $UR(q)$ is always non-empty (because it always contains q), whereas $UR(q, y)$ may be empty. □

3.4 Comments and Further Reading

The case of labeled nondeterministic finite automata (LNFA) in Definition 3.26 is widely studied in the literature of detectability, fault diagnosis, and opacity (which are the subjects of study in Chaps. 6, 7 and 8, respectively). One of the reasons is that notation becomes a bit easier, since the observation sequence only depends on the event sequence and is decoupled from the matching state sequence(s). In fact, as mentioned earlier, a very common assumption in the literature is for each observable event to have a unique label, which leads to the natural projection in Eq. (3.11).

Several books provide interesting discussions on finite automata, including Hopcroft (2008), Cassandras and Lafortune (2007). Algebraic approaches toward the analysis of finite automata and languages can be found in Ginzburg (1968), Arbib (1968), Arbib (1969). Observation models, finite automata reductions, and decompositions can be found in Booth (1968), Kohavi (1978).

References

Arbib MA (ed) (1968) Algebraic theory of machines, languages, and semigroups. Academic Press, New York

Arbib MA (1969) Theories of abstract automata. Prentice-Hall, Englewood Cliffs, New Jersey

Booth TL (1968) Sequential machines and automata theory. Wiley, New York

Cassandras CG, Lafortune S (2007) Introduction to discrete event systems. Springer
Ginzburg A (1968) Algebraic theory of automata. Academic Press, New York
Hopcroft JE (2008) Introduction to automata theory, languages, and computation. Pearson Education
Kohavi Z (1978) Switching and finite automata theory. McGraw-Hill, New York
Rosen KH (2011) Discrete mathematics and its applications. McGraw-Hill, New York

Chapter 4
State Estimation

4.1 Introduction and Motivation

In many applications of interest, such as supervisory control (Ramadge and Wonham 1987, 1989; Moody and Antsaklis 1997, 1998) and fault diagnosis (Sampath et al. 1995; Hashtrudi Zad et al. 2003), one frequently needs to estimate the state of a finite automaton that models a given system of interest. Typically, the model of the underlying automaton and its initial state (or set of possible initial states) *before* any observations are made are assumed known; based on this knowledge, the task is to characterize the set of states that are possible at a given time epoch during the operation of the system, following the observation of a sequence of outputs $y_0^k \equiv y[0], y[1], y[2], \ldots, y[k]$. In other words, one needs to enumerate all states that are compatible with the system model, the given set of initial states, and the observed sequence of outputs. Depending on the point in time at which the set of possible states needs to be obtained, we can classify state estimation tasks into the following three categories:

Current-State Estimation: This task requires the enumeration of the states that are possible in the system after the observation of the last output $y[k]$, taking into account the system model, the prior set of possible initial states (before any observations are made), and the fact that the system subsequently generated the sequence of outputs y_0^k. We denote this set of states by $\hat{q}_{k+1}(y_0^k)$ or $\hat{q}_{y[k]}(y_0^k)$. [Note that the subscript $k+1$ indicates an estimate of the state at (observation) time epoch $k+1$ and the subscript $y[k]$ indicates an estimate of the state immediately following the kth output. If silent transitions are present, observation time epochs do not necessarily coincide with activity (input) time epochs; for simplicity we will use the term "time epoch" to refer to observation time epochs.]

Smoothing: This requires the enumeration of the states that the system was possibly in after it generated observation $y[i]$ ($0 \leq i < k$), taking into account the system model, the prior set of possible initial states, and the facts that (i) the sequence of observations $y_0^i \equiv y[0], y[1], \ldots, y[i]$ was generated in order to reach this state (or states), and (ii) the system subsequently generated the sequence of observations $y_{i+1}^k \equiv y[i+1], y[i+2], \ldots, y[k]$. We denote this set of states by $\hat{q}_{i+1}(y_0^k)$ or

C. N. Hadjicostis, *Estimation and Inference in Discrete Event Systems*, Communications and Control Engineering,
https://doi.org/10.1007/978-3-030-30821-6_4

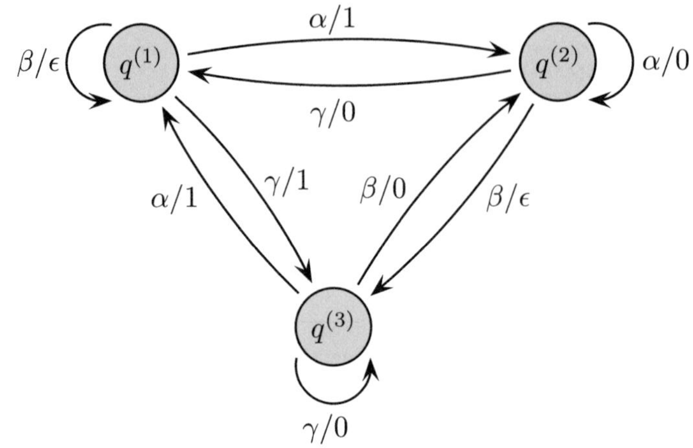

Fig. 4.1 Finite automaton with outputs used in Example 4.1 to illustrate the various state estimation tasks

$\hat{q}_{y[i]}(y_0^k)$ (again, the subscript $i+1$ indicates an estimate of the state at time epoch $i+1$ and the subscript $y[i]$ indicates an estimate of the state immediately following the ith output). A common special case of this scenario is when one is interested in refining the estimate of the states of the system at some past point in time by incorporating the knowledge of a fixed number of D observations that occur after that point in time. In such case, one aims to obtain the estimate $\hat{q}_{k-D+1}(y_0^k)$ or $\hat{q}_{y[k-D]}(y_0^k)$, which is referred to as *D-delayed- state estimation* or *smoothing with delay D*.

Initial-State Estimation: This is a special case of smoothing that requires the enumeration of the set of possible initial states (from which the system might have started from); this is essentially the prior set of initial states refined by the knowledge that the system subsequently generated the sequence of observations y_0^k. In other words, states that were thought to be possible at system start-up but cannot possibly have generated the observed sequence of outputs y_0^k are eliminated from consideration. We denote the set of initial states that are compatible with the sequence of observations y_0^k by $\hat{q}_0(y_0^k)$.

The following example illustrates the ideas of current-state estimation, smoothing, and initial-state estimation.

Example 4.1 Consider the deterministic finite automation (DFA) with outputs $\mathcal{DFA} = (Q, \Sigma, Y, \delta, \lambda, Q_0)$ shown in Fig. 4.1, where $Q = \{q^{(1)}, q^{(2)}, q^{(3)}\}$, $\sigma = \{\alpha, \beta, \gamma\}$, $Y = \{0, 1\}$, δ and λ are as described in the figure, and $Q_0 = \{q^{(1)}, q^{(3)}\}$. We illustrate the estimation tasks described above for the sequence of observations 110 (i.e., $y_0^2 = 110$ with $y[0] = 1$, $y[1] = 1$ and $y[2] = 0$).

Regarding current-state estimation, it is not hard to see that, following the sequence of observations $y_0^2 = 110$, we have

$$\hat{q}_{y[2]}(y_0^2) = \{q^{(1)}, q^{(2)}, q^{(3)}\} .$$

For instance, if we start in state $q^{(3)}$, then with $\alpha\alpha\gamma$ the output sequence is 110 and we end up in state $q^{(1)}$; with $\alpha\alpha\alpha$ the output sequence is 110 and we end up in state $q^{(2)}$; with $\alpha\gamma\gamma$ the output sequence is 110 and we end up in state $q^{(3)}$. Thus, all three states are possible current states following the observation sequence 110.

Regarding initial-state estimation, the discussion above makes it clear that $q^{(3)}$ is a possible initial state. Since we know that $Q_0 = \{q^{(1)}, q^{(3)}\}$, the only other possible initial state is $q^{(1)}$. However, if we start in state $q^{(1)}$ there is no sequence of inputs that generates the observed sequence of outputs (110); thus, we conclude that

$$\hat{q}_0(y_0^2) = \{q^{(3)}\} .$$

Regarding smoothing, let us focus on the possible states of the system following $y[1]$ but before $y[2]$, i.e., $\hat{q}_{y[1]}(y_0^2)$. We already know that we can only start in $q^{(3)}$; we observe the following:

- With $\alpha\alpha\gamma$ the output sequence is 110 and we go through the sequence of states $q^{(3)} \to q^{(1)} \to q^{(2)} \to q^{(1)}$; thus, $q^{(2)} \in \hat{q}_{y[1]}(y_0^2)$. [Note that we arrive in the same conclusion with $\alpha\alpha\alpha$.]
- With $\alpha\gamma\gamma$ the output sequence is 110 and we go through the sequence of states $q^{(3)} \to q^{(1)} \to q^{(3)} \to q^{(3)}$; thus, $q^{(3)} \in \hat{q}_{y[1]}(y_0^2)$.

It is not hard to see that $q^{(1)} \notin \hat{q}_{y[1]}(y_0^2)$ since there is no way that the last output ($y[2] = 0$) can be produced via a transition from state $q^{(1)}$ (including sequences of transitions from state $q^{(1)}$ that produce observation $y[2] = 0$). Thus, we conclude that

$$\hat{q}_{y[1]}(y_0^2) = \{q^{(2)}, q^{(3)}\} .$$

Note that this example aimed at illustrating the various state estimation concepts that we will be interested in. In the remainder of this chapter, we will be deriving systematic ways for performing these estimation tasks. □

An implicit (but not crucial) assumption in our discussions thus far is that the sequence of observations y_0^k is generated by actual activity in the given system; this implies that the sets of states described in the various estimation tasks described before Example 4.1 are always non-empty. Obviously, if y_0^k cannot be generated by the given system for any of the possible initial states, then *all* of the above sets of state estimates will be empty. Clearly, however, as long as our knowledge of the system

Fig. 4.2 State estimation under unknown inputs versus state estimation under (partially) known inputs

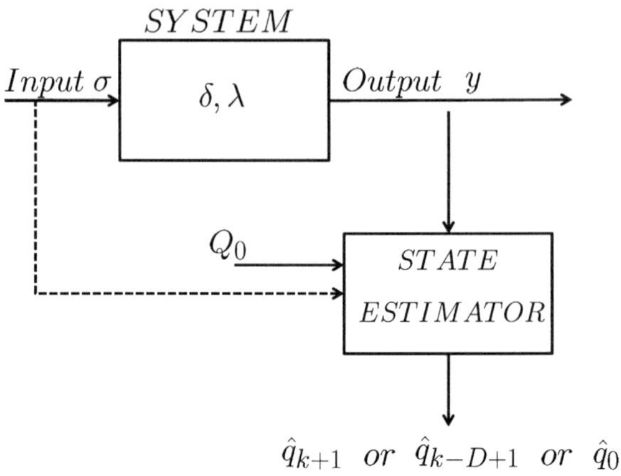

model and possible initial states is correct, this situation cannot arise. In certain cases, an empty set of possible current states could be an indication of a hardware fault or a communication error (e.g., when a sensor has a fault or when the communication from a sensor to a central monitor is unreliable).

Remark 4.1 We have assumed that state estimation needs to be determined based solely on output information. This formulation essentially treats the inputs as *unknown*, and is reminiscent of the construction of unknown input observers in the control literature, see for example Frank (1990). More generally, however, some or all of the inputs to the system might also be available to the state estimator. The distinction between the two cases is shown graphically in Fig. 4.2 where the dotted arrow indicates that some of the inputs might be available to the state estimator. Note that the latter case, in which some (or all of the) inputs to the system may be known, essentially reduces to the case of unknown inputs if one simply annotates the output label with the known input that is applied to the system. For instance, if from state q, an observable input σ causes a transition to state q' and generates output y, we can capture the knowledge of the input by replacing the output y with the output (σ, y). In this way, the observable output contains the information regarding the known input (and this can be used by the mechanism that performs state estimation if necessary). For this reason, we will not address the case of (partially) known inputs separately, although, when appropriate, we will make pertinent remarks about this case.

Another interesting observation to make at this point is that the functionality of the finite automaton *as observed through the available outputs* is essentially equivalent—as far as state estimation is concerned—to a labeled nondeterministic finite automaton (LNFA). We elaborate on this issue in Sect. 4.7 when we construct the observation-equivalent nondeterministic finite automaton for a given finite automaton with outputs. □

Note that in many cases, it will be of interest to perform the above tasks recursively, i.e., to be able to update the set of possible states as additional observations become available. This will be important, for example, when performing state estimation or fault diagnosis online. Thus, we will also be discussing recursive ways for obtaining such estimates. We next establish some notation and formulate the problem for the case of DFA with outputs; we start with automata without silent transitions and then discuss the case of automata with silent transitions. Our development will focus on the latter case, which is more general.

4.2 Problem Formulation

In this section we formulate state estimation problems for DFA with outputs, starting initially with the simplest case, in which automata are not allowed to take silent transitions, and then moving to the more general case of DFA with silent transitions.

4.2.1 *State Estimation in DFA Without Silent Transitions*

We now consider DFA with outputs and without silent transitions (as described in Definition 3.15) for which the initial state is only partially known. More specifically, we assume that we are given a DFA with outputs and without silent transitions $\mathcal{DFA} = (Q, \Sigma, Y, \delta, \lambda, Q_0)$ where $Q = \{q^{(1)}, q^{(2)}, \ldots, q^{(N)}\}$ is a finite set of states, $\Sigma = \{\sigma^{(1)}, \sigma^{(2)}, \ldots, \sigma^{(K)}\}$ is a finite set of inputs, $Y = \{y^{(1)}, y^{(2)}, \ldots, y^{(R)}\}$ is a finite set of outputs, $\delta : Q \times \Sigma \rightarrow Q$ is the (possibly partially defined) next-state transition function, $\lambda : Q \times \Sigma \rightarrow Y$ is the (possibly partially defined) output function, and $Q_0 \subseteq Q$ is the set of prior possible initial states. Furthermore, we assume that we observe a sequence of outputs y_0^k that is generated by underlying unknown activity in the system.

Remark 4.2 As mentioned earlier, the case of (partially) known inputs can be handled easily by taking the set of outputs to be $\tilde{Y} = Y \cup (\Sigma \times Y)$ and by setting the observation $\tilde{y}[i]$ at time epoch i to $\tilde{y}[i] = (\sigma[i], y[i])$ when $\sigma[i]$ (the input applied at observation time epoch i) is observable, or to $\tilde{y}[i] = y[i]$ otherwise. Note that if input availability depends on the state of the system (i.e., a particular input is not always available), then the transformation will be a bit more complex (though the set of possible outputs $\tilde{Y}$ remains the same): in such case, we can modify the system model so as to include as a second output the input symbol when that is available. Naturally, we also need to modify the observation sequence y_0^k to $\tilde{y}_0^k$ (where $\tilde{y} \in \tilde{Y}$) to also reflect any inputs that are observed. □

We assume that the given system (DFA with outputs and without silent transitions) starts operating at time epoch 0, at which point an unknown sequence of inputs σ_0^k is applied (at time epochs $0, 1, 2, \ldots, k$), causing the system to transition to a new state (at time epochs $1, 2, 3, \ldots, k+1$, respectively) and produce the sequence of outputs $y_0, y_1, \ldots, y_k$. The state estimation problems we are interested in are based on the observed sequence of outputs y_0^k.

Problem 4.1 (*Current-State Estimation in DFA without Silent Transitions*) Given $\mathcal{DFA} = (Q, \Sigma, Y, \delta, \lambda, Q_0)$ and a sequence of observations $y_0^k \equiv y[0], y[1], \ldots, y[k]$, determine the set of states that the automaton could reside in at time epoch $k+1$, i.e., determine

$$\hat{q}_{k+1}(y_0^k) = \{q \in Q \mid \exists q_0 \in Q_0,\ \exists \sigma_0^k \in \Sigma^{k+1} \text{ such that } \lambda_{seq}(q_0, \sigma_0^k) = y_0^k \text{ and } \delta(q_0, \sigma_0^k) = q\}\ .$$

In other words, $\hat{q}_{k+1}(y_0^k)$ is the set of states that can be reached (at time epoch $k+1$) from a possible initial state (in the set Q_0) via a sequence of inputs that generates the observed sequence of outputs. (Note that $\lambda_{seq}(q_0, \sigma_0^k)$ was defined in Eq. (3.6) and its existence implies that $\delta(q_0, \sigma_0^k)$ is defined.)

After having observed the output sequence y_0^k, we can also define smoothing problems for states that are possible at earlier time epochs, namely at some time epoch i, $i = 1, 2, \ldots, k$.

Problem 4.2 (*Smoothing in DFA without Silent Transitions*) Given $\mathcal{DFA} = (Q, \Sigma, Y, \delta, \lambda, Q_0)$ and a sequence of observations $y_0^k \equiv y[0], \ldots, y[k]$, determine the set of states that the automaton could reside in at time epoch i, $1 \leq i \leq k$, i.e., determine

$$\hat{q}_i(y_0^k) = \{q \in Q \mid \exists q_0 \in Q_0,\ \exists \sigma_0^k \in \Sigma^{k+1} \text{ such that } \lambda_{seq}(q_0, \sigma_0^k) = y_0^k \text{ and } \delta(q_0, \sigma_0^{i-1}) = q\}\ .$$

In other words, $\hat{q}_i(y_0^k)$ is the set of states that is visited at time epoch i from an initial state (in the set Q_0) via at least one sequence of inputs in a way that generates the observed sequence of outputs.

The initial-state estimation problem is concerned with the set of possible states at the initialization of the system. Before the observation of the sequence of outputs, the set of possible initial states is, of course, given by Q_0; however, this prior estimate can be refined once observations become available.

Problem 4.3 (*Initial-State Estimation in DFA without Silent Transitions*) Given $\mathcal{DFA} = (Q, \Sigma, Y, \delta, \lambda, Q_0)$ and a sequence of observations $y_0^k \equiv y[0], y[1], \ldots, y[k]$, determine the set of states that the automaton could have started from, i.e., determine

$$\hat{q}_0(y_0^k) = \{q_0 \in Q_0 \mid \exists \sigma_0^k \text{ such that } \lambda_{seq}(q_0, \sigma_0^k) = y_0^k\}\ .$$

Remark 4.3 A special case of the setting described above is that of a DFA with a unique initial state $q[0]$. Clearly, the initial-state estimation problem, in this case, becomes trivial since the initial state is obviously $q[0]$ for all valid sequences of outputs y_0^k (i.e., for all sequences of outputs that can be generated by the system starting from the initial state $q[0]$ via sequences of inputs σ_0^k). Even though the initial state is unique and known, state estimates at other points in time (as in the current-state estimation and smoothing problems) will not necessarily be singleton sets because identical outputs maybe produced by multiple transitions (from a given state to another). □

4.2.2 State Estimation in DFA with Silent Transitions

We now consider DFA with outputs and possibly silent transitions for which the initial state may be only partially known (as described in Sect. 3.3.2). More specifically, we assume that we are given the model of a DFA with outputs and silent transitions described by a six-tuple $\mathcal{DFA} = (Q, \Sigma, Y \cup \{\epsilon\}, \delta, \lambda, Q_0)$ where $Q = \{q^{(1)}, q^{(2)}, \ldots, q^{(N)}\}$ is a finite set of states, $\Sigma = \{\sigma^{(1)}, \sigma^{(2)}, \ldots, \sigma^{(K)}\}$ is a finite set of inputs, $Y = \{y^{(1)}, y^{(2)}, \ldots, y^{(R)}\}$ is a finite set of outputs, ϵ is the empty output, $\delta : Q \times \Sigma \to Q$ is the (possibly partially defined) next-state transition function, $\lambda : Q \times \Sigma \to Y \cup \{\epsilon\}$ is the (possibly partially defined) output function, and $Q_0 \subseteq Q$ is the prior set of possible initial states. Furthermore, we assume that we observe a sequence of outputs y_0^k that is generated by underlying unknown activity in the system.

Remark 4.4 As also discussed in Remark 4.1, the case of (partially) known inputs can be handled easily by taking the set of outputs to be $\tilde{Y} = Y \cup (\Sigma \times Y) \cup (\Sigma \times \epsilon)$ and by setting the observation $\tilde{y}[i]$ at time epoch i, $0 \leq i \leq k$, to be $\tilde{y}[i] = (\sigma[i], y[i])$ when $\sigma[i]$ (the input applied at observation time epoch i) is known and output $y[i]$ is observed, or to $\tilde{y}[i] = (\sigma[i], \epsilon)$ if $\sigma[i]$ is known and no output is observed, or to $\tilde{y}[i] = y[i]$ otherwise. We will not discuss this case explicitly since the aforementioned transformation of the automaton (its set of inputs and its output function λ), together with the transformation of the sequence of observations $y_0^k = y[0], y[1], \ldots, y[k]$ to $\tilde{y}_0^k = \tilde{y}[0], \tilde{y}[1], \ldots, \tilde{y}[k]$, results in completely equivalent state estimation problems. Note that if input availability depends on the state of the system (i.e., a particular input is not always available), then the transformation will be a bit more complex: in such case, we can modify the system model so as to include as a second output the input symbol when that is available. □

State estimation in DFA with outputs and silent transitions resembles state estimation in such automata without silent transitions (which was described in the previous section). The main difference is that, after the observation of a sequence of outputs y_0^i, one no longer has knowledge of the actual number of events (transitions) that have occurred in the system, since this number could be any integer greater or equal to $i + 1$ (depending on the unknown number of silent transitions). To circumvent this problem we consider state estimation with respect to the observed sequence of outputs. More specifically, we take $\hat{q}_{y[k]}(y_0^k)$ to be the set of states that can be reached from some initial state in Q_0 with a sequence of events that (can have length $k+1, k+2, \ldots,$ and) generates the sequence of observations y_0^k; this can also be seen as the set of states that the system could be in after it generated the last observation $y[k]$. Similarly, we define $\hat{q}_{y[i]}(y_0^k)$ to be the set of states that (i) can be reached from some initial state in Q_0 with a sequence of events that (can have length $i+1, i+2, \ldots,$ and) generates the sequence of observations y_0^i, *and* (ii) allow the execution of a sequence of events that (can have length $k-i, k-i+1, \ldots,$ and) generates the sequence of observations y_{i+1}^k; this can also be interpreted as the set of states that the system could reside in *after* it generated observation $y[i]$ and *before* it generated observation $y[i+1]$, given of course that the sequence of observations y_0^k was observed.

With the above notation in hand, we can easily redefine the state estimation problems of interest in the context of DFA with outputs and silent transitions.

Problem 4.4 (*Current-State Estimation in DFA with Silent Transitions*) Given $\mathcal{DFA} = (Q, \Sigma, Y \cup \{\epsilon\}, \delta, \lambda, Q_0)$ and a sequence of observations $y_0^k \equiv y[0], y[1], \ldots, y[k]$, determine the set of states that the automaton could reside in after it generated its last observation $y[k]$, i.e., determine

$$\hat{q}_{y[k]}(y_0^k) = \{q \in Q \mid \exists q_0 \in Q_0,\ \exists \sigma_0^{k'} \in \Sigma^* \text{ such that } \lambda_{seq}(q_0, \sigma_0^{k'}) = y_0^k \\ \text{and } \delta(q_0, \sigma_0^{k'}) = q\}\ .$$

Note that k' in the above definition would necessarily have values in $\{k, k+1, k+2, \ldots\}$ so that $\sigma_0^{k'} \in \Sigma^{k+1}\Sigma^*$.

Problem 4.5 (*Smoothing in DFA with Silent Transitions*) Given $\mathcal{DFA} = (Q, \Sigma, Y \cup \{\epsilon\}, \delta, \lambda, Q_0)$ and a sequence of observations $y_0^k \equiv y[0], y[1], \ldots, y[k]$, determine the set of states that the automaton could reside in after it generated observation $y[i]$, $0 \leq i < k$, i.e., determine

$$\hat{q}_{y[i]}(y_0^k) = \{q \in Q \mid \exists q_0 \in Q_0,\ \exists \sigma_0^{k'}, \sigma_0^{k''} \in \Sigma^* \text{ such that} \\ \text{(i) } \lambda_{seq}(q_0, \sigma_0^{k'}) = y_0^i \text{ and } \delta(q_0, \sigma_0^{k'}) = q, \\ \text{and (ii) } \lambda_{seq}(q, \sigma_0^{k''}) = y_{i+1}^k \text{ and } \delta(q, \sigma_0^{k''}) \text{ is defined}\}\ .$$

In other words, $\hat{q}_{y[i]}(y_0^k)$ is the set of states that is reached from an initial state (in the set Q_0) *after* the ith observation (and *before* the $(i+1)$st observation) via a sequence of inputs that generates the observed sequence of outputs. Note that the statement "$\delta(q, \sigma_0^{k''})$ is defined" is not really needed in condition (ii) of the above definition because it is implied by the fact that $\lambda_{seq}(q, \sigma_0^{k''}) = y_{i+1}^k$. Also note that in the above definition we necessarily have $k' \geq i$ and $k'' \geq k - i$.

Problem 4.6 (*Initial-State Estimation in DFA with Silent Transitions*) Given $\mathcal{DFA} = (Q, \Sigma, Y \cup \{\epsilon\}, \delta, \lambda, Q_0)$ and a sequence of observations $y_0^k \equiv y[0], y[1], \ldots, y[k]$, determine the set of states that the automaton could have started from, i.e., determine

$$\hat{q}_0(y_0^k) = \{q_0 \in Q_0 \mid \exists \sigma_0^{k'} \in \Sigma^* \text{ such that } \lambda_{seq}(q_0, \sigma_0^{k'}) = y_0^k\} .$$

In the above definition, k' satisfies $k' \geq k$.

Remark 4.5 Note that state estimation problems define sets of possible states with respect to the observation time epochs, i.e., time epochs at which *observations* were made. In the previous section, since there were no silent transitions, the index of an observation time epoch coincided with the index of an input time epoch. □

Remark 4.6 In an automaton without silent transitions, the set of possible states before the observation of any output activity (i.e., when $y_0^k = \epsilon$) satisfies

$$\hat{q}_0(\epsilon) = Q_0 ,$$

where Q_0 is the set of initial states. However, in an automaton with silent transitions, the set of possible states before the observation of any output activity (i.e., when $y_0^k = \epsilon$) is captured by

$$\hat{q}_\epsilon(\epsilon) = UR(Q_0) ,$$

where $UR(Q_0)$ denotes the unobservable reach of the set of initial states Q_0 and necessarily includes the set Q_0 (recall Definition 3.27). Note that according to the definition of $\hat{q}_0(\cdot)$ given above, we have $\hat{q}_0(\epsilon) = Q_0$ even for an automaton with silent transitions. Therefore, it is important to keep in mind that in the case of automata with silent transitions, we have

$$Q_0 = \hat{q}_0(\epsilon) \subseteq \hat{q}_\epsilon(\epsilon)$$

with the inclusion potentially being strict.

This discrepancy between $\hat{q}_0(y_0^k)$ and $\hat{q}_\epsilon(y_0^k)$ continues as more observations become available because the latter set might include states in the unobservable reach of Q_0 which are not necessarily included in $\hat{q}_0(y_0^k)$. In general, we have

$$\hat{q}_0(y_0^k) \subseteq \hat{q}_\epsilon(y_0^k) .$$

Many researchers overcome this nuisance by assuming that Q_0 is a set that is equal to its unobservable reach.

Given a sequence of observations y_0^k, we also have

$$\hat{q}_0(y_0^{i+1}) \subseteq \hat{q}_0(y_0^i) ,$$
$$\hat{q}_\epsilon(y_0^{i+1}) \subseteq \hat{q}_\epsilon(y_0^i) ,$$

for $0 \leq i \leq k-1$, i.e., the estimate of the initial state can only get refined as more observations become available.

Finally, we point out that given a sequence of observations y_0^k, we also have

$$\hat{q}_{y[i]}(y_0^j) \subseteq \hat{q}_{y[i]}(y_0^i) ,$$

for $0 \leq i < j \leq k$, i.e., the estimate of the set of possible states after the observation of $y[i]$ can only get refined as more observations become available.

In particular, the above discussion implies the following *monotonic refinement property* for initial-state estimation: given a sequence of observations y_0^k, we have

$$\hat{q}_0(y_0^j) \subseteq \hat{q}_0(y_0^i) ,$$

for $0 \leq i < j \leq k$, i.e., the estimate of the set of possible initial states can only get refined as more observations become available. □

4.3 Intuitive Discussion on Current-State Estimation

In order to gain some intuition about state estimation problems, we start with the current-state estimation problem, which is the one most commonly discussed in the literature. To keep the notation simple, we consider current-state estimation for a DFA with outputs and without silent transitions denoted by $G = (Q, \Sigma, Y, \delta, \lambda, Q_0)$; however, the analysis can be easily modified to apply to a DFA with outputs and silent transitions (without significantly affecting the complexity of the approach). Given a sequence of observations $y_0^k \equiv y[0], y[1], \ldots, y[k]$, we would like to determine (for $i = 0, 1, \ldots, k$), the set of states $\hat{q}_{i+1}(y_0^i)$ that the automaton could reside in at time epoch $i + 1$, given the observation of the sequence of outputs y_0^i up to that point.

A straightforward way to solve the above problem is to maintain the set of possible current states Q_C of the system ($Q_C \subseteq Q$) and update this set each time an observation becomes available. In other words, we can start by setting $Q_C = Q_0$ and, once the first observation $y[0]$ becomes available, we can update Q_C as follows:

$$\begin{aligned} Q_C^{new} &= \{q \in Q \mid \exists q_c \in Q_C, \exists \sigma \in \Sigma \text{ such that } y[0] = \lambda(q_c, \sigma) \\ &\qquad (\text{and } \delta(q_c, \sigma) \text{ is defined})\} \\ Q_C &= Q_C^{new} . \end{aligned}$$

As $y[i]$ becomes available for $i = 1, 2, \ldots, k$, we can continue updating Q_C in the same way, i.e., for $i = 1, 2, \ldots, k$ we can iterate as follows:

$$\begin{aligned} Q_C^{new} &= \{q \in Q \mid \exists q_c \in Q_C, \exists \sigma \in \Sigma \text{ such that } y[i] = \lambda(q_c, \sigma) \\ &\qquad (\text{and } \delta(q_c, \sigma) \text{ is defined})\} \\ Q_C &= Q_C^{new} . \end{aligned}$$

At the end of each update during this iterative process, we have $\hat{q}_{i+1}(y_0^i) = Q_C$.

The above approach requires $O(N)$ storage, to be able to store the set of current states ($N = |Q|$). Of course, it also assumes that the system model is available (stored), which requires $O(NK)$ storage where $K = |\Sigma|$. The complexity of performing each update is also $O(NK)$.

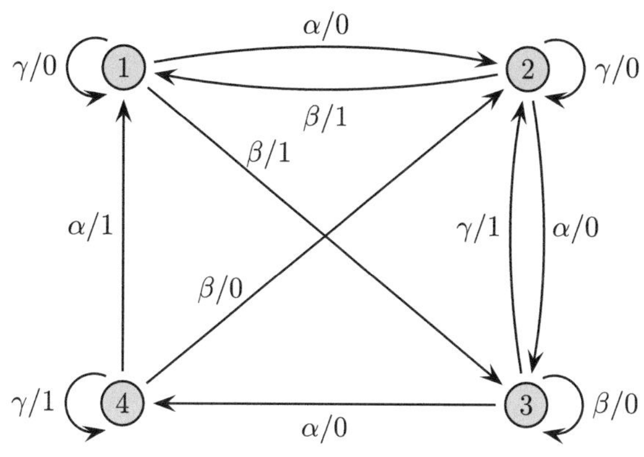

Fig. 4.3 Deterministic finite automaton with outputs (and without silent transitions) used to illustrate the recursive approach for current-state estimation

Example 4.2 To illustrate the above ideas, consider the DFA with outputs (and without silent transitions) $\mathcal{DFA} = (Q, \Sigma, Y, \delta, \lambda, Q_0)$ shown in Fig. 4.3, where $Q = \{1, 2, 3, 4\}$, $\sigma = \{\alpha, \beta, \gamma\}$, $Y = \{0, 1\}$, δ and λ are as described in the figure, and $Q_0 = \{1\}$. [Note that, for simplicity, we denote states by 1, 2, 3, 4 as opposed to $q^{(1)}, q^{(2)}, q^{(3)}, q^{(4)}$; the distinction between state 1 and output 1 should be obvious from context.] We illustrate the estimation tasks described above for the sequence of observations 1010 (i.e., $y_0^3 = 1010$ with $y[0] = 1$, $y[1] = 0$, $y[2] = 1$, and $y[3] = 0$).

The sequence of current state estimates for y_0^3 is as follows:

$$
\begin{aligned}
Q_C &= Q_0 = \{1\} \\
1 : Q_C &= \{3\} \\
10 : Q_C &= \{3, 4\} \\
101 : Q_C &= \{1, 2, 4\} \\
1010 : Q_C &= \{1, 2, 3\}
\end{aligned}
$$

To follow this recursive updating, we note the following:
(1) From state 1 with output 1 we can only go to state 3 (via β); thus, following $y[0] = 1$, the set of current state estimates is $Q_C = \{3\}$.
(2) From state 3 with output 0 we remain in state 3 (via β) or go to state 4 (via α). Thus, following $y[1] = 0$, the set of current state estimates is $Q_C = \{3, 4\}$.
(3) From state 3 with output 1 we can go to state 2 (via γ), whereas from state 4 with output 1 we can remain in state 4 (via γ) or go to state 1 (via α). Thus, following $y[2] = 1$, the set of current state estimates is $Q_C = \{1, 2, 4\}$.
(4) From state 1 with output 0 we remain in state 1 (via γ) or go to state 2 (via α); from state 2 with output 0 we remain in state 2 (via γ) or go to state 3 (via α); finally, from state 4 with output 0 we go to state 2 (via β). Thus, following $y[2] = 1$, the set of current state estimates is $Q_C = \{1, 2, 3\}$.

A similar approach can be followed for any sequence of observations. Note that at any given time we will need to maintain at most four current states. When the subsequent output becomes available, we can update each such state estimate separately by considering all possible transitions out of that state that produce the observed output. In this example, there are at most three possible transitions from each state; thus, for each updating, we have to consider at most 12 cases (4 states, each with 3 possible inputs). □

Note that the number of possible sets of current state estimates Q_C is finite (because Q_C is a subset of Q, i.e., $Q_C \in 2^Q$ and there are 2^N different such subsets). Since the update of Q_C^{new} only depends on Q_C and the particular observation $y \in Y$ (one out of R possible observations) that becomes available, the above online estimation strategy can also be implemented using a DFA without outputs, called *observer*, that is driven by the output sequence generated by G such that its state at any given time represents the set of possible current states for G. The observer for G is denoted by

$$G_{obs} := (Q_{obs}, \Sigma_{obs}, \delta_{obs}, Q_{obs,0}) = AC(2^Q, Y, \delta_{obs}, Q_0) ;$$

it is initialized at Q_0 (before any observations are made, Q_0 represents the set of possible current states) and its next-state transition function is defined for $Q_C \in 2^Q$, $y \in Y$ as

$$\delta_{obs}(Q_C, y) := Q_C^{new} = \{q \in Q \mid \exists q_c \in Q_C, \exists \sigma \in \Sigma \text{ such that } y = \lambda(q_c, \sigma) \\ (\text{and } \delta(q_c, \sigma) \text{ is defined})\} .$$

Note that AC denotes the accessible part of the automaton (refer to Definition 3.3) and simply means that Q_{obs} only retains the set of states in 2^Q that are accessible in G_{obs} from the initial state Q_0. Note that the empty subset $\emptyset$ could be a state in G_{obs} if there exist sequences of observations that cannot possibly be generated by G: this state is an absorbing state (once reached, G_{obs} remains in that state) and it is reached by sequences of outputs that do not correspond to any valid pair of an initial state and a sequence of inputs. It is worth pointing out that the construction of the observer for a given DFA with outputs resembles the process of determining a nondeterministic finite automaton (NFA) as described in Sect. 3.2.1.2.

A formal description of the observer is provided in Sect. 5.2, where we also discuss how it can be used to verify various properties of interest.

Example 4.3 The observer $G_{obs} = AC(2^Q, Y, \delta_{obs}, Q_0)$ for the DFA with outputs (and without silent transitions) $\mathcal{DFA} = (Q, \Sigma, Y, \delta, \lambda, Q_0)$ in Fig. 4.3 is shown in Fig. 4.4. It has 12 states which are denoted as subsets of $Q = \{1, 2, 3, 4\}$. (If we exclude the empty subset, there are $2^4 - 1 = 15$ such subsets, but three of them are not reachable and are not indicated in the construction in Fig. 4.4.) The observer is driven by the outputs $Y = \{0, 1\}$ of the automaton and its initial state is the set $Q_0 = \{1\} \subseteq Q$. The next-state transition function is determined by the observed output. For example, from state $\{3, 4\}$ with input 1 we transition to state $\{1, 2, 4\}$ because (i) from state 3 with output 1 we can go to state 2 (via γ), and (ii) from state 4 with output 1 we can remain in state 4 (via γ) or go to state 1 (via α). The transition functionality of δ_{obs} can be determined from each state in a similar fashion.

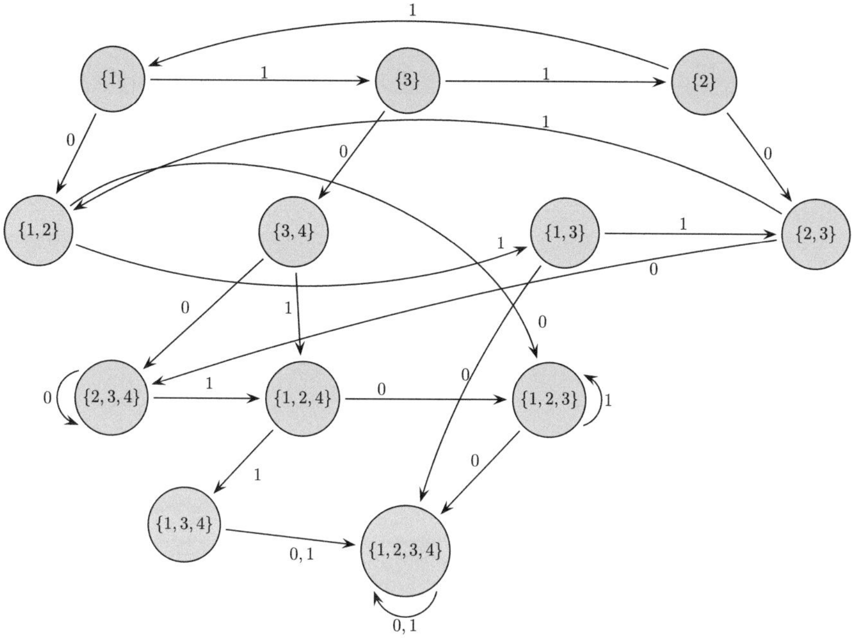

Fig. 4.4 Observer G_{obs} for the deterministic finite automaton with outputs (and without silent transitions) in Fig. 4.3

Note that once the observer is constructed, it can be used to perform current-state estimation in the given finite automaton in Fig. 4.3 in a straightforward way. For instance, if the sequence of observations is 1010 (i.e., $y_0^3 = 1010$ with $y[0] = 1$, $y[1] = 0$, $y[2] = 1$ and $y[3] = 0$) as in Example 4.2, then by applying this sequence as inputs to the observer (starting from Q_0), we obtain the sequence of states

$$\begin{aligned} q_{obs}[0] &= \{1\} , \\ y[0] = 1,\ q_{obs}[1] &= \{3\} , \\ y[1] = 0,\ q_{obs}[2] &= \{3, 4\} , \\ y[2] = 1,\ q_{obs}[3] &= \{1, 2, 4\} , \\ y[3] = 0,\ q_{obs}[4] &= \{1, 2, 3\} , \end{aligned}$$

which match exactly the sequence of state estimates we obtained in Example 4.2 using the recursive approach for current-state estimation. □

Remark 4.7 Building the observer G_{obs} for a given finite automaton G requires space complexity of $O(R2^N)$. In contrast, the online approach that maintains the set of possible current states Q_C would require significantly smaller complexity: $O(N)$ storage complexity is needed to store the possible states after each observation, $O(NK)$ storage complexity is needed to store the transition model of finite automaton G, and $O(NK)$ computational complexity is required to perform the update of the set of possible states after each observation. Clearly, the online approach should be preferred in most cases; nevertheless, the construction of G_{obs} is useful in cases where we are interested in capturing the set of possible current states in the system *under all possible behaviors* that might be generated by G. For example, if one is interested in checking whether the observer always knows the current state exactly or within an uncertainty set that involves at most one state in addition to the actual state that the system resides in, then one can build the observer G_{obs} and verify that all of its states correspond to subsets of Q that have cardinality at most two, i.e., $Q_{obs} \subseteq \{\emptyset, \{q^{(1)}\}, \{q^{(2)}\}, \ldots, \{q^{(N)}\}, \{q^{(1)}, q^{(2)}\}, \{q^{(1)}, q^{(3)}\}, \ldots, \{q^{(N-1)}, q^{(N)}\}\}$. These questions are related to detectability and K-detectability (see Chap. 5 where we discuss the verification of state-based properties and Chap. 6 where we discuss variations of the property of detectability). □

4.4 Mathematical Preliminaries

4.4.1 State Mappings and State Trajectories

In our discussions of various state estimation problems based on sequences of observations, we will find it useful to track sequences of states that are compatible with a given sequence of observations. To do that we will rely on the notions of state mappings and state trajectories.

Definition 4.1 (*State Mapping*) Given a set of states $Q = \{q^{(1)}, q^{(2)}, \ldots, q^{(N)}\}$, a state mapping M is a subset of $Q^2 \equiv Q \times Q$.

Clearly, given the set of states Q of cardinality $N = |Q|$, there are at most 2^{N^2} different state mappings. Note that M is a finite set of ordered pairs of states of the form (q_0, q_1) or (q_i, q_f), where $q_0, q_1 \in Q$ or $q_i, q_f \in Q$. The leftmost component is referred to as the 0th component (or the initial state) and the rightmost component is referred to as the 1st component (or the final state). To retrieve the components of a state mapping, we define projection operations with respect to the 0th (initial) and 1st (final) components; we also define a composition operation for state mappings.

Definition 4.2 (*State Mapping Projections*) Given a state mapping M, we define the projections Π_0 and Π_1 as follows:

$$\Pi_0(M) = \{q_0 \mid (q_0, q_1) \in M\} ,$$
$$\Pi_1(M) = \{q_1 \mid (q_0, q_1) \in M\} .$$

Definition 4.3 (*State Mapping Composition*) Given two state mappings M_1 and M_2, we define their composition $\odot$ as another state mapping

$$M = M_1 \odot M_2 = \{(q_{01}, q_{12}) \mid \exists q_{11} = q_{02} \text{ such that } (q_{01}, q_{11}) \in M_1 \text{ and } (q_{02}, q_{12}) \in M_2\} .$$

Remark 4.8 One can easily show that operation $\odot$ is associative so that the set 2^{Q^2} forms a semigroup under operation $\odot$ (in fact, it forms a monoid with identity element $M_{ident} = \{(q^{(1)}, q^{(1)}), (q^{(2)}, q^{(2)}), \ldots, (q^{(N)}, q^{(N)})\}$). □

Definition 4.4 (*State L-Trajectory and Related Operations*) Given a set of states $Q = \{q^{(1)}, q^{(2)}, \ldots, q^{(N)}\}$, a state L-trajectory $M^{(L)}$ (for $L = 1, 2, 3, 4, \ldots$) is a subset of $Q^L \equiv Q \times Q \times \cdots \times Q$ (L times). We define four operations on state L-trajectories: (i) projection Π_i for $i = 0, 1, \ldots, L-1$, (ii) concatenation $\cdot$, (iii) shift $>>$, and (iv) i-trimming $trim_i$, as follows:

1. Given a state L-trajectory $M^{(L)}$, we define the projections Π_i for $i = 0, 1, \ldots, L-1$ as
$$\begin{aligned} \Pi_0(M^{(L)}) &:= \{q_0 \mid (q_0, q_1, \ldots, q_{L-1}) \in M^{(L)}\} , \\ \Pi_1(M^{(L)}) &:= \{q_1 \mid (q_0, q_1, \ldots, q_{L-1}) \in M^{(L)}\} , \\ &\vdots \\ \Pi_{L-1}(M^{(L)}) &:= \{q_{L-1} \mid (q_0, q_1, \ldots, q_{L-1}) \in M^{(L)}\} . \end{aligned}$$

2. Given a state L_1-trajectory $M^{(L_1)}$ and a state L_2-trajectory $M^{(L_2)}$ with $L_1 \geq 1$ and $L_2 \geq 1$, we define their concatenation $\cdot$ as a state $(L_1 + L_2 - 1)$-trajectory $M^{(L_1+L_2-1)}$ defined as
$$\begin{aligned} M^{(L_1+L_2-1)} &= M^{(L_1)} \cdot M^{(L_2)} \\ &:= \{(q_0, q_1, \ldots, q_{L_1-1}, q'_1, \ldots, q'_{L_2-1}) \mid \exists q_{L_1-1} = q'_0 \\ &\qquad \text{such that } (q_0, q_1, \ldots, q_{L_1-1}) \in M^{(L_1)} \text{ and} \\ &\qquad (q'_0, q'_1, \ldots, q'_{L_2-1}) \in M^{(L_2)}\} . \end{aligned}$$

   [Note that given a state 1-trajectory $M_1^{(1)}$ and a state 1-trajectory $M_2^{(1)}$, their concatenation is a state 1-trajectory $M^{(1)} = \{(q_i) \mid \exists q_i \text{ such that } (q_i) \in M_1^{(1)} \text{ and } (q_i) \in M_2^{(1)}\}$.]
3. Given a state L-trajectory $M^{(L)}$ and a state 2-trajectory $M^{(2)}$, we define the shift operation $>>$ as a state L-trajectory
$$\begin{aligned} M^{(L)} >> M^{(2)} := \{(q_1, q_2, \ldots, q_{L-1}, q_L) \mid (q_0, q_1, \ldots, q_{L-1}) \in M^{(L)} \\ \text{and } (q_{L-1}, q_L) \in M^{(2)}\} . \end{aligned}$$

4. Given a state L-trajectory $M^{(L)}$ where $L \geq 2$, the i-trimmed version of it for $i = 0, 1, 2, \ldots, L-1$ is a state $(L-i)$-trajectory $M^{(L-i)}$ defined as
$$M^{(L-i)} = trim_i(M^{(L)}) := \{(q_i, q_{i+1}, \ldots, q_{L-1}) \mid (q_0, q_1, \ldots, q_{L-1}) \in M^{(L)}\} .$$

   Note that $trim_0(M^{(L)}) = M^{(L)}$. Also, if we slightly abuse notation, we can say that $trim_{L-1}(M^{(L)}) = \Pi_{L-1}(M^{(L)})$. [The problem with the latter equation is that the trim operation returns a state 1-trajectory, whereas the projection operation returns a subset of states; however, if we think of each state in the subset of states returned by the projection operation as the corresponding element of the state 1-trajectory, then we easily arrive at the latter equality.]

Note that a state L-trajectory could be the empty set. Clearly, given the set of states Q of cardinality $N = |Q|$, there are at most 2^{N^L} different state L-trajectories. One can easily show that operation $\cdot$ is associative, i.e., for arbitrary state trajectories M_1, M_2, and M_3, we have

$$M_1 \cdot (M_2 \cdot M_3) = (M_1 \cdot M_2) \cdot M_3 \ .$$

Also, note that the shift operation can be expressed via the composition operation followed by 1-trimming, i.e., given a state L-trajectory $M^{(L)}$ and a state 2-trajectory $M^{(2)}$, we have

$$M^{(L)} >> M^{(2)} = trim_1(M^{(L)} \cdot M^{(2)}) \ .$$

In fact, for $L \geq 2$, we also have

$$M^{(L)} >> M^{(2)} = trim_1(M^{(L)}) \cdot M^{(2)} \ .$$

Example 4.4 Consider the set of states $\{1, 2, 3\}$ and the following three state trajectories, one of length 3, one of length 4, and one of length 2:

$$\begin{aligned} M_1^{(3)} &= \{(1,2,3), (1,3,3), (3,1,2)\} \ , \\ M_2^{(4)} &= \{(1,2,3,3), (2,1,1,2), (3,2,1,1), (1,1,1,1)\} \ , \\ M_3^{(2)} &= \{(1,2), (2,1), (3,2)\} \ . \end{aligned}$$

Below we illustrate some of the operations and properties for state trajectories that were introduced in this section:

- Projection:

$$\begin{aligned} \Pi_0(M_1^{(3)}) &= \{1,3\} \ , \\ \Pi_2(M_2^{(4)}) &= \{1,3\} \ , \\ \Pi_1(M_3^{(2)}) &= \{1,2\} \ . \end{aligned}$$

- Concatenation:

$$\begin{aligned} M_4^{(6)} := M_1^{(3)} \cdot M_2^{(4)} &= \{(1,2,3,2,1,1), (1,3,3,2,1,1), (3,1,2,1,1,2)\} \ , \\ M_5^{(5)} := M_2^{(4)} \cdot M_3^{(2)} &= \{(1,2,3,3,2), (2,1,1,2,1), (3,2,1,1,2), \\ & \quad (1,1,1,1,2)\} \ . \end{aligned}$$

- Shift:

$$\begin{aligned} M_1^{(3)} >> M_3^{(2)} &= \{(2,3,2), (3,3,2), (1,2,1)\} \ , \\ M_2^{(4)} >> M_3^{(2)} &= \{(2,3,3,2), (1,1,2,1), (2,1,1,2), (1,1,1,2)\} \ . \end{aligned}$$

- Trim:

$$\begin{aligned} trim_1(M_1^{(3)}) &= \{(2,3),(3,3),(1,2)\}\ , \\ trim_2(M_2^{(4)}) &= \{(3,3),(1,2),(1,1)\}\ , \\ trim_1(M_2^{(4)} \cdot M_3^{(2)}) &= trim_1(M_5^{(5)}) \\ &= \{(2,3,3,2),(1,1,2,1),(2,1,1,2),(1,1,1,2)\}\ . \end{aligned}$$

Note that indeed $M_2^{(4)} >> M_3^{(2)} = trim_1(M_2^{(4)} \cdot M_3^{(2)})$.

Also, it is easy to verify in this example that concatenation is an associative operation. For instance,

$$\begin{aligned} (M_1^{(3)} \cdot M_2^{(4)}) \cdot M_3^{(2)} &= M_4^{(6)} \cdot M_3^{(2)} \\ &= \{(1,2,3,2,1,1,2),(1,3,3,2,1,1,2),(3,1,2,1,1,2,1)\}\ , \end{aligned}$$

which is the same as

$$\begin{aligned} M_1^{(3)} \cdot (M_2^{(4)} \cdot M_3^{(2)}) &= M_1^{(3)} \cdot M_5^{(5)} \\ &= \{(1,2,3,2,1,1,2),(1,3,3,2,1,1,2),(3,1,2,1,1,2,1)\}\ . \end{aligned}$$

Graphical ways of representing state trajectories, as well as ways of performing the above operations graphically, will be presented in Example 4.8 (at least for a class of state mappings and trajectories). □

Remark 4.9 Note that for $L = 2$, a state 2-trajectory is identical to a state mapping in the sense that both are subsets of Q^2. However, the concatenation operation $\cdot$ on state 2-trajectories is quite different from the composition operation $\odot$ on state mappings: the former results in a state 3-trajectory $M^{(3)}$, whereas the latter results in a state 2-trajectory (state mapping) M; nevertheless, the resulting state 3-trajectory can be used to obtain the corresponding state mapping by eliminating the intermediate states in the trajectories, i.e., one can easily show that for arbitrary $M_1, M_2 \subseteq Q^2$, the set

$$M = \{(q_1, q_3) \mid \exists q_2 \text{ such that } (q_1, q_2, q_3) \in M^{(3)}\}\ ,$$

where $M^{(3)} = M_1 \cdot M_2$, satisfies $M = M_1 \odot M_2$.

In Example 4.4, the composition of $M_3^{(2)}$ with itself results in

$$M_3^{(2)} \odot M_3^{(2)} = \{(1,1),(2,2),(3,1)\}\ ,$$

whereas the concatenation with itself results in

$$M_3^{(2)} \cdot M_3^{(2)} = \{(1,2,1),(2,1,2),(3,2,1)\}\ .$$

Clearly, we can obtain $M_3^{(2)} \odot M_3^{(2)}$ by ignoring the intermediate states in $M_3^{(2)} \cdot M_3^{(2)}$ (and getting rid of any replicas—which was not necessary in this example). □

An important property of state trajectories is the fact that the continuation of a state L_1-trajectory by another state L_2-trajectory via the concatenation operation only depends on its last component. This property follows easily from the definition of the concatenation operation, but it is formalized in the theorem below because it is important in simplifying the recursive state estimation algorithms we develop later on.

Theorem 4.1 *Given a state L_1-trajectory $M^{(L_1)}$ and a state L_2-trajectory $M^{(L_2)}$ with $L_1 \geq 1$ and $L_2 \geq 1$, and their concatenated state $(L_1 + L_2 - 1)$-trajectory $M^{(L_1+L_2-1)} = M^{(L_1)} \cdot M^{(L_2)}$, we have*

$$trim_{L_1-1}(M^{(L_1+L_2-1)}) = P_{L_1-1}(M^{(L_1)}) \cdot M^{(L_2)}$$

or, equivalently,

$$trim_{L_1-1}(M^{(L_1)} \cdot M^{(L_2)}) = P_{L_1-1}(M^{(L_1)}) \cdot M^{(L_2)} .$$

Note that in the above theorem $\Pi_{L_1-1}(M^{(L_1)})$ is treated as a state 1-trajectory that gets concatenated with a state L_2-trajectory (resulting in a state L_2-trajectory—just like the trim operation on the left of the equation).

Proof From the definition of the *trim* and concatenation operations we get that the left hand side satisfies

$$\begin{aligned} trim_{L_1-1}(M^{(L_1+L_2-1)}) &= trim_{L_1-1}(M^{(L_1)} \cdot M^{(L_2)}) \\ &= \{(q_{L_1-1}, q'_1, \ldots, q'_{L_2-1}) \mid \exists q_{L_1-1} = q'_0, \\ &\quad \exists q_0, q_1, \ldots, q_{L_1-2} \in Q \text{ such that} \\ &\quad\quad (q_0, q_1, \ldots, q_{L_1-1}) \in M^{(L_1)} \text{ and} \\ &\quad\quad (q'_0, q'_1, \ldots, q'_{L_2-1}) \in M^{(L_2)}\} . \end{aligned}$$

From the definition of the projection operation we have on the right hand side

$$\begin{aligned} P_{L_1-1}(M^{L_1}) = \{q_{L_1-1} \mid \exists q_0, q_1, \ldots, q_{L_1-2} \in Q \\ \text{such that } (q_0, q_1, \ldots, q_{L_1-1}) \in M^{(L_1)}\} . \end{aligned}$$

Combining the above with the definition of the concatenation operation we see that the right hand side satisfies

$$\begin{aligned} P_{L_1-1}(M^{(L_1)}) \cdot M^{(L_2)} = \{(q_{L_1-1}, q'_1, \ldots, q'_{L_2-1}) \mid \\ \exists q_0, q_1, \ldots, q_{L_1-2} \in Q \text{ such that} \\ (q_0, q_1, \ldots, q_{L_1-1}) \in M^{(L_1)} \text{ and} \\ \exists q'_0 = q_{L_1-1} \text{ such that } (q'_0, q'_1, \ldots, q'_{L_2-1}) \in M^{(L_2)}\} , \end{aligned}$$

which is exactly equal to the left side. □

Theorem 4.1 can be generalized easily to the following theorem whose proof is omitted because it is very similar to the previous proof (note that for $i = L_1 - 1$, the theorem statement reduces to Theorem 4.1).

Theorem 4.2 *Given a state L_1-trajectory $M^{(L_1)}$ and a state L_2-trajectory $M^{(L_2)}$ with $L_1 \geq 1$ and $L_2 \geq 1$, and their concatenated state $(L_1 + L_2 - 1)$-trajectory $M^{(L_1+L_2-1)} = M^{(L_1)} \cdot M^{(L_2)}$, we have*

$$trim_i(M^{(L_1+L_2-1)}) = trim_i(M^{(L_1)}) \cdot M^{(L_2)}$$

or, equivalently,

$$trim_i(M^{(L_1)} \cdot M^{(L_2)}) = trim_i(M^{(L_1)}) \cdot M^{(L_2)} .$$

for $i = 0, 1, 2, \ldots, L_1 - 1$.

4.4.2 Induced State Mappings

Given a DFA with outputs and possibly silent transitions $(Q, \Sigma, Y \cup \{\epsilon\}, \delta, \lambda, Q_0)$, we are interested in using state mappings to track the sequences of states that are compatible with the observation of the sequence y_0^k. We first start by defining the notion of an induced state mapping for a single observation, we then extend this concept for a sequence of observations, and finally, we discuss how information about the prior set of possible initial states can be incorporated. We choose to deal directly with automata with silent transitions since the special case of automata without silent transitions can be handled in exactly the same way.

Definition 4.5 (*Induced State Mapping*) Given the DFA with outputs and (possibly) silent transitions $\mathcal{DFA} = (Q, \Sigma, Y \cup \{\epsilon\}, \delta, \lambda, Q_0)$, we define for each label $y \in Y$, the induced state mapping M_y to be

$$\begin{aligned} M_y = \{(q_i, q_f) \in Q^2 \mid \exists m' \geq 0, \sigma_k^{k+m'} \in \Sigma^* \text{ such that } q_f = \delta(q_i, \sigma_k^{k+m'}) \\ \text{and } y = \lambda_{seq}(q_i, \sigma_k^{k+m'})\} . \end{aligned}$$

Remark 4.10 Notice that the notion of a state mapping induced by observation y can also be captured by an $N \times N$ matrix A_y whose (j, i)th entry satisfies $a_y(j, i) = 1$ iff (if and only if) there exists a sequence of inputs that takes us from state $q^{(i)}$ to state $q^{(j)}$ while generating output y. This is reminiscent of the transition matrix notation that was introduced (for inputs rather than outputs) in Sect. 3.2.1.3. □

The notion of an induced state mapping also extends very naturally to induced state mappings over sequences of observations.

Definition 4.6 (*Induced State Mapping over a Sequence of Observations*) Given the DFA with outputs and (possibly) silent transitions $\mathcal{DFA} = (Q, \Sigma, Y \cup \{\epsilon\}, \delta, \lambda, Q_0)$, we define for a sequence of observations y_k^{k+m}, $m \geq 0$, the induced state mapping $M_{y_k^{k+m}}$ to be

$$M_{y_k^{k+m}} = \{(q_i, q_f) \in Q^2 \mid \exists m' \geq 0, \sigma_k^{k+m'} \in \Sigma^* \text{ such that } q_f = \delta(q_i, \sigma_k^{k+m'}) \\ \text{and } y_k^{k+m} = \lambda_{seq}(q_i, \sigma_k^{k+m'})\} \ .$$

An important property of induced state mappings is captured by the following theorem:

Theorem 4.3 *Consider a DFA with outputs and (possibly) silent transitions* $\mathcal{DFA} = (Q, \Sigma, Y \cup \{\epsilon\}, \delta, \lambda, Q_0)$. *Given sequence of outputs* $y_k^{k'}$ *and any* $k \leq m < k'$, *we have*

$$M_{y_k^{k'}} = M_{y_k^m} \odot M_{y_{m+1}^{k'}} \ .$$

Proof From the definition of induced state mapping, the left hand side satisfies

$$M_{y_k^{k'}} = \{(q_i, q_f) \in Q^2 \mid \exists m' \geq 0, \sigma_k^{k+m'} \in \Sigma^* \text{ such that } q_f = \delta(q_i, \sigma_k^{k+m'}) \\ \text{and } y_k^{k'} = \lambda_{seq}(q_i, \sigma_k^{k+m'})\} \ .$$

Similarly, for the right hand side we have

$$M_{y_k^m} = \{(q_i, q_{f1}) \in Q^2 \mid \exists m_1' \geq 0, \sigma_k^{k+m_1'} \in \Sigma^* \text{ such that} \\ q_{f1} = \delta(q_i, \sigma_k^{k+m_1'}) \text{ and } y_k^m = \lambda_{seq}(q_i, \sigma_k^{k+m_1'})\}$$

and

$$M_{y_{m+1}^{k'}} = \{(q_{i2}, q_f) \in Q^2 \mid \exists m_2' \geq 0, \sigma_{m+1}^{m+1+m_2'} \in \Sigma^* \text{ such that} \\ q_f = \delta(q_{i2}, \sigma_{m+1}^{m+1+m_2'}) \text{ and } y_{m+1}^{k'} = \lambda_{seq}(q_{i2}, \sigma_{m+1}^{m+1+m_2'})\} \ .$$

The composition operation on $M_{y_k^m}$ and $M_{y_{m+1}^{k'}}$ is by definition

$$\begin{aligned} M_{y_k^m} \odot M_{y_{m+1}^{k'}} &= \{(q_i, q_f) \in Q^2 \mid \exists q_{f1} = q_{i2} \text{ such that} \\ &\quad (q_i, q_{f1}) \in M_{y_k^m} \text{ and } (q_{i2}, q_f) \in M_{y_{m+1}^{k'}}\} \\ &= \{(q_i, q_f) \in Q^2 \mid \exists q_{f1} = q_{i2}, \ \exists m_1' \geq 0, \sigma_k^{k+m_1'} \in \Sigma^*, \\ &\quad \exists m_2' \geq 0, \sigma_{m+1}^{m+1+m_2'} \in \Sigma^* \text{ such that} \\ &\quad q_{f1} = \delta(q_i, \sigma_k^{k+m_1'}) \text{ and } y_k^m = \lambda_{seq}(q_i, \sigma_k^{k+m_1'}) \text{ and} \\ &\quad q_f = \delta(q_{i2}, \sigma_{m+1}^{m+1+m_2'}) \text{ and } y_{m+1}^{k'} = \lambda_{seq}(q_{i2}, \sigma_{m+1}^{m+1+m_2'})\} \\ &= \{(q_i, q_f) \in Q^2 \mid \exists \ m' \geq 0, \sigma_k^{k+m'} \in \Sigma^* \\ &\quad \text{such that } q_f = \delta(q_i, \sigma_k^{k+m'}) \text{ and } y_k^{k'} = \lambda_{seq}(q_i, \sigma_k^{k+m'})\} \ , \end{aligned}$$

where in the last equality $\sigma_k^{k+m'}$ is the concatenation of $\sigma_k^{k+m_1'}$ and $\sigma_{m+1}^{m+1+m_2'}$. □

The following corollary follows directly from the above theorem using the fact that operation $\odot$ is associative.

Corollary 4.1 *Consider a DFA with outputs and (possibly) silent transitions $\mathcal{DFA} = (Q, \Sigma, Y \cup \{\epsilon\}, \delta, \lambda, Q_0)$. Given sequence of outputs $y_k^{k'}$, we have*

$$M_{y_k^{k'}} = M_{y_k} \odot M_{y_{k+1}} \odot \cdots \odot M_{y_{k'}} .$$

Example 4.5 Consider the finite automaton G with outputs (and without silent transitions) in Fig. 4.3. There are two possible outputs ($Y = \{0, 1\}$) and their corresponding induced mappings are

$$M_0 = \{(1, 1), (1, 2), (2, 2), (2, 3), (3, 3), (3, 4), (4, 2)\} ,$$
$$M_1 = \{(1, 3), (2, 1), (3, 2), (4, 1), (4, 4)\} .$$

If we consider a sequence of outputs (observations) 01 (i.e., $y_0^1 = 01$ with $y[0] = 0$ and $y[1] = 1$), then M_{01} contains exactly the pairs of states of the form (q_i, q_f) such that there exists a sequence of inputs that takes us from q_i to q_f and generates the sequence of outputs 01. It is not hard to confirm that

$$M_{01} = \{(1, 1), (1, 3), (2, 1), (2, 2), (3, 1), (3, 2), (3, 4), (4, 1)\} .$$

For example, from state 1 we can go to state 1 (via $\alpha\beta$) or state 3 (via $\gamma\beta$). Similarly, we can obtain M_{010} for sequence of outputs $y_0^2 = 010$ as

$$M_{010} = \{(1, 1), (1, 2), (1, 3), (1, 4), (2, 1), (2, 2), (2, 3), (3, 1), (3, 2), (3, 3), (4, 1), (4, 2)\} .$$

For example, from state 1 we can go to state 1 (via $\alpha\beta\gamma$), or state 2 (via $\alpha\beta\alpha$), or state 3 (via $\gamma\beta\beta$), or state 4 (via $\gamma\beta\alpha$).

Notice that if we compose $M_0 \odot M_1$ we obtain exactly M_{01}; similarly, we have $M_{010} = M_{01} \odot M_0$. □

4.4.3 Induced State Trajectories

The definitions in the previous section for induced state mappings can be extended to induced state L-trajectories as follows.

Definition 4.7 (*Induced State 2-Trajectory over a Single Observation*) Given the DFA with outputs and (possibly) silent transitions $\mathcal{DFA} = (Q, \Sigma, Y \cup \{\epsilon\}, \delta, \lambda, Q_0)$, we define for each output $y \in Y$, the induced state 2-trajectory $M_y^{(2)}$ to be

$$M_y^{(2)} = \{(q_i, q_f) \in Q^2 \mid \exists m' \geq 0, \sigma_k^{k+m'} \in \Sigma^* \text{ such that } q_f = \delta(q_i, \sigma_k^{k+m'}) \\ \text{and } y = \lambda_{seq}(q_i, \sigma_k^{k+m'})\} .$$

Clearly, for a single output $y \in Y$, the induced state 2-trajectory is identical to the corresponding induced state mapping. However, the notion of an induced state trajectory is quite different from an induced state mapping in the case of a sequence of observations of length greater than one.

Definition 4.8 (*Induced State L-Trajectory over a Sequence of Observations*) Given the DFA with outputs and (possibly) silent transitions $\mathcal{DFA} = (Q, \Sigma, Y \cup \{\epsilon\}, \delta, \lambda, Q_0)$, we define for each sequence of outputs $y_k^{k+m} \equiv y[k], y[k+1], \ldots, y[k+m]$, the induced state $(m+2)$-trajectory to be

$$M_{y_k^{k+m}}^{(m+2)} = \{(q_0, q_1, \ldots, q_m, q_{m+1}) \in Q^{m+2} | \exists\, 0 \leq k_0 = k < k_1 < \cdots < k_{m+1}, \\ \exists\, \sigma_{k_0}^{k_1-1}, \sigma_{k_1}^{k_2-1}, \ldots, \sigma_{k_m}^{k_{m+1}-1} \in \Sigma^* \text{ such that } \forall i = 0, 1, \ldots, m, \\ q_{i+1} = \delta(q_i, \sigma_{k_i}^{k_{i+1}-1}) \text{ and } y[k+i] = \lambda_{seq}(q_i, \sigma_{k_i}^{k_{i+1}-1})\} .$$

The proof of the following theorem is similar to the proof of Theorem 4.3 for induced state mappings and is omitted.

Theorem 4.4 *Consider a DFA with outputs and (possibly) silent transitions* $\mathcal{DFA} = (Q, \Sigma, Y \cup \{\epsilon\}, \delta, \lambda, Q_0)$. *Given sequence of outputs* $y_k^{k'} \equiv y[k], y[k+1], \ldots, y[k'] \in Y^{k'-k+1}$, *we have for any* $k \leq m < k'$

$$M_{y_k^{k'}}^{(k'-k+2)} = M_{y_k^m}^{(m-k+2)} \cdot M_{y_{m+1}^{k'}}^{(k'-m+1)} .$$

The following corollary, which can be obtained from the above theorem using the fact that the concatenation operation on state trajectories is associative, identifies a property of induced state L-trajectories that is key for allowing us to recursively perform state estimation in the sequel. [Recall that operation $\cdot$ is an associative operation and thus there is no need for parenthesis (e.g., $(M_{y[0]}^{(2)} \cdot M_{y[1]}^{(2)}) \cdot M_{y[2]}^{(2)} = M_{y[0]}^{(2)} \cdot (M_{y[1]}^{(2)} \cdot M_{y[2]}^{(2)})$).]

Corollary 4.2 *Consider a DFA with outputs and (possibly) silent transitions* $\mathcal{DFA} = (Q, \Sigma, Y \cup \{\epsilon\}, \delta, \lambda, Q_0)$. *Given sequence of outputs* $y_k^{k+m} \equiv y[k], y[k+1], \ldots, y[k+m] \in Y^{m+1}$, *we have*

$$M_{y_k^{k+m}}^{(m+2)} = M_{y[k]}^{(2)} \cdot M_{y[k+1]}^{(2)} \cdot \cdots \cdot M_{y[k+m]}^{(2)} .$$

Example 4.6 Consider again the finite automaton G with outputs (and without silent transitions) in Fig. 4.3. The induced state 4-trajectory given the sequence of observations 010 (i.e., $y_0^2 = 010$ with $y[0] = 0$, $y[1] = 1$, and $y[2] = 0$) is captured by

$$M_{010}^{(4)} = \{(1,1,3,3), (1,1,3,4), (1,2,1,1), (1,2,1,2), (2,2,1,1), (2,2,1,2), (2,3,2,2), (2,3,2,3), (3,3,2,2), (3,3,2,3), (3,4,1,1), (3,4,1,2), (3,4,4,2), (4,2,1,1), (4,2,1,2)\} .$$

Notice that one can obtain $M_{010}^{(4)}$ by performing the concatenation

$$M_{010}^{(4)} = M_0 \cdot M_1 \cdot M_0 ,$$

where M_0 and M_1 were defined in Example 4.5. Also, if we only consider initial and final states in each 4-tuple of states (i.e., if we ignore the intermediate states), we obtain the set of pairs

$$\{(1,1), (1,2), (1,3), (1,4), (2,1), (2,2), (2,3), (3,1), (3,2), (3,3), (4,1), (4,2)\} ,$$

which is exactly the state mapping M_{010} we obtained in Example 4.5. □

4.4.4 *Tracking Induced State Trajectories via Trellis Diagrams*

An L-dimensional trellis diagram is an L-partite (directed) graph for which the L partitions, indexed from P_0 to P_{L-1} and drawn from left to right, have equal numbers of nodes. More specifically, we assume that each partition has N nodes, indexed from 1 to N and drawn in a vertical slice from top to bottom; connections exist only between pairs of nodes that belong in two partitions with consecutive indices: more specifically, connections can only exist from a node in partition P_0 to a node in partition P_1, or from a node in partition P_1 to a node in partition P_2, ..., or from a node in partition P_{L-2} to a node in partition P_{L-1}. Clearly, an L-dimensional trellis diagram has LN nodes and, as shown in Fig. 4.5 for the case when $L = 4$ and $N = 4$, it can be drawn by arranging the L partitions into vertical columns (indexed by P_0, P_1, ..., P_{L-1}), each of which is a vertical slice with N nodes (indexed by 1, 2, ..., N).

A key property of a trellis diagram is that each node in a non-boundary partition P_i $(1 \leq i \leq L-2)$ is either isolated or has connections from (at least) one node

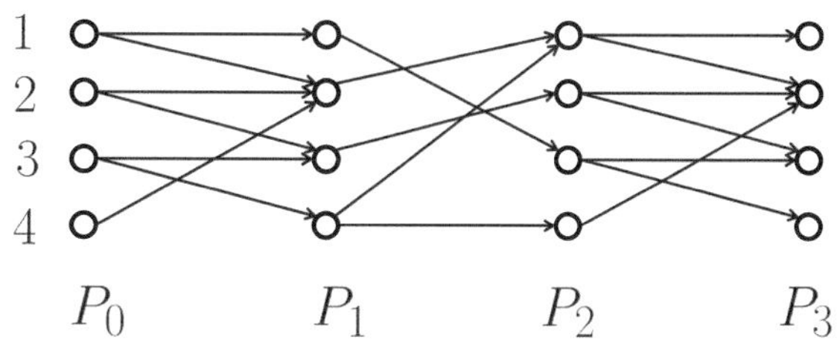

Fig. 4.5 An example of a trellis diagram with $L = 4$ and $N = 4$

in partition P_{i-1} *and* to (at least) one node in partition P_{i+1}; furthermore, nodes in each of the two boundary partitions (P_0 or P_{L-1}) are either isolated or have connections to/from (at least) one node in their neighboring partition (P_1 or P_{L-2}, respectively). This requirement (that nodes be either isolated or connected from/to at least one node in each of the two neighboring partitions, unless the node belongs to a boundary partition) implies that one can start from any non-isolated node in the leftmost partition (partition P_0) and find at least one path to some node in the rightmost partition (partition P_{L-1}); similarly, one can start from any non-isolated node in the rightmost partition (partition P_{L-1}) and find at least one reverse path to some node in the leftmost partition (partition P_0); in fact, one can start from any non-isolated node in any non-boundary partition and find at least one path to a node in the rightmost partition (partition P_{L-1}) and at least one reverse path to a node in the leftmost partition (partition P_0). This will always be possible, unless one deals with the trivial trellis diagram which has no edges between nodes.

Note that if we (1) associate each partition of an L-dimensional trellis diagram with an observation time (so that partition P_i, $0 \leq i \leq L-1$, corresponds to observation time i), and (2) associate each of the N nodes in a partition with a state in the set $Q = \{q^{(1)}, q^{(2)}, \ldots, q^{(N)}\}$, then any given state L-trajectory $M^{(L)}$ can be associated with a unique trellis diagram. More specifically, we can construct the trellis diagram by taking each $(q_0, q_1, \ldots, q_{L-1}) \in M^{(L)}$ and including edges between node (q_0, P_0) (i.e., the node that corresponds to q_0 in partition P_0) and node (q_1, P_1) (i.e., the node that corresponds to q_1 in partition P_1), between node (q_1, P_1) and node (q_2, P_2) (i.e., the node that corresponds to q_2 in partition P_2), ..., and between node (q_{L-2}, P_{L-2}) (i.e., the node that corresponds to q_{L-2} in partition $L-2$) and node (q_{L-1}, P_{L-1}) (i.e., the node that corresponds to q_{L-1} in partition P_{L-1}). This implies that for each $(q_0, q_1, \ldots, q_{L-1}) \in M^{(L)}$, we include a path $(q_0, P_0), (q_1, P_1), \ldots, (q_{L-1}, P_{L-1})$ in the corresponding trellis diagram; this trivially satisfies the requirement that nodes in the trellis diagram are either isolated or connected to at least one node in each of the two neighboring partitions.

Note that a trellis diagram is *not* a multigraph and cannot have multiple edges between the same pair of nodes: an L-dimensional trellis diagram includes an edge from a node (q, P_i) to a node (q', P_{i+1}) in consecutive partitions P_i and P_{i+1} as long as there is at least one element in the original state L-trajectory $M^{(L)}$ that includes state q at its i component and state q' at its $(i+1)$ component. It should be clear from the above discussion that when we restrict ourselves to induced state L-trajectories, the induced state L-trajectory that corresponds to an L-dimensional trellis diagram is unique. The reason is that in an induced state L-trajectory $M^{(L)}_{y_0^{L-1}}$, continuations from a given state in the trellis diagram have to be the *same* for all trajectories leading to that state. This claim is illustrated in the example below and formalized in Lemma 4.1.

Example 4.7 Consider again DFA G with outputs (and without silent transitions) in Fig. 4.3. As argued in Example 4.6, the induced state 4-trajectory under the sequence of observations 010 is given by

$$M_{010}^{(4)} = \{(1,1,3,3),(1,1,3,4),(1,2,1,1),(1,2,1,2),(2,2,1,1),(2,2,1,2),\\ (2,3,2,2),(2,3,2,3),(3,3,2,2),(3,3,2,3),(3,4,1,1),(3,4,1,2),\\ (3,4,4,2),(4,2,1,1),(4,2,1,2)\}$$

and it can easily be verified that its corresponding trellis diagram is actually the one in Fig. 4.5.

In general, given an L-dimensional trellis diagram, there could be multiple state L-trajectories that could generate it; for instance, the trellis diagram in Fig. 4.5 could have been generated not only by $M_{010}^{(4)}$ above, but also by the state 4-trajectory

$$M^{(4)} = \{(1,1,3,3),(1,1,3,4),(1,2,1,1),(2,2,1,2),(2,3,2,2),(3,3,2,3),\\ (3,4,1,2),(3,4,4,2),(4,2,1,2)\}\ ,$$

or others. As mentioned before the example, however, the induced state L-trajectory that corresponds to an L-dimensional trellis diagram is unique (this claim is formalized in Lemma 4.1 below). This property ensures a one-to-one correspondence between an induced state L-trajectory and an L-dimensional trellis diagram (for the trellis in Fig. 4.5 the only possible induced state 4-trajectory would be $M_{010}^{(4)}$ as specified above). □

The precise reasoning for the one-to-one correspondence between an L-dimensional trellis diagram and an induced state L-trajectory is provided in the lemma below, whose proof follows directly from the definition of the concatenation operation and is omitted.

Lemma 4.1 *Consider a DFA with outputs and (possibly) silent transitions* $\mathcal{DFA} = (Q, \Sigma, Y \cup \{\epsilon\}, \delta, \lambda, Q_0)$. *Given a sequence of outputs* $y_0^{L-1} \equiv y[0], y[1], \ldots, y[L-1] \in Y^L$, *we use* $M_{y_0^{L-1}}^{(L+1)}$ *to denote the induced state trajectory. If the following is true:*

$$(q_0, q_1, \ldots, q_{i-1}, q_i, q_{i+1}, \ldots, q_L) \in M_{y_0^{L-1}}^{(L+1)}\ ,$$
$$(q_0', q_1', \ldots, q_{i-1}', q_i, q_{i+1}', \ldots, q_L') \in M_{y_0^{L-1}}^{(L+1)}\ ,$$

then we can conclude that

$$(q_0', q_1', \ldots, q_{i-1}', q_i, q_{i+1}, \ldots, q_L) \in M_{y_0^{L-1}}^{(L+1)}\ ,$$
$$(q_0, q_1, \ldots, q_{i-1}, q_i, q_{i+1}', \ldots, q_L') \in M_{y_0^{L-1}}^{(L+1)}\ .$$

Example 4.8 Using trellis diagrams, we can easily represent state mappings and state trajectories. For instance, the top of Fig. 4.6 shows the induced state mappings $M_0 = m_1$ and $M_1 = m_2$ (under output 0 and output 1, respectively) for DFA with outputs (and without silent transitions) G in Fig. 4.3, which were first discussed in Example 4.5. Specifically, m_1 and m_2 in the figure represent

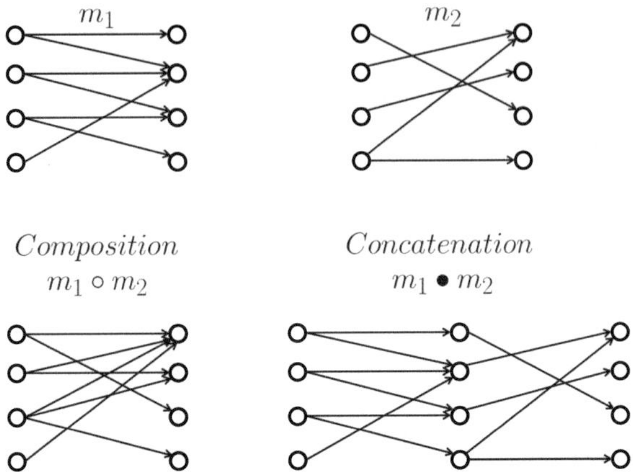

Fig. 4.6 Graphical representation of state mappings and state trajectories, and the operations of composition and concatenation

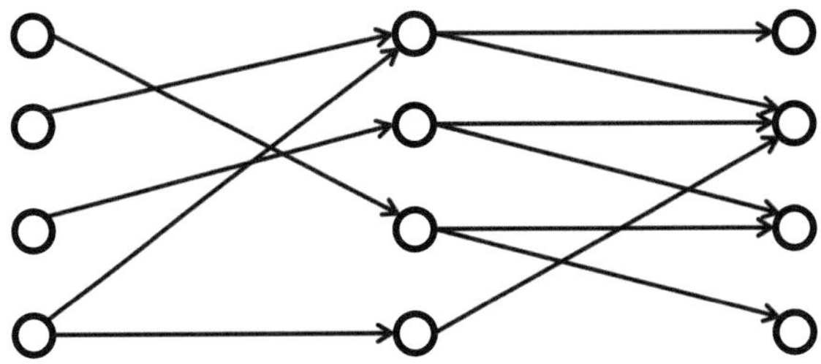

Fig. 4.7 Trellis diagram of induced state trajectory for observation sequence 10 (i.e., $y_0^1 = 10$ with $y[0] = 1$ and $y[1] = 0$) for deterministic finite automaton G in Fig. 4.3

$$m_1 = M_0 = \{(1,1), (1,2), (2,2), (2,3), (3,3), (3,4), (4,2)\} ,$$
$$m_2 = M_1 = \{(1,3), (2,1), (3,2), (4,1), (4,4)\} .$$

At the bottom of Fig. 4.6 we show graphically, the result of composing/ concatenating m_1 and m_2. Specifically, we have

$$m_1 \odot m_2 = \{(1,1), (1,3), (2,1), (2,2), (3,1), (3,2), (3,4), (4,1)\} ,$$

which is identical to M_{01} obtained in Example 4.5. Also, the concatenation operation is given by

$$m_1 \cdot m_2 = \{(1,1,3), (1,2,1), (2,2,1), (2,3,2), (3,3,2), (3,4,1), (3,4,4), (4,2,1)\} .$$

Note that it is rather straightforward to graphically perform operations on state mappings: for instance, Fig. 4.7 shows the induced state 3-trajectory for the sequence of observations 10 (i.e., $y_0^1 = 10$ with $y[0] = 1$ and $y[1] = 0$) for DFA G in Fig. 4.3. This is essentially the concatenation $m_2 \cdot m_1$ of the state 2-trajectories m_1 and m_2 in Fig. 4.6. □

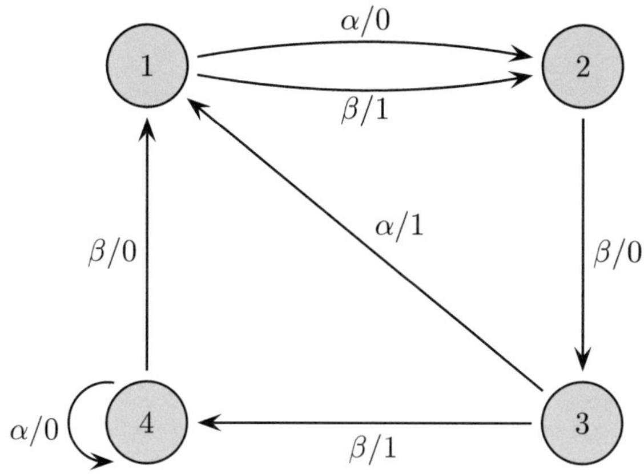

Fig. 4.8 Deterministic finite automaton with outputs considered in Example 4.9

Example 4.9 Consider the DFA in Fig. 4.8 where the initial state is unknown (i.e., we take $Q_0 = \{1, 2, 3, 4\}$). There are two possible outputs that generate two induced state mappings, namely for output $y = 0$ the induced state mapping is

$$M_0 = \{(1, 2), (2, 3), (4, 1), (4, 4)\} ,$$

whereas for output 1, the induced state mapping is

$$M_1 = \{(1, 2), (3, 1), (3, 4)\} .$$

These two state mappings are shown graphically in Fig. 4.9.

Let us now consider the sequence of observations $y[0] = 0$ followed by $y[1] = 1$. The induced state trajectory can be captured by the concatenation of $M_0 \cdot M_1$ as

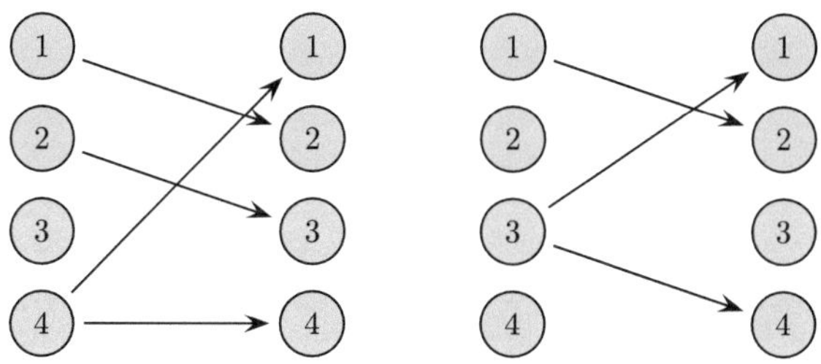

Fig. 4.9 Induced state mappings associated with the deterministic finite automaton in Fig. 4.8 discussed in Example 4.9: on the left we have the induced state mapping associated with output $y = 0$ (i.e., M_0) and on the right, we have the induced state mapping associated with output $y = 1$ (i.e., M_1)

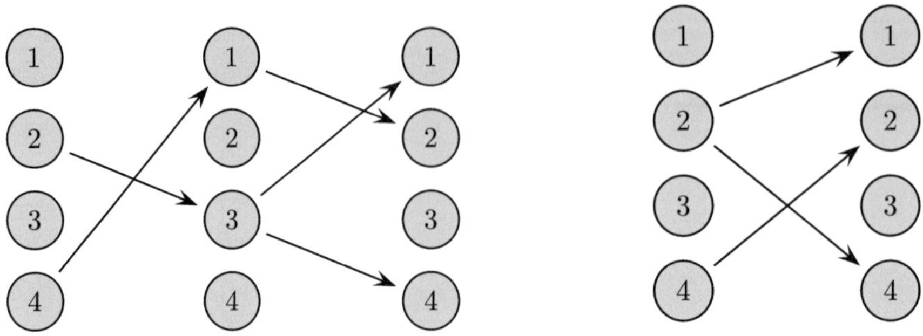

Fig. 4.10 Trellis diagram of induced state trajectory (left) and trellis diagram of induced state mapping (right) associated with the sequence of observations 01 (i.e., $y[0] = 0$ and $y[1] = 1$) for the deterministic finite automaton in Fig. 4.8 discussed in Example 4.9

$$M_{01}^{(3)} := M_0 \cdot M_1 = \{(2, 3, 1), (2, 3, 4), (4, 1, 2)\} ,$$

which is shown on the left of Fig. 4.10. Graphically, it can be obtained by matching the states on the right of M_0 with the states on the left of M_1, and eliminating trajectories that cannot be extended from the starting states to the ending states.

The induced state mapping associated with the sequence of observations 01 can be captured by the composition of M_0 and M_1 as

$$M_{01} := M_0 \odot M_1 = \{(2, 1), (2, 4), (4, 2)\} ,$$

which is shown graphically on the right of Fig. 4.10. Note that this composition can also be obtained by eliminating the intermediate states on the trellis diagram of the induced state trajectory on the left of the figure.

Consider now one additional observation $y[2] = 0$. The induced state trajectory following the sequence of observations 010 can be captured by the concatenation of $M_0 \cdot M_1 \cdot M_0$ as

$$\begin{aligned} M_{010}^{(4)} &= M_0 \cdot M_1 \cdot M_0 \\ &= (M_0 \cdot M_1) \cdot M_0 \\ &= \{(2, 3, 1, 2), (2, 3, 4, 1), (2, 3, 4, 4), (4, 1, 2, 3)\} , \end{aligned}$$

which is shown graphically on the left of Fig. 4.11. It can be obtained by matching the states on the right of state trajectory $M_{01}^{(3)}$ (shown on the left of Fig. 4.10) with the states on the left of M_0, and eliminating trajectories that cannot be extended from the starting states to the ending states.

The induced state mapping associated with sequence of observations 010 can be captured by the composition of M_0, M_1, and M_0 as

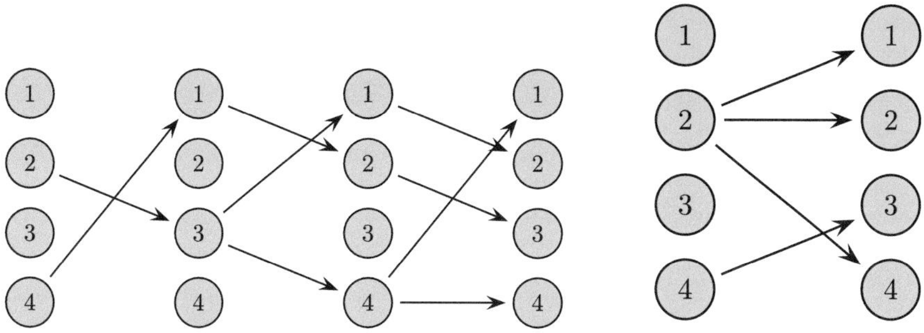

Fig. 4.11 Trellis diagram of induced state trajectory (left) and trellis diagram of induced state mapping (right) associated with the sequence of observations 010 (i.e., $y[0] = 0$, $y[1] = 1$, and $y[2] = 0$) for the deterministic finite automaton in Fig. 4.8 discussed in Example 4.9

$$\begin{aligned} M_{010} &= M_0 \odot M_1 \odot M_0 \\ &= (M_0 \odot M_1) \odot M_0 \\ &= \{(2, 1), (2, 2), (2, 4), (4, 3)\} , \end{aligned}$$

which is shown graphically on the right of Fig. 4.11. Note that this composition can also be obtained by eliminating the intermediate states on the trellis diagram of the induced state trajectory on the left of the figure or by composing the state mapping M_{01} (shown on the right of Fig. 4.10) with M_0, i.e., $M_{010} = M_{01} \odot M_0$. □

4.5 State Estimation

In this section, we are interested in performing state estimation in a DFA with outputs and (possibly) silent transitions $\mathcal{DFA} = (Q, \Sigma, Y \cup \{\epsilon\}, \delta, \lambda, Q_0)$, given a sequence of observed outputs y_0^k. This task can be achieved in a number of different ways, some of which might be preferable to others. The most general methodology would be to simply track all state trajectories that are compatible with the underlying finite automaton and the sequence of observations. This means that we can first obtain the induced state $(k+2)$-trajectory $M^{(k+2)}_{y_0^k}$ associated with the sequence of observations y_0^k (this can be done, for example, using the result in Corollary 4.2), and then also incorporate our knowledge of the set of possible initial states (by eliminating trajectories that originate from states outside the set Q_0), so that we obtain

$$M^{(k+2)}_{y_0^k, Q_0} = \{(q_{j_0}, q_{j_1}, \ldots, q_{j_{k+1}}) \in Q^{(k+2)} | (q_{j_0}, q_{j_1}, \ldots, q_{j_{k+1}}) \in M^{(k+2)}_{y_0^k}, q_{j_0} \in Q_0\}.$$

If the state $(k+2)$-trajectory $M^{(k+2)}_{y_0^k, Q_0}$ is available, we can easily perform the various state estimation tasks (current-state estimation, smoothing, and initial-state estimation) as follows:

$$\hat{q}_{y[k]}(y_0^k) = \{q_{j_{k+1}} \mid (q_{j_0}, q_{j_1}, \ldots, q_{j_{k+1}}) \in M^{(k+2)}_{y_0^k, Q_0}\} ,$$

$$\hat{q}_{y[i]}(y_0^k) = \{q_{j_{i+1}} \mid (q_{j_0}, q_{j_1}, \ldots, q_{j_{k+1}}) \in M^{(k+2)}_{y_0^k, Q_0}\} ,$$

$$\hat{q}_0(y_0^k) = \{q_{j_0} \mid (q_{j_0}, q_{j_1}, \ldots, q_{j_{k+1}}) \in M^{(k+2)}_{y_0^k, Q_0}\} .$$

What is important to keep in mind here is that from Corollary 4.2 we have

$$M^{(k+2)}_{y_0^k} = M^{(2)}_{y[0]} \cdot M^{(2)}_{y[1]} \cdot \cdots \cdot M^{(2)}_{y[k]} ,$$

which implies that we can obtain the induced state trajectories recursively as observations are coming in: first we calculate $M^{(2)}_{y_0^0} = M^{(2)}_{y[0]}$ and then set

$$M^{(i+2)}_{y_0^i} = M^{(i+1)}_{y_0^{i-1}} \cdot M^{(2)}_{y[i]}$$

for $i = 1, 2, \ldots, k$. In fact, we can simplify the process of incorporating initial-state information if we initialize this iteration with the state trajectory

$$M^{(2)}_{y_0^0, Q_0} = \{(q_{j_0}, q_{j_1}) \mid (q_{j_0}, q_{j_1}) \in M^{(2)}_{y_0^0},\ q_{j_0} \in Q_0\}$$

and then perform the iteration in the same fashion, i.e., perform

$$M^{(i+2)}_{y_0^i, Q_0} = M^{(i+1)}_{y_0^{i-1}, Q_0} \cdot M^{(2)}_{y[i]}$$

for $i = 1, 2, \ldots, k$.

Assuming a streaming sequence of observations $y[0], y[1], \ldots, y[k]$, a pseudocode describing the above recursive approach is found below.

Initialization: Set $InducedStateTrajectory = \{(q_0) \mid q_0 \in Q_0\}$
For $i = 0$ to k: Do
  $CurrentInducedStateMapping = M^{(2)}_{y[i]}$
  $Temp = InducedStateTrajectory \cdot CurrentInducedStateMapping$
  $InducedStateTrajectory = Temp$

Note that the induced state trajectory at the end of the ith iteration is the state $(i+2)$-trajectory $M^{(i+2)}_{y_0^i, Q_0}$. With the above notation in hand, we get the following answers to the state estimation problems we might be interested in after observing the sequence of observations y_0^k:

$$\hat{q}_{y[k]}(y_0^k) = \Pi_{k+1}(M^{(k+2)}_{y_0^k, Q_0}) ,$$

$$\hat{q}_{y[i]}(y_0^k) = \Pi_{i+1}(M^{(k+2)}_{y_0^k, Q_0}) \text{ for } 0 \le i < k , \qquad (4.1)$$

$$\hat{q}_0(y_0^k) = \Pi_0(M^{(k+2)}_{y_0^k, Q_0}) ,$$

where Π is the projection operation of the corresponding state $(k+2)$-trajectory.

Though the above approach to state estimation is rather universal, one potential problem with it is that it (perhaps unnecessarily) tracks all possible state trajectories. As k increases, the amount of information that might have to be stored increases as well. In cases when it is not necessary to maintain the set of all compatible state trajectories (e.g., when interested in current-state estimation), it is possible to do better and this is what we discuss in the remainder of this chapter. The following example illustrates the universal procedure outlined above.

Example 4.10 We consider again the DFA with outputs (and without silent transitions) G in Fig. 4.3 and illustrate the above approach for recursively calculating the induced state trajectory assuming that the set of initial states is $Q_0 = \{1\}$ and that the sequence of observations is 010 (i.e., $y_0^2 = 010$ with $y[0] = 0$, $y[1] = 1$, and $y[2] = 0$). The sequence of induced state trajectories $M^{(i+2)}_{y_0^i, Q_0}$ after each observation is made ($i = 0, 1, 2$) is shown in Fig. 4.12.

- Initially, we start with the set of states $Q_0 = \{1\}$ (which can be thought of as a state 1-trajectory and is not shown in the figure).
- Once the first observation (0) is made, we concatenate this state 1-trajectory with the induced state mapping M_0 corresponding to the observation (this is mapping m_1 shown at the top of Fig. 4.6 which was discussed in Example 4.8). The result is a state 2-trajectory which essentially eliminates from M_0 any trajectories that do not start from a state in $Q_0 = \{1\}$; its corresponding trellis is shown at the top of Fig. 4.12.
- Once the second observation (1) is made, we concatenate the previous induced state 2-trajectory with the induced state mapping M_1 corresponding to the observation (this is mapping m_2 shown at the top of Fig. 4.6). The result is a state 3-trajectory whose trellis is shown in the middle of Fig. 4.12.
- Once the third observation (0) is made, we concatenate the previous induced state 3-trajectory with M_0 and the result is the state 4-trajectory whose trellis is shown at the bottom of Fig. 4.12.

Note that the induced state 4-trajectory $M_{y_0^2, Q_0}$ resulting after three observations can also be obtained from the induced state mapping M_{010} (whose trellis is shown in Fig. 4.5) by eliminating all trajectories that do no start from a state in $Q_0 = \{1\}$. □

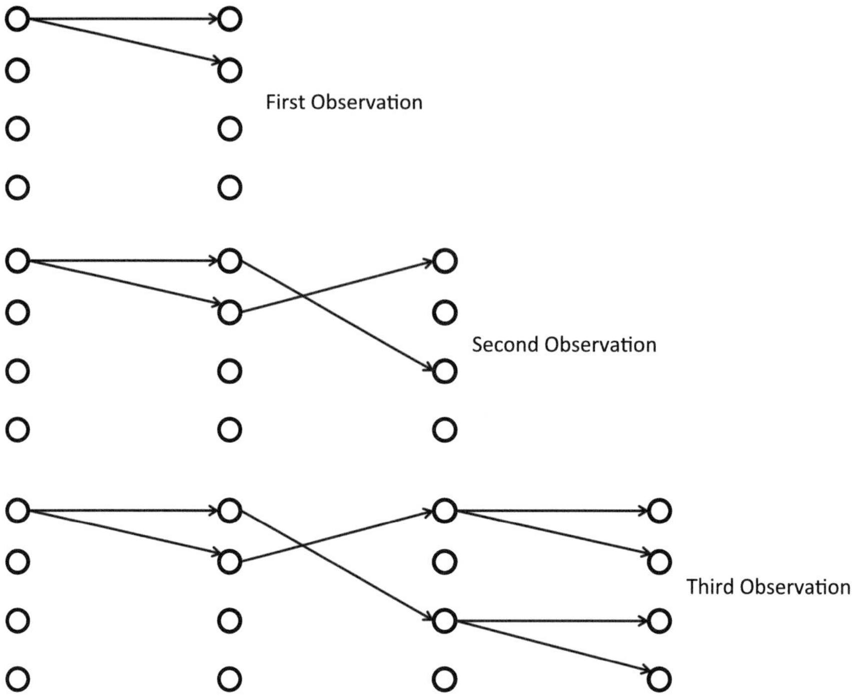

Fig. 4.12 Trellis diagrams illustrating the evolution of induced state trajectories $M^{(i+2)}_{y_0^i, Q_0}$, $i = 0, 1, 2,$ at the end of each observation in the observation sequence 010 (i.e., $y_0^2 = 010$ with $y[0] = 0$, $y[1] = 1$, and $y[2] = 0$) for deterministic finite automaton G in Fig. 4.3

4.5.1 Current-State Estimation

With the machinery introduced in the previous section, we now revisit the problem of current-state estimation. What is important to note in this case is that, given a streaming sequence of observations y_0^k, we are only interested at each i, $0 \leq i \leq k$, at reconstructing the set of possible current states. This set can be obtained as the projection of the last component of the state trajectory induced by the set of initial states Q_0 and the sequence of observations:

$$\hat{q}_{y[i]}(y_0^i) = \Pi_{i+1}(M^{(i+2)}_{y_0^i, Q_0}) \ .$$

In other words, we are only interested in the set of states that are possible from the set of initial states Q_0 given the set of observations y_0^i seen so far. Perhaps not surprisingly, we show next that, by simply keeping track of this set of states, one can recursively obtain the estimate of the set of possible current states at the next time step. The key observation is formalized in the corollary below.

Corollary 4.3 *Consider a DFA with outputs and (possibly) silent transitions* $\mathcal{DFA} = (Q, \Sigma, Y \cup \{\epsilon\}, \delta, \lambda, Q_0)$. *Given a sequence of outputs* $y_0^k \equiv y[0], y[1], \ldots, y[k] \in Y^{k+1}$, *we have*

$$\Pi_{i+1}\left(M^{(i+2)}_{y_0^i, Q_0}\right) = \Pi_1\left(\Pi_i(M^{(i+1)}_{y_0^{i-1}, Q_0}) \cdot M^{(2)}_{y[i]}\right)$$

for $i = 1, 2, \ldots, k$.

Proof Note that $\Pi_i(M^{(i+1)}_{y_0^{i-1}, Q_0})$ is treated as a state 1-trajectory that gets composed with the state 2-trajectory $M^{(2)}_{y[i]}$ (resulting in a state 2-trajectory). The statement of the corollary follows directly from Theorem 4.1 by realizing that $trim_i(M^{(i+2)}_{y_0^i, Q_0}) = \Pi_i(M^{(i+1)}_{y_0^{i-1}, Q_0}) \cdot M^{(2)}_{y[i]}$ and that $\Pi_{i+1}(M^{(i+2)}_{y_0^i, Q_0}) = \Pi_1(trim_i(M^{(i+2)}_{y_0^i, Q_0})) = \Pi_1(\Pi_i(M^{(i+1)}_{y_0^{i-1}, Q_0}) \cdot M^{(2)}_{y[i]})$. □

With the above corollary in hand, the recursive approach for obtaining the set of current states reduces to the following:

Initialization: Set $CurrentStates = Q_0$
For $i = 0$ to k: Do
  $CurrentInducedStateMapping = M^{(2)}_{y[i]}$
  $Temp = \Pi_1(CurrentStates \cdot CurrentInducedStateMapping)$
  $CurrentStates = Temp$

At this point, it is not difficult to realize that the above recursion is essentially the intuitive approach we developed in Sect. 4.3 when we had a preliminary discussion on current-state estimation. However, we reached this conclusion starting from an approach that keeps track of all information (and which can be generated recursively using Corollary 4.2) and then realized (via Corollary 4.3) that it is sufficient to keep track of the set of estimates for the current state (i.e., simply keeping track of the current states allows us to perform the recursion and also maintain the information we need). Note that our approach here also establishes the correctness of the approach in Sect. 4.3; of course, the correctness of that approach could also be established by other means (e.g., by showing that continuations of state trajectories only depend on the set of current states, which the algorithm explicitly keeps track of).

Example 4.11 Consider again the finite automaton G with outputs (and without silent transitions) in Fig. 4.3 assuming that the set of initial states is $Q_0 = \{1\}$ and that the sequence of observations is 010 (i.e., $y_0^2 = 010$ with $y[0] = 0$, $y[1] = 1$, and $y[2] = 0$). Figure 4.13 shows the evolution of the induced state trajectories when keeping track of only the ending states. Specifically, the figure shows the evolutions of $\Pi_{i+1}(M^{(i+2)}_{y_0^i, Q_0})$ for $i = 0, 1, 2$. To make the comparisons with the corresponding (full) induced state trajectories shown in Fig. 4.12, the diagram puts a shaded box over the part of the trajectory that is eliminated due to the projection operation. One

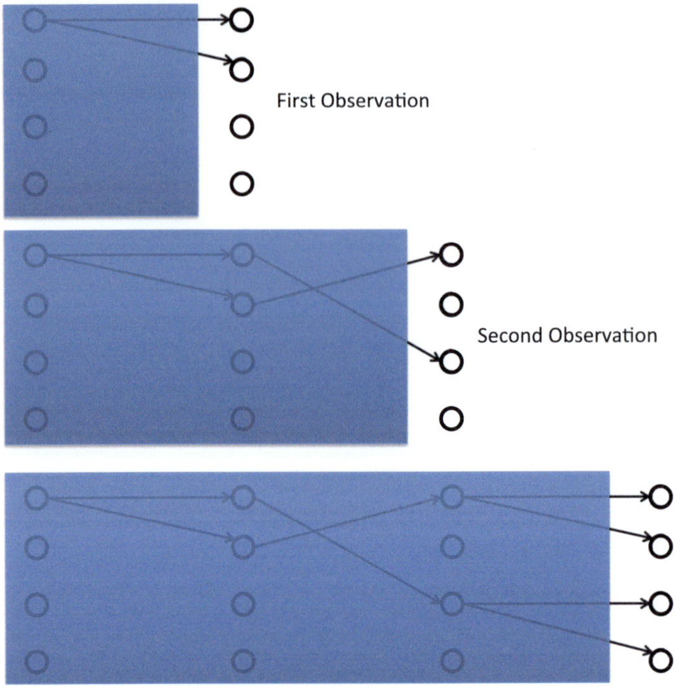

Fig. 4.13 Trellis diagrams illustrating the evolution of induced state trajectories $\Pi_{i+1}(M^{(i+2)}_{y_0^i, Q_0})$, $i = 0, 1, 2$, when projecting on the ending states at the end of each observation (for current-state estimation)

easily sees that the ending states of each induced state trajectory (which are *not* eliminated by the projection) are sufficient to perform the concatenation with the state mapping induced by the new observation. □

4.5.2 Delayed-State Estimation—Smoothing

The discussion in this section thus far (in particular, Eq. (4.1)) can also be applied toward smoothing, in order to obtain delayed state estimates for some fixed delay D. Under this scenario, given a streaming sequence of observations y_0^k, $k = 0, 1, \ldots$, we are interested to reconstruct, for any given i, $D \leq i \leq k$, the set of possible states at the instant immediately following the $(i - D)$th observation. This set of possible states is given by the projection on the corresponding component of the state trajectory induced by the set of initial states Q_0 and the sequence of observations y_0^i seen so far.

Clearly, we can obtain the D-delayed state estimates if we have access to the last $D+1$ components of the state $(i+2)$-trajectory $M^{(i+2)}_{y_0^i, Q_0}$ that corresponds to the sequence of observations y_0^i seen up to that point and the initial states Q_0. Similarly, once observation $y[i+1]$ becomes available, what we need are the last $D+1$ components of the state $(i+3)$-trajectory $M^{(i+3)}_{y_0^{i+1}, Q_0}$ that corresponds to y_0^{i+1} and the initial states Q_0. From the discussion in Sect. 4.4 (in particular, from Theorem 4.2), we know that the last $D+1$ components of this state $(i+2)$-trajectory are sufficient to determine the last $D+1$ components of the subsequent state $(i+3)$-trajectory (in fact, they are sufficient to determine its last $D+2$ components if so desired).

The key observation is formalized in the corollary below which follows directly from Theorem 4.2.

Corollary 4.4 *Consider a DFA with outputs and (possibly) silent transitions* $\mathcal{DFA} = (Q, \Sigma, Y \cup \{\epsilon\}, \delta, \lambda, Q_0)$. *Given a sequence of outputs* $y_0^k \equiv y[0], y[1], \ldots, y[k] \in Y^{k+1}$ *where* $k \geq D+1$, *we have for* $i = D, D+1, \ldots, k-1$

$$trim_{i-D}\left(M^{(i+2)}_{y_0^i, Q_0}\right) = trim_1\left(trim_{i-D}(M^{(i+1)}_{y_0^{i-1}, Q_0}) \cdot M^{(2)}_{y[i]}\right),$$

where $trim_0(M) = M$.

The corollary establishes that, by keeping track of the last $D+1$ components of the state trajectory that corresponds to a given sequence of observations and a set initial states, one can recursively maintain the last $D+1$ components of subsequent state trajectories. Since, at any given instant, we are interested in the set of states that are possible at the instant immediately following the Dth-to-last observation, these last $D+1$ components are sufficient for obtaining the set of possible states of interest.

With the above theorem in hand, the recursive approach for obtaining the set of D-delayed state estimates reduces to the following algorithm, which essentially has a separate part to deal with the first D observations.

Initialization: Set
$D - DelayedStateEstimates = Q_0$,
$CurrentStateTrajectory = Q_0$
For $i = 0$ to $D-1$: Do
$CurrentInducedStateMapping = M^{(2)}_{y[i]}$
$Temp = CurrentStateTrajectory \cdot CurrentInducedStateMapping$
$CurrentStateTrajectory = Temp$
$D - DelayedStateEstimates = \Pi_0(CurrentStateTrajectory)$
For $i = D$ to k: Do
$CurrentInducedStateMapping = M^{(2)}_{y[i]}$
$Temp = CurrentStateTrajectory \cdot CurrentInducedStateMapping$
$CurrentStateTrajectory = trim_1(Temp)$
$D - DelayedStateEstimates = \Pi_0(CurrentStateTrajectory)$

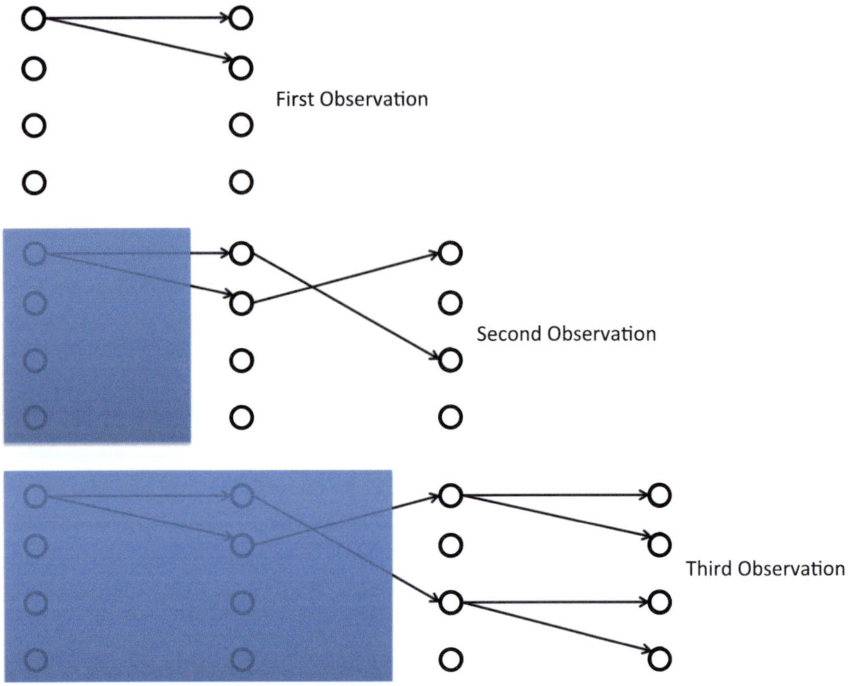

Fig. 4.14 Trellis diagrams illustrating the evolution of induced state trajectories $trim_{i-1}(M^{(i+2)}_{y_0^i, Q_0})$, $i = 0, 1, 2$, when retaining only the last two stages of the trajectories (for 1-delayed estimation)

Though not as intuitive as the recursion we obtained for current-state estimation, the correctness of the above recursion can be established following mathematical steps that are very similar to the ones we used to prove the correctness of the recursion for current-state estimation. What is important to keep in mind here is that continuations of state trajectories only depend at the current (latest) state, however, subsequent observations might influence previous state estimates by discontinuing (i.e., invalidating) certain sequences of states.

Example 4.12 Consider again the DFA with outputs (and without silent transitions) G in Fig. 4.3 assuming that the set of initial states is $Q_0 = \{1\}$ and that the sequence of observations is 010 (i.e., $y_0^2 = 010$ with $y[0] = 0$, $y[1] = 1$, and $y[2] = 0$). Figure 4.14 shows the evolution of the induced state trajectories when keeping only the last two stages of the induced state trajectories (which can be seen in Fig. 4.12). Specifically, the figure shows the evolutions of $trim_{i-1}(M^{(i+2)}_{y_0^i, Q_0})$ for $i = 0, 1, 2$. Clearly, the ending states of each induced state trajectory allow one to concatenate the last part of the induced state trajectory so far, with the state mapping induced by the new observation.

At any given time, by taking the projection on the next to last stage, one can obtain precisely the set of 1-delayed state estimates. For example, by projecting on the next to last state, we obtain the 1-delayed state estimates after the observation sequence 010 as $\hat{q}_{y[1]}(y_0^2) = \{1, 3\}$. □

Remark 4.11 The above recursive approach for D-delayed-state estimation requires significantly more storage than the recursion for current-state estimation analyzed in the previous section. Clearly, both approaches need to maintain/store the induced state mappings for each possible output, which requires $O(N^2 R)$ storage where $N = |Q|$ and $R = |Y|$; however, unlike the previous approach that only needed to maintain the set of current states (which required storage of only $O(N)$), the tracking of state $(D + 1)$-trajectories, appears at first glance to be significantly more complicated: each state $(D + 1)$-trajectory could have as many as N^{D+1} elements (sequences of states), each requiring storage $D + 1$, for a total storage of $O((D + 1)N^{D+1})$. In reality, the storage needed can be reduced significantly if one realizes that the sequence of the last D observations together with the set of possible states D observations ago uniquely determine the state $(D + 1)$-trajectory under consideration. In fact, Theorem 4.5 below shows that this is true for any induced state trajectory. Based on this discussion, the storage required by the recursive algorithm for performing D-delayed-state estimation is $O(N + D)$ where N is the storage needed to maintain the (at most N) states that are possible D observations ago and D is the storage needed to maintain the sequence of the last D observations. Overall, the storage complexity needed by the recursive algorithm is $O(NR + D)$; each recursive step requires computation of $O(N^2 D)$ to construct the induced state trajectories (since these state trajectories are not explicitly stored). □

The following theorem establishes that any induced mapping can be reconstructed *from a specific observation point onwards* using the set of states that are possible at this specific observation point (due to the sequence observations leading to this point) and the sequence of observations that have been seen from this point onwards.

Theorem 4.5 *Consider a DFA with outputs and (possibly) silent transitions* $\mathcal{DFA} = (Q, \Sigma, Y \cup \{\epsilon\}, \delta, \lambda, Q_0)$. *Given sequence of outputs* $y_0^k \equiv y[0], y[1], \ldots, y[k] \in Y^{k+1}$, *we have for* $i = 1, 2, \ldots, k$

$$trim_i \left(M^{(k+2)}_{y_0^k, Q_0} \right) = M^{(k+2-i)}_{y_i^k, Q_i} ,$$

where $Q_i = \Pi_i \left(M^{(i+1)}_{y_0^{i-1}, Q_0} \right)$.

Proof Note that $M^{(k+2)}_{y_0^k, Q_0} = M^{(1)}_{Q_0} \cdot M^{(k+2)}_{y_0^k}$, where $M^{(1)}_{Q_0} = \{(q_0) \mid q_0 \in Q_0\}$. Moreover, from Theorem 4.4 we have $M^{(k+2)}_{y_0^k} = M^{(i+1)}_{y_0^{i-1}} \cdot M^{(k+2-i)}_{y_i^k}$. Putting these together, we have

$$M^{(k+2)}_{y_0^k, Q_0} = \underbrace{\left(M^{(1)}_{Q_0} \cdot M^{(i+1)}_{y_0^{i-1}} \right)}_{M^{(i+1)}_{y_0^{i-1}, Q_0}} \cdot M^{(k+2-i)}_{y_i^k} .$$

Using Theorem 4.2, we have

$$\begin{aligned} trim_i\left(M^{(k+2)}_{y_0^k,Q_0}\right) &= trim_i\left(M^{(i+1)}_{y_0^{i-1},Q_0}\right)\cdot M^{(k+2-i)}_{y_i^k} \\ &= \Pi_i\left(M^{(i+1)}_{y_0^{i-1},Q_0}\right)\cdot M^{(k+2-i)}_{y_i^k} \\ &= M^{(k+2-i)}_{y_i^k,Q_i}\ , \end{aligned}$$

which completes the proof. □

4.5.3 Initial-State Estimation

The discussion in Sect. 4.5 (in particular, Eq. (4.1)) can also be applied toward initial-state estimation. In this case, given a streaming sequence of observations y_0^k, we are interested to reconstruct (for any given i, $0 \leq i \leq k$) the set of possible initial states which is given by

$$\hat{q}_0(y_0^i) = \Pi_0(M^{(i+2)}_{y_0^i,Q_0})\ , 0 \leq i \leq k\ .$$

Alternatively, this set of state estimates can be obtained via the set of initial states of the induced state mapping

$$\hat{q}_0(y_0^i) = \Pi_0(M_{y_0^i,Q_0})\ ,$$

where $M_{y_0^i,Q_0} = \{(q_i, q_f) \in M_{y_0^i} \mid q_i \in Q_0\}$.

By repeated application of Theorem 4.3, we easily obtain the following:

$$M_{y_0^i,Q_0} = M_{y[0],Q_0} \odot M_{y[1]} \odot M_{y[2]} \odot \cdots \odot M_{y[i]}\ ,$$

which immediately leads to the recursive methodology below that can be used for tracking the set of initial states $\hat{q}_0(y_0^i)$ for $i = 0, 1, \ldots, k$.

Initialization: Set
  $CurrentStateMapping = \{(q_0, q_0) \mid q_0 \in Q_0\}$
  $InitialStateEstimates = \Pi_0(CurrentStateMapping)$
For $i = 0$ to k: Do
  $CurrentInducedStateMapping = M^{(2)}_{y[i]}$
  $Temp = CurrentStateMapping \odot CurrentInducedStateMapping$
  $CurrentStateMapping = Temp$
  $InitialStateEstimates = \Pi_0(CurrentStateMapping)$

Remark 4.12 The main difference between the recursive approach described above and the previously described recursive methodologies for current and D-delayed-state estimation is that the previous methods maintained a sliding window of the set

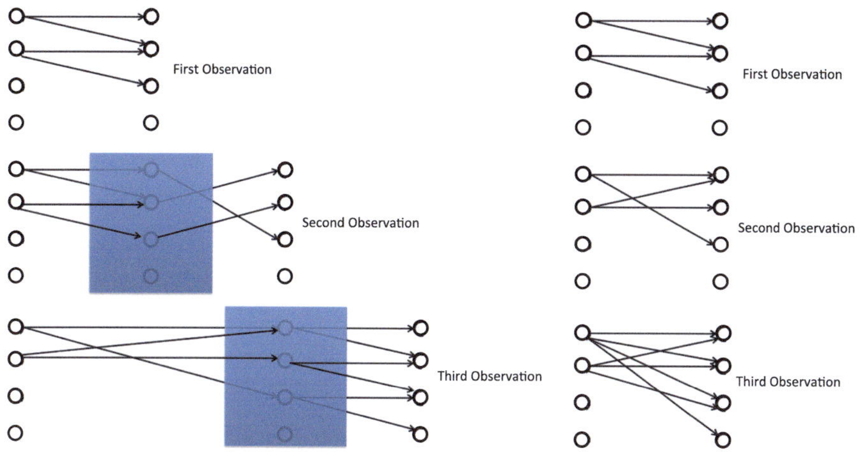

Fig. 4.15 Trellis diagrams illustrating the evolution of the induced state trajectories $M^{(i+2)}_{y_0^i, Q_0})$, $i = 0, 1, 2$ (left), and the evolution of the corresponding state mappings $M_{y_0^i, Q_0}$, $i = 0, 1, 2$, when retaining only the initial and final stages of the trajectories (for initial-state estimation)

of possible estimates for each (observation) time index corresponding to this location. The above algorithm, however, maintains the endpoints (initial and current (observation) time index) of a window whose size keeps increasing as more observations become available. □

Example 4.13 Consider again the DFA with outputs (and without silent transitions) G in Fig. 4.3 assuming that the set of initial states is $Q_0 = \{1, 2\}$ and that the sequence of observations is 010 (i.e., $y_0^2 = 010$ with $y[0] = 0$, $y[1] = 1$, and $y[2] = 0$). Figure 4.15 shows the evolution of all stages of the induced state trajectories $M^{(i+2)}_{y_0^i, Q_0})$, $i = 0, 1, 2$ (left), and the evolution of the first and last stages of the corresponding induced state mappings $M_{y_0^i, Q_0}$, $i = 0, 1, 2$ (right). The shaded regions correspond to parts of induced state trajectories that are removed (and replaced by appropriate connections between starting and ending states at that corresponding point in time).

Clearly, since the ending states of each induced state trajectory are retained in the corresponding state mapping, one has sufficient information to perform the composition of the given induced state trajectory with the state mapping induced by the new observation. At any given time, by taking the projection on the first stage, one can obtain precisely the set of initial state estimates. For instance, the initial state estimates after the observation sequence 010 can be obtained as $\hat{q}_0(y_0^2) = \{1, 2\}$. □

4.6 Extensions to Nondeterministic Finite Automata

Our development in this chapter so far assumed that the underlying system is a DFA with outputs and (possibly) silent transitions $\mathcal{DFA} = (Q, \Sigma, Y \cup \{\epsilon\}, \delta, \lambda, Q_0)$. Nevertheless, the recursive algorithms that we developed for state estimation (current, D-delayed, and initial) given a streaming sequence of observations $y_0^k = y[0], y[1], \ldots, y[k]$ rely solely on the induced state mappings $\{M_y^{(2)} \mid y \in Y\}$ (and, of course, on the set of initial states Q_0). Thus, as long as we can appropriately extend the notions of induced state mappings and trajectories to NFA, we can readily obtain the corresponding state estimation algorithms.

Note that state mappings, state trajectories, and their operations were defined in Sect. 4.4.1 independently from the underlying system; however, induced states mappings/trajectories, which were established in Sects. 4.4.2 and 4.4.3, relied in an underlying deterministic systems and need to be modified appropriately.

Without any loss of generality, we choose to deal directly with automata with silent transitions. Given an NFA with outputs and (possibly) silent transitions $(Q, \Sigma, Y \cup \{\epsilon\}, \delta, \lambda, Q_0)$, we are interested in using state mappings to track the sequences of states that are compatible with the observation of the sequence y_0^k. The key problem is that an input (or an input sequence) can generate multiple state trajectories with corresponding output sequences that may differ. As a result, we need to make sure that the state trajectories that we consider indeed generate the desired output (in the case of induced state mappings) or sequence of outputs (in the case of induced state trajectories). The definition below makes use of the erase function E which was defined in Eq. (3.8).

Definition 4.9 (*Induced State Mapping for Nondeterministic Automaton*) Given the NFA with outputs and (possibly) silent transitions $\mathcal{NFA} = (Q, \Sigma, Y \cup \{\epsilon\}, \delta, \lambda, Q_0)$, we define for each label $y \in Y$, the induced state mapping M_y to be

$$\begin{aligned} M_y = \{(q_i, q_f) \in Q^2 \mid \exists m' \geq 0, \sigma_k^{k+m'} \in \Sigma^*, q_{j_k} = q_i, q_{j_{k+1}}, \ldots, q_{j_{k+m'}}, \\ q_{j_{k+m'+1}} = q_f \in Q, \text{ such that } q_{j_{k+\ell+1}} \in \delta(q_{j_{k+\ell}}, \sigma_{k+\ell}) \\ \text{for } \ell = 0, 1, \ldots, m' \text{ and } E(y_k^{k+m'}) = y \text{ where} \\ y_{k+\ell} = \lambda(q_{j_{k+\ell}}, \sigma_{k+\ell}, q_{j_{k+\ell+1}}) \text{ for } \ell = 0, 1, \ldots, m'\} . \end{aligned}$$

Remark 4.13 The key change in the above definition, compared to the definition of induced state mappings in Sect. 4.4.2, is that we explicitly ensure that the state trajectory that leads from q_i to q_f generates a sequence of outputs y_k^{k+m} that is equivalent (after the removal of empty outputs) to the output y. If we did not explicitly track the output generated by this state sequence, it is possible that we would erroneously include in our state mapping a pair (q_i, q_f') simply because q_f' is reachable from q_i via the sequence of inputs $\sigma_k^{k+m'}$ (and, of course, the fact that output y can be generated from this input sequence starting at state q_i); the problem, however, is that q_f' might be reachable only with state sequences that generate a *different* sequence of outputs (that is not equivalent to y). □

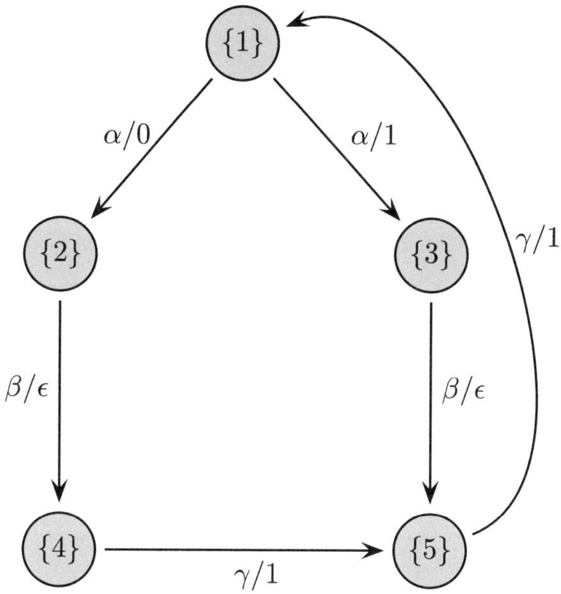

Fig. 4.16 Nondeterministic finite automaton used in Example 4.14 to illustrate the notion of induced state mappings and trajectories

Example 4.14 Consider the NFA in Fig. 4.16. There are two possible outputs that we need to consider. Under output 0, the induced state mapping is

$$M_0 = \{(1, 2), (1, 4)\} ,$$

whereas under output 1, the induced state mapping is

$$M_1 = \{(1, 3), (1, 5), (2, 5), (3, 1), (4, 5), (5, 1)\} .$$

As an illustration of the point made in the previous remark, note that the sequence of inputs α from state 1 *can* generate the sequence of observations 0; however, not all states reachable from state 1 under this input sequence can be included in M_0. Specifically, even though state 3 is reachable from state 1 via input sequence α, the pair $(1, 3) \notin M_0$ because the state trajectory leading to it does not generate output 0. Similarly, the sequence of inputs $\alpha\beta$ from state 1 *can* generate the sequence of observations 0; however, even though state 5 is reachable from state 1 via input sequence $\alpha\beta$, the pair $(1, 5) \notin M_0$ because the state trajectory leading to it does not generate output 0. □

Definition 4.10 (*Induced State Mapping over a Sequence of Observations for NFA*) Given the NFA with outputs and (possibly) silent transitions $\mathcal{NFA} = (Q, \Sigma, Y \cup \{\epsilon\}, \delta, \lambda, Q_0)$, we define for a sequence of observations y_k^{k+m}, the induced state mapping $M_{y_k^{k+m}}$ to be

$$M_{y_k^{k+m}} = \{(q_i, q_f) \in Q^2 \mid \exists m' \geq 0, \sigma_k^{k+m'} \in \Sigma^*, q_{j_k} = q_i, q_{j_{k+1}}, \ldots, q_{j_{k+m'}}, q_{j_{k+m'+1}} = q_f \in Q, \text{ such that } q_{j_{k+\ell+1}} \in \delta(q_{j_{k+\ell}}, \sigma_{k+\ell}) \text{ for } \ell = 0, 1, \ldots, m' \text{ and } E(y_k^{k+m'}) = y_k^{k+m} \text{ where } y_{k+\ell} = \lambda(q_{j_{k+\ell}}, \sigma_{k+\ell}, q_{j_{k+\ell+1}}) \text{ for } \ell = 0, 1, \ldots, m'\} .$$

The important property captured by Theorem 4.3 for induced state mappings over deterministic automata also extends to the case of induced state mappings over nondeterministic automata as discussed below. The proof is straightforward and is omitted.

Theorem 4.6 *Consider an NFA with outputs and (possibly) silent transitions $\mathcal{NFA} = (Q, \Sigma, Y \cup \{\epsilon\}, \delta, \lambda, Q_0)$. Given sequence of outputs $y_k^{k'}$ and any $k \leq m < k'$, we have*

$$M_{y_k^{k'}} = M_{y_k^m} \odot M_{y_{m+1}^{k'}} .$$

The definitions for induced state mappings can be extended to induced state L-trajectories as follows.

Definition 4.11 (*Induced State 2-Trajectory over a Single Observation for NFA*) Given the NFA with outputs and (possibly) silent transitions $\mathcal{NFA} = (Q, \Sigma, Y \cup \{\epsilon\}, \delta, \lambda, Q_0)$, we define for each label $y \in Y$, the induced state 2-trajectory $M_y^{(2)}$ to be

$$M_y^{(2)} = \{(q_i, q_f) \in Q^2 \mid \exists m' \geq 0, \sigma_k^{k+m'} \in \Sigma^*, q_{j_k} = q_i, q_{j_{k+1}}, \ldots, q_{j_{k+m'}}, q_{j_{k+m'+1}} = q_f \in Q, \text{ such that } q_{j_{k+\ell+1}} \in \delta(q_{j_{k+\ell}}, \sigma_{k+\ell}) \text{ for } \ell = 0, 1, \ldots, m' \text{ and } E(y_k^{k+m'}) = y \text{ where } y_{k+\ell} = \lambda(q_{j_{k+\ell}}, \sigma_{k+\ell}, q_{j_{k+\ell+1}}) \text{ for } \ell = 0, 1, \ldots, m'\} .$$

Definition 4.12 (*Induced State L-Trajectory over a Sequence of Observations for an NFA*) Given the NFA with outputs and (possibly) silent transitions $\mathcal{NFA} = (Q, \Sigma, Y \cup \{\epsilon\}, \delta, \lambda, Q_0)$, we define for each sequence of outputs $y_k^{k+m} \equiv y[k], y[k+1], \ldots, y[k+m]$, the induced state $(m+2)$-trajectory to be

$$M^{(m+2)}_{y_k^{k+m}} = \{(q_0, \ldots, q_m, q_{m+1}) \in Q^{m+2} \mid (q_i, q_{i+1}) \in M_{y[k+i]} \ \forall i = 0, 1, \ldots, m\}.$$

As was the case for a DFA, the induced state 2-trajectory for a single output of an NFA is identical to the corresponding induced state mapping. The important property captured by Corollary 4.2 for induced state L-trajectories over deterministic automata (which was important in allowing us to recursively perform state estimation), also holds for induced state trajectories over nondeterministic automata. The proof follows from the definition and is omitted.

Theorem 4.7 *Consider an NFA with outputs and (possibly) silent transitions $\mathcal{NFA} = (Q, \Sigma, Y \cup \{\epsilon\}, \delta, \lambda, Q_0)$. Given sequence of outputs $y_k^{k+m} \equiv y[k], y[k+1], \ldots, y[k+m] \in Y^{m+1}$, we have*

$$M^{(m+2)}_{y_k^{k+m}} = M^{(2)}_{y[k]} \cdot M^{(2)}_{y[k+1]} \cdot \cdots \cdot M^{(2)}_{y[k+m]} .$$

At this point, it is clear that once the notions of state mappings/trajectories and induced state mappings/trajectories are properly adjusted for nondeterministic automata, the approaches that were described in Sects. 4.5, 4.5.1, 4.5.2, and 4.5.3 can be used without any changes.

4.7 Observation Equivalence

Thus far, our discussion in this chapter revolved around various estimation problems (namely, current estimation, D-delayed estimation, and initial-state estimation). Specifically, given the DFA with outputs and (possibly) silent transitions $\mathcal{DFA} = (Q, \Sigma, Y \cup \{\epsilon\}, \delta, \lambda, Q_0)$ and a sequence of observations y_0^k, the answers to all of the above state estimation problems relied on the induced state $(k+2)$-trajectory

$$M_{y_0^k}^{(k+2)} = M_{y[0]}^{(2)} \cdot M_{y[1]}^{(2)} \cdot \dots \cdot M_{y[k]}^{(2)} ,$$

which, together with the set of initial states Q_0, uniquely determine the set of all compatible state trajectories.

It is important to observe that $M_{y_0^k}^{(k+2)}$ is a state trajectory that is obtained via the concatenation of induced state 2-trajectories from the set $\{M_y^{(2)} \mid y \in Y\}$. This set includes at most $|Y| = R$ different induced state 2-trajectories that get concatenated according to the sequence y_0^k that is observed.

Consider now another (possibly nondeterministic) finite automaton $\mathcal{NFA}' = (Q, \Sigma', Y, \delta', \lambda', Q_0)$, with the same set of states and outputs but with no unobservable transitions, such that its set of induced state mappings $\{M_y'^{(2)} \mid y \in Y\}$ is identical to the induced states mappings for $\mathcal{DFA}$. More specifically, we assume that

$$M_y'^{(2)} = M_y^{(2)} \text{ for all } y \in Y .$$

Since the induced state mappings are identical for the two automata, then the induced state $(k+2)$-trajectories will be identical for any sequence of observations y_0^k; this, together with the fact that the sets of initial states are identical, implies that all state estimation problems we considered in this chapter will result in the same set of state estimates (in fact, the same set of state trajectories) for the two automata. For this reason, we say that automata $\mathcal{DFA}$ and $\mathcal{NFA}'$ are *observation-equivalent*.

Example 4.15 In Fig. 4.17, we see NFA G' that is observation equivalent to the DFA with outputs in Fig. 4.1. Note that G' is nondeterministic (e.g., from state $q^{(1)}$ with input α one can end up in state $q^{(2)}$ or $q^{(3)}$). Also, note that G' has functionality that is seemingly not present in the original automaton. For example, from $q^{(2)}$ there is a transition that leads to state $q^{(3)}$ and generates output 0; similarly, from $q^{(2)}$ there is a transition that leads to state $q^{(1)}$ and generates output 1.

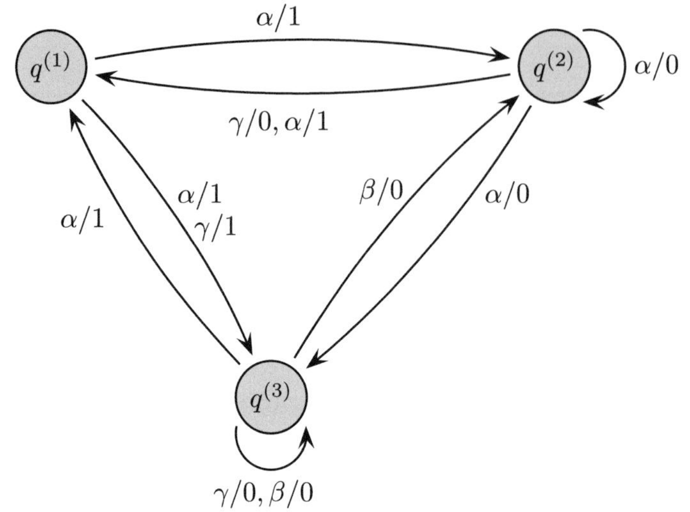

Fig. 4.17 Nondeterministic finite automaton G' that is observation equivalent to the finite automaton of Fig. 4.1

The only requirement for observation equivalence is that, for each output $y \in Y$, the induced state mappings M_y for the two automata are identical. In particular, for both aforementioned automata, we have

$$M_0 = \{(q^{(2)}, q^{(1)}), (q^{(2)}, q^{(2)}), (q^{(2)}, q^{(3)}), (q^{(3)}, q^{(2)}), (q^{(3)}, q^{(3)})\} ,$$
$$M_1 = \{(q^{(1)}, q^{(2)}), (q^{(1)}, q^{(3)}), (q^{(2)}, q^{(1)}), (q^{(3)}, q^{(1)})\} ,$$

which establishes that they are observation equivalent. □

From the above discussion, it follows that given a DFA with outputs and (possibly) silent transitions $\mathcal{DFA} = (Q, \Sigma, Y \cup \{\epsilon\}, \delta, \lambda, Q_0)$, we can perform state estimation using (instead of $\mathcal{DFA}$) any (possibly nondeterministic) finite automaton with outputs (but without silent transitions) $\mathcal{NFA}'$, as long as $\mathcal{NFA}'$ is observation-equivalent to it. One such automaton is the labeled nondeterministic finite automaton $\mathcal{LNFA}_{ur}$ defined below. Note that $\mathcal{LNFA}_{ur}$ has no silent transitions. The above construction can also be done for an NFA (instead of the DFA we consider here) but the definition of δ_{ur} and λ_{ur} would have to take into account the non-determinism of the underlying automaton (much like the way we handled non-determinism in the definition of induced state mappings for nondeterministic systems).

Definition 4.13 (*Observation-Equivalent Labeled Nondeterministic Finite Automaton*) Given a DFA with outputs and (possibly) silent transitions $\mathcal{DFA} = (Q, \Sigma, Y \cup \{\epsilon\}, \delta, \lambda, Q_0)$, the LNFA with outputs $\mathcal{LNFA}_{ur} = (Q, Y, Y, \delta_{ur}, \lambda_{ur}, Q_0)$ is defined with the following δ_{ur} and λ_{ur}:

- For $q_i \in Q$ and $y \in Y$, we set $\delta_{ur} : Q \times Y \to 2^Q$ to be

$$\delta_{ur}(q_i, y) = \{q_f \mid \exists m \geq 0, \sigma_0^m \in \Sigma^* \text{ such that } q_f = \delta(q_i, \sigma_0^m) \\ \text{and } y = \lambda_{seq}(q_i, \sigma_0^m)\} .$$

- For $y \in Y$ and for $q_f \in \delta_{ur}(q_i, y)$ we set $\lambda_{ur} : Q \times Y \times Q :\to Y$ to be

$$\lambda_{ur}(q_i, y, q_f) = \lambda_{ur}(y) = y .$$

Note that when constructing automaton $\mathcal{LNFA}_{ur}$ the input set Σ of the original automaton is not important; this is hardly surprising since, when performing the various state estimation tasks, the inputs do not appear explicitly in the induced state mappings. The fact that the labeled automaton $\mathcal{LNFA}_{ur}$ is observation-equivalent to the given DFA can be established easily: the state mappings for the labeled automaton $\mathcal{LNFA}_{ur}$ are seen to satisfy the following: for $y \in Y$

$$M_y^{ur,(2)} = \{(q_i, q_f) \mid \lambda_{ur}(q_i, y, q_f) = y \text{ and } q_f \in \delta_{ur}(q_i, y)\ \} ,$$

which is equivalent to

$$M_y^{ur,(2)} = \{(q_i, q_f) | \exists m \geq 0, \sigma_0^m \in \Sigma^*, \{q_f = \delta(q_i, \sigma_0^m) \text{ and } y = \lambda_{seq}(q_i, \sigma_0^m)\}\}.$$

Therefore, the state mappings associated with $\mathcal{LNFA}_{ur}$ are identical to the state mappings associated with the original DFA (with outputs and possibly silent transitions) $\mathcal{DFA}$ we started off with.

Example 4.16 On the top of Fig. 4.18, we see automaton $\mathcal{LNFA}_{ur}$ (top) that is observation equivalent to the DFA of Fig. 4.1. At the bottom we see automaton $\mathcal{LNFA}'_{ur}$ that is observation equivalent to the DFA in Fig. 4.3. As discussed earlier, the induced state mapping for each output of $\mathcal{LNFA}_{ur}$ (or $\mathcal{LNFA}'_{ur}$) is identical to the induced state mapping of the automaton in Fig. 4.1 (or Fig. 4.3). For example, for $\mathcal{LNFA}_{ur}$, we have

$$M_0 = \{(q^{(2)}, q^{(1)}), (q^{(2)}, q^{(2)}), (q^{(2)}, q^{(3)}), (q^{(3)}, q^{(2)}), (q^{(3)}, q^{(3)})\} ,$$
$$M_1 = \{(q^{(1)}, q^{(2)}), (q^{(1)}, q^{(3)}), (q^{(2)}, q^{(1)}), (q^{(3)}, q^{(1)})\} .$$

Similarly, for $\mathcal{LNFA}'_{ur}$ at the bottom of Fig. 4.18, we have

$$M_0 = \{(1, 1), (1, 2), (2, 2), (2, 3), (3, 3), (3, 4), (4, 2)\} ,$$
$$M_1 = \{(1, 3), (2, 1), (3, 2), (4, 1), (4, 4)\} .$$

It is worth noting some key differences in automaton $\mathcal{LNFA}_{ur}$. For instance, under input (observation) 1, we can go from state $q^{(2)}$ to state $q^{(1)}$ due to the unobservable transition from $q^{(2)}$ to $q^{(3)}$ (under input β) in the finite automaton of Fig. 4.1. □

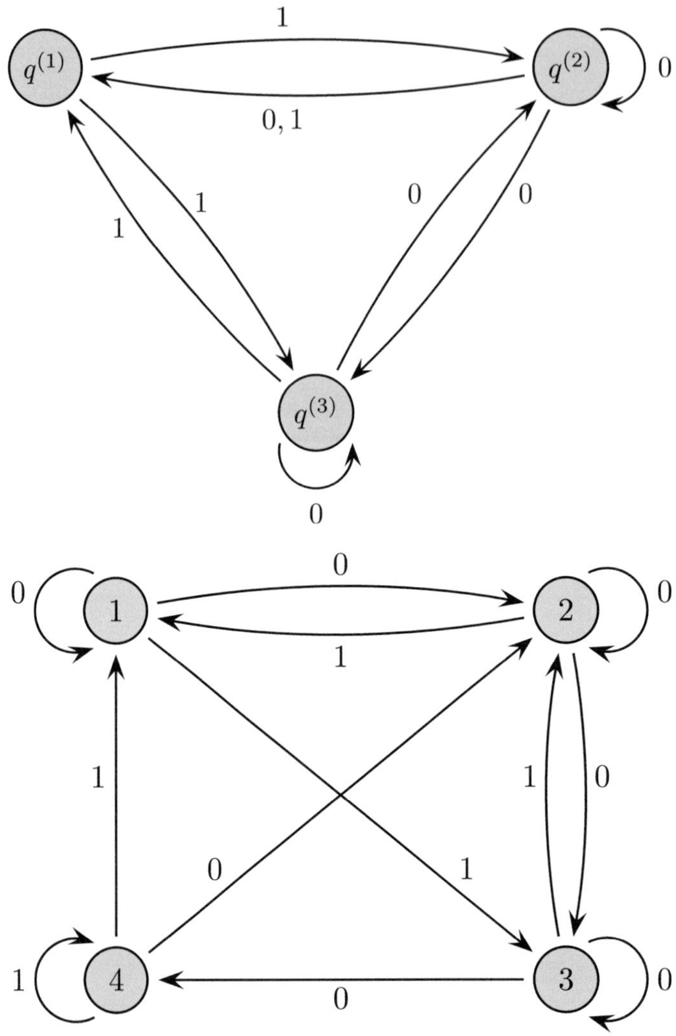

Fig. 4.18 Labeled nondeterministic finite automaton $\mathcal{LNFA}_{ur}$ (top) that is observation equivalent to the deterministic finite automaton in Fig. 4.1 and labeled nondeterministic finite automaton $\mathcal{LNFA}'_{ur}$ (bottom) that is observation equivalent to the deterministic finite automaton in Fig. 4.3

4.8 Complexity of Recursive State Estimation

This section discusses briefly the complexity of the recursive algorithms that were presented for state estimation in earlier sections. All algorithms assume knowledge of the system model, which imposes a memory requirement of either (i) $O(NK)$ for deterministic automata, or (ii) $O(N^2K)$ for nondeterministic automata, or (iii) $O(N^2R)$ when storing induced state mappings for each output (where N is the number of states, K is the number of inputs, and R is the number of outputs of the underlying finite automaton). We also have the following additional memory and computational requirements for each estimator.

Current-State Estimator: As already mentioned in Remark 4.7, the current-state estimator requires $O(N)$ memory to store the set of current states. Each time a new observation is obtained, the complexity of updating the set of possible current states is $O(N^2)$.

Initial-State Estimator: The initial-state estimator requires $O(N^2)$ memory to store the set of pairs of possible initial and current states. Each time a new observation is obtained, the complexity of updating state mappings can be bounded by $O(N^2)$ set operations (to check paths from each initial state to a possible final state via matching intermediate states).

Delayed-State Estimator: The D-delayed-state estimator requires complexity $O(N^2 D)$ to store the set of possible states and transitions out of them at each of the D stages of the trimmed trellis diagram that is maintained. The complexity of updating the last stage of the trellis diagram is $O(N^2)$ (as in the case of current-state estimation), whereas the complexity of updating the trellis diagram (i.e., removing trajectories that have become infeasible and getting rid of the initial stage) can be bounded by $O(N^2 D)$ operations.

4.9 Comments and Further Reading

State estimation is a key task in system monitoring and control (Simon 2006). As such, it has been the focus of study in a variety of settings (e.g., Luenberger observers in linear dynamic systems (Luenberger 1964) or Kalman filtering (Kailath 1981; Kumar and Varaiya 2015). In the context of DES that are modeled by partially observable finite automata, state estimation and related notions (e.g., observability or input inversion) have also been studied quite extensively, starting with the works of Lin and Wonham (1988), Özveren and Willsky (1990), Caines et al. (1991), Özveren and Willsky (1992). These initial ideas on state estimation (more specifically, observability and input inversion) were subsequently shown to play a critical role in the verification of important system properties, such as detectability, diagnosability, and opacity, which are discussed in more detail in later chapters of this book (respectively, in Chaps. 6, 7 and 8).

State estimation utilizing Petri net models is another direction that has been studied heavily. Starting with the work in Giua (1997), several researchers considered Petri net modes in which the initial marking (state) is partially known and events are observed via some labeling function, which for some transitions could even be the empty label (unobservable transitions). Later on, partial marking observations were also incorporated in this type of approaches. For example, Giua and Seatzu (2002) considered the estimation of the initial marking (observability analysis), whereas Li and Hadjicostis (2010), Li and Hadjicostis (2012) considered, respectively, least cost estimation of the initial marking (observability analysis) and the applied sequence of events (input invertibility), under minimum cost criteria. The authors of Giua et al. (2005, 2007) considered current marking estimation in the absence/presence of unobservable transitions. Bounds on the number of markings consistent with a sequence of observations are obtained in Ru and Hadjicostis (2009), whereas the authors of Jiroveanu et al. (2008) consider aspects of online monitoring and estimation. Extensions to time Petri nets were considered in Declerck and Bonhomme (2013), Basile et al. (2014), Bonhomme (2014). A nice survey of state estimation results using Petri net models can be found in Giua (2011).

The state estimation algorithms that were developed in this chapter for NFA were primarily concerned with whether a state is feasible or not. For example, given a sequence of observations, the algorithm that performs current-state estimation returns a set of states, each of which is a feasible final state (reached under one or more executions that generate the given sequence observations). In stochastic settings (e.g., probabilistic finite automata), one may have additional information pertaining to the likelihood of being in a particular current state, i.e., the posterior probability that the system is in a particular state, conditioned on the given sequence of observations. This additional likelihood information can be important when trying to assess which is the most likely state (among possible system states), particularly when different events have unequal probabilities (this should be contrasted against the techniques described in this chapter, which in a stochastic setting reduce to simply tracking whether or not the posterior probability of a state is nonzero). State estimation in probabilistic finite automata is related to treatments that deal with Markov chains (Meyn and Tweedie 2012; Brémaud 2013) and hidden Markov models (HMM) (Rabiner and Juang 1986; Rabiner 1989), in which state estimation is well studied (see, for example, Elliott et al. 2008).

References

Basile F, Cabasino MP, Seatzu C (2014) State estimation and fault diagnosis of labeled time Petri net systems with unobservable transitions. IEEE Trans Autom Control 60(4):997–1009

Bonhomme P (2014) Marking estimation of P-time Petri nets with unobservable transitions. IEEE Trans Syst Man Cybern Syst 45(3):508–518

Brémaud P (2013) Markov chains: Gibbs fields, Monte Carlo simulation, and queues, vol 31. Springer Science & Business Media

Caines PE, Greiner R, Wang S (1991) Classical and logic-based dynamic observers for finite automata. IMA J Math Control Inf 8(1):45–80

Declerck P, Bonhomme P (2013) State estimation of timed labeled Petri nets with unobservable transitions. IEEE Trans Autom Sci Eng 11(1):103–110

Elliott RJ, Aggoun L, Moore JB (2008) Hidden Markov models: estimation and control, vol 29. Springer Science & Business Media

Frank PM (1990) Fault diagnosis in dynamic systems using analytical and knowledge-based redundancy: a survey and some new results. Automatica 26(3):459–474

Giua A (1997) Petri net state estimators based on event observation. In: Proceedings of 36th IEEE conference on decision and control (CDC), vol 4, pp 4086–4091

Giua A (2011) State estimation and fault detection using Petri nets. In: Proceedings of international conference on application and theory of Petri nets and concurrency, pp 38–48
Giua A, Seatzu C (2002) Observability of place/transition nets. IEEE Trans Autom Control 47(9):1424–1437
Giua A, Corona D, Seatzu C (2005) State estimation of λ-free labeled Petri nets with contact-free nondeterministic transitions. Discrete Event Dyn Syst 15(1):85–108
Giua A, Seatzu C, Corona D (2007) Marking estimation of Petri nets with silent transitions. IEEE Trans Autom Control 52(9):1695–1699
Hashtrudi Zad S, Kwong R, Wonham W (2003) Fault diagnosis in discrete-event systems: framework and model reduction. IEEE Trans Autom Control 48(7):1199–1212
Jiroveanu G, Boel RK, Bordbar B (2008) On-line monitoring of large Petri net models under partial observation. Discrete Event Dyn Syst 18(3):323–354
Kailath T (1981) Lectures on Wiener and Kalman filtering. In: Lectures on Wiener and Kalman filtering. Springer, pp 1–143
Kumar PR, Varaiya P (2015) Stochastic systems: estimation, identification, and adaptive control, vol 75. SIAM
Li L, Hadjicostis CN (2010) Least-cost transition firing sequence estimation in labeled Petri nets with unobservable transitions. IEEE Trans Autom Sci Eng 8(2):394–403
Li L, Hadjicostis CN (2012) Minimum initial marking estimation in labeled Petri nets. IEEE Trans Autom Control 58(1):198–203
Lin F, Wonham WM (1988) On observability of discrete-event systems. Inf Sci 44(3):173–198
Luenberger DG (1964) Observing the state of a linear system. IEEE Trans Mil Electron 8(2):74–80
Meyn SP, Tweedie RL (2012) Markov chains and stochastic stability. Springer Science & Business Media
Moody JO, Antsaklis PJ (1997) Supervisory control using computationally efficient linear techniques: a tutorial introduction. In: Proceedings of 5th IEEE mediterranean conference on control and systems (MED)
Moody JO, Antsaklis PJ (1998) Supervisory control of discrete event systems using Petri nets. Springer Science & Business Media
Özveren CM, Willsky AS (1990) Observability of discrete event dynamic systems. IEEE Trans Autom Control 35(7):797–806
Özveren CM, Willsky AS (1992) Invertibility of discrete-event dynamic systems. Math Control Signals Syst 5(4):365–390
Rabiner LR (1989) A tutorial on Hidden Markov Models and selected applications in speech recognition. Proc IEEE 77(2):257–286
Rabiner LR, Juang BH (1986) An introduction to Hidden Markov Models. IEEE ASSP Mag 3(1): 4–16
Ramadge PJ, Wonham WM (1987) Supervisory control of a class of discrete event processes. SIAM J Control Optim 25(1):206–230
Ramadge PJ, Wonham WM (1989) The control of discrete event systems. Proc IEEE 77(1):81–97
Ru Y, Hadjicostis CN (2009) Bounds on the number of markings consistent with label observations in Petri nets. IEEE Trans Autom Sci Eng 6(2):334–344
Sampath M, Sengupta R, Lafortune S, Sinnamohideen K, Teneketzis D (1995) Diagnosability of discrete-event systems. IEEE Trans Autom Control 40(9):1555–1575
Simon D (2006) Optimal state estimation: Kalman, H infinity, and nonlinear approaches. Wiley

Chapter 5
Verification of State Isolation Properties

5.1 Introduction and Motivation

Chapter 4 discussed ways to perform state estimation in discrete event systems that can be modeled as deterministic finite automata (DFA) with outputs and (possibly) silent transitions. In particular, given a sequence of observations, we formulated and solved (using rather efficient recursive algorithms) the problems of current-state estimation, D-delayed estimation (or smoothing), and initial-state estimation. In many applications of interest, we might not simply be satisfied with solving such estimation problems; rather, we might want to determine a priori (i.e., before the system starts operating and before any observations are made) what we should expect when performing these estimation tasks. Examples of some relevant questions that might arise in such contexts are given below.

Current-State Estimation: Given a DFA with outputs and possibly silent transitions, will we always be in position to precisely pinpoint its current state (assuming, for example, we know its initial state exactly)? In other words, regardless of the sequence of observations that might be observed, will the corresponding current-state estimate (set of possible current states given the sequence of observations) be a singleton set? If not, will we at least be in position to eventually (i.e., after a finite number of observations, bounded perhaps by a constant that can be precisely calculated) know the state of the system exactly? If the latter case is not possible, will we be in position to precisely pinpoint the state of the system periodically? All of the above questions can also be rephrased (i) by relaxing the requirement that we know the state of the system exactly to a requirement that we know whether or not the state of the system belongs to some specific subset of states, and/or (ii) by requiring that we will be in position to pinpoint the current state for at least one sequence of observations (as opposed to *all* sequences of observations).

D-Delayed-State Estimation: We can also ask the above questions in a delayed-state estimation setting: for instance, given a DFA with outputs and possibly silent

C. N. Hadjicostis, *Estimation and Inference in Discrete Event Systems*,
Communications and Control Engineering,
https://doi.org/10.1007/978-3-030-30821-6_5

transitions, will we always be in position to precisely pinpoint its D-delayed state for some given finite delay D? In other words, regardless of the sequence of observations that might be observed, will the corresponding set of D-delayed-state estimates (i.e., the set of possible states D observations ago, given a sequence of observations of length greater or equal to D) be a singleton set? If not, will we at least be in position to eventually (i.e., after a finite number of observations) or periodically know precisely the D-delayed state of the system? As in the case of current-state estimation, all of the above questions can also be rephrased (i) by relaxing the requirement that we know the D-delayed state of system exactly to a requirement that we know whether or not it belonged to some specific subset of states, and/or (ii) by requiring that we will be in position to pinpoint the set of possible D-delayed states for at least one sequence of observations (as opposed to *all* sequences of observations).

Initial-State Estimation: Similar questions also arise in the context of initial-state estimation. For example, we might be interested in answering the following questions: (i) Will we (eventually) be in position to pinpoint the initial state of the system for any (or at least one) sequence of observations? (ii) Will we (eventually) be in position to determine whether or not the system started from a certain subset of states for any (or at least one) sequence of observations?

As it turns out, the answers to the above questions may involve different degrees of difficulty in terms of computational complexity. The complexity of the solutions that we consider in this chapter will, in general, be exponential in the size of the given finite automaton (this should be contrasted to the polynomial complexity of the recursive algorithms for solving state estimation problems in Chap. 4). The difficulty in the type of questions we answer in this chapter is that they need to *verify* properties for all possible behaviors in the system (and not simply track what happens following a particular sequence of observations). In this chapter, we are content to obtain answers to the questions above using various types of estimators. The reliance on state estimator (observer) constructions provides a universal methodology for answering this type of questions; however, we will see that, in certain cases, this universal approach is not the most efficient in terms of computational and/or storage complexity.

After briefly motivating such questions via some application examples, we focus on describing ways to systematically obtain the answers. Before we do that, we formalize the questions posed above, which requires us to establish some notation. For the remainder of this chapter, we assume that we are dealing with a DFA with outputs and possibly silent transitions for which the initial state may only be partially known (as described in Sect. 3.3.2). However, one should keep in mind that the techniques that we described in Chap. 4 (and which are the basis for our analysis here) can easily handle nondeterministic finite automata (NFA), as long as one is willing to invest in heavier notation (as explained in Sect. 4.6 of Chap. 4); thus, the discussions in this chapter can be easily extended to NFA.

The questions we are interested in can be formally described as follows: we are given the model of a DFA with outputs and possibly silent transitions $G = (Q, \Sigma, Y \cup \{\epsilon\}, \delta, \lambda, Q_0)$, with state set Q, input set Σ, output set Y, next-state transition function $\delta : Q \times \Sigma \rightarrow Q$, output function $\lambda : Q \times \Sigma \rightarrow Y \cup \{\epsilon\}$ (with ϵ representing the empty output), and set of possible initial states $Q_0 \subseteq Q$. Furthermore, we are given an arbitrary subset of states $S \subseteq Q$.

Current-State Isolation: A sequence of outputs y_0^k such that the set of possible states (following this sequence of observations) is non-empty and satisfies $\hat{q}_{y[k]}(y_0^k) \subseteq S$ is said to "isolate the current state of the DFA G to be within the set S." Given DFA G, we can ask a number of questions that relate to current-state isolation. For example, one may be interested to determine whether all sequences of observations (or at least one sequence of observations or no sequence of observations) allow us to isolate the current state of G to be within the set S. Similarly, one may be interested to determine whether all sequences of observations *of sufficient length* (or at least one sequence of observations of sufficient length or no sequence of observations of sufficient length) allow(s) us to isolate the current state of G to be within the set S. Note that these questions can have various degrees of complexity and we will use the umbrella term *current-state isolation* to refer to them. In this chapter, we answer such questions by constructing a *current-state estimator.*

D-Delayed-State Isolation: A sequence of outputs y_0^k such that the set of possible states (following this sequence of observations) is non-empty and satisfies $\hat{q}_{y[k-D]}(y_0^k) \subseteq S$ is said to "isolate the D-delayed state of the given DFA (i.e., the state of the automaton D observations ago) to be within the set S." [By convention, if $k - D < 0$, we take $\hat{q}_{y[k-D]}(y_0^k) = \hat{q}_0(y_0^k)$.] Given the DFA G, we can ask questions similar to the ones we asked above for current-state isolation: do all sequences of observations of sufficient length (or at least one sequence of observations of sufficient length or no sequence of observations of sufficient length) allow(s) us to isolate the D-delayed state of the system to be within the set S? We will use the umbrella term *D-delayed-state isolation* to refer to this type of questions and we will be addressing them by constructing a *delayed-state estimator.*

Initial-State Isolation: A sequence of outputs y_0^k such that the set of possible initial states (following this sequence of observations) is non-empty and satisfies $\hat{q}_0(y_0^k) \subseteq S$ is said to "isolate the initial state of the given DFA to be within the set S." Note that the problem remains unchanged if the set of states S is changed to $S \cap Q_0$ and becomes trivial if $S \cap Q_0 = Q_0$ or $S \cap Q_0 = \emptyset$. Again, one may be interested to determine whether all sequences of observations of sufficient length (or at least one sequence of observations of sufficient length or no sequence of observations of sufficient length) allow(s) us to isolate the initial state of G to be within the set S. We will use the umbrella term *initial-state isolation* to refer to this type of questions and we will be addressing them by constructing an *initial-state estimator.*

Questions like the above (or variations of them) arise in many applications and have appeared in various forms in the literature. Below, we briefly discuss some representative examples.

5.1.1 Detectability of Discrete Event Systems

The notion of *detectability* (e.g., Shu et al. 2007) asks whether an arbitrary sequence of observations, generated by a finite automaton with outputs and (possibly) silent transitions, will "eventually" or "almost always" allow the observer to infer the exact current state of the system. "Eventually" in this case means that there exists a finite number of events/observations, after which we are guaranteed to know *exactly* the current state of the system, regardless of the actual activity in the system (and the corresponding sequence of observations that it generates). An equivalent question is whether there exists at least one sequence of observations (of unbounded length) that does *not* allow, not even from a certain point onwards, the current state of the system to be always isolated within a singleton set. "Almost always" in this case means that for any (infinitely extensible) sequence of observations, the fraction of times the observer will be able to know *exactly* the current state of the system tends to unity (as the length of the observation sequence increases). A related question of being able to pinpoint the system's state was asked in Caines et al. (1988, 1991) and was answered rather efficiently for a class of finite automata. Clearly, detectability relates to the notion of current-state isolation described earlier when the set of states estimates can be isolated within one of the singleton subsets of the set of states. An illustration of how detectability can be analyzed and verified by constructing a current-state estimator can be found in Example 5.5. Formal definitions for various notions of detectability and related extensions are discussed in more detail in Chap. 6.

5.1.2 Testing of Digital Circuits

When testing a circuit implementation of a finite-state machine, the initial state might be unknown and thus a common question that arises is to identify the initial and/or current state of the machine. Testing and verification of digital sequential circuits deal with a variety of difficult questions, including finite-state machine (FSM) identification (originally posed in the seminal work Moore 1956) and the related problem of conformance testing which is important for verifying functional properties of a given digital circuit. Conformance testing aims to verify whether a given "gray-box" implementation of an FSM conforms to specified state transition and output functionalities. Conformance testing is nontrivial because (i) the machine may be in an unknown initial state and (ii) a solution will have to verify that all machine states exist and that all state transitions execute according to the specified next-state functionality. The complexity (and tractability) of this problem depends on whether the machine is strongly connected, whether the change that may have occurred belongs to a known class of changes, and whether the machine has a *distinguishing* sequence, i.e. a sequence of inputs that allows one to uniquely identify the initial state by observing the output sequence. (Later on, such sequences were referred to as *preset* sequences to distinguish them from *adaptive* sequences, which were introduced to allow the choice of the kth input of the sequence to depend on the sequence of outputs observed up to time step $k-1$.) In other words, a distinguishing sequence is an input sequence that produces a unique output sequence for each possible starting state, thus allowing the observer to differentiate among possible starting states. The pioneering work in Hennie (1964, 1968) showed that, subject to certain assumptions about the machine structure, conformance testing can take time polynomial in the size of the machine and in the possibly exponential length of its distinguishing sequence (if one exists). Following the work by Moore and Hennie, many researchers studied related topics, managing to refine these techniques (e.g., by improving the bounds on the length of checking sequences Kime 1966; Gönenc 1970; Hsieh 1971; Vasilevskii 1973; Chow 1978) and to demonstrate how a machine may be augmented with additional inputs and/or outputs in order to facilitate testing (e.g., allowing an FSM that initially has no distinguishing sequence to posses one Kohavi and Lavalee 1967; Murakami et al. 1970; Sheppart and Vranesic 1974; Fujiwara and Kinoshita 1978; Pradhan 1983; Bhattacharyya 1983). The work in Yannakakis and Lee (1994) has shown that it is PSPACE-complete to determine whether or not a DFA has a distinguishing sequence (as there exist machines whose shortest distinguishing sequence is exponential in length). However, they have also shown that one can determine in polynomial time whether a DFA has an "adaptive" distinguishing sequence, and, if that is the case, one can find such a sequence (whose length is argued to be $O(|Q|^2)$) in polynomial time. An "adaptive" distinguishing sequence (distinct from a "preset" distinguishing sequence as explained earlier in this paragraph) is not really a sequence at all, but rather a decision tree whose branches correspond to the outputs generated by the system. Further discussions and examples of such types of sequences can be found in Chap. 6.

5.1.3 Fault Diagnosis

Fault diagnosis will be discussed explicitly in Chap. 7, but we make a quick reference here in order to relate it to the notion of state isolation. One of the most frequently used fault diagnosis formulations was established in Sampath et al. (1995) and considers a finite automaton setting where (the automaton is typically labeled and observations adhere to a natural projection mapping, and) certain events, called *fault events*, need to be detected and/or identified and/or classified ("diagnosed") by the diagnoser. The diagnoser is an external observer who has access to the sequence of outputs generated by the system (in response to an unknown sequence of inputs/events) and is also assumed to have full knowledge of the system model and its possible initial state(s).

The simplest question one can ask is whether the occurrence of a fault f_i from a given set of fault events $F=\{f_1, f_2, \ldots, f_{|F|}\}$ can be detected or even identified after a finite number of events or observations. Specifically, the detection task aims to determine whether *at least one* fault from the set F has occurred, whereas the identification task is concerned with precisely pinpointing the fault(s) $f_i, f_i \in F$, that has (have) occurred. Clearly, the task becomes trivial if event $f_i, f_i \in F$, generates an output that is uniquely associated with it, because the diagnoser will immediately be able to conclude that f_i has occurred. For this reason, in the frequently studied labeled automaton setting (Sampath et al. 1995), fault events are assumed (without loss of generality) to be unobservable. (Otherwise, in the labeled automaton setting, the fault event $f_i, f_i \in F$, will be associated with the unique label f_i, which will allow us to immediately detect and, in fact, identify its occurrence.) In the more general setting we consider in this book (where an event maybe associated with an output label that is not necessarily unique), this assumption is not necessary and we do not impose it.

For fault classification, the setting is slightly more general: the set of fault events $F \subseteq \Sigma$ is partitioned into C mutually exclusive fault classes (types) with respect to which faults need to be classified. In other words, we have C types (sets) of faults $F_1, F_2, \ldots, F_C$, where $F_c \subseteq F$ for $c = 1, 2, \ldots, C$; $F_{c_i} \cap F_{c_j} = \emptyset$ for $c_i, c_j \in \{1, 2, \ldots, C\}$, $c_i \neq c_j$; and $F = F_1 \dot{\cup} F_2 \dot{\cup} \cdots \dot{\cup} F_C$. Under this more general formulation, fault classification only needs to identify whether faults from one or more classes have occurred. In certain cases, when multiple faults occur, it might be desirable to also determine the order in which faults from different classes have occurred.

The fault detection problem can easily be converted to a state isolation problem by the procedure illustrated in the example below.

Example 5.1 Consider the DFA with outputs $G = (Q, \Sigma, Y \cup \{\epsilon\}, \delta, \lambda, Q_0)$ shown on the top of Fig. 5.1, where

$$Q = \{1, 2, 3, 4, 5, 6, 7, 8, 9\} ,$$

$\Sigma = \{\alpha, \beta, \gamma, f_1, f_2\}$, $Y = \{0, 1\}$, $Q_0 = \{1\}$, and δ and λ are as shown at the top of the figure. The *detection* of faults in the set $F = \{f_1, f_2\}$ can be converted to a state isolation problem by considering the automaton $G_F = (Q \dot{\cup} Q_F, \Sigma, Y \cup \{\epsilon\}, \delta_F, \lambda_F, Q_0)$ where

$$Q_F = \{1F, 2F, 3F, 4F, 5F, 6F, 7F, 8F, 9F\} ,$$

and δ_F and λ_F are as shown at the bottom of the figure (note that the transition from state 9 to state 1 under input γ generates no observation). Then, it can be easily shown that faults in F are detectable (i.e., their occurrence can be detected after a finite number of events/observations) if and only if the states Q_F in automaton G_F can be isolated following the occurrence of a sequence of inputs with one or more faults from the set F. More precisely, if after a finite number of observations following the occurrence of a fault, the set of state estimates is a subset of Q_F, then we are certain that some fault from the set F has occurred. □

The procedure described in the above example can be summarized as follows. Given a finite automaton with outputs and possibly silent transitions $G = (Q, \Sigma, Y \cup \{\epsilon\}, \delta, \lambda, Q_0)$ with states $Q = \{q^{(1)}, q^{(2)}, \ldots, q^{(N)}\}$ and fault events $F \subseteq \Sigma$ that

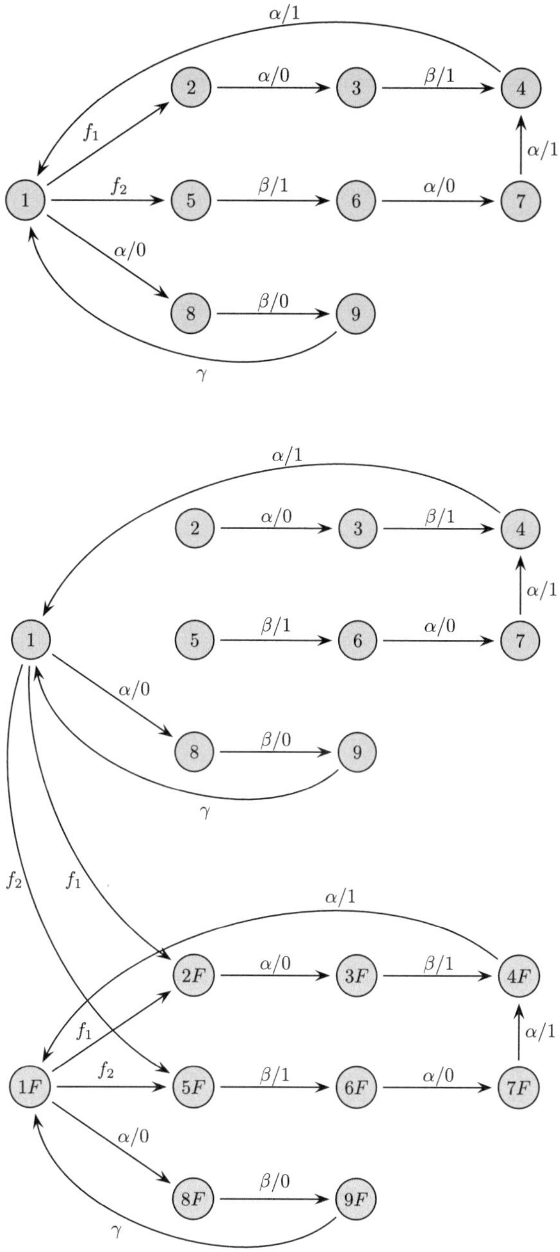

Fig. 5.1 Finite automaton G (top) and corresponding finite automaton G_F (bottom) illustrating the conversion of the fault detection problem into a state isolation problem

need to be detected, one can generate a new automaton $G_F = (Q\dot{\cup}Q_F, \Sigma, Y \cup \{\epsilon\}, \delta_F, \lambda_F, Q_0)$, where $Q_F = \{q_F^{(1)}, q_F^{(2)}, \ldots, q_F^{(N)}\}$, and δ_F and λ_F are defined so that the restriction to states Q_F essentially yields a copy of the original automaton (with state $q^{(i)}$ replaced by state $q_F^{(i)}$), whereas the restriction to states Q yields the transition functionality of the original automaton with the only difference being that fault events, instead of taking us to states in Q, they take us to the corresponding states in Q_F (i.e., under fault event $f \in F$, instead of a transition from state $q^{(i)}$ to state $q^{(j)}$ that generates output $y \in Y \cup \{\epsilon\}$ in G, we transition in G_F from state $q^{(i)}$ to state $q_F^{(j)}$ generating the same, possibly empty, output y). It is clear from this construction that states Q_F are trapping states and if, following a sequence of observations, we can isolate the current state of the system to be within the set $S = Q_F$ (at least for all sequences of inputs that include one or more faults), then we can safely conclude that a fault has definitely taken place. Therefore, the requirement that faults in the set F will be detected within a finite number of at most D events/observations is equivalent to a requirement that, following the occurrence of any fault in F and regardless of the ensuing sequence of events, the observer will always be in position to isolate the state of the system within the set Q_F, after at most D events/observations. More details about the construction of automaton Q_F and the reduction of fault detection to state isolation are provided in Chap. 7.

To transform the fault classification problem to a state isolation problem, we first need to state the goal of fault classification in case more than one faults have occurred. For instance, if a fault from a class of faults F_1 was subsequently followed by a fault from class F_2, does one aim to identify (after a finite number of at most D events/observations) the fault class of the first fault that occurred, or does one aim to identify (after a finite number of at most D events/observations) that faults from both of these fault classes have definitely occurred (perhaps also indicating the order in which faults from different classes have occurred)? We will see that different fault diagnosis goals can be translated to different state isolation problems. The goal of separately identifying faults from each particular class F_i, $i = 1, 2, \ldots, C$, can be converted to C separate state isolation problems of the type discussed earlier for fault detection. [In the ith state isolation problem, class F_i is taken to be the class of faults ($F = F_i$) and all other fault events (in other fault classes, i.e., in $F \backslash F_i$) are treated as normal events.]

When one is interested in identifying with certainty that a particular class of faults has occurred first, the conversion to a corresponding state isolation problem needs to be slightly modified. Specifically, suppose we are given a finite automaton with outputs and possibly silent transitions $G = (Q, \Sigma, Y \cup \{\epsilon\}, \delta, \lambda, Q_0)$ with states $Q = \{q^{(1)}, q^{(2)}, \ldots, q^{(N)}\}$ and fault events $F \subseteq \Sigma$ that need to be classified according to mutually exclusive fault classes $F_1, F_2, \ldots, F_C$, where $F_c \subseteq F$, $c = 1, 2, \ldots, C$; $F_{c_i} \cap F_{c_j} = \emptyset$ for $c_i, c_j \in \{1, 2, \ldots, C\}$, $c_i \neq c_j$; and $F = F_1 \dot{\cup} F_2 \dot{\cup} \cdots \dot{\cup} F_C$. We can generate a new automaton $G_{\mathcal{F}'} = (Q\dot{\cup}Q_{F_1}\dot{\cup}Q_{F_2}\dot{\cup}\cdots\dot{\cup}Q_{F_C}, \Sigma, Y \cup \{\epsilon\}, \delta_{\mathcal{F}'}, \lambda_{\mathcal{F}'}, Q_0)$ where $Q_{F_c} = \{q_{F_c}^{(1)}, q_{F_c}^{(2)}, \ldots, q_{F_c}^{(N)}\}$ for $c = 1, 2, \ldots, C$, and $\delta_{\mathcal{F}'}$ and $\lambda_{\mathcal{F}'}$ are defined so that the restriction to states Q_{F_c} essentially yields a copy of the original automaton (with state $q^{(i)}$ represented by state $q_{F_c}^{(i)}$), whereas the restriction to states Q yields the transition functionality of the original automaton with the only difference being that fault events in F_c, instead of taking us to states in Q, they take us to the corresponding states in Q_{F_c} (i.e., under fault event $f_c \in F_c$, instead of a transition from state $q^{(i)}$ to state $q^{(j)}$ that generates output $y \in Y \cup \{\epsilon\}$ in G, we transition in $G_{\mathcal{F}'}$ from state $q^{(i)}$ to state $q^{(j)}_{F_c}$ generating the same, possibly empty, output y). It is clear from this construction that each of the sets of states $Q_{F_c}, c = 1, 2, \ldots, C$, are trapping states and if, following a sequence of observations, we can isolate the current state of the system to be within the set $S = Q_{F_c}$, then we know that a fault in class F_c was the first fault to occur (though it is possible that subsequent faults from the same or other classes also occurred). Therefore, the requirement that the class (type) of the first fault $f_c, f_c \in F_c$, that occurs can be identified after a finite number of at most D events/observations, regardless of the ensuing sequence of events, is equivalent to the observer always being in position to isolate the state of the system within the set Q_{F_c} after a finite number of at most D events/observations (following the occurrence of the fault f_c). Note that, using automaton $G_{\mathcal{F}'}$, fault detection is equivalent to the observer always being in position to isolate the state of the system within the set $Q_{F_1} \dot\cup Q_{F_2} \dot\cup \cdots \dot\cup Q_{F_C}$ after a finite number of at most D events/observations (in such case, the observer is certain that a fault has occurred, but it cannot determine which class of faults occurred). An illustration of this approach is found in the following example.

Example 5.2 Consider again the DFA with outputs $G = (Q, \Sigma, Y \cup \{\epsilon\}, \delta, \lambda, Q_0)$ shown at the top of Fig. 5.1, where

$$Q = \{1, 2, 3, 4, 5, 6, 7, 8, 9\} ,$$

$\Sigma = \{\alpha, \beta, \gamma, f_1, f_2\}$, $Y = \{0, 1\}$, $Q_0 = \{1\}$, and δ and λ are as shown at the top of the figure. The faults classes are $F_1 = \{f_1\}$ and $F_2 = \{f_2\}$ so that the fault classification task essentially reduces to the identification of faults f_1 and f_2. If the goal is to simply determine which fault occured first, the fault classification problem can be converted to a state isolation problem by considering the automaton $G_{\mathcal{F}'} = (Q \dot\cup Q_{F_1} \dot\cup Q_{F_2}, \Sigma, Y \cup \{\epsilon\}, \delta_{\mathcal{F}'}, \lambda_{\mathcal{F}'}, Q_0)$ where

$$Q_{F_1} = \{1F_1, 2F_1, 3F_1, 4F_1, 5F_1, 6F_1, 7F_1, 8F_1, 9F_1\} ,$$
$$Q_{F_2} = \{1F_2, 2F_2, 3F_2, 4F_2, 5F_2, 6F_2, 7F_2, 8F_2, 9F_2\} ,$$

and $\delta_{\mathcal{F}'}$ and $\lambda_{\mathcal{F}'}$ are as shown in Fig. 5.2. Then, the first fault that occurs (either fault f_1 or f_2) can be identified if and only if the states Q_{F_1} in automaton $G_{\mathcal{F}'}$ can be isolated when the first fault occurring is f_1, and the states Q_{F_2} in automaton $G_{\mathcal{F}'}$ can be isolated when the first fault occurring is f_2.

Note that if the sequence of events that occurs involves both f_1 and f_2, then the state isolation problem in $G_{\mathcal{F}'}$ will be able to determine that fault f_1 or fault f_2 occurred (depending on which one occurred first), but it will not be able to clarify whether both types of faults occurred. If one wanted to isolate f_1 and f_2 (without regards to their ordering), then one could use two separate state isolation problems,

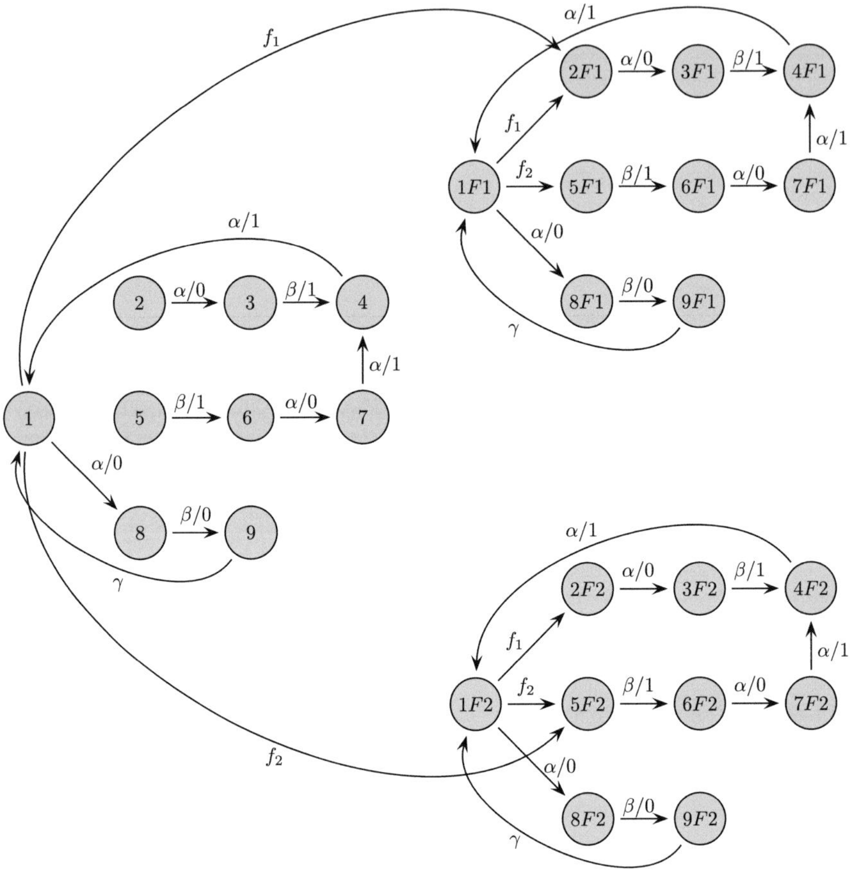

Fig. 5.2 Finite automaton $G_{\mathcal{F}'}$ illustrating the conversion of the fault classification problem for the automaton G at the top of Fig. 5.1 to a state isolation problem

one for f_1 and one for f_2. In each case, a reduction of the type shown at the bottom of Fig. 5.1 would be used: in one case, f_1 would be treated as the (only) fault and f_2 as a regular event (i.e., in automaton $G_{\mathcal{F}}$ at the bottom of Fig. 5.1, f_2 would be a transition from state 1 to state 5 and not to state $5F$); in the other case, f_2 would be treated as the (only) fault and f_1 as a regular event (i.e., in automaton $G_{\mathcal{F}}$ at the bottom of Fig. 5.1, f_1 would be a transition from state 1 to state 2 and not to state $2F$). □

Note that a similar reduction with 2^C trapping sets of states can be used to translate the problem of identifying faults from one or more different classes to an equivalent state isolation problem. Formal definitions of various fault diagnosis formulations, as well as details about their verification can be found in Chap. 7 where fault diagnosis is discussed explicitly.

5.1.4 State-Based Notions of Opacity

The proliferation of digital technologies and interconnectivity has led to the emergence of complex DES in numerous applications, ranging from automated manufacturing systems and autonomous vehicles to traffic networks and healthcare/information systems. Important security and privacy concerns arise in such applications when shared (typically untrustworthy) infrastructures are used as the communication backbone to exchange information. In particular, the non-dedicated nature of the network implies the possible presence of eavesdroppers, which may try to learn important details about various operational parameters of the system (thus, compromising privacy), or even malicious adversaries, which may try to affect the functionality of the system (thus, compromising security and potentially causing safety-critical violations). A simple motivational example was presented in Chap. 1.

Various notions of opacity have been proposed in an effort to characterize the information flow from the system to an intruder. In these settings, the intruder is typically an entity that observes activity in the system, has full or partial knowledge of the system model, and aims to infer certain important (e.g., critical or private) information, such as passwords, account balances, and others (Focardi and Gorrieri 1994). In general, opacity aims to determine whether a given system's secret or private behavior (i.e., a subset of the behavior of the system that is considered critical and is usually represented by a predicate) is kept opaque to outsiders; this means the intruder (modeled as an active or passive observer of the system's behavior) will never be able to establish the truth of the predicate.

There are many formulations for opacity in finite automata and they are discussed in detail in Chap. 8. Here, we make a quick reference to *current-state opacity* in order to relate it to the notion of state isolation. A system is said to be current-state opaque if the entrance of the system state to a specific secret state (or to a set of secret states) remains opaque (uncertain) to an intruder—at least until the system leaves the set of secret state(s) (Saboori and Hadjicostis 2007). For a system that can be modeled as an NFA, the notion of current-state opacity is related to the current-state isolation problem, where the set S is taken to be the set of secret states. In other words, if for all possible sequence of observations, the corresponding (non-empty) set of state estimates does not fall entirely within the set of secret states S, then the system is deemed current-state opaque. The notion of current-state opacity has also been extended to the notion of initial-state opacity (which is equivalent to the initial-state isolation problem with the set S taken to be the set of secret initial states), D-delayed-state opacity (which is equivalent to the D-delayed-state isolation problem with the set S taken to be the set of secret D-delayed states Saboori and Hadjicostis 2011b), and to infinite step opacity (Saboori and Hadjicostis 2012) (which is equivalent to the limit of D-delayed opacity when D goes to infinity).

There are numerous examples where the above state-based notions of opacity arise naturally as the obvious way of defining security/privacy properties in a given system; these include tracking and coverage of mobile agents in sensor networks (Saboori and Hadjicostis 2011a), and encryption guarantees of pseudorandom generators (Saboori and Hadjicostis 2007), both of which are discussed in Chap. 8. The verification of state-based notions of opacity in these and other applications is equivalent to the verification of state isolation properties, as discussed in the remainder of this chapter. Note that, unlike the fault diagnosis case described in the previous section, the set of secret states S in the case of state-based opacity is not necessarily a trapping set of states; for this and other reasons, the verification of fault diagnosis is not necessarily of the same complexity as the verification of current-state opacity. An illustration of how current-state opacity can be analyzed and verified using a current-state estimator can be found in Example 5.5 in this chapter; formal definitions of various notions of opacity, as well as ways to verify them and enforce them, can be found in Chap. 8.

5.2 Current-State Isolation Using the Current-State Estimator

In Sect. 4.3 of Chap. 4, we discussed how, given a streaming sequence of observations, the current state of a DFA (with outputs and possibly silent transitions) G can be estimated online using a recursive algorithm that keeps track of the set of possible current states, starting from the set of initial states and obtaining the new set of possible current states, each time using the previous states and the new observation that becomes available. As discussed in Sect. 4.6 of Chap. 4, these ideas can be extended to NFA as long as one is willing to invest in heavier notation (specifically, one needs to rely on more complex definitions of state mappings).

In Sect. 4.3, we also talked about an alternative approach that uses an observer (or a current-state estimator) G_{obs}, i.e., a DFA without outputs, constructed and initialized so that the state it reaches (following a sequence of observations generated by activity in G) represents the set of possible current states for G. Thus, the observer (whose states can be viewed as subsets of states of G) is an automaton that is driven by the set of outputs of G; its construction for the case of an NFA (with outputs and silent transitions) as defined in 3.24 is formally described below.

Definition 5.1 (*Observer, Current-State Estimator*) Given an NFA with outputs and (possibly) silent transitions $G = (Q, \Sigma, Y \cup \{\epsilon\}, \delta, \lambda, Q_0)$, its observer is the DFA without outputs $G_{obs} = AC(2^Q, Y, \delta_{obs}, Q_{0,obs}) =: (Q_{obs}, Y, \delta_{obs}, Q_{0,obs})$, where

1. The set of states Q_{obs} is (the accessible part of) 2^Q (the set of subsets of the set Q);
2. For $q_{obs} \in 2^Q$, $y \in Y$, the mapping $\delta_{obs} : 2^Q \times Y \mapsto 2^Q$ is defined as

$$\delta_{obs}(q_{obs}, y) = q'_{obs} := \{q_f \in Q \mid \exists q_i \in q_{obs} \text{ such that } (q_i, q_f) \in M_y\} ,$$

   where $q'_{obs} \in 2^Q$ and M_y is the induced state mapping under input y for NFA G (refer to Definition 4.9);
3. $Q_{0,obs}$ is the unobservable reach of the set of possible initial states of G (i.e., the set of states Q_0 together with any states that can be reached from a state in Q_0

via a sequence of transitions that does not generate any observation—refer to Definition 3.27);
4. AC denotes the accessible part of the automaton (i.e., the part of the automaton G_{obs} that can be reached from its initial state $Q_{0,obs}$).

The construction of G_{obs} in the above definition essentially considers all subsets of Q as potential states of G_{obs}; for each pair of two such subsets, q_{obs} and q'_{obs}, it adds a transition from q_{obs} to q'_{obs} under input y (which is an output of automaton G) if and only if the set of states that can be reached in automaton G from states in the set q_{obs} while generating output y is exactly captured by the set q'_{obs} (note that this ensures that G_{obs} is a DFA). After constructing the next-state transition function of G_{obs} in this fashion, the states in 2^Q that cannot be reached from the initial state $Q_{0,obs}$ can be safely ignored, and this is exactly what the operation AC does. It is clear from the above construction that G_{obs} has at most 2^N states because that is the number of distinct subsets of the set Q (where $N = |Q|$).

Remark 5.1 Note that the next-state transition mapping $\delta_{obs}(q_{obs}, y)$ for the observer G_{obs} can also be written as

$$\delta_{obs}(q_{obs}, y) = \Pi_1(q_{obs} \cdot M_y) \ ,$$

where q_{obs} is treated as an one-dimensional state mapping and M_y is treated as a two-dimensional state mapping. Also, note that the empty subset $\emptyset$ could be a state in G_{obs} if there exist sequences of observations that cannot possibly be generated by G: this state is an absorbing state (once reached, G_{obs} remains in that state) and it is reached by sequences of outputs that cannot possibly be generated in G from a valid initial state (in the set Q_0) and a valid sequence of inputs. It is typical to draw G_{obs} without including this absorbing state or the transitions leading to it. □

The following example clarifies the construction of G_{obs}.

Example 5.3 Consider the automaton $G = (Q, \Sigma, Y \cup \{\epsilon\}, \delta, \lambda, Q_0)$ at the top of Fig. 5.3, which describes a labeled deterministic finite automaton (LDFA) with silent transitions under the natural projection mapping. Specifically, we have $Q = \{1, 2, 3, 4, 5\}$, $\Sigma = \{a, b, c\}$, $Y = \{a, b\}$, $Q_0 = \{1\}$, and δ as shown in the figure. The set of observable events is $\Sigma_{obs} = \{a, b\}$ and the set of unobservable events is $\Sigma_{uo} = \{c\}$ (so that the output mapping λ simply implements the natural projection, i.e., $\lambda(a) = a$, $\lambda(b) = b$, $\lambda(c) = \epsilon$).

The observer (current-state estimator) for automaton G is the DFA (without outputs) $G_{obs} = (Q_{obs}, \Sigma_{obs}, \delta_{obs}, Q_{0,obs})$ shown at the bottom of Fig. 5.3. The observer has the following properties:

1. Its states are subsets of Q (thus, $Q_{obs} \subseteq 2^Q$);
2. Its inputs are the observable outputs of G (namely Σ_{obs});
3. Its next-state transition function δ_{obs} is as shown at the bottom of Fig. 5.3;

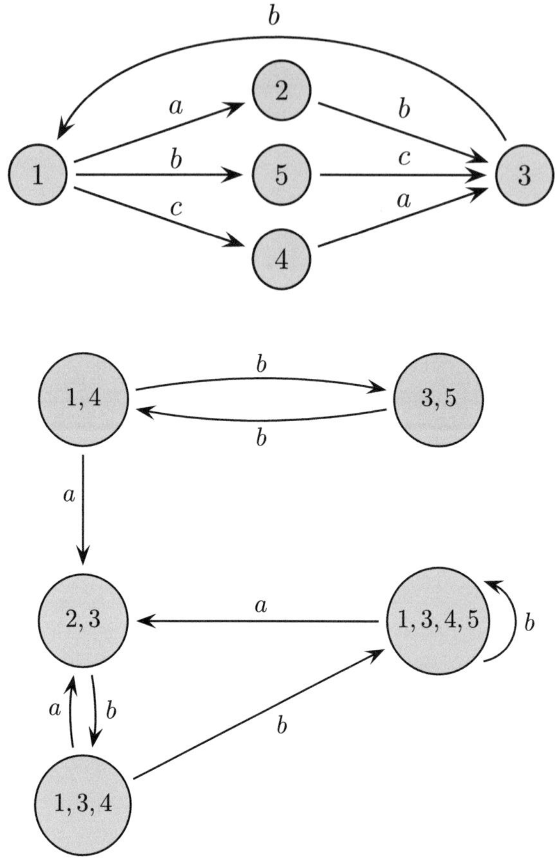

Fig. 5.3 Finite automaton G (top) and its corresponding observer G_{obs} (bottom)

4. Its initial state $Q_{0,obs}$ is the set of states in Q_0 (namely, state 1) and its unobservable reach (namely, state 1 together with state 4—the latter can be reached from state 1 via c which is unobservable).

To better understand the construction of G_{obs}, consider the following. Since we know that G starts in state 1, the initial state of G_{obs} includes not only state 1 but also all states that can be reached from state 1 via silent sequences of events. Thus, the initial state of G_{obs} is state $Q_{0,obs} = \{1, 4\}$. From that observer state, there are two possible observations, namely a and b. Observation a implies that a sequence of inputs of the form c^*ac^* has occurred, which can take system G to either state 2 (from state 1) or to state 3 (from state 4); this implies the next observer state from $Q_{0,obs}$ under observation a is state $\{2, 3\}$. Observation b implies that a sequence of inputs of the form c^*bc^* has occurred, which can take system G to either state 5 (from state 1) or to state 3 (from state 5, which can be reached from state 1, via the unobservable event c); this implies the next observer state from $Q_{0,obs}$ under observation b is observer state $\{3, 5\}$.

From each of the two observer states generated above, there are two possible observations to consider, namely a and b. If we consider observer state $\{3, 5\}$, we have the following:
(i) observation a implies that a sequence of inputs of the form c^*ac^* has occurred, which is not possible from either state 3 or state 5. In such case, these transitions are typically not drawn in the observer diagram, and this is what we have done in the figure. An alternative approach would have been to include in the observer an absorbing state (associated with the empty set of states $\emptyset$) and have all such transitions lead to this state.
(ii) observation b implies that a sequence of inputs of the form c^*bc^* has occurred, which can take system G to either state 1 (from state 3 or state 5) or to state 4 (from state 3 or state 5); this implies that from observer state $\{3, 5\}$ under observation b, the next observer state is $\{1, 4\}$.

The construction of the observer G_{obs} can be completed by continuing in this fashion. □

Automaton G_{obs} can be utilized to obtain the set of possible current states of automaton G as follows: given a sequence of observations $y_0^k \in Y^*, k \geq 0$, generated by underlying activity in G, the set $\hat{q}_{y[k]}(y_0^k)$ of possible current states of G is given by $\hat{q}_{y[k]}(y_0^k) = \delta_{obs}(Q_{0,obs}, y_0^k)$, i.e., the set of possible current states in G following the sequence of observations y_0^k is the set of states represented in G_{obs} by the observer state reached, starting from the initial state $Q_{0,obs}$ and applying $y[0]$, then $y[1]$, ..., and finally $y[k]$. The example below illustrates this idea, whereas the theorem that follows establishes this property formally.

Example 5.4 If we consider the sequence of observations $y_0^3 = abba$ (i.e., $y[0] = a$, $y[1] = b$, $y[2] = b$, and $y[3] = a$) in the finite automaton G shown on the top of Fig. 5.3, we can easily obtain the corresponding sequence of current-state estimates as

$$\begin{aligned} \hat{q}_{y[0]}(y_0^0) &= \{2, 3\} , \\ \hat{q}_{y[1]}(y_0^1) &= \{1, 3, 4\} , \\ \hat{q}_{y[2]}(y_0^2) &= \{1, 3, 4, 5\} , \\ \hat{q}_{y[3]}(y_0^3) &= \{2, 3\} . \end{aligned}$$

This follows simply from the fact that, starting from initial state $Q_{0,obs} = \{1, 4\}$, the observer G_{obs} at the bottom of Fig. 5.3 under the sequence of inputs (observations) *abba*, follows the sequence of states $\{2, 3\}$, $\{1, 3, 4\}$, $\{1, 3, 4, 5\}$, and $\{2, 3\}$. □

The following theorem states the property illustrated in the above example formally.

Theorem 5.1 *Consider an NFA with outputs and (possibly) silent transitions* $G = (Q, \Sigma, Y \cup \{\epsilon\}, \delta, \lambda, Q_0)$ *and its observer* $G_{obs} = AC(2^Q, Y, \delta_{obs}, Q_{0,obs}) =: (Q_{obs}, Y, \delta_{obs}, Q_{0,obs})$ *constructed as described in Definition 5.1. For any integer* $k \geq 0$ *and for any* $y_0^k \in Y^*$, *we have*

$$\hat{q}_{y[k]}(y_0^k) = \delta_{obs}(Q_{0,obs}, y_0^k) .$$

Proof We will establish that

$$\hat{q}_{y[i]}(y_0^i) = \Pi_{i+1}\left(M^{(i+2)}_{y_0^i,Q_0}\right) = \delta_{obs}(Q_{0,obs}, y_0^i) \ ,$$

where $M^{(i+2)}_{y_0^i,Q_0}$ was defined in Sect. 4.5 of Chap. 4. The proof is by induction: for $k = 0$, we have that

$$\begin{aligned}\hat{q}_{y[0]}(y[0]) &= \Pi_1\left(M^{(2)}_{y[0],Q_0}\right)\\ &= \{q_f \in Q \mid \exists q_i \in Q_0 \text{ such that } (q_i, q_f) \in M_{y[0]}\}\\ &= \delta_{obs}(Q_{0,obs}, y[0]) \ ,\end{aligned}$$

where the first equality follows from the discussions in Sect. 4.5.1 (the discussion there is stated for a DFA but also holds—pending appropriate definitions of state mappings—for an NFA), the second equality follows from the definition of $M^{(2)}_{y[0],Q_0}$, and the last equality follows from the definition of δ_{obs}. [Note that $\delta_{obs}(Q_{0,obs}, y[0]) = \delta_{obs}(Q_0, y[0])$, i.e., whether or not we use the unobservable reach makes no difference in the state estimation process.]

For the induction hypothesis, we assume that

$$\hat{q}_{y[k-1]}(y_0^{k-1}) = \Pi_k\left(M^{(k+1)}_{y_0^{k-1},Q_0}\right) = \delta_{obs}(Q_{0,obs}, y_0^{k-1}) \ ,$$

and we have to show that

$$\hat{q}_{y[k]}(y_0^k) = \Pi_{k+1}\left(M^{(k+2)}_{y_0^k,Q_0}\right) = \delta_{obs}(Q_{0,obs}, y_0^k) \ .$$

From Corollary 4.3, we have

$$\Pi_{k+1}\left(M^{(k+2)}_{y_0^k,Q_0}\right) = \Pi_1\left(\Pi_k(M^{(k+1)}_{y_0^{k-1},Q_0}) \cdot M^{(2)}_{y[k]}\right) \ ,$$

which together with the induction hypothesis implies

$$\hat{q}_{y[k]}(y_0^k) = \Pi_1\left(\delta_{obs}(Q_{0,obs}, y_0^{k-1}) \cdot M^{(2)}_{y[k]}\right) \ .$$

If we set $q_{obs} = \delta_{obs}(Q_{0,obs}, y_0^{k-1})$, then we have

$$\begin{aligned}\hat{q}_{y[k]}(y_0^k) &= \Pi_1\left(\{(q_i, q_f) \in M^{(2)}_{y[k]} \mid q_i \in q_{obs}\}\right)\\ &= \{q_f \in Q \mid \exists q_i \in q_{obs} \text{ such that } (q_i, q_f) \in M_{y[k]}\}\\ &= \delta_{obs}(q_{obs}, y[k]) \ ,\end{aligned}$$

where the last equality follows from the definition of δ_{obs}. □

Having constructed the observer $G_{obs} = AC(2^Q, Y, \delta_{obs}, Q_{0,obs}) =: (Q_{obs}, Y, \delta_{obs}, Q_{0,obs})$ for a given NFA G, it is rather straightforward to verify current-state isolation with respect to a set of states S: if there is an accessible state $q_{obs} \in Q_{obs}$ such that $q_{obs} \subseteq S$, $q_{obs} \neq \emptyset$, then we have a sequence of observations that allows the observer to determine with certainty that the set of possible current states is within the set S. Note that one has to exclude the case when $q_{obs} = \emptyset$ because that state is reached only via sequences of observations (outputs) that cannot be generated by an underlying sequence of events (inputs) in the given system G; however, all other states in Q_{obs} are reached by sequences of observations that can be generated in the given system. The following corollary states this formally.

Corollary 5.1 (Aid for Current-State Isolation) *We consider an NFA with outputs and (possibly) silent transitions* $G = (Q, \Sigma, Y \cup \{\epsilon\}, \delta, \lambda, Q_0)$ *and its observer* $G_{obs} = AC(2^Q, Y, \delta_{obs}, Q_{0,obs}) =: (Q_{obs}, Y, \delta_{obs}, Q_{0,obs})$. *Consider a sequence of events in G that generates a sequence of observations* y_0^k *that drives* G_{obs} *to a state* $q_{obs} = \delta_{obs}(Q_{0,obs}, y_0^k)$. *The sequence* y_0^k *isolates the current state of G to be within the set S,* $S \subseteq Q$, *iff* $q_{obs} \subseteq S$ *and* $q_{obs} \neq \emptyset$.

The above corollary can be applied to many of the settings described in the previous section. For example, if S is the set of secret states, then system G is current-state opaque with respect to S if and only if its observer G_{obs} does not have a reachable state that is a non-empty subset of S (Saboori and Hadjicostis 2007). Similarly, if one is interested in verifying that, regardless of the sequence of events, the resulting sequence of observations allows the observer to exactly determine the current state of system G, at least almost always, then one can equivalently verify that all reachable cycles of states in G_{obs} consist solely of singleton subsets of Q.

Example 5.5 In this example, we briefly illustrate how state isolation relates to opacity and detectability using two different automata. Detectability is discussed in more detail in Chap. 6, whereas notions of opacity are elaborated upon in Chap. 8.

Consider first the finite automaton G shown on the top of Fig. 5.3 and its observer G_{obs} shown at the bottom of the same figure. Let $S_1 = \{3, 5\}$ and $S_2 = \{1, 2\}$ be the sets of states of interest. From G_{obs}, it is relatively easy to make the following inferences:

1. When the sequence of inputs generates the sequence of observations $b(bb)^*$ (namely, when the sequence of inputs is $b(cbb)^*$), one can isolate the current state of G with respect to set S_1 (because under these sequences of observations the observer ends in observer state $\{3, 5\}$).
2. No observation sequence (and thus no input sequence) allows us to isolate the current state to be within the set S_2 (because no state of the observer is contained within S_2). In the terminology of Saboori and Hadjicostis (2011b), one says that system G is *current-state opaque* with respect to S_2.
3. All sequences of inputs generate a corresponding sequence of observations that does not allow us to pinpoint the state of the system with certainty (for each possible sequence of inputs, the corresponding sequence of observations is associated with a set of states of cardinality at least two).

Consider now the system shown at the top of Fig. 5.4 whose observer is shown at the bottom of the same figure (this system is discussed later in this chapter in Example 5.6 in the context of delayed-state estimation). From the observer at the bottom of Fig. 5.4 we can easily reach the following conclusions:

1. The system is *not detectable*. This is because the infinite sequence of observations $a(ba)^*$ (which can be generated by the infinite sequence of events $a((c + \epsilon)ba)^*$) leads to an observer state that has cardinality more than one; note that this is the only problematic state in the observer (as the other states that have cardinality more than one are states that can only be visited once).
2. The system is *periodically detectable*. This is because for all sufficiently long sequences of observations, we will periodically be able to determine the state of the system exactly (all loops in G_{obs} go through state $\{4\}$ and/or state $\{1\}$.

Note that the above inferences regarding state isolation, opacity and detectability properties can be made solely based on G_{obs}. □

Remark 5.2 It is worth mentioning at this point that the use of a current-state estimator is not necessarily the best method for verifying current-state isolation; depending on the underlying system and objectives, it may be possible to verify a current-state isolation problem using more efficient techniques. For instance, the work in Özveren and Willsky (1990), Shu et al. (2007) aims to verify whether the current state of the system will eventually be isolated to be within a set of cardinality one; as it turns out, this property of a system can be verified with polynomial complexity by constructing a product automaton called a *detector* (which is discussed in Chap. 6). Also, note that in Chap. 7 we show that the notion of diagnosability can be verified with polynomial complexity using a related construction called a *verifier*. □

Remark 5.3 In some cases, simply looking at cycles that are present in G_{obs} might not be enough. For instance, if one is interested in verifying that following the occurrence of an unobservable event f, the state of the system can be isolated to be within some set S, then one should keep in mind that some of the cycles that are present in G_{obs} might be "broken" once event f (which is not explicitly present in G_{obs}) occurs. This is the case in the setting of fault diagnosis (where one is interested in identifying that fault f has occurred, once a finite number of events follows the occurrence of f). We revisit this issue in Chap. 7 (in particular, Example 7.5) where we discuss fault diagnosis explicitly. □

5.3 Delayed-State Isolation Using the Delayed-State Estimator

In Sect. 4.5.2 of Chap. 4, we discussed how, given a streaming sequence of observations y_0^k, the state of a DFA G (with outputs and possibly silent transitions) D observations ago can be estimated using a recursive algorithm that keeps track of the last $D+1$ components of the state trajectory $M^{(i+2)}_{y_0^i, Q_0}$, starting from the set of possible initial states and updating the set of possible trajectories by appending the possible current states each time a new observation becomes available. The key to the recursive algorithm in Sect. 4.5.2 was the observation in Corollary 4.4: keeping track of only the last $D+1$ components of state trajectory $M^{(i+2)}_{y_0^i, Q_0}$, does not inhibit our ability to obtain the last $D+1$ components of state trajectory $M^{(i+3)}_{y_0^{i+1}, Q_0}$, once the new observation $y[i+1]$ becomes available.

An alternative to the approach in Sect. 4.5.2 is to use a D-delayed- state estimator G_{Dobs} to accomplish this task. Much like the current-state estimator presented in the previous section, the D-delayed-state estimator is a DFA without outputs, constructed and initialized so that the state it reaches following a sequence of observations y_0^k (generated by underlying activity in G) represents the last $D+1$ components of the state trajectory $M^{(k+2)}_{y_0^k, Q_0}$. Once this property is established, it is easy to verify that the D-delayed-state estimator G_{Dobs} can be used to perform delayed-state estimation. Note that G_{Dobs} has states that can be viewed as subsets of

$$Q^{D+1} \equiv \underbrace{Q \times Q \times \cdots \times Q}_{(D+1)\text{ times}}$$

(where Q is the set of states of G and the product is taken $D+1$ times), and its inputs consist of the outputs of G; its construction is formally described below.

Definition 5.2 (*D-Delayed-State Estimator*) Given an NFA with outputs and (possibly) silent transitions $G = (Q, \Sigma, Y \cup \{\epsilon\}, \delta, \lambda, Q_0)$, its D-delayed-state estimator is the DFA without outputs

$$G_{Dobs} = AC(2^{Q^{D+1}}, Y, \delta_{Dobs}, Q_{0D}) =: (Q_{Dobs}, Y, \delta_{Dobs}, Q_{0D}) ,$$

where

1. The set of states Q_{Dobs} is (the accessible part of) $2^{Q^{D+1}}$ (all subsets of the set $Q^{D+1} \equiv Q \times Q \times \cdots \times Q$, where the product is taken $D+1$ times);
2. For $q_{Dobs} \in 2^{Q^{D+1}}$, $y \in Y$, the mapping $\delta_{Dobs} : 2^{Q^{D+1}} \times Y \mapsto 2^{Q^{D+1}}$ is defined as

$$\delta_{Dobs}(q_{Dobs}, y) = q'_{Dobs} := trim_1(q_{Dobs} \cdot M_y) ,$$

   where $q'_{Dobs} \in 2^{Q^{D+1}}$ and M_y is the induced state mapping under y, which is an output of the NFA G (refer to Definition 4.9);
3. The initial state of G_{Dobs} is given by $Q_{0D} = \{(q_0, q_0, \ldots, q_0) \in Q^{D+1} \mid q_0 \in Q_0\}$, where Q_0 is the set of possible initial states of G (and for ease of notation is assumed to include its unobservable reach);
4. AC denotes the accessible part of the automaton (i.e., the part of the automaton G_{Dobs} that can be reached from its initial state Q_{0D}).

The construction of G_{Dobs} in the above definition essentially considers all subsets of Q^{D+1} as potential states of G_{Dobs}; for each pair of two such subsets (i.e., for each pair of subsets of state $(D+1)$-trajectories) q_{Dobs} and q'_{Dobs}, it adds a transition from q_{Dobs} to q'_{Dobs} under input y (which is an output of automaton G) if and only if the last $D+1$ components of the state $(D+2)$-trajectories that can be reached in automaton G by extending the state $(D+1)$-trajectories in q_{Dobs} while generating output y are exactly captured by the set q'_{Dobs} (note that this ensures that G_{Dobs} is a DFA and that we have a mathematically well-defined structure). After constructing the next-state transition function of G_{Dobs} in this fashion, the states in G_{Dobs} that cannot be reached from the initial state Q_{0D} can be safely ignored, and this is exactly what the operation AC does. It is clear from the above construction that G_{Dobs} has at most $2^{N^{D+1}}$ states because that is the number of distinct subsets of the set Q^{D+1} (where $N = |Q|$). In reality, as we will argue later, the number of states of G_{Dobs} could be significantly smaller than this bound.

Remark 5.4 The empty state trajectory could be a state in G_{Dobs} if there exist sequences of observations that cannot be generated by G. The empty state trajectory is an absorbing state (once reached, G_{Dobs} remains in that state). Typically, one draws G_{Dobs} without including this absorbing state and the transitions leading to it. □

The following example clarifies the construction of G_{Dobs}.

Example 5.6 Consider the automaton $G = (Q, \Sigma, Y \cup \{\epsilon\}, \delta, \lambda, Q_0)$ at the top of Fig. 5.4, which describes a labeled nondeterministic finite automaton (LNFA) with silent transitions under the natural projection mapping. Specifically, we have $Q = \{1, 2, 3, 4\}$, $\Sigma = \{a, b, c\}$, $Y = \{a, b\}$, $Q_0 = \{1, 2, 3, 4\}$, and δ as shown in the figure. The set of observable events is $\Sigma_{obs} = \{a, b\}$ and the set of unobservable events is $\Sigma_{uo} = \{c\}$ (so that the output mapping λ simply implements the natural projection, i.e., $\lambda(a) = a$, $\lambda(b) = b$, $\lambda(c) = \epsilon$). For completeness, the observer G_{obs} for automaton G (which can also be thought of as a 0-delayed- state estimator) is included at the bottom of Fig. 5.4. [For details on the construction of the observer refer to Example 5.3; note that again the state corresponding to the empty set of states $\emptyset$ has not been included for simplicity.]

Figure 5.5 shows the 2-delayed-state estimator G_{2obs} for LNFA G. The top of the figure shows the structure (state transition mechanism) of G_{2obs}, whereas the bottom of the figure shows graphically (using trellis diagrams) the state 3-trajectory corresponding to each state (note that the states of the 2-delayed-state estimator will be denoted as $1_{2obs}, 2_{2obs}, \ldots, 10_{2obs}$ to avoid confusion with states 1, 2, 3, 4 of the finite automaton G). The 2-delayed-state estimator is a DFA $G_{2obs} = (Q_{2obs}, \Sigma_{obs}, \delta_{2obs}, Q_{0,2obs})$ whose

- States are denoted as state 3-trajectories (note that $Q_{2obs} \subseteq 2^{Q \times Q \times Q}$);
- Inputs are the observable outputs of G (namely Σ_{obs});
- Next-state transition function δ_{2obs} is as shown at the top of Fig. 5.5;
- Initial state $Q_{0,2obs} = 1_{2obs}$ is the state 3-trajectory

$$\{(1, 1, 1), (2, 2, 2), (3, 3, 3), (4, 4, 4)\}$$

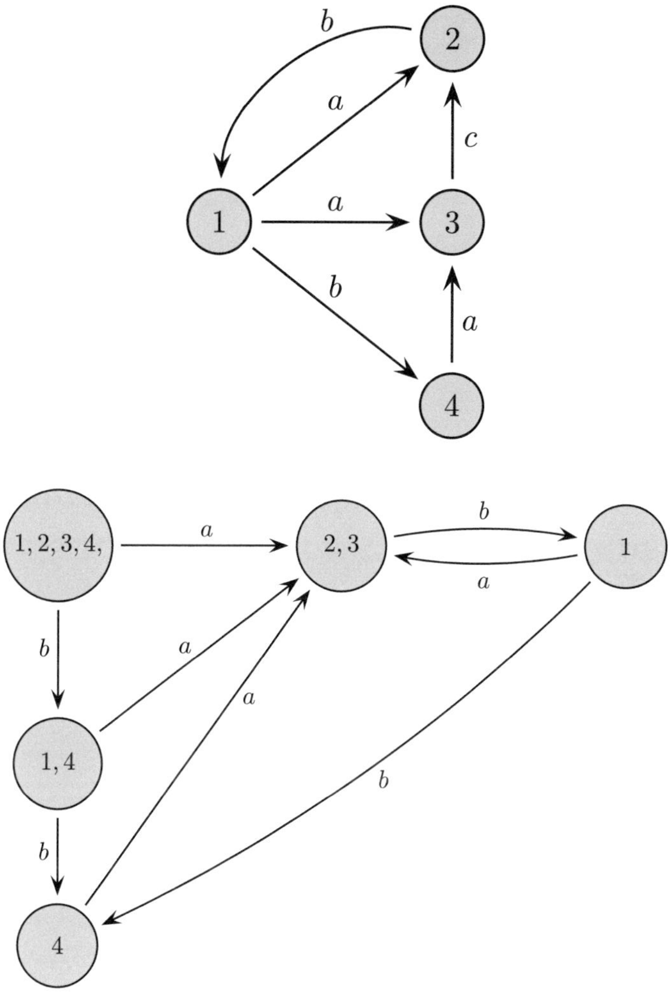

Fig. 5.4 Finite automaton G (top) and its corresponding observer G_{obs} (bottom)

(namely, all states in Q_0, each triplicated in a sequence).

[Note that if the unobservable reach of the initial states of system G included states outside Q_0, we would have to replace Q_0 with its unobservable reach in the above construction.]

To better understand the construction of G_{2obs}, consider the following. Since we know that G starts in any state in $\{1, 2, 3, 4\}$, the initial state of G_{2obs} includes the state 3-trajectories $\{(1, 1, 1), (2, 2, 2), (3, 3, 3), (4, 4, 4)\}$. From this initial state $Q_{0,2obs} = 1_{2obs}$, there are two possible observations, namely a and b. Observation a implies that an event sequence of the form c^*ac^* has occurred, which can take system G to either state 2 (from either state 1 or state 4) or to state 3 (from either state 1 or

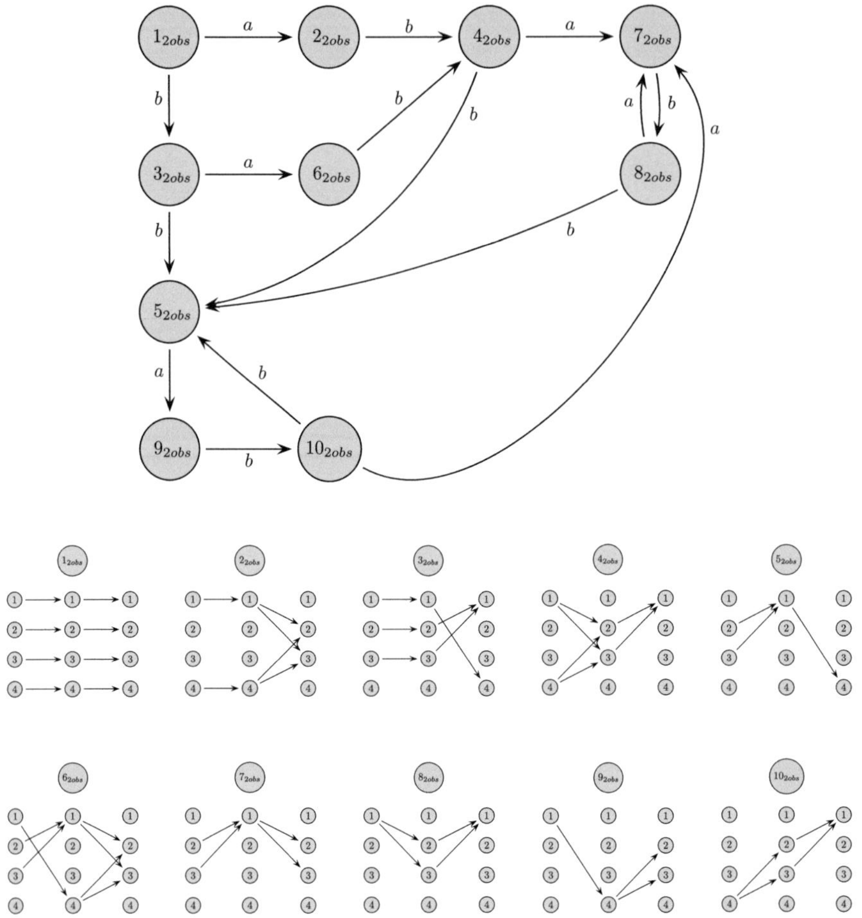

Fig. 5.5 2-delayed-state estimator G_{2obs} for automaton G at the top of Fig. 5.4 (top) and the state 3-trajectories associated with each of its states (bottom)

state 4); this implies the next estimator state under observation a includes the state 3-trajectories $\{(1, 1, 2), (1, 1, 3), (4, 4, 2), (4, 4, 3)\} = 2_{2obs}$. This can be obtained systematically by *concatenating* the induced state mapping corresponding to observation a (given by $M_a = \{(1, 2), (1, 3), (4, 2), (4, 3)\}$) with the state 3-trajectory associated with estimator state 1_{2obs}, and then trimming the starting stage of the resulting state 4-trajectory. Similarly, from the initial estimator state $Q_{0,2obs} = 1_{2obs}$, observation b implies that an event sequence of the form c^*bc^* has occurred, which can take system G to either state 1 (from state 2 or state 3) or to state 4 (from state 1); this implies the next estimator state from $Q_{0,2obs} = 1_{2obs}$ under observation b includes the state 3-trajectories $\{(1, 1, 4), (2, 2, 1), (3, 3, 1)\} = 3_{2obs}$. Again, this can be obtained systematically by *concatenating* the induced state mapping corresponding to observation b (given by $M_b = \{(1, 4), (2, 1), (3, 1)\}$) with the state

3-trajectory associated with estimator state $Q_{0,2obs} = 1_{2obs}$, and then trimming the starting stage of the resulting state 4-trajectory.

For each of the two 2-delayed-state estimator states obtained above, there are two possible observations to consider, namely a and b. For instance, if we consider state 2_{2obs}, we have the following:
(i) Observation a implies that a sequence of inputs of the form c^*ac^* has occurred, which is not possible in system G from either state 2 or state 3 (which are the ending states in the state 3-trajectory associated with estimator state 2_{2obs}). Again, this can be obtained systematically by concatenating the state 3-trajectory associated with observer state 2_{2obs} with the induced state mapping corresponding to observation a, which would result in the empty state 4-trajectory. Transitions that lead to the empty state trajectory are typically not drawn in the estimator diagram and this is what we have done in the figure (an alternative approach would have been to include in the 2-delayed-state estimator an absorbing state, associated with the empty state 3-trajectory $\emptyset$, and have all such transitions lead to this state).
(ii) Observation b implies that a sequence of inputs of the form c^*bc^* has occurred, which can take system G to state 1 (from state 2 or state 3); this implies the next estimator state from 2_{2obs} under observation b includes the state 3-trajectories $\{(1, 2, 1), (1, 3, 1), (4, 2, 1), (4, 3, 1)\} = 4_{2obs}$. Again, this can be obtained systematically by concatenating the state 3-trajectory associated with estimator state 2_{2obs} with the induced state mapping corresponding to observation b (given by $M_b = \{(1, 4), (2, 1), (3, 1)\}$), and then trimming the starting stage of the resulting state 4-trajectory.

The construction of the 2-delayed-state estimator G_{2obs} can be completed by continuing in this fashion. The end result is shown in Fig. 5.5. □

Automaton G_{Dobs} can be utilized to obtain the set of possible delayed-state estimates for automaton G as follows: given a sequence of observations $y_0^k \in Y^*, k \geq 0$, generated by underlying activity in G, the set $\hat{q}_{y[k-d]}(y_0^k)$, $d = 0, 1, \ldots, D$, of possible d-delayed-state estimates of G is given by $\hat{q}_{y[k-d]}(y_0^k) = \Pi_{D-d}(\delta_{Dobs}(Q_{0D}, y_0^k))$. In other words, the set of possible d-delayed-state estimates of G $(d = 0, 1, \ldots, D)$ following the sequence of observations y_0^k is captured by an appropriate projection of the components of the state $(D+1)$-trajectory reached in G_{Dobs} starting from the initial state Q_{0D} and applying y_0^k. Note that, under the convention that $\hat{q}_{y[k-d]}(y_0^k) = \hat{q}_0(y_0^k)$ for $k < d$, one does not need to treat the case $k < D$ separately due to the particular way in which Q_{0D} is chosen. The example below illustrates this idea, whereas the theorem that follows establishes this property formally.

Example 5.7 If we consider the sequence of observations $y_0^4 = abbab$ (i.e., $y[0] = a$, $y[1] = b$, $y[2] = b$, $y[3] = a$, and $y[4] = b$) in the finite automaton G shown on the top of Fig. 5.4, we can easily obtain the corresponding sequence of 2-delayed-state estimates as

$$\begin{aligned}
\hat{q}_{y[-2]}(y_0^0) &= \{1,4\}\\
\hat{q}_{y[-1]}(y_0^1) &= \{1,4\}\\
\hat{q}_{y[0]}(y_0^2) &= \{2,3\}\\
\hat{q}_{y[1]}(y_0^3) &= \{1\}\\
\hat{q}_{y[2]}(y_0^4) &= \{4\}\ .
\end{aligned}$$

This follows simply from the fact that, starting from the initial state $Q_{0,2obs}$, the 2-delayed-state estimator G_{2obs} in Fig. 5.5 under the sequence of inputs (observations) *abbab*, follows the sequence of states 1_{2obs}, 2_{2obs}, 4_{2obs}, 5_{2obs}, 9_{2obs}, and 10_{2obs}. By projecting on the starting stage of the state 3-trajectories associated with these states of the 2-delayed-state estimator, one obtains the 2-delayed-state estimates described above. □

Remark 5.5 Using the 2-delayed-state estimator, one can also perform current-state estimation. For instance, in the example above, the sequence of observations $y_0^4 =$ *abbab* (i.e., $y[0] = a$, $y[1] = b$, $y[2] = b$, $y[3] = a$, and $y[4] = b$) takes us through the estimator states 1_{2obs}, 2_{2obs}, 4_{2obs}, 5_{2obs}, 9_{2obs}, and 10_{2obs}. We can easily obtain the corresponding sequence of current-state estimates by projecting on the last stage of the state 2-trajectories associated with these estimator states:

$$\begin{aligned}
\hat{q}_{\epsilon}(\epsilon) &= \{1,2,3,4\}\\
\hat{q}_{y[0]}(y_0^0) &= \{2,3\}\\
\hat{q}_{y[1]}(y_0^1) &= \{1\}\\
\hat{q}_{y[2]}(y_0^2) &= \{4\}\\
\hat{q}_{y[3]}(y_0^3) &= \{2,3\}\\
\hat{q}_{y[4]}(y_0^4) &= \{1\}\ .
\end{aligned}$$

Not surprisingly, this is in agreement with the sequence of current-state estimates one obtains using the observer G_{obs} at the bottom of Fig. 5.4. This means that if we focus on the last stage of each of the state 2-trajectories associated with states of the 2-delayed (or, more generally, D-delayed) estimator, one obtains a version of the observer G_{obs}, which is likely redundant but equivalent to the one we obtained earlier at the bottom of Fig. 5.4. For instance, the sequence *abab* takes us to state 8_{2obs} which is associated with current-state estimates in the set $\{1\}$. Similarly, in G_{obs} sequence *abab* takes us to state $\{1\}$ (i.e., the current-state estimates are the same). □

Theorem 5.2 *Consider an NFA with outputs and (possibly) silent transitions* $G = (Q, \Sigma, Y \cup \{\epsilon\}, \delta, \lambda, Q_0)$ *and its D-delayed-state estimator* $G_{Dobs} = AC(2^{Q^{D+1}}, Y, \delta_{Dobs}, Q_{0D}) =: (Q_{Dobs}, Y, \delta_{Dobs}, Q_{0D})$ *constructed as in Definition 5.2. For any integer* $k \geq 0$ *and for any* $y_0^k \in Y^*$*, we have*

$$\hat{q}_{y[k-d]}(y_0^k) = \Pi_{D-d}(\delta_{Dobs}(Q_{0D}, y_0^k))\ , d = 0, 1, \ldots, D\ ,$$

with the convention that $\hat{q}_{y[k-d]}(y_0^k) = \hat{q}_0(y_0^k)$ *when* $d > k$.

Proof Note that Q_{0D} is a state $(D+1)$-trajectory and thus one can think of $M^{(i+2+D)}_{y_0^i, Q_{0D}} \equiv Q_{0D} \cdot M^{(i+2)}_{y_0^i}$ as a state $(i+2+D)$-trajectory that also includes the states of the system at the fictitious time instants $-1, -2, \ldots, -D$. With this understanding, we will establish that

$$\delta_{Dobs}(Q_{0D}, y_0^i) = trim_{i+1}\left(M^{(i+2+D)}_{y_0^i, Q_{0D}}\right) ,$$

from which it is clear that $\delta_{Dobs}(Q_{0D}, y_0^i)$ is a state $(D+1)$-trajectory that captures the sequences of the last $D+1$ states that are possible in system G given the observation y_0^i; once this is shown, the theorem statement follows easily.

The proof of the above statement is by induction: for $i=0$, we have (from the definition of δ_{Dobs}) that

$$\delta_{Dobs}(Q_{0D}, y_0^0) = trim_1\left(Q_{0D} \cdot M_{y[0]}\right) .$$

For the induction hypothesis, we assume that

$$\delta_{Dobs}(Q_{0D}, y_0^i) = trim_{i+1}\left(Q_{0D} \cdot M_{y_0^i}\right)$$

and we have to show that

$$\delta_{Dobs}(Q_{0D}, y_0^{i+1}) = trim_{i+2}\left(Q_{0D} \cdot M_{y_0^{i+1}}\right) .$$

The construction of G_{Dobs} ensures that

$$\begin{aligned}
\delta_{Dobs}(Q_{0D}, y_0^{i+1}) &= trim_1\left(\delta_{Dobs}(Q_{0D}, y_0^i) \cdot M_{y[i+1]}\right) \\
&= trim_1\left(trim_{i+1}(Q_{0D} \cdot M_{y_0^i}) \cdot M_{y[i+1]}\right) \\
&= trim_1\left(trim_{i+1}(Q_{0D} \cdot M_{y_0^{i+1}})\right) \\
&= trim_{i+2}\left(Q_{0D} \cdot M_{y_0^{i+1}}\right) ,
\end{aligned}$$

where the second equality follows from the induction hypothesis, the third equality follows from Theorem 4.2, and the last equality follows from the definition of the *trim* operation.

At this point the proof of the theorem is complete. □

Having constructed the D-delayed-state estimator $G_{Dobs} = AC(2^{Q^{D+1}}, Y, \delta_{Dobs}, Q_{0D}) =: (Q_{Dobs}, Y, \delta_{Dobs}, Q_{0D})$ for a given NFA G, it is rather straightforward to verify D-delayed-state isolation with respect to a set of states S: if there is an accessible state $q_{Dobs} \in Q_{Dobs}$ such that $\Pi_0(q_{Dobs}) \subseteq S$, $\Pi_0(q_{Dobs}) \neq \emptyset$, then we have at least one sequence of observations that allows the observer to determine with certainty that the set of possible states of the system D observation steps ago is within the set S. Note that one has to exclude the case when $q_{Dobs} = \emptyset$ (which is equivalent to $\Pi_0(q_{Dobs}) = \emptyset$) because that state is reached only via sequences of observations that cannot be generated by an underlying sequence of events in the given system G; however, all other states in Q_{Dobs} are reached by sequences of observations that can indeed be generated by sequences of events in the given system. The following corollary states this formally.

Corollary 5.2 (Aid for D-Delayed-State Isolation) *Consider an NFA with outputs and (possibly) silent transitions $G = (Q, \Sigma, Y \cup \{\epsilon\}, \delta, \lambda, Q_0)$ and its D-delayed-state estimator $G_{Dobs} = AC(2^{Q^{D+1}}, Y, \delta_{Dobs}, Q_{0D}) =: (Q_{Dobs}, Y, \delta_{Dobs}, Q_{0D})$. Also consider a sequence of events in G that generates a sequence of observations y_0^k that drives G_{Dobs} to state $q_{Dobs} = \delta_{Dobs}(Q_{0D}, y_0^k)$; sequence y_0^k isolates the D-delayed-state estimate of G (i.e., the estimate of the state of the given system D observations ago) to be within the set S, $S \subseteq Q$, iff $\Pi_0(q_{Dobs}) \subseteq S$, $q_{Dobs} \neq \emptyset$.*

The above corollary can be applied to many of the settings described in the previous section. For example, if S is the set of secret states, then system G is D-delayed-state opaque with respect to S if and only if its D-delayed-state estimator G_{Dobs} does not have a reachable (non-empty) state whose zeroth projection (corresponding to the D-delayed-state estimates) is a subset of S (Saboori and Hadjicostis 2011b).

Example 5.8 Consider again the finite automaton G shown on the top of Fig. 5.4. Consider the set of states $S_1 = \{1, 2\}$ and $S_2 = \{3\}$. We see that G allows us to isolate the 2-delayed state with respect to set S_1 (e.g., when the sequence of inputs generates the sequence of observations $abab(ab)^*$, as under these sequences of observations the 2-delayed- state estimator ends in state 8_{2obs}). [Note that we are guaranteed that these sequences of observations can be generated by sequences of events that are possible in system G; in this particular case, $(a(c+\epsilon)ba(c+\epsilon)b(a(c+\epsilon)b)^*$ are possible strings in the system that generate such sequences of observations.] On the contrary, no observation sequence (and thus no input sequence) allows us to isolate the 2-delayed state to be within the set S_2 (because no state of the 2-delayed-state estimator is associated with a state 2-trajectory whose starting stage involves states that are contained within S_2). □

Remark 5.6 As mentioned earlier, the number of states of G_{Dobs} is clearly bounded by $2^{N^{D+1}}$ (since there are that many different state $(D+1)$-trajectories for an automaton with N states). Another way to obtain a (most likely tighter) bound on the number of states of G_{Dobs} is to incorporate the fact that its state $(D+1)$-trajectories are generated according to the R different output symbols in the set Y. Specifically, given a sequence of observations y_0^k, there is a corresponding state $(k+2)$-trajectory; the last $(D+1)$ components of this state trajectory, given by $trim_{k-D+1}(M^{(k+2)}_{y_0^k, Q_{0D}})$, are completely defined if one knows the last D outputs y_{k-D+1}^k and the states $\Pi_{k-D+1}(M^{(k+2)}_{y_0^k, Q_{0D}}) = \Pi_0\left(trim_{k-D+1}(M^{(k+2)}_{y_0^k, Q_{0D}})\right)$, a conclusion that follows easily from Theorem 4.2. Therefore, there are at most $2^N \times R^{D+1}$ different states (state trajectories) that might be reached in G_{Dobs}. □

5.4 Initial-State Isolation Using the Initial-State Estimator

In Sect. 4.5.3 of Chap. 4, we discussed how, given a streaming sequence of observations y_0^k, the initial state of a DFA G (with outputs and possibly silent transitions) can be estimated using a recursive online algorithm. This algorithm keeps track of the pairs of initial and final states that are possible following the sequence of observations y_0^i for $i = 0, 1, 2, \ldots, k$. Specifically, the algorithm in Sect. 4.5.3 tracks the state mapping $M^{(2)}_{y_0^i, Q_0}$, starting with the state mapping $Q_{0I} = \{(q_0, q_0) \mid q_0 \in Q_0\}$ and composing each time the current-state mapping with the mapping induced by the new observation, via the recursion

$$M^{(2)}_{y_0^{i+1}, Q_0} = M^{(2)}_{y_0^i, Q_0} \odot M^{(2)}_{y[i+1]}, \; i = -1, 0, 1, \ldots, k-1,$$

where $M^{(2)}_{y_0^{-1}, Q_0} = Q_{0I} = \{(q_0, q_0) \mid q_0 \in Q_0\}$.

An alternative to the approach in Sect. 4.5.3 is to use an initial-state estimator G_{Iobs} to accomplish this task. Much like the current-state estimator and the D-delayed-state estimator presented in the previous sections, the initial-state estimator is a DFA without outputs, constructed and initialized so that the state it reaches following a sequence of observations y_0^k (generated by underlying activity in G) allows us to recover the possible initial states associated with the state trajectory $M^{(k+2)}_{y_0^k, Q_0}$. Once this property is established, it is easy to verify that the initial-state estimator G_{Iobs} can be used to perform initial-state estimation. Note that G_{Iobs} has states that are state mappings (i.e., subsets of $Q^2 \equiv Q \times Q$ where Q is the set of states of G) and its inputs consist of the outputs of G; its construction is formally described below.

Definition 5.3 (*Initial-State Estimator*) Given an NFA with outputs and (possibly) silent transitions $G = (Q, \Sigma, Y \cup \{\epsilon\}, \delta, \lambda, Q_0)$, its initial-state estimator is the DFA without outputs $G_{Iobs} = AC(2^{Q^2}, Y, \delta_{Iobs}, Q_{0I}) =: (Q_{Iobs}, Y, \delta_{Iobs}, Q_{0I})$, where

1. The set of states Q_{Iobs} is (the accessible part of) 2^{Q^2} (all subsets of the set $Q^2 \equiv Q \times Q$);
2. For $q_{Iobs} \in 2^{Q^2}$, $y \in Y$, the mapping $\delta_{Iobs} : 2^{Q^2} \times Y \mapsto 2^{Q^2}$ is defined as

$$\delta_{Iobs}(q_{Iobs}, y) = q'_{Iobs} := q_{Iobs} \odot M_y \; ,$$

   where $q'_{Iobs} \in 2^{Q^2}$ and M_y is the induced state mapping under input y for NFA G (refer to Definition 4.9);
3. The initial state of G_{Iobs} is given by $Q_{0I} = \{(q_0, q_0) \in Q^2 \mid q_0 \in Q_0\}$, where Q_0 is the set of possible initial states of G;
4. AC denotes the accessible part of the automaton (i.e., the part of the automaton G_{Iobs} that can be reached from its initial state Q_{0I}).

The construction of G_{Iobs} in the above definition essentially considers all subsets of Q^2 as potential states of G_{Iobs}; for each pair of two such subsets (i.e., for each pair of state mappings), q_{Iobs} and q'_{Iobs}, it adds a transition from q_{Iobs} to q'_{Iobs} under input y (which is an output of automaton G) if and only if the pairs of initial and final states (q_i, q_f) in q'_{Iobs} are exactly the ones obtained by taking a pair (q_i, q_{int}) (with the initial state q_i and an intermediate state q_{int}) from q_{Iobs} and a pair (q_{int}, q_f) (with the same intermediate state q_{int} and the final state q_f) from M_y. This should be interpreted that the previous sequence of observations could be explained by a sequence of events (or multiple sequences of events) that start from the initial state q_i and end in the state q_{int}, whereas the last observation could be generated by one or more sequences of events that take us from state q_{int} to the final state q_f. Note that the construction above ensures that G_{Iobs} is a DFA with at most 2^{N^2} states (where $N = |Q|$ is the number of states of the original automaton G). After constructing the next-state transition function of G_{Iobs} in this fashion, the states in G_{Iobs} that cannot be reached from the initial state Q_{0I} can be safely ignored, and this is exactly what the operation AC does.

Remark 5.7 If there exist sequences of observations that cannot possibly be generated by G, then the above construction implies that the empty state mapping could be a state in G_{Iobs}. As in the case of the current-state estimator and the delayed-state estimator, the empty state mapping is an absorbing state (once reached, G_{Iobs} remains in that state). One typically draws G_{Iobs} without including this absorbing state or the transitions leading to it. □

The following example clarifies the construction of G_{Iobs}.

Example 5.9 Consider again the automaton $G = (Q, \Sigma, Y \cup \{\epsilon\}, \delta, \lambda, Q_0)$ at the top of Fig. 5.4, which describes a labeled nondeterministic finite automaton (LNFA) with silent transitions under the natural projection mapping. Specifically, we have $Q = \{1, 2, 3, 4\}$, $\Sigma = \{a, b, c\}$, $Y = \{a, b\}$, $Q_0 = \{1, 2, 3, 4\}$, and δ as shown in the figure. The set of observable events is $\Sigma_{obs} = \{a, b\}$ and the set of unobservable events is $\Sigma_{uo} = \{c\}$ (so that the output mapping λ simply implements the natural projection, i.e., $\lambda(a) = a$, $\lambda(b) = b$, $\lambda(c) = \epsilon$).

Figure 5.6 shows the initial-state estimator G_{Iobs} for finite automaton G. The top of the figure shows the structure (state transition mechanism) of G_{Iobs}, whereas the bottom of the figure shows graphically (using trellis diagrams) the state mappings corresponding to each state (note that the states of the initial-state estimator will be denoted as $1_{Iobs}, 2_{Iobs}, \ldots, 11_{Iobs}$ to avoid confusion with the states 1, 2, 3, 4 of the finite automaton G). The initial-state estimator is a DFA $G_{Iobs} = (Q_{Iobs}, \Sigma_{obs}, \delta_{Iobs}, Q_{0I})$ whose

1. States are denoted as state mappings (note that $Q_{Iobs} \subseteq Q \times Q$);
2. Inputs are the observable outputs of G (namely Σ_{obs});
3. Next-state transition function δ_{Iobs} is as shown at the top of Fig. 5.6;
4. Initial state $Q_{0I} = 1_{Iobs}$ is the state mapping $\{(1, 1), (2, 2), (3, 3), (4, 4)\}$.

To better understand the construction of G_{Iobs}, consider the following. Since we know that G starts in any state in $\{1, 2, 3, 4\}$, the initial state of G_{Iobs} is the state mapping $\{(1, 1), (2, 2), (3, 3), (4, 4)\} = 1_{Iobs}$. From this initial state $Q_{0I} = 1_{Iobs}$,

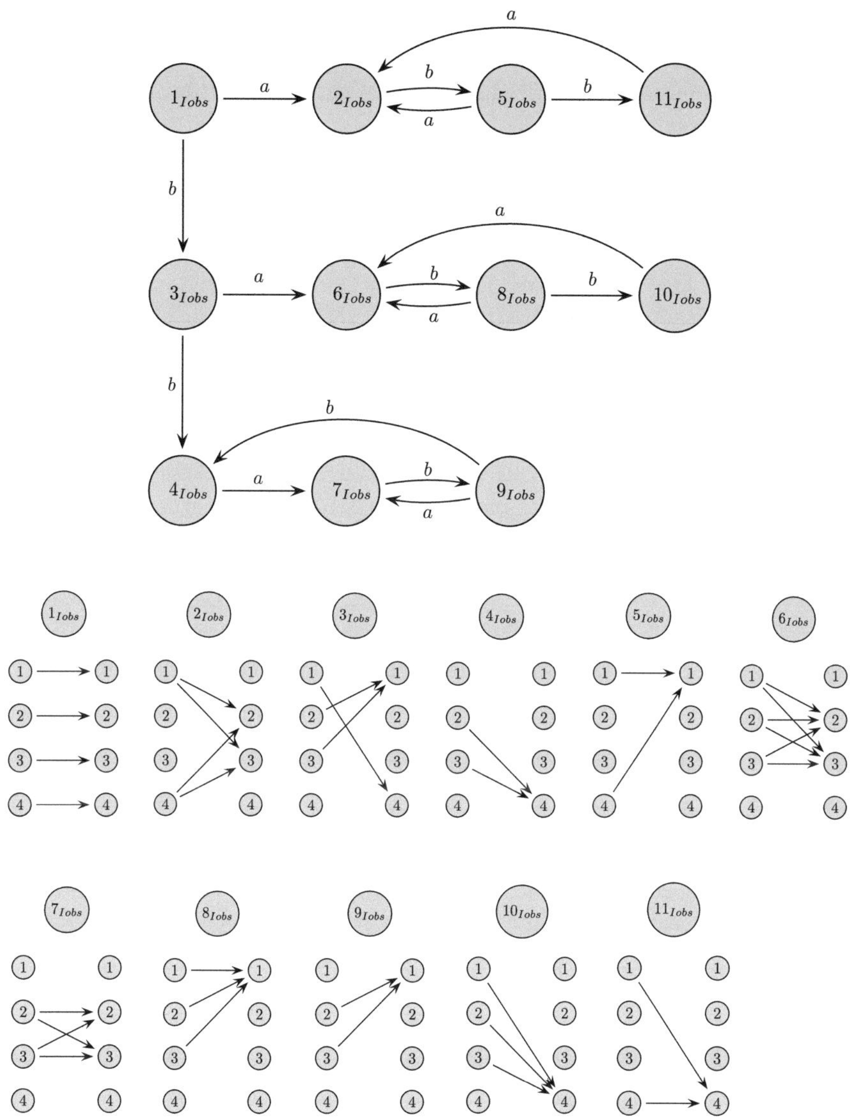

Fig. 5.6 Initial-state estimator G_{Iobs} for automaton G at the top of Fig. 5.4 (top) and state mappings associated with each of its states (bottom)

there are two possible observations, namely a and b. Observation a implies that an event sequence of the form c^*ac^* has occurred, which can take system G to either state 2 (from either state 1 or state 4) or to state 3 (from either state 1 or state 4); this implies that the next state from Q_{Iobs} under observation a is the state mapping $\{(1, 2), (1, 3), (4, 2), (4, 3)\} = 2_{Iobs}$. This can be obtained systematically by *composing* the induced state mapping corresponding to observation a (given by $M_a = \{(1, 2), (1, 3), (4, 2), (4, 3)\}$) with the state mapping associated with state 1_{Iobs}. Similarly, from the initial state $Q_{0I} = 1_{Iobs}$, observation b implies that an event sequence of the form c^*bc^* has occurred, which can take system G to either state 1 (from state 2 or state 3) or to state 4 (from state 1); this implies that the next state from $Q_{0I} = 1_{Iobs}$ under observation b is the state mapping $\{(1, 4), (2, 1), (3, 1)\} = 3_{Iobs}$. This can be obtained systematically by *composing* the induced state mapping corresponding to observation b (given by $M_b = \{(1, 4), (2, 1), (3, 1)\}$) with the state mapping associated with state $Q_{0I} = 1_{Iobs}$.

For each of the two initial-state estimator states obtained above, there are two possible observations to consider, namely a and b. For instance, if we consider state 2_{Iobs}, we have the following:

(i) Observation a implies that a sequence of inputs of the form c^*ac^* has occurred, which is not possible in system G from either state 2 or state 3 (which are the ending states in the state mapping associated with estimator state 2_{Iobs}). In such cases, these transitions are typically not drawn in the estimator diagram and this is what we have done in the figure (an alternative approach would have been to include in the initial-state estimator an absorbing state, associated with the empty state mapping $\emptyset$, and have all such transitions lead to this state).

(ii) Observation b implies that a sequence of inputs of the form c^*bc^* has occurred, which can take system G to state 1 (from state 2 or state 3); this implies the next estimator state from 2_{Iobs} under observation b is the state mapping $\{(1, 1), (4, 1)\} = 5_{Iobs}$. Again, this can be obtained systematically by *composing* the induced state mapping corresponding to observation b (given by $M_b = \{(1, 4), (2, 1), (3, 1)\}$) with the state mapping associated with estimator state 2_{Iobs}.

The construction of the initial-state estimator G_{Iobs} can be completed by continuing in this fashion. The end result is shown in Fig. 5.6. □

The initial-state estimator G_{Iobs} can be utilized to obtain the set of possible initial states of automaton G as follows: given a sequence of observations $y_0^k \in Y^*, k \geq 0$, generated by underlying activity in G, the set $\hat{q}_0(y_0^k)$ of possible initial-state estimates of G is given by $\hat{q}_0(y_0^k) = \Pi_0(\delta_{Iobs}(Q_{0I}, y_0^k))$. In other words, the set of possible initial-state estimates of G following the sequence of observations y_0^k is captured by the initial states associated with the pairs in the state mapping reached in G_{Iobs}, starting from the initial state Q_{0I} and applying y_0^k. The example below illustrates this idea, whereas the theorem that follows establishes this property formally.

Example 5.10 If we consider the sequence of observations $y_0^4 = abbab$ (i.e., $y[0] = a$, $y[1] = b$, $y[2] = b$, $y[3] = a$, and $y[4] = b$) in the finite automaton G shown on the top of Fig. 5.4, one can easily obtain the corresponding sequence of initial-state estimates as

$$\begin{aligned}\hat{q}_0(y_0^0) &= \{1,4\}\\ \hat{q}_0(y_0^1) &= \{1,4\}\\ \hat{q}_0(y_0^2) &= \{1,4\}\\ \hat{q}_0(y_0^3) &= \{1,4\}\\ \hat{q}_0(y_0^4) &= \{1,4\}\ .\end{aligned}$$

This follows simply from the fact that, starting from the initial state Q_{0I}, the initial-state estimator G_{Iobs} in Fig. 5.6 under the sequence of inputs (observations) *abbab* follows the sequence of states $1_{Iobs}, 2_{Iobs}, 5_{Iobs}, 11_{Iobs}, 2_{Iobs}$, and 5_{Iobs}. By projecting on the starting stage of the state mappings associated with these states of the initial-state estimator, one obtains the initial-state estimates described above. □

Remark 5.8 As in the case of the 2-delayed-state estimator, one can also use the initial-state estimator to perform current-state estimation. For example, if we consider the sequence of observations $y_0^4 = abbab$ (i.e., $y[0] = a$, $y[1] = b$, $y[2] = b$, $y[3] = a$, and $y[4] = b$) in the finite automaton G shown on the top of Fig. 5.4, we can obtain the corresponding sequence of current-state estimates by projecting on the ending stage of the state mappings associated with the states 1_{Iobs}, 2_{Iobs}, 5_{Iobs}, 11_{Iobs}, 2_{Iobs}, and 5_{Iobs} that are visited in the initial-state estimator:

$$\begin{aligned}\hat{q}_\epsilon(\epsilon) &= \{1,2,3,4\}\\ \hat{q}_{y[0]}(y_0^0) &= \{2,3\}\\ \hat{q}_{y[1]}(y_0^1) &= \{1\}\\ \hat{q}_{y[2]}(y_0^2) &= \{4\}\\ \hat{q}_{y[3]}(y_0^3) &= \{2,3\}\\ \hat{q}_{y[4]}(y_0^4) &= \{1\}\ .\end{aligned}$$

Not surprisingly, this is in agreement with the sequence of current-state estimates one obtains if one uses the observer G_{obs} at the bottom of Fig. 5.4. This means that if we focus on the ending stage of each of the state mappings associated with states of the initial-state estimator, we obtain a version of the observer G_{obs}, which is likely redundant but equivalent to the one in Fig. 5.4. □

Theorem 5.3 *Consider an NFA with outputs and (possibly) silent transitions $G = (Q, \Sigma, Y \cup \{\epsilon\}, \delta, \lambda, Q_0)$ and its initial-state estimator $G_{Iobs} = AC(2^{Q^2}, Y, \delta_{Iobs}, Q_{0I}) =: (Q_{Iobs}, Y, \delta_{Iobs}, Q_{0I})$ as constructed in Definition 5.3. For any integer $k \geq 0$ and for any $y_0^k \in Y^*$, we have*

$$\hat{q}_0(y_0^k) = \Pi_0(\delta_{Iobs}(Q_{0I}, y_0^k))\ .$$

Proof We will establish that

$$\delta_{Iobs}(Q_{0I}, y_0^i) = M^{(2)}_{y_0^i, Q_0}\ ,$$

where $M^{(2)}_{y_0^i, Q_0} = \{(q_i, q_f) \in M_{y_0^i} \mid q_i \in Q_0\}$; from this, the proof of the theorem follows easily.

The proof of the above statement is by induction: for $i = 0$, we have from the definition of δ_{Iobs} that

$$\delta_{Iobs}(Q_{0I}, y_0^0) = Q_{0I} \odot M_{y[0]} = M_{y_0^0, Q_0} \ .$$

From the induction hypothesis, we assume that

$$\delta_{Iobs}(Q_{0I}, y_0^i) = M^{(2)}_{y_0^i, Q_0}$$

and we have to show that

$$\delta_{Iobs}(Q_{0I}, y_0^{i+1}) = M^{(2)}_{y_0^{i+1}, Q_0} \ .$$

The construction of G_{Iobs} ensures that

$$\begin{aligned}\delta_{Iobs}(Q_{0I}, y_0^{i+1}) &= \delta_{Iobs}(Q_{0I}, y_0^i) \odot M_{y[i+1]} \\ &= M_{y_0^i, Q_0} \odot M_{y[i+1]} \\ &= M_{y_0^{i+1}, Q_0} \ ,\end{aligned}$$

where the second equality follows from the induction hypothesis and the third equality follows from the discussion in Sect. 4.5.3. □

Having constructed the initial-state estimator $G_{Iobs} = AC(2^{Q^2}, Y, \delta_{Iobs}, Q_{0I}) =: (Q_{Iobs}, Y, \delta_{Iobs}, Q_{0I})$ for a given NFA G, it is rather straightforward to verify initial-state isolation with respect to a set of states S: if there is an accessible state $q_{Iobs} \in Q_{Iobs}$ such that $\Pi_0(q_{Iobs}) \subseteq S$, $\Pi_0(q_{Iobs}) \neq \emptyset$, then we have a sequence of observations that allows the observer to determine with certainty that the set of possible initial states of the system is within the set S. Note that one has to exclude the case when $q_{Iobs} = \emptyset$ (which is equivalent to $\Pi_0(q_{Iobs}) = \emptyset$) because that observer state is reached only via sequences of observations that cannot be generated by an underlying sequence of events in the given system G; however, all other states in Q_{Iobs} are reached by sequences of observations that can indeed be generated by sequences of events in the given system. The following corollary states this formally.

Corollary 5.3 (Aid for Initial-State Isolation) *We consider an NFA with outputs and (possibly) silent transitions $G = (Q, \Sigma, Y \cup \{\epsilon\}, \delta, \lambda, Q_0)$ and initial-state estimator $G_{Iobs} = AC(2^{Q^2}, Y, \delta_{Iobs}, Q_{0I}) =: (Q_{Iobs}, Y, \delta_{Iobs}, Q_{0I})$. Also consider a sequence of events in G that generates a sequence of observations y_0^k that drives G_{Iobs} to a state $q_{Iobs} = \delta_{Iobs}(Q_{0I}, y_0^k)$; sequence y_0^k isolates the initial state of G to be within the set S, $S \subseteq Q$, iff $\Pi_0(q_{Iobs}) \subseteq S$, $q_{Iobs} \neq \emptyset$.*

The above corollary can be applied to many of the settings described in the previous section. For example, if S is the set of secret states, then system G is initial-state opaque with respect to S if and only if its initial-state estimator G_{Iobs} does not have a reachable (non-empty) state whose 0th projection (corresponding to the initial-state estimates) is a subset of S (Saboori and Hadjicostis 2009).

Example 5.11 Consider again the finite automaton G shown at the top of Fig. 5.4 and the sets of states $S_1 = \{2, 3\}$ and $S_2 = \{1, 2\}$. We see that G allows us to isolate the initial state with respect to set S_1 (e.g., when the sequence of inputs generates the sequence of observations $bb(ab(ab)^*b)^*$ as under these sequences of observations the initial-state estimator ends in state 4_{Iobs} (whose state mapping has starting states 2 and 3). On the contrary, no observation sequence (and thus no input sequence) allows us to isolate the initial state to be within the set S_2 (because no state of the initial-state estimator is associated with a state mapping whose starting states are contained within S_2). □

Remark 5.9 The number of states of G_{Iobs} is clearly bounded by 2^{N^2} (since there are that many different state mappings for an automaton with N states). A tighter bound on the number of states of G_{Iobs} can be obtained by realizing that the initial set of states Q_0 might be significantly smaller than Q; since the set of possible initial states is always going to be a subset of Q_0, we obtain the bound $2^{|Q_0|\times N}$, because the pairs (q_i, q_f) in the reachable state mappings can only involve a state $q_i \in Q_0$ (the final state q_f can, of course, be any state in Q). □

Remark 5.10 The above corollary gives necessary and sufficient conditions for the verification of initial-state isolation properties using the initial-state estimator. However, the verification of initial-state isolation does not necessarily have to be done using an initial-state estimator. If one is interested in only verifying initial-state isolation with respect to a specific set S, other methods may be more efficient. For example, instead of using state mappings induced by a sequence of observations y_0^k, one can use state-status mappings: given the sequence of observations y_0^k, the corresponding state-status mapping SM associates with each state $q \in Q$ a label from the set $\{\mathcal{N}, \mathcal{S}, \mathcal{M}, \mathcal{X}\}$; this label indicates that state q can be reached (via a sequence of events that generates the sequence of observations y_0^k) only from states in the set $Q \backslash S$ (label $\mathcal{N}$), or q can be reached only from states in the set S (label $\mathcal{S}$), or q can be reached from states from both sets S and $Q \backslash S$ (label $\mathcal{M}$), or q cannot be reached at all (label $\mathcal{X}$). In other words, the state-status mapping is a set of N pairs of the form $\{(label_i, q^{(i)}) \mid q^{(i)} \in Q\}$ where $label_i \in \{\mathcal{N}, \mathcal{S}, \mathcal{M}, \mathcal{X}\}$. Once a new observation $y[k+1]$ is obtained, the state-status mapping for the sequence of observations y_0^{k+1}, which will be denoted by $SM_{y_0^{k+1}}$, can be obtained recursively using the state-status mapping associated with the sequence of observations y_0^k, denoted by $SM_{y_0^k}$, and the state mapping $M_{y[k+1]}$ induced by the new observation $y[k+1]$ as follows: for each possible ending state $q_f \in Q$,

1. Find the set $Q_{starting}(q_f) = \{q_i \mid (q_i, q_f) \in M_{y[k+1]}\}$ which represents all states from which q_f can be reached while producing output $y[k+1]$;
2. Consider the set of labels $Q_{label}(q_f) = \{label_i \mid (label_i, q_i) \in SM_{y_0^k},\ q_i \in Q_{starting}(q_f)\}$ which represents the set of labels of the states via which q_f can be reached;
3. Assign a label to state q_f according to the following rules:

- If $Q_{label}(q_f) = \{\mathcal{N}\}$ or $Q_{label}(q_f) = \{\mathcal{N}, \mathcal{X}\}$, then set the label of q_f in the state-status mapping $SM_{y_0^{k+1}}$ to $\mathcal{N}$;
- If $Q_{label}(q_f) = \{\mathcal{S}\}$ or $Q_{label}(q_f) = \{\mathcal{S}, \mathcal{X}\}$, then set the label of q_f in the state-status mapping $SM_{y_0^{k+1}}$ to $\mathcal{S}$;
- If $Q_{label}(q_f) = \emptyset$ (because q_f is not a final state in $M_{y[k+1]}$) or if $Q_{label}(q_f) = \{\mathcal{X}\}$, then set the label of q_f in the state-status mapping $SM_{y_0^{k+1}}$ to $\mathcal{X}$;
- Otherwise, set the label of q_f in the state-status mapping $SM_{y_0^{i+1}}$ to $\mathcal{M}$.

The above method ensures the proper propagation of labels: for example, if two or more trajectories with different labels from the set $\{\mathcal{N}, \mathcal{S}, \mathcal{M}\}$ merge at the same final state, then that state receives label $\mathcal{M}$; similarly, final states that cannot be reached at all receive label $\mathcal{X}$.

The use of state-status mappings simplifies the verification of initial-state isolation with respect to the set S because the number of possible states (different state-status mappings) is reduced from 2^{N^2} (or, more accurately, $2^{|Q_0| \times N}$) to 4^N where N is the number of states of the original automaton. More details about state-status mappings, and their construction and manipulation, can be found in Saboori and Hadjicostis (2009). □

5.5 Comments and Further Reading

In this chapter, we have studied the verification of state isolation properties using standard observer constructions (current, delayed, or initial-state estimators). More generally, beyond the verification of properties related to state isolation, these observers can be used to check properties that relate to the estimates/inferences an observer can make, based on observations that are generated by underlying activity in a given system. The main disadvantage of these standard observer constructions is their exponential complexity (with respect to the size of the underlying finite automaton); their main advantage is that they are universal in terms of the ability to verify state isolation properties.

Though state isolation itself has not been studied heavily, certain problems that relate to it have been heavily studied by the DES research community. For example, fault diagnosis (discussed in Chap. 7) has attracted the attention of many researchers, whereas opacity analysis and enforcement (see Chap. 8) recently emerged as a DES topic of interest. Also related is the notion of detectability, which is discussed in Chap. 6. We will see that in certain cases, properties of interest can be verified with less complexity, without resorting to the construction of an observer. These discussions, as well as relevant references, are provided in the corresponding chapters.

References

Bhattacharyya A (1983) On a novel approach of fault detection in an easily testable sequential machine with extra inputs and extra outputs. IEEE Trans Comput 32(3):323–325
Caines PE, Greiner R, Wang S (1988) Dynamical logic observers for finite automata. In: Proceedings of IEEE conference on decision and control (CDC), pp 226–233
Caines P, Greiner R, Wang S (1991) Classical and logic-based dynamic observers for finite automata. IMA J Math Control Inf 8(1):45–80
Chow TS (1978) Testing software design modeled by finite-state machines. IEEE Trans Softw Eng 4(3):178–187
Focardi R, Gorrieri R (1994) A taxonomy of trace–based security properties for CCS. In: Proceedings of the 7th workshop on computer security foundations, pp 126–136
Fujiwara H, Kinoshita K (1978) On the complexity of system diagnosis. IEEE Trans Comput 27(10):881–885
Gönenc G (1970) A method for the design of fault detection experiments. IEEE Trans Comput 19(6):551–558
Hennie FC (1964) Fault detecting experiments for sequential circuits. In: Proceedings of the 5th annual symposium on switching circuit theory and logical design, pp 95–110
Hennie FC (1968) Finite state models for logical machines. Wiley, New York
Hsieh EP (1971) Checking experiments for sequential machines. IEEE Trans Comput 20(10):1152–1166
Kime CR (1966) An organization for checking experiments on sequential circuits. IEEE Trans Electron Comput 15(1):113–115
Kohavi Z, Lavalee P (1967) Design of sequential machines with fault detection capabilities. IEEE Trans Electron Comput 16(4):473–484
Moore EF (1956) Gedanken-experiments on sequential machines. Automata studies, vol 34. Annual mathematical studies. Princeton University Press, Princeton, pp 129–153
Murakami SI, Kinoshita K, Ozaki Z (1970) Sequential machines capable of fault diagnosis. IEEE Trans Comput 19(11):1079–1085
Özveren CM, Willsky AS (1990) Observability of discrete event dynamic systems. IEEE Trans Autom Control 35(7):797–806
Pradhan DK (1983) Sequential network design using extra inputs for fault detection. IEEE Trans Comput 32(3):319–323
Saboori A, Hadjicostis CN (2007) Notions of security and opacity in discrete event systems. In: Proceedings of 46th IEEE conference on decision and control (CDC), pp 5056–5061
Saboori A, Hadjicostis CN (2009) Verification of infinite-step opacity and analysis of its complexity. In: Proceedings of dependable control of discrete systems (DCDS), vol 2, pp 46–51
Saboori A, Hadjicostis CN (2011a) Coverage analysis of mobile agent trajectory via state-based opacity formulations. Control Eng Pract 19(9):967–977
Saboori A, Hadjicostis CN (2011b) Verification of K-step opacity and analysis of its complexity. IEEE Trans Autom Sci Eng 8(3):549–559
Saboori A, Hadjicostis CN (2012) Verification of infinite-step opacity and complexity considerations. IEEE Trans Autom Control 57(5):1265–1269
Sampath M, Sengupta R, Lafortune S, Sinnamohideen K, Teneketzis D (1995) Diagnosability of discrete-event systems. IEEE Transn Autom Control 40(9):1555–1575
Sheppart DA, Vranesic ZG (1974) Fault detection of binary sequential machines. IEEE Trans Comput 23(4):352–358
Shu S, Lin F, Ying H (2007) Detectability of discrete event systems. IEEE Trans Autom Control 52(12):2356–2359
Vasilevskii MP (1973) Failure diagnosis in automata. Kybernetika 9(4):98–108
Yannakakis M, Lee D (1994) Testing finite-state machines: state identification and verification. IEEE Trans Comput 43(3):209–227

Chapter 6
Detectability

6.1 Introduction and Motivation

As mentioned in Chap. 5, there are many variations of the notion of *detectability*. For example, Shu et al. (2007) asks whether an arbitrary sequence of observations, generated by an underlying known finite automaton will allow an external observer to eventually (after a finite number of observations) determine exactly the (current) state of the system. More specifically, *strong detectability* (or simply *detectability*) requires that, regardless of the actual activity that may occur in the system and after a finite number of observations, the external observer is guaranteed to be able to determine *exactly* the current state of the system as well as all future (subsequent) states (by analyzing the sequence of observations that is generated). Equivalently, one can ask whether there exists at least one sequence of observations (of unbounded length) that does *not* allow the current state of the system to be isolated eventually within singleton sets (even if we ignore any uncertainty at the beginning of the observation sequence). Clearly, if an observer construction is available for the given finite automaton, strong detectability can be verified by checking for problematic observer states, i.e., observer states that are associated with sets of system states of cardinality more than unity. In particular, the presence of a cycle of observer states that involves at least one such problematic state would indicate that the system is not detectable (an illustration of how detectability can be analyzed and verified by constructing a current-state estimator was discussed in Example 5.5 in Chap. 5). This approach would require polynomial (linear) complexity on the size of the observer; however, the observer will, in general, have exponential complexity in the size of the given finite automaton.

In this chapter, we define and discuss the verification of several variations of detectability for a given finite automaton, generally nondeterministic, with outputs and possibly silent transitions. We discuss how to verify detectability notions using observer constructions, but we also elaborate on more efficient *detector* constructions that allow us to verify detectability with polynomial complexity. Most of this

C. N. Hadjicostis, *Estimation and Inference in Discrete Event Systems*,
Communications and Control Engineering,
https://doi.org/10.1007/978-3-030-30821-6_6

chapter focuses on detectability notions that concern the current state of the system. However, our analysis can be extended to D-delayed-state detectability and initial-state detectability in a straightforward manner, using respectively D-delayed-state estimators (or detectors) and initial-state estimators (or detectors). We choose not to address these cases in detail to avoid cluttering the presentation. It is worth pointing out that most of the existing literature on detectability focuses on the ability to determine the current state of a system exactly or within a set of at most K possible states. The notions of D-delayed- and initial-state detectability have also been addressed more recently (e.g., Hadjicostis 2012; Shu and Lin 2012).

This chapter also discusses the traditional notions of distinguishing sequences, synchronizing sequences, and unique input–output (UIO) sequences that have appeared in the context of testing of digital circuits (and were briefly mentioned in Chap. 5). These types of sequences are related to (current-state or initial-state) detectability in the sense that they aim at determining exactly the (current or initial) state of the system by choosing the input sequence and by observing the output sequence that is generated. One big difference from the discussions in Chap. 5 (as well as the discussion on detectability in this chapter) is that distinguishing/synchronizing sequences allow the user to not only observe the outputs, but also choose the inputs in a way that aids the identification of the current or initial state. This should be contrasted with the discussions in Chap. 5 where the external observer had no control over the input to the system (even if we consider the case where the observer is aware of the input as clarified in Remark 4.1, the ability to apply a specific input was absent in Chap. 5—and will also be absent in the discussions on detectability in this chapter).

Note that the existence of distinguishing, synchronizing, and UIO sequences imply that certain detectability-like properties may hold for some sequences of inputs, but not necessarily all. For example, a synchronizing sequence is a sequence of inputs that, when applied, allows the observer to determine the current state of the system exactly (based on the sequence of outputs that is generated); in fact, an adaptive synchronizing sequence allows the user (which is simultaneously the observer and the controller) to choose the input adaptively based on the sequence of observations seen thus far, so as to eventually reach a point where the current state of the system is known exactly. However, the existence of an (adaptive) synchronizing sequence does not imply detectability, which is a requirement that concerns all possible behavior in the system.

6.2 Notions of Detectability

In this section, we describe different notions of detectability in finite automata, which have received attention recently (see, for example, Shu et al. 2007). In its most basic form, detectability considers the following question: if an observer is willing to wait for an arbitrarily large (but finite) number of events to take place in the given finite automaton, will this observer be able to eventually determine and maintain *exact* knowledge of the current state of the system based on the sequence of observations that has been generated? If that is the case for *all* possible behaviors, the system is called *strongly detectable* (or sometimes simply detectable). Related notions of eventually being able to pinpoint exactly the state of the system were first studied in Caines et al. (1988, 1991), Özveren and Willsky (1990) under different nomenclature.

We will be interested in four notions of detectability,[1] which are described informally below (and more formally later in this chapter) for an underlying system that is captured by a nondeterministic finite automaton (NFA) G with outputs and possibly silent transitions. In the informal descriptions below, we consider arbitrarily long sequences of observations, denoted by y, that can be generated by activity in the given system. For simplicity, in our discussion we assume that: (i) from each state of the underlying system there exists at least one defined transition, which implies that each sequence of events can be extended indefinitely, and (ii) there are no *unobservable cycles*, i.e., cycles of events that do not generate any observation.[2] Note that the two assumptions together imply that any sequence of observations can be extended arbitrarily.

Strong Detectability or Detectability: Given the sequence of observations y, the external observer needs to *almost always* be able to determine *exactly* (i.e., uniquely) the current state of the system. Almost always in this context means that, as the length of y goes to infinity, the fraction of time instances at which the observer does *not* know the state of the system exactly goes to zero. For (strong) detectability to hold true, this property should hold for *all* infinitely extensible sequences of observations y that can be generated by the system.

Weak detectability: Again, given the sequence of observations y, the external observer needs to *almost always* be able to determine *exactly* (i.e., uniquely) the current state of the system. For weak detectability to hold true, this property should hold for *at least one* infinitely extensible sequence of observations y that can be generated by the system.

(Strong) Periodic Detectability: Given the sequence of observations y, the external observer needs to *periodically* be able to determine *exactly* (i.e., uniquely) the current state of the system. Periodically in this context means that the number of observations between instances of time when the observer can determine *exactly* (i.e., uniquely) the state of the system is no larger than a bound p. For (strong) periodic detectability to hold true, we need to be able to find a finite value of p for which this property holds for *all* infinitely extensible sequences of observations y that can be generated by the system.

Weak Periodic Detectability: Again, given the sequence of observations y, the external observer needs to *periodically* be able to determine *exactly* (i.e., uniquely) the current state of the system. For weak periodic detectability to hold true, this

---

[1]Our definitions are slightly different from Shu et al. (2007) to avoid having to deal with issues associated with the transient behavior of the system G.

[2]The presence of cycles of events that do not generate any observations would imply that it is possible for the system to follow an arbitrarily long sequence of events without generating any observation.

property should hold for *at least one* infinitely extensible sequence of observations y that can be generated by the system.

6.2.1 Detectability

In this section, we formulate the detectability problem assuming that the underlying system is a nondeterministic finite automaton (NFA) with outputs and (possibly) silent transitions $G = (Q, \Sigma, Y \cup \{\epsilon\}, \delta, \lambda, Q_0)$ (this is the most general case of the observation models discussed in Chap. 3—refer to Definition 3.24).

We start the section with an example to illustrate important properties that may or may not hold following a long sequence of observations generated by underlying activity in the given automaton. We then formalize the various notions of detectability.

Example 6.1 Consider $G = (Q, \Sigma, Y \cup \{\epsilon\}, \delta, \lambda, Q_0)$ in Fig. 6.1, which describes an labeled *nondeterministic* finite automaton (LNFA) with silent transitions under the natural projection mapping. Specifically, we have $Q = \{1, 2, 3, 4, \ldots, 11, 12\}$, $\Sigma = \{a, b, c, d\}$, $Y = \{a, b, c\}$, $Q_0 = \{1\}$, and δ as shown in the figure. The set of observable events is $\Sigma_{obs} = \{a, b, c\}$ and the set of unobservable events is $\Sigma_{uo} = \{d\}$ (so that the output mapping λ simply implements the natural projection $P_{\Sigma_{obs}}$ (also denoted by P below), i.e., $\lambda(a) = a$, $\lambda(b) = b$, $\lambda(c) = c$, and $\lambda(d) = \epsilon$). [The observer G_{obs} for automaton G can be found in Fig. 6.2 later in this chapter when we discuss the verification of detectability.]

The language of automaton G contains the following infinitely extensible sequences of events where n is any nonnegative integer (below, we also indicate the corresponding sequence of observations):

$$\begin{aligned} adc^n &\text{ with } P(adc^n) = ac^n \\ ac^n &\text{ with } P(ac^n) = ac^n \\ dc^n &\text{ with } P(dc^n) = c^n \\ b(aca)^n &\text{ with } P(b(aca)^n) = b(aca)^n . \end{aligned}$$

It is not hard to realize that if we observe ac^n (i.e., either adc^n or ac^n occur), we will *not* be able to determine exactly the state of the system, as states 3 and 5 are both possible (for any $n \geq 1$). Based on the sequence of observations $y = ac^*$, we can conclude that G is neither strongly detectable nor strongly periodically detectable.

On the other hand, if we observe c^n (i.e., dc^n occurs), we will know that the system is in state 7 (for any $n \geq 1$); thus, based on the sequence of observations $y = c^*$, we can conclude that the system *is* weakly detectable (and also weakly periodically detectable).

Finally, if we consider the sequence $b(aca)^n$, we realize that we will perfectly know the state of the system (namely, state 8) for any $n \geq 0$. However, in between, we will not know the state of the system exactly: (i) when we observe $b(aca)^*a$, both states 9 and 11 are possible, whereas (ii) when we observe $b(aca)^*ac$, both states

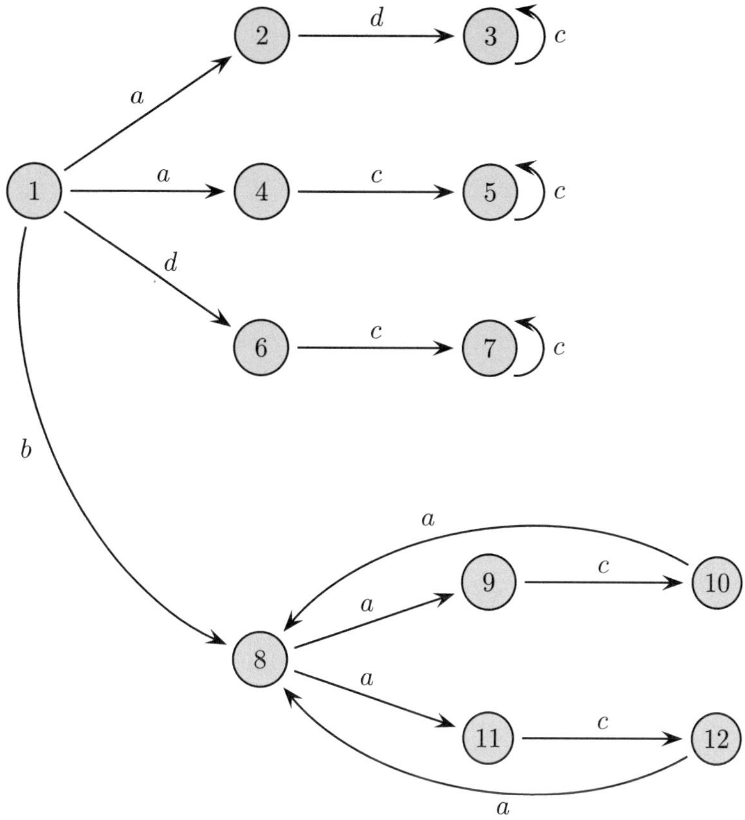

Fig. 6.1 Labeled nondeterministic finite automaton discussed in Example 6.1

10 and 12 are possible. Based on the sequence of observations $y = b(aca)^*$, we can conclude that the system is weakly periodically detectable.

To determine if the system is (strongly) periodically detectable, we need to check whether all possible sequences of events allow the observer to pinpoint the exact state of the system periodically; since this is not the case for sequence ac^n, we can conclude that the system is not (strongly) periodically detectable. □

In the remainder of this section we formulate the various notions of detectability. We assume that we are given an NFA with outputs and (possibly) silent transitions $G = (Q, \Sigma, Y \cup \{\epsilon\}, \delta, \lambda, Q_0)$, as described in Definition 3.24 in Chap. 3. To simplify the development and the notation, we will make the following (rather common) assumptions for automaton G:

1. Liveness: Automaton G is live, i.e., for each state $q \in Q$, there exists $\sigma \in \Sigma$ such that $\delta(q, \sigma)$ is non-empty.
2. Absence of unobservable cycles: There do not exist cycles of events in G that do not generate any observation, i.e., for all $q \in Q$ and all $\sigma_k^{k+m} \in \Sigma^*$, $m \geq 0$, if

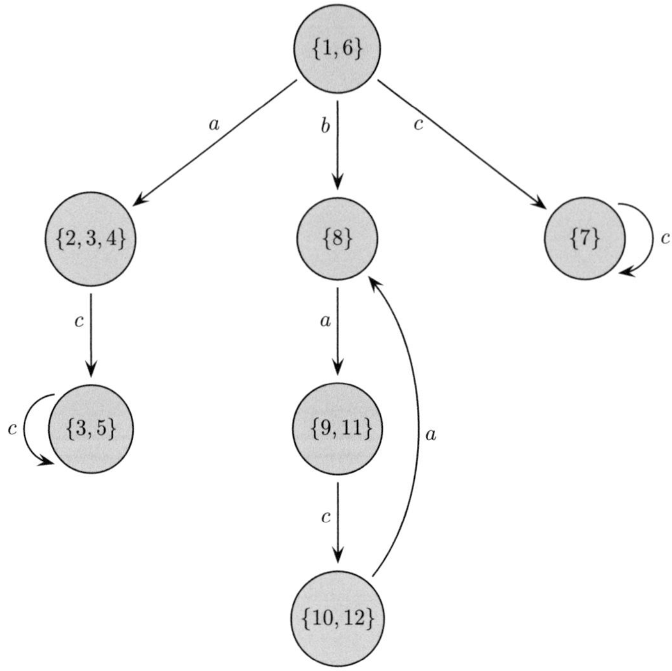

Fig. 6.2 Observer for labeled nondeterministic finite automaton G in Fig. 6.1 discussed in Example 6.1

we can find a sequence of states $q_k = q, q_{k+1}, \ldots, q_{k+m}, q_{k+m+1} = q$ such that $q_{k+i+1} \in \delta(q_{k+i}, \sigma[k+i])$ for $i = 0, 1, \ldots m$, then the corresponding outputs

$$\lambda(q_{k+i}, \sigma[k+i], q_{k+i+1}),\ i = 0, 1, \ldots m ,$$

cannot all be empty.

Liveness is needed so as to be able to always consider an arbitrary long sequence of observations (if this assumption is removed, one would have to consider how to adjust the definitions of detectability below to account for the existence of possibly finite sequences of observations). Also, absence of unobservable cycles is needed to ensure that the system cannot remain active indefinitely without generating any observations.

Given a live NFA with outputs and (possibly) silent transitions $G = (Q, \Sigma, Y \cup \{\epsilon\}, \delta, \lambda, Q_0)$, we use $Y(G)$ to denote the set of all possible sequences of observations that can be generated by G, i.e.,

$$Y(G) = \cup_{q_0 \in Q_0, s \in L(G, q_0)} E(\lambda_{seq}(q_0, s)) , \tag{6.1}$$

where $E(\lambda_{seq}(q, s))$ was defined in (3.9) in Chap. 3). Note that any sequence of observations $y \in Y(G)$ can be extended indefinitely under the two assumptions we made above (liveness and absence of unobservable cycles). Given an infinitely extensible sequence of observations $y \in Y(G)$, we will use $y_0^k := y[0], y[1], \ldots, y[k]$ to denote its prefix of length $k+1$, and $\hat{q}_{y[k]}(y_0^k)$ to denote the set of (current) state estimates following observation $y[k]$ (as defined in Problem 4.4 in Chap. 4). For a given y_0^k, we can define the set of indices κ, $\kappa \in \{0, 1, 2, \ldots, k\}$ for which the state of system G can be estimated exactly by the observer as

$$\mathcal{D}_k(y) = \{\kappa \in \{0, 1, \ldots, k\} \mid |\hat{q}_{y[\kappa]}(y_0^\kappa)| = 1\} . \tag{6.2}$$

Also, we denote the cardinality of set $\mathcal{D}_k(y)$ by

$$d_k(y) = |\mathcal{D}_k(y)| . \tag{6.3}$$

Note that $d_k(y)$ satisfies $0 \leq d_k(y) \leq k+1$ for all $y \in Y(G)$ and all k.

Definition 6.1 (*(Strong) Detectability*) A live NFA with outputs and (possibly) silent transitions $G = (Q, \Sigma, Y \cup \{\epsilon\}, \delta, \lambda, Q_0)$ is (strongly) detectable if for all infinite sequences of observations $y \in Y(G)$ (generated by unknown underlying activity in automaton G), we have that

$$\lim_{k \to \infty} \frac{d_k(y)}{k+1} = 1 ,$$

where $d_k(y)$ is defined in (6.3) for each prefix y_0^k of the sequence of observations y.

In words, (strong) detectability requires that, for any infinite sequence of observations y that can be generated by the given system, the corresponding set of current-state estimates (following the last observation $y[k]$ along a prefix y_0^k of this observation sequence) is a singleton set, perhaps with the exception of a small set of indices (whose cardinality relative to $k+1$ decreases toward zero as k goes to infinity). Note that a singleton set of state estimates consists of a single state estimate, which obviously has to coincide with the true underlying state of the system. [Since the infinite sequence of observations $y[0], y[1], \ldots, y[k], \ldots$ is assumed to have been generated by unknown underlying activity in automaton G, then clearly we have $|\hat{q}_{y[k]}(y_0^k)| \geq 1$ for all $k = 0, 1, 2, \ldots$ (because the true underlying state is included in the set of possible state estimates).]

Weak detectability relaxes the above requirement considerably.

Definition 6.2 (*Weak Detectability*) A live NFA with outputs and (possibly) silent transitions $G = (Q, \Sigma, Y \cup \{\epsilon\}, \delta, \lambda, Q_0)$ is weakly detectable if there exists at least one infinite sequence of observations y (generated by unknown underlying activity in automaton G) for which

$$\lim_{k \to \infty} \frac{d_k(y)}{k+1} = 1 ,$$

where $d_k(y)$ is defined in (6.3) for each prefix y_0^k of the sequence of observations y.

Weak detectability requires that we can find at least one infinite sequence of observations y that can be generated by the given system, for which the corresponding set of current-state estimates (following the last observation $y[k]$ along a prefix y_0^k of this trajectory) is a singleton set, perhaps with the exception of a small set of indices (whose cardinality relative to $k+1$ decreases toward zero as k goes to infinity). In other words, along this particular trajectory, one can almost always pinpoint the state of the system exactly.

Definition 6.3 ((*Strong*) *Periodic Detectability*) A live NFA with outputs and (possibly) silent transitions $G = (Q, \Sigma, Y \cup \{\epsilon\}, \delta, \lambda, Q_0)$ is (strongly) periodically detectable if there exists an integer p, $0 < p < \infty$, such that for all infinite sequences of observations y (generated by unknown underlying activity in automaton G), we have that the following property: if we consider all time indices $k_1 < k_2 < \cdots < k_j < \cdots$ for which $|\hat{q}_{y[k_j]}(y_0^{k_j})| = 1$, then we have an infinite set of indices and

$$k_{j+1} - k_j \leq p, \quad \forall j = 0, 1, 2, \ldots$$

where $k_0 = 0$.

In words, (strong) periodic detectability requires that for any infinite trajectory in the system, we will be able to pinpoint exactly the state of the system, at least once every p observations. Weak periodic detectability relaxes this requirement by only requiring the existence of at least one infinite trajectory that behaves in this manner.

Definition 6.4 (*Weak Periodic Detectability*) A live NFA with outputs and (possibly) silent transitions $G = (Q, \Sigma, Y \cup \{\epsilon\}, \delta, \lambda, Q_0)$ is (strongly) periodically detectable if there exists an integer p, $0 < p < \infty$, such that we can find at least one infinite sequence of observations y (generated by unknown underlying activity in automaton G) such that the following property holds: if we consider all time indices $k_1 < k_2 < \cdots < k_j < \cdots$ for which $|\hat{q}_{y[k_j]}(y_0^{k_j})| = 1$, then we have an infinite set of indices and

$$k_{j+1} - k_j \leq p, \quad \forall j = 0, 1, 2, \ldots$$

where $k_0 = 0$.

Example 6.2 Having formulated the various notions of opacity, we revisit the LNFA G in Fig. 6.1 (discussed earlier in Example 6.1). What we can say about G is that

1. G is not strongly detectable;
2. G is weakly detectable;
3. G is not strongly periodically detectable;
4. G is weakly periodically detectable.

Since the sequence of observations ac^n creates confusion about the state (3 or 5) for arbitrary n, the system is neither strongly detectable nor strongly periodically detectable. In fact, if we let $y = ac^*$, the limit

$$\lim_{k\to\infty} \frac{d_k(y)}{k+1} = 0 \; .$$

On the other hand, since the sequence of observations c^n allows us to determine the state exactly (namely, state 7) for every n, we can conclude that the system is weakly detectable. More specifically, if we let $y' = c^*$, then we have

$$\frac{d_k(y')}{k+1} = 1 \; ,$$

for all $k \geq 1$ (and thus the limit is also equal to unity). Using this same sequence of observations, we can also see that the system is weakly periodically detectable (with $p = 1$).

Note that if states 6 and 7 were not present in system G (along with any transitions in and out of them), the system would not be weakly detectable as there are two infinite sequences of observations, namely $y = ac^*$ analyzed above and $y'' = b(aca)^*$, both of which do not satisfy the limit requirement (sequence y has $\lim_{k\to\infty} \frac{d_k(y)}{k+1} = 0$, whereas sequence y'' has $\lim_{k\to\infty} \frac{d_k(y'')}{k+1} = 1/3$). However, the system would still be weakly periodically detectable with $p = 3$, because of the presence of the sequence of observations $b(aca)^*$. In this sequence, we can determine the state of the system exactly once every three observations (namely, after we observe b, $b(bab)$, $b(bab)^2$, $b(bab)^3, \ldots$, and so forth). □

6.2.2 Initial-State and D-Delayed-State Detectability

Before closing this section, it is worth pointing out that one could also extend the above definitions of detectability to the case where the focus is on the initial-state (or some delayed state) estimate rather than the current state. For example, one could define the notions of *(strong) initial-state detectability* and *weak initial-state detectability* as described below. Note that periodic initial-state detectability is not interesting due to the monotonic refinement property of the initial-state estimate (discussed in Remark 4.6 of Chap. 4): given a sequence of observations y_0^k, we have

$$\hat{q}_0(y_0^{i+1}) \subseteq \hat{q}_0(y_0^i) \; ,$$

for $0 \leq i \leq k-1$, i.e., the estimate of the initial state can only get refined as more observations become available. In the definitions below we use $\hat{q}_0(y_0^k)$ to denote the set of initial-state estimates following observation sequence y_0^k, as defined in Problem 4.6 in Chap. 4.

Remark 6.1 One could certainly define (strong) initial-state detectability and weak initial-state detectability by taking the corresponding definitions of (strong) current-state detectability and weak current-state detectability, and replacing $\hat{q}_{y[k]}(y_0^k)$ with

$\hat{q}_0(y_0^k)$ (also redefining $\mathcal{D}_k(y)$ and $d_k(y)$ with respect to $\hat{q}_0(y_0^k)$). The definitions below take advantage of the monotonic refinement property of initial-state estimation and are a bit simpler (and closer to the notions of *I-detectability* and *weak I-detectability* defined in Shu and Lin 2012). □

Definition 6.5 ((*Strong*) *Initial-State Detectability*) A live NFA with outputs and (possibly) silent transitions $G = (Q, \Sigma, Y \cup \{\epsilon\}, \delta, \lambda, Q_0)$ is (strongly) initial-state detectable if there exists an integer k_0, $0 \leq k_0 < \infty$, such that for all infinite sequences of observations $y \in Y(G)$ (generated by unknown underlying activity in automaton G), we have that

$$k \geq k_0 \Rightarrow |\hat{q}_0(y_0^k)| = 1 \ .$$

Definition 6.6 (*Weak Initial-State Detectability*) A live NFA with outputs and (possibly) silent transitions $G = (Q, \Sigma, Y \cup \{\epsilon\}, \delta, \lambda, Q_0)$ is weakly initial-state detectable if there exists an integer k_0, $0 \leq k_0 < \infty$, such that there exists at least one infinite sequence of observations $y \in Y(G)$ (generated by unknown underlying activity in automaton G) for which we have that

$$k \geq k_0 \Rightarrow |\hat{q}_0(y_0^k)| = 1 \ .$$

Note that definitions can also be stated for D-delayed-state (strong) detectability, D-delayed-state weak detectability, D-delayed-state (strong) periodic detectability, and D-delayed-state weak periodic detectability (by taking the corresponding definitions of current-state detectability, and replacing $\hat{q}_{y[k]}(y_0^k)$ with $\hat{q}_{[k-D]}(y_0^k)$ (also redefining $\mathcal{D}_k(y)$ and $d_k(y)$ with respect to $\hat{q}_{[k-D]}(y_0^k)$).

6.3 Verification of Detectability

The verification of all notions of (current-state) detectability mentioned in the previous section can be achieved by constructing the current-state estimator (observer) of the given finite automaton. The following theorem can be proved easily if one realizes that any sequence of observations $y \in Y(G)$ that can be generated by a given NFA G corresponds to a trajectory (sequence of states) in the observer G_{obs} and vice versa. Given this property of the observer and the fact that its set of states is finite, one direction of the proof becomes trivial because an infinitely extensible sequence of observations will correspond to a state trajectory that enters one or more loops in the observer (and does not visit observer states that are not part of a loop more than once); the other direction can be easily established by contradiction.

Theorem 6.1 *Consider an NFA with outputs and (possibly) silent transitions $G = (Q, \Sigma, Y \cup \{\epsilon\}, \delta, \lambda, Q_0)$ that is live and has no cycles of events that do not generate any observation (unobservable cycles). Construct the corresponding current-state estimator, denoted by*

$$G_{obs} = AC(2^Q, Y, \delta_{obs}, Q_{0,obs}) =: (Q_{obs}, Y, \delta_{obs}, Q_{0,obs}) ,$$

as described in Definition 5.1 in Chap. 5. Let $Q_{obs} \subseteq 2^Q$ *denote the states of* G_{obs} *and let*

$$Q_{m,obs} = \{q_{obs} \in Q_{obs} \mid |q_{obs}| = 1\}$$

denote the set of states in the current-state estimator ($Q_{m,obs} \subseteq Q$) *that are singleton subsets of* Q. *Then, the following hold:*

1. *G is (strongly) detectable if and only if $Q_{m,obs}$ is non-empty and all loops in G_{obs} are entirely within $Q_{m,obs}$ (i.e., no loop exists that involves a state in $Q_{obs} \setminus Q_{m,obs}$);*
2. *G is weakly detectable if and only if $Q_{m,obs}$ is non-empty and there exists at least one loop that is entirely within $Q_{m,obs}$;*
3. *G is (strongly) periodically detectable if and only if all loops in G_{obs} involve at least one state in $Q_{m,obs}$ (i.e., no loop exists that involves states exclusively in $Q_{obs} \setminus Q_{m,obs}$);*
4. *G is weakly periodically detectable if and only if there exists at least one loop in G_{obs} that involves at least one state in $Q_{m,obs}$.*

Remark 6.2 From the above theorem, it becomes evident that (strong) detectability implies (strong) periodic detectability and that weak detectability implies weak periodic detectability. □

Example 6.3 In this example, we revisit the LNFA G in Fig. 6.1, which was discussed earlier in Example 6.1. The observer G_{obs} for G is shown in Fig. 6.2. Once we have constructed the observer, we can reach the following conclusions:

1. G is not strongly detectable (e.g., because of the presence of the cycle involving the state $\{3, 5\}$ and event/observation c in the observer);
2. G is weakly detectable (because of the presence of the cycle involving the state $\{7\}$ and event/observation c in the observer);
3. G is not strongly periodically detectable (e.g., because of the presence of the cycle involving the state $\{3, 5\}$ and event/observation c in the observer);
4. G is weakly periodically detectable (e.g., because of the presence of the cycle involving the states $\{8\}$, $\{9, 11\}$, and $\{10, 12\}$ and events/observations a, c, and a in the observer).

These conclusions are in agreement with what we concluded about system G in our discussions in Example 6.2. □

In a similar manner, one can describe the verification of initial-state detectability using the initial-state estimator. More specifically, one can prove the theorem below.

Theorem 6.2 *Consider an NFA with outputs and (possibly) silent transitions* $G = (Q, \Sigma, Y \cup \{\epsilon\}, \delta, \lambda, Q_0)$ *that is live and has no cycles of events that do not generate any observation (unobservable cycles). Construct the corresponding initial-state estimator, denoted by*

$$G_{Iobs} = AC(2^{Q^2}, Y, \delta_{Iobs}, Q_{0I}) =: (Q_{Iobs}, Y, \delta_{Iobs}, Q_{0I}) ,$$

as described in Definition 5.3 in Chap. 5. Let

$$Q_{m,Iobs} = \{q_{Iobs} \in Q_{Iobs} \mid \exists q_0 \in Q_0, \forall (q_i, q_f) \in q_{Iobs} \{q_i = q_0\}\}$$

denote the set of states in the initial-state estimator ($Q_{m,Iobs} \subseteq Q_{Iobs} \subseteq Q^2$) *that are associated with pairs of states (of the form* $(q_i, q_f) \in Q^2$*) that involve the* same *initial state (for all pairs* (q_i, q_f)*, the initial state* q_i *equals some* $q_0 \in Q_0$*). Then, the following hold:*

1. *G is (strongly) initial-state detectable if and only if* $Q_{m,Iobs}$ *is non-empty and all loops in* G_{Iobs} *are entirely within* $Q_{m,Iobs}$ *(i.e., no loop exists that involves a state in* $Q_{Iobs} \backslash Q_{m,Iobs}$*);*
2. *G is weakly initial-state detectable if and only if* $Q_{m,Iobs}$ *is non-empty and there exists at least one loop that is entirely within* $Q_{m,Iobs}$*.*

6.4 Verification of Strong Detectability Using the Detector

The previous section argued that we can determine whether or not a given NFA is detectable by building its current-state estimator and checking for loops that involve singleton states. The construction of the current-state estimator could involve, in the worst case, an exponential number of states (potentially 2^N states where $N = |Q|$ is the number of states of the given automaton). In this section, we describe the construction of a *detector*, which allows us to determine whether or not a given system is (strongly) detectable or (strongly) periodically detectable.

Unlike the current-state estimator, the detector is an NFA and has complexity that is polynomial in N (its number of states is $O(N^2)$). The detector was introduced in Shu et al. (2007); related ideas also appeared in Caines et al. (1988, 1991) (the authors of the latter works were interested in eventually being able to pinpoint the system's state in a class of finite automata, and were able to answer this question rather efficiently).

Definition 6.7 (*Detector*) Given an NFA with outputs and (possibly) silent transitions $G = (Q, \Sigma, Y \cup \{\epsilon\}, \delta, \lambda, Q_0)$, its detector is the *nondeterministic* finite automaton without outputs $G_d = AC(X, Y, \delta_d, X_{0,d})$, where

1. $X_{0,d} = UR(Q_0)$ is the unobservable reach of the set of possible initial states of G (i.e., the set of states Q_0 together with any states that can be reached from a state in Q_0 via a sequence of transitions that does not generate any observation—refer to Definition 3.27 in Chap. 3);
2. The set of states is (the accessible part of) X which is defined as

$$X = X_p \cup X_s \cup \{X_{0,d}\} ,$$

where $X_p = \{\{q_i, q_j\} \mid q_i, q_j \in Q,\ q_i \neq q_j\}$ (is the set of subsets of Q of cardinality 2) and $X_s = \{\{q_i\} \mid q_i \in Q\}$ (is the set of subsets of Q of cardinality 1);

3. For $x_d \in X$, $y \in Y$, the mapping $\delta_d : X \times Y \mapsto X$ is defined as follows: let $\hat{x}_d = UR(x_d, y) := \cup_{q \in x_d} UR(q, y)$ where the unobservable reach $UR(q, y)$ from state $q, q \in Q$, is taken with respect to an output y in the given finite automaton G, as defined in Definition 3.28 in Chap. 3; then, we have

$$\delta_d(x_d, y) = \begin{cases} \{x_p \in X_p \mid x_p \subseteq \hat{x}_d\}\ , & \text{if } |\hat{x}_d| \geq 2\ , \\ \{x_s \mid x_s \in X_s\}\ , & \text{if } \hat{x}_d = \{x_s\} \text{ (i.e., if } |\hat{x}_d| = 1)\ , \\ \emptyset\ , & \text{if } |\hat{x}_d| = 0 \text{ (i.e., if } \hat{x}_d = \emptyset)\ . \end{cases}$$

4. AC denotes the accessible part of the automaton (i.e., the part of the automaton G_d that can be reached from its initial state $X_{0,d}$).

It is not hard to prove the following theorem.

Theorem 6.3 *Consider an NFA with outputs and (possibly) silent transitions $G = (Q, \Sigma, Y \cup \{\epsilon\}, \delta, \lambda, Q_0)$ that is live and has no cycles of events that do not generate any observation (unobservable cycles). Construct the corresponding detector $G_d = AC(X, Y, \delta_d, X_{0,d})$ with $X = X_p \cup X_s \cup \{X_{0,d}\}$ as described in Definition 6.7 above. Then, the following hold true:*

1. *G is (strongly) detectable if and only if all loops in G_d are entirely within X_s (i.e., no loop exists that involves a state in X_p);*
2. *G is (strongly) periodically detectable if and only if all loops in G_d involve at least one state in X_s (i.e., no loop exists that involves states exclusively in X_p).*

Note that the detector can be used to verify (strong) detectability and (strong) periodic detectability, but cannot be used to determine weak detectability and weak periodic detectability. Checking whether the detector possesses at least one loop that involves one or more states from the set X_p can be accomplished with complexity that is polynomial in the size of the state transition graph of the detector (which, in turn, is polynomial in the number of states $N = |Q|$ of the given NFA).

Example 6.4 We revisit the LNFA G in Fig. 6.1, whose verification using the observer was discussed in Example 6.3. The detector for G is shown in Fig. 6.3. Once we have constructed the detector, we can reach the following conclusions:

1. G is not strongly detectable (because of the presence of the cycle involving the state $\{3, 5\}$ and event/observation c in the detector);
2. G is not strongly periodically detectable (because of the presence of the cycle involving the state $\{3, 5\}$ and event/observation c in the detector).

The above statements are in agreement with the conclusions we reached regarding strong detectability and strong periodic detectability for system G in our earlier discussions. It is also interesting to notice the differences between the observer (in Fig. 6.2) and the detector (in Fig. 6.3): starting from the same initial state as the observer, the detector states and next-state transition function resemble the ones for

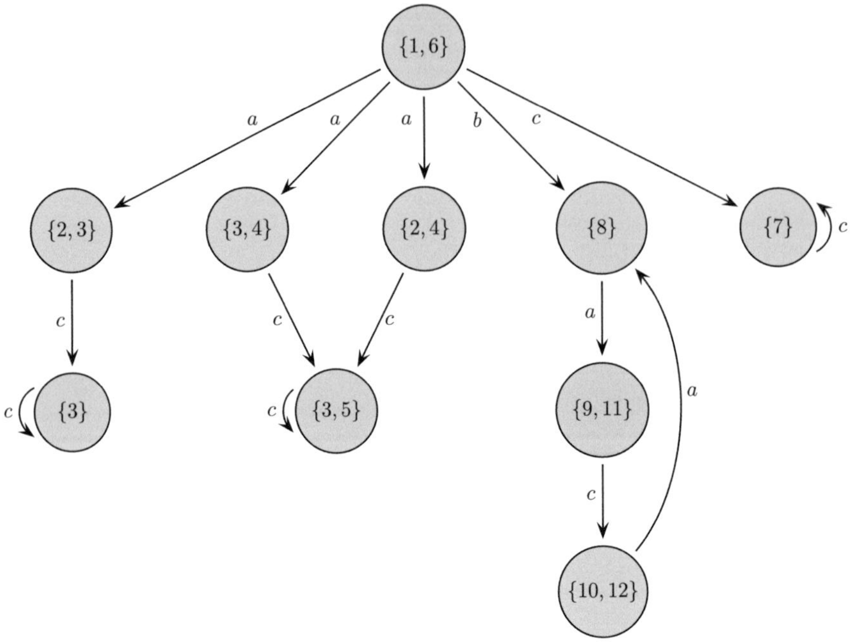

Fig. 6.3 Detector for labeled nondeterministic finite automaton G in Fig. 6.1, discussed in Example 6.1

the observer, as long as the observer states involve less than one or two system states (see, for example, the two rightmost branches from state $\{1, 6\}$ of the detector); however, when the observer involves more than two states, we get different behavior (see, for example, the three leftmost branches from state $\{1, 6\}$ of the detector). □

Before closing this section, we should also point out that the notion of (strong) initial-state detectability can also be verified with polynomial complexity using detector-like constructors that focus on tracking the pairwise uncertainty associated with possible initial states. The initial-state detector (referred to as I-detector in Shu and Lin 2012) is a construction that resembles the detector described in this section, except that it also tracks the initial state.

6.5 Extensions to K-Detectability and Verification Using the K-Detector

In this section, we formalize the notion of K-detectability, where K is a positive integer, and describe its verification using the K-detector (Hadjicostis and Seatzu 2016). Roughly speaking, given an NFA with outputs and (possibly) silent transitions (which, as in earlier sections, is assumed to be live and posses no unobservable cycles), variations of K-detectability are concerned with whether the sets of state estimates, at least after a long enough sequence of observations, have cardinality smaller or equal to K. When $K = 1$, we recover the standard notions of detectability described in the previous section. The formal definition of (strong) K-detectability is provided below.

6.5.1 *K-Detectability*

As in the case of detectability, we assume that we are given an NFA with outputs and (possibly) silent transitions $G = (Q, \Sigma, Y \cup \{\epsilon\}, \delta, \lambda, Q_0)$, as described in Definition 3.24 in Chap. 3. Again, we assume that G is live and that there are no unobservable cycles in G (refer to Sect. 6.2.1).

Given a live NFA with outputs and (possibly) silent transitions $G = (Q, \Sigma, Y \cup \{\epsilon\}, \delta, \lambda, Q_0)$, we use $Y(G)$ to denote the set of all possible sequences of observations that can be generated by G (see (6.1)) and, for any infinitely extensible $y \in Y(G)$ we use $y_0^k := y[0], y[1], \ldots, y[k]$ to denote its prefix of length $k+1$, and $\hat{q}_{y[k]}(y_0^k)$ to denote the set of (current) state estimates following observation $y[k]$ (as defined in Problem 4.4 in Chap. 4). For a given y_0^k, we can define the set of indices κ, $\kappa \in \{0, 1, 2, \ldots, k\}$ for which the state of system G can be estimated within an uncertainty set of cardinality at most K as follows:

$$\mathcal{D}_{k,K}(y) = \{\kappa \in \{0, 1, \ldots, k\} \mid |\hat{q}_{y[\kappa]}(y_0^\kappa)| \leq K\} . \tag{6.4}$$

Also, we denote the cardinality of set $\mathcal{D}_{k,K}(y)$ by

$$d_{k,K}(y) = |\mathcal{D}_{k,K}(y)| . \tag{6.5}$$

Note that $d_{k,K}(y)$ satisfies $0 \leq d_{k,K}(y) \leq k+1$ for all $y \in Y(G)$ and all k.

Definition 6.8 ((*Strong*) *K-Detectability*) A live NFA with outputs and (possibly) silent transitions $G = (Q, \Sigma, Y \cup \{\epsilon\}, \delta, \lambda, Q_0)$ is (strongly) K-detectable if for all infinite sequences of observations $y \in Y(G)$ (generated by unknown underlying activity in automaton G), we have that

$$\lim_{k \to \infty} \frac{d_{k,K}(y)}{k+1} = 1 ,$$

where $d_{k,K}(y)$ is defined in (6.5) for each prefix y_0^k of the sequence of observations y.

In words, (strong) K-detectability requires that, for any infinite sequence of observations that can be generated by the given system, the corresponding set of current-state estimates (following the last observation along this trajectory) is a set of cardinality at most K, perhaps with the exception of a small set of indices k (whose cardinality relative to $k+1$ decreases toward zero as k goes to infinity). Note that the set of state estimates (of cardinality at most K) has to include the underlying state of the system, but for $K > 1$ it may also include other possible states. Though not discussed explicitly here, weak K-detectability, (strong) periodic K-detectability, and weak periodic K-detectability can be defined in a similar manner: one can take the corresponding definitions in the previous section and simply allow the set of state estimates to have cardinality smaller or equal to K. Similarly, one can talk about initial-state K-detectability and D-delayed-state K-detectability.

6.5.2 Verification of K-Detectability

Clearly, K-detectability can be verified using a current-state estimator. A proof for a version of the theorem below can be found in Hadjicostis and Seatzu (2016).

Theorem 6.4 *Consider an NFA with outputs and (possibly) silent transitions $G = (Q, \Sigma, Y \cup \{\epsilon\}, \delta, \lambda, Q_0)$ that is live and has no cycles of events that do not generate any observation (unobservable cycles). Construct the corresponding current-state estimator, denoted by*

$$G_{obs} = AC(2^Q, Y, \delta_{obs}, Q_{0,obs}) =: (Q_{obs}, Y, \delta_{obs}, Q_{0,obs}) ,$$

as described in Definition 5.1 in Chap. 5. Let $Q_{obs} \subseteq 2^Q$ denote the states of G_{obs} and let

$$Q_{m,obs} = \{q_{obs} \in Q_{obs} \mid |q_{obs}| \leq K\}$$

denote the set of states in the current-state estimator ($Q_{m,obs} \subseteq Q_{obs}$) that are subsets of Q of cardinality at most K. Then, the following hold:

1. *G is (strongly) K-detectable if and only if $Q_{m,obs}$ is non-empty and all loops in G_{obs} are entirely within $Q_{m,obs}$ (i.e., no loop exists that involves a state in $Q_{obs} \setminus Q_{m,obs}$);*
2. *G is weakly K-detectable if and only if $Q_{m,obs}$ is non-empty and there exists at least one loop that is entirely within $Q_{m,obs}$;*
3. *G is (strongly) periodically K-detectable if and only if all loops in G_{obs} involve at least one state in $Q_{m,obs}$ (i.e., no loop exists that involves states exclusively in $Q_{obs} \setminus Q_{m,obs}$);*
4. *G is weakly periodically K-detectable if and only if there exists at least one loop in G_{obs} that involves at least one state in $Q_{m,obs}$.*

As in the case of detectability, the verification of variations of K-detectability using the observer can be accomplished with complexity that is linear in the number of states of the observer (which nevertheless may have an exponentially many states with respect to the size of the given finite automaton). Next, we describe the K-detector, which is a generalization of the detector and allows us to verify K-detectability with complexity $O(N^{K+1})$ where $N = |Q|$ is the number of states of the underlying system.

Definition 6.9 (*K-Detector*) Given an NFA with outputs and (possibly) silent transitions $G = (Q, \Sigma, Y \cup \{\epsilon\}, \delta, \lambda, Q_0)$, its K-detector is the *nondeterministic* finite automaton without outputs, denoted by

$$G_{Kd} = AC(X, Y, \delta_{Kd}, Q_{0,Kd}) =: (Q_{Kd}, Y, \delta_{Kd}, Q_{0,Kd}) ,$$

where

1. $Q_{0,Kd} = UR(Q_0)$ is the unobservable reach of the set of possible initial states of G (i.e., the set of states Q_0 together with any states that can be reached from a state in Q_0 via a sequence of transitions that does not generate any observation—refer to Definition 3.27 in Chap. 3);
2. The set of states is (the accessible part of) X which is defined as

$$X = X_p \cup X_s \cup \{Q_{0,Kd}\} ,$$

with (i) $Q_{0,Kd} = UR(Q_0)$ being the set of all possible initial states before any observation is made, (ii) $X_s = \{T_s \mid T_s \subseteq Q \text{ and } |T_s| \leq K\}$, and (iii) $X_p = \{T_p \mid T_p \subseteq Q \text{ and } |T_p| = K + 1\}$;
3. For $x_{Kd} \in X$, $y \in Y$, the mapping $\delta_{Kd} : X \times Y \mapsto X$ is defined as follows: let $\hat{x}_{Kd} = UR(x_{Kd}, y) := \cup_{q \in x_{Kd}} UR(q, y)$ where the unobservable reach $UR(q, y)$ from state q with respect to an output y is defined for the given automaton G as in Definition 3.28 in Chap. 3; then, we have

$$\delta_{Kd}(x_{Kd}, y) = \begin{cases} \{T_s \in X_s \mid T_s = \hat{x}_{Kd}\}, & \text{if } |\hat{x}_{Kd}| \leq K, \\ \{T_p \in X_p \mid T_p \subseteq \hat{x}_{Kd}\}, & \text{if } |\hat{x}_{Kd}| > K. \end{cases}$$

4. AC denotes the accessible part of the automaton (i.e., the part of the automaton G_{Kd} that can be reached from its initial state $Q_{0,Kd}$).

The following theorem is proven in Hadjicostis and Seatzu (2016).

Theorem 6.5 *Consider an NFA with outputs and (possibly) silent transitions $G = (Q, \Sigma, Y \cup \{\epsilon\}, \delta, \lambda, Q_0)$ that is live and has no cycles of events that do not generate any observation (unobservable cycles). Construct the corresponding K-detector $G_{Kd} = AC(X, Y, \delta_{Kd}, X_{0,Kd})$ with $X = X_p \cup X_s \cup \{Q_{0,Kd}\}$ as described in Definition 6.9 above. Then, the following hold true:*

1. *G is (strongly) K-detectable if and only if all loops in G_{Kd} are entirely within X_s (i.e., no loop exists that involves a state in X_p);*
2. *G is (strongly) periodically K-detectable if and only if all loops in G_{Kd} involve at least one state in X_s (i.e., no loop exists that involves states exclusively in X_p).*

Note that the K-detector can be used to verify (strong) K-detectability and (strong) periodic K-detectability, but cannot be used to determine weak K-detectability and weak periodic K-detectability. Checking whether the detector possesses at least one loop that involves one or more states from the set X_p can be accomplished with complexity that is polynomial in the size of the state transition graph of the K-detector (which, in turn, is polynomial in the number of states $N = |Q|$ of the given NFA).

Example 6.5 Consider the automaton $G = (Q, \Sigma, Y \cup \{\epsilon\}, \delta, \lambda, Q_0)$ at the top of Fig. 6.4, which was also considered in Example 5.3 in Chap. 5 but is reproduced here for convenience. System G is an LDFA with silent transitions under the natural projection mapping, with $Q = \{1, 2, 3, 4, 5\}$, $\Sigma = \{a, b, c\}$, $Y = \{a, b\}$, $Q_0 = \{1\}$, and δ as shown in the figure. The set of observable events is $\Sigma_{obs} = \{a, b\}$ and the set of unobservable events is $\Sigma_{uo} = \{c\}$ (so that the output mapping λ simply implements the natural projection $P_{\Sigma_{obs}}$, i.e., $\lambda(a) = a$, $\lambda(b) = b$, $\lambda(c) = \epsilon$). Note that the observer for G is the deterministic finite automation (DFA) without outputs $G_{obs} = (Q_{obs}, \Sigma_{obs}, \delta_{obs}, Q_{0,obs})$ shown at the bottom of Fig. 6.4.

The 2-detector for system G is shown in Fig. 6.5. Notice that each state of the 2-detector is associated with at most three system states (state estimates). Unlike the observer, the 2-detector is a nondeterministic system (for example, from state $\{1, 3, 4\}$ with observation b, we go to four different states, namely $\{1, 3, 4\}$, $\{1, 3, 5\}$, $\{1, 4, 5\}$, $\{3, 4, 5\}$).

Note that, for $K \geq 3$, the K-detector for system G becomes identical to the observer shown at the bottom of Fig. 6.4 (since observer states are associated with at most four system states). □

6.6 Synchronizing, Homing, and Distinguishing Sequences

As mentioned in Chap. 5, a common problem that arises in the context of testing of digital circuits is that, at start-up, the initial state might be unknown. In order to verify the functionality of the system (or, in some cases, in order to simply utilize the system), it is often useful to first find a way to either drive the system to a specified (or, at least, known) state or simply identify the initial state of the system (before the application of any sequence of inputs) and/or the current state of the system (after the application of a specific sequence of inputs).

A number of works have dealt with the notions of *synchronizing* sequences, *homing* sequences, and *distinguishing* sequences, which we define and discuss in this section. Our goal is primarily to introduce these notions and connect them to the state estimation discussions/constructions in this book; for more details, the interested reader is referred to works that deal explicitly with these topics (e.g., the surveys in Lee and Yannakakis 1996; Sandberg 2005).

Due to the nature of the motivating application (testing of digital circuits), most works in this area have considered DFA with outputs (finite-state machines), whose

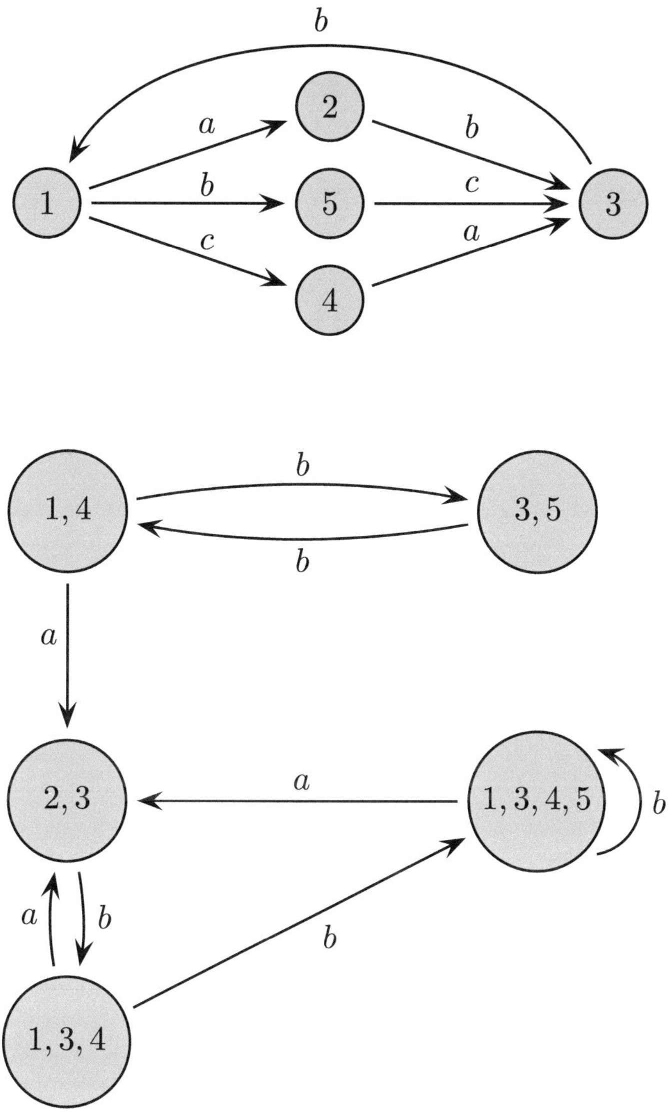

Fig. 6.4 Labeled deterministic finite automaton G (top) and its corresponding observer G_{obs} (bottom)

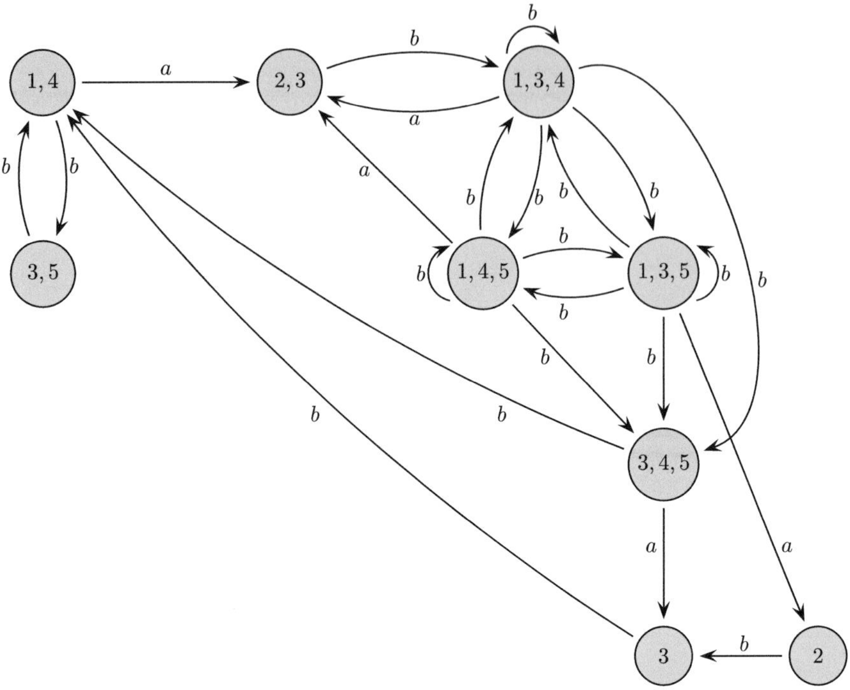

Fig. 6.5 2-detector for labeled deterministic finite automaton G at the top of Fig. 6.4

initial state is unknown (and is typically taken to be the set of all states in the given automaton) and whose inputs can be chosen at each step (by the user/observer); however, many of the concepts can also be defined for NFA. Apart from focusing on DFA with outputs (finite-state machines), other typical assumptions are that the automaton is *minimal* (i.e., we cannot have an automaton with fewer states and the same input–output behavior), *connected* (any state is accessible from any other state via a sequence of inputs), and *completely specified* (each input is defined at each state). Another issue that deserves attention is that in our estimation analysis so far we have dealt with finite automata with outputs, where we assume that the inputs are unknown (see Fig. 4.2 and the discussion in Chap. 4). Thus, the presentation in this section differs in the sense that inputs are controllable and, of course, known to the user/observer.

Remark 6.3 Given a DFA with outputs $G = (Q, \Sigma, Y, \delta, \lambda, Q_0)$, a reset input is an input $\sigma_r \in \Sigma$ that satisfies, for all $q \in Q$, $\delta(q, \sigma_r) = q_r$ for some state $q_r \in Q$. Clearly, in such case, $s = \sigma_r$ drives the system to a known state (in terms of the definitions in this section, we will see that s is both a synchronizing sequence and a homing sequence). For machines that do not have reset inputs, one typically resorts to more complex synchronizing or homing sequences, in order to drive the system to a known state. □

Definition 6.10 (*Synchronizing Sequence*) Given a DFA with outputs $G = (Q, \Sigma, Y, \delta, \lambda, Q_0)$, a sequence of inputs $s \in \Sigma^*$ is called synchronizing if

$$|\delta(Q_0, s)| = 1 .$$

Remark 6.4 In words, when a synchronizing sequence is applied to the system, then we know that the end state (current state) of the system is the same regardless of which initial state (from the set Q_0) the system started from. Note that synchronizing sequences are independent of the outputs of the given finite automaton. Typically, the set of initial states is taken to be the set of all states (i.e., $Q_0 = Q$). If one makes the typical assumption that the finite automaton is completely specified and uses $f_{\sigma^{(i)}} : Q \to Q$, $i = 1, 2, \ldots, K$, to denote the state transition mappings (i.e., $f_{\sigma^{(i)}} = \delta(\cdot, \sigma^{(i)})$ for $\sigma^{(i)} \in \Sigma$), then $s = \sigma[0]\sigma[1]\ldots\sigma[m]$ being a synchronizing sequence is equivalent to the composed mapping $f_{\sigma[m]}(f_{\sigma[m-1]}(\ldots f_{\sigma[1]}(f_{\sigma[0]}(\cdot))\ldots))$ being a constant mapping (Sandberg 2005). □

We can easily find a synchronizing sequence (if one exists) using the current estimator (observer) construction: more specifically, if we ignore the outputs and use the inputs Σ as the observations (see Fig. 4.2 and the discussion in Chap. 4), we can build a current-state estimator of the form

$$G_{obs} = AC(2^Q, \Sigma, \delta_{obs}, Q_{0,obs}) =: (Q_{obs}, \Sigma, \delta_{obs}, Q_{0,obs}) ,$$

where $Q_{0,obs} = Q_0$ (there are no silent transitions, thus no unobservable reach) and δ_{obs} is defined, for $q_{obs} \in 2^Q$ and $\sigma \in \Sigma$, as

$$q'_{obs} = \delta_{obs}(q_{obs}, \sigma) = \delta(q_{obs}, \sigma) .$$

Then, if we look for a state $q_{f,obs} \in 2^Q$ that is a singleton subset of Q (one may not exist), we can obtain a synchronizing sequence by finding a path in the observer (sequence of inputs) that takes us from $Q_{0,obs}$ to $q_{f,obs}$: specifically, sequence $s = \sigma[0]\sigma[1]\ldots\sigma[m]$ is synchronizing if the following transitions are present in the observer:

$$Q_{0,obs} \overset{\sigma[0]}{\to} q_{1,obs} \overset{\sigma[1]}{\to} q_{2,obs} \to \cdots \overset{\sigma[m-1]}{\to} q_{m,obs} \overset{\sigma[m]}{\to} q_{f,obs}$$

for some intermediate observer states $q_{1,obs}, q_{2,obs}, \ldots q_{m,obs}$. This is illustrated in the following example.

Example 6.6 Consider the system $G = (Q, \Sigma, Y, \delta, \lambda, Q_0)$ shown in Fig. 6.6. System G is a DFA without silent transitions, with $Q = \{1, 2, 3, 4\}$, $\Sigma = \{\alpha, \beta\}$, $Y = \{0, 1\}$, $Q_0 = \{1, 2, 3, 4\}$, and δ as shown in the figure. Notice that δ is fully defined (for each input, from each state, there is a defined transition and output).

It is not hard to verify that the sequence $\beta\beta\beta$ drives the system to state 3, regardless of what the initial state is. More specifically, from state 1, we go to state 3, and then stay there; from state 2, we go to state 1, then to state 3, and then stay at state 3; from

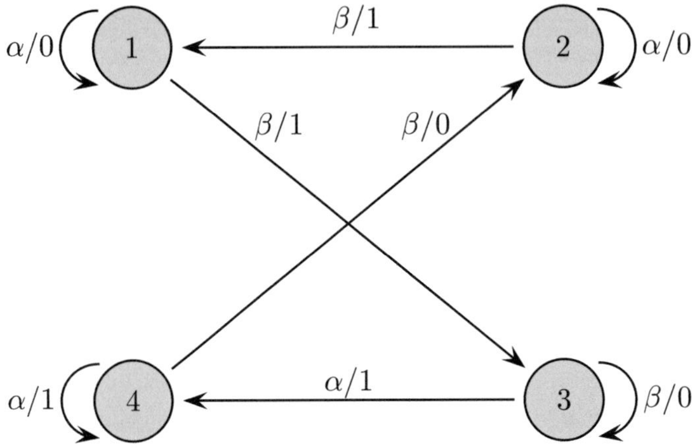

Fig. 6.6 Deterministic finite automaton discussed in Example 6.6

state 3, we stay at state 3 for all subsequent β; and from state 4, we go to state 2, then state 1, and then state 3. We conclude that $\beta\beta\beta$ is a synchronizing sequence.

To understand how to systematically obtain synchronizing sequences (or determine that one does not exist), consider the construction of the "observer" for system G shown in Fig. 6.7. This "observer" ignores the output of the system (0 or 1) and is driven by the inputs to the system (i.e., what is observed is α and β). From this construction, we can see that $\beta\beta\beta$ and its continuations (i.e., $\beta\beta\beta(\alpha+\beta)^*$) are the only synchronizing sequences. Clearly, for synchronizing sequence to exist, we need the "observer" construction in Fig. 6.7 to have at least one state associated with a singleton subset of states of G. [Note that once the "observer" construction reaches a state associated with a singleton subset of states of G, it will remain in states that are associated with singleton subsets of states of G regardless of the subsequent sequence of inputs that is applied (because the system G is taken to be deterministic).] □

Finding a synchronizing sequence using an observer (as illustrated in the above example) would require complexity that is linear in the size of the observer (which, in turn, is exponential in the size of the given finite automaton). This is not necessarily the most efficient approach to find a synchronizing sequence. An algorithm with complexity $O(N^3 + N^2|s|)$ (where $N = |Q|$ and $|s|$ is the length of the synchronizing sequence s) is discussed in Sandberg (2005). It is worth pointing out that, if a synchronizing sequence exists, then at least one such sequence has length at most $\frac{N^3}{3}$; however, finding such a sequence is generally NP-hard (Sandberg 2005).

Definition 6.11 (*Homing Sequence*) Given a DFA with outputs $G = (Q, \Sigma, Y, \delta, \lambda, Q_0)$, a sequence of inputs $s \in \Sigma^*$ is called homing if, for every pair of possible initial states $q_{i_1}, q_{i_2} \in Q_0$, we have

$$\delta(q_{i_1}, s) \neq \delta(q_{i_2}, s) \Rightarrow \lambda_{seq}(q_{i_1}, s) \neq \lambda_{seq}(q_{i_2}, s) .$$

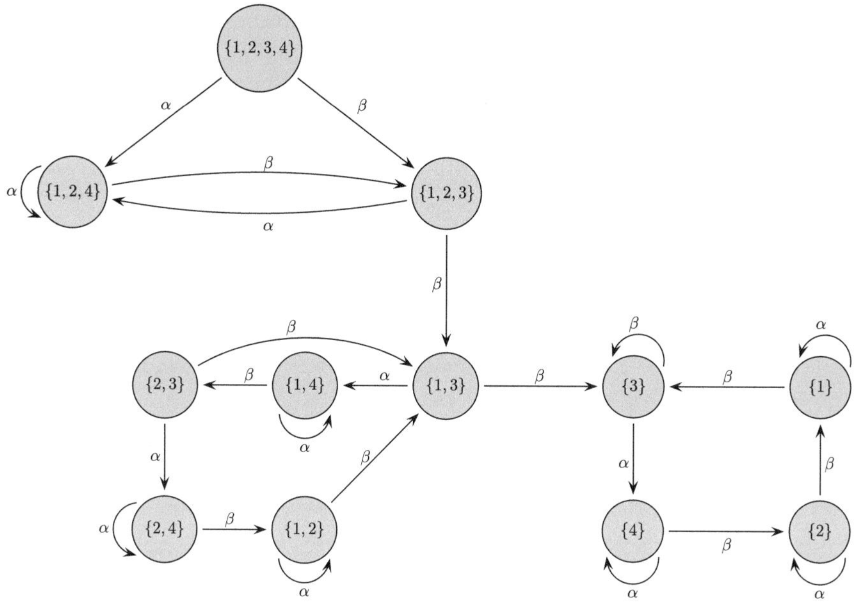

Fig. 6.7 Observer driven by observation of the inputs (and not the outputs) of the deterministic finite automaton in Fig. 6.6 discussed in Example 6.6

A homing sequence is a concept that is related to a synchronizing sequence, in the sense that it allows us to identify the end state of the machine (current state). The main difference is that, in order to identify the final state of the machine, a homing sequence also relies on the output sequence. Another difference is that the homing sequence could be either *preset* or *adaptive*. A preset homing sequence is one that is chosen ahead of time (at start-up), whereas an adaptive sequence is one that can be chosen during the operation of the system, utilizing also the observed sequence of outputs. In other words, for an adaptive homing sequence $s = \sigma[0]\sigma[1]\ldots\sigma[m]$, we are allowed to choose $\sigma[k+1]$, $k = 0, 1, \ldots, m-1$, based on $\sigma[0]\sigma[1]\ldots\sigma[k]$ (the previous inputs) and $\lambda_{seq}(q_0, \sigma[0]\sigma[1]\ldots\sigma[k])$ (i.e., the outputs seen so far, which depend on the unknown initial state q_0). It should be clear based on the above definition, that a synchronizing sequence is always a homing sequence; however, the reverse is not necessarily true.

Finding an adaptive homing sequence is more like building a decision tree: given the DFA with outputs $G = (Q, \Sigma, Y, \delta, \lambda, Q_0)$ (and assuming the inputs are also observable), we can build its current-state estimator

$$G_{obs} = AC(2^Q, \Sigma \times Y, \delta_{obs}, Q_{0,obs}) =: (Q_{obs}, \Sigma \times Y, \delta_{obs}, Q_{0,obs}) ,$$

where $Q_{0,obs} = Q_0$ (there are no silent transitions). The current-state estimator is driven by a pair (σ, y) of an input σ and an output y. Effectively, what we need to do is to start at $Q_{0,obs}$, choose an input σ, observe the corresponding output y, use (σ, y) to drive the observer to the next state and then repeat. The goal is to eventually reach a state in the observer that is a singleton set. This can be treated as a game involving two players, where, at each round, the first player (user) chooses the input σ and the second player (system) chooses a possible output y. Unlike synchronizing sequences, which (as mentioned earlier) do not always exist, adaptive homing sequence(s) always exist, at least if the given automaton is minimal (Sandberg 2005).

The above process is illustrated in the following example.

Example 6.7 Consider again the system $G = (Q, \Sigma, Y, \delta, \lambda, Q_0)$ shown in Fig. 6.6. System G is a DFA without silent transitions, with $Q = \{1, 2, 3, 4\}$, $\Sigma = \{\alpha, \beta\}$, $Y = \{0, 1\}$, $Q_0 = \{1, 2, 3, 4\}$, and δ as shown in the figure. Notice that δ is fully defined (i.e., for each input, from each state, there is a defined transition and output).

The process of obtaining an adaptive homing sequence is illustrated in Fig. 6.8.

- Starting from the set of initial states $\{1, 2, 3, 4\}$ we apply input α. There are two possibilities as we can either observe "0" or "1". If we observe "1" then we know

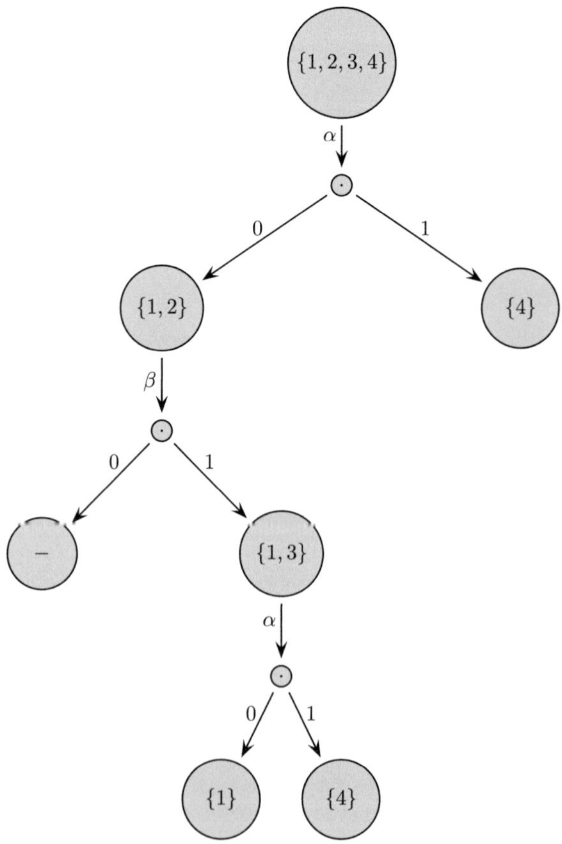

Fig. 6.8 Decision tree used in obtaining an adaptive homing sequence for the deterministic finite automaton in Fig. 6.6 discussed in Example 6.6

that the state of the system is state 4 (at which point we know the state exactly and we are done), whereas if we observe "0" then we know that the state of the system is either 1 or 2 (indicated by $\{1, 2\}$ in the figure).

- If observe "0", we can subsequently apply β; we will necessarily observe "1" and we will know that the state of the system is either 1 or 3 (indicated by $\{1, 3\}$ in the figure.
- Finally, we can apply α and we again have two possibilities as we can either observe "0" or "1". If we observe "0", then we know the system state is 1, whereas if we observe "1" we know that the system state is 4. Regardless of what we observe, we will know the state of the system exactly and will be done.

The above choices resulted in an adaptive homing sequence of length three in the worst-case scenario. We can actually obtain an adaptive homing sequence of length at most two in the worst-case scenario as follows:

- Step 1: Starting from the set of initial states $\{1, 2, 3, 4\}$ we apply input β. There are two possibilities as we can either observe "0" or "1". If we observe "0" then we know that the state of the system is either state 2 or 3, whereas if we observe "1" then we know that the state of the system is either 1 or 3.
- Step 2a: If we observe "0" after Step 1, we can subsequently apply α. There are two possibilities as we can either observe "0" or "1". If we observe "0" then we know that the state of the system is state 2, whereas if we observe "1" then we know that the state of the system is state 4.
- Step 2b: If we observe "1" after Step 1, we can subsequently apply α. There are two possibilities as we can either observe "0" or "1". If we observe "0" then we know that the state of the system is state "1", whereas if we observe "1" then we know that the state of the system is state 4.

Thus, there are many different adaptive sequences, some of which may be preferable than others. □

Finding a homing sequence using an observer (as illustrated in the above example) would require complexity that is polynomial in the size of the observer (which, in turn, is exponential in the size of the given finite automaton). This is not necessarily the most efficient approach to find a homing sequence (see the algorithms in Sandberg 2005).

Remark 6.5 A *distinguishing* sequence is a sequence of inputs that allows one to uniquely identify the initial state by observing the output sequence. Just like homing sequences, distinguishing sequences could be *preset* or *adaptive*. In other words, a distinguishing sequence is an input sequence that produces a unique output sequence for each possible starting state, thus allowing the observer to differentiate among possible starting states. Since the underlying finite automaton is assumed to be deterministic, a distinguishing sequence is also a homing sequence (and, thus, a synchronizing sequence). □

6.7 Comments and Further Reading

Detectability is a very active research topic with researchers exploiting extensions in various directions. For instance, Shu and Lin (2012), Yin (2017) consider *I-detectability* with respect to the initial state of the system, whereas Shu and Lin (2013a) considers *delayed detectability*, a notion that allows future observations to be used to refine the estimate about a previous state of the system, and requires that, after a finite number of observations, this delayed (smoothed) estimate of the state of the system remains a singleton set or becomes a singleton set periodically (depending on the variant of delayed detectability that is used). Other works consider sensor activation to achieve detectability (see, for example, Shu and Lin 2010; Shu et al. 2013) or supervisory control strategies to enforce detectability (see, for example, Shu and Lin 2013b). The complexity of verifying different versions of detectability is considered in Zhang (2017), Masopust (2018). Researchers have also started looking at notions of detectability in labeled Petri nets, with some initial results appearing in Zhang and Giua (2018), Masopust and Yin (2019).

Notions of detectability have been extended to stochastic settings, starting with Shu et al. (2008) and continuing with Keroglou and Hadjicostis (2015, 2017), Zhao et al. (2019). The work in Zhao et al. (2019) introduced and analyzed a measure of detectability in probabilistic finite automata (PFA), whereas the works in Keroglou and Hadjicostis (2015, 2017) study different stochastic versions of detectability in PFA. Below, we discuss two such versions.

- *A-detectability* considers problematic sequences of events (i.e., sequences of events that generate observations that do not allow the external observer to eventually determine the exact state of the system or continue to know it exactly), and requires that the total (prior) probability of such sequences goes to zero as the number of observed events increases. Effectively, this means that if one builds a (logical) observer (using the NFA that results from the given PFA when one ignores probabilities), the resulting observer can have problematic cycles (that involve sets of state estimates of cardinality larger than one), as long as these cycles are *transient*, i.e., there are events (with nonzero probability) that will eventually drive us out of these problematic cycles. This approach for checking A-detectability relies on the construction of an observer, which means that it has complexity exponential in the size of the given NFA (its complexity is linear in the size of its observer, which is in turn exponential in the size of the given PFA).
- *AA-detectability* relaxes A-detectability by considering the posterior probability on the various states, conditioned on the observed sequence. This means that even though a particular sequence of observations may not allow perfect state estimation in the logical sense (because there is a nonzero posterior probability for more than one possible states), one may consider whether the (posterior) probabilities of the various state estimates allow *almost* exact state estimation, with increasing certainty as more information is acquired from observing the behavior of the given PFA. Specifically, AA-detectability requires that the prior probability of sequences of events that allow the external observer to determine the exact state of the system,

with probability arbitrarily close to unity, goes to zero as the number of observed events increases. The work in Keroglou and Hadjicostis (2017) established necessary and sufficient conditions for AA-detectability, and showed that it can be verified with complexity that is polynomial in the size of the given PFA.

Conformance testing, distinguishing sequences, checking sequences, and homing sequences were topics that were heavily researched in the early days of digital system design. The pioneering work in Hennie (1964, 1968) showed that, subject to certain assumptions about the machine structure, conformance testing (i.e., testing whether a machine conforms to its specification) can take time polynomial in the size of the machine and in the possibly exponential length of its distinguishing sequence (if one exists). As mentioned in Chap. 5, following the work by Moore and Hennie, many researchers studied related topics, managing to refine these techniques (e.g., by improving the bounds on the length of checking sequences Kime 1966; Gönenc 1970; Hsieh 1971; Vasilevskii 1973; Chow 1978) and to demonstrate how a machine may be augmented with additional inputs and/or outputs in order to facilitate testing (e.g., allowing an FSM that initially has no distinguishing sequence to posses one Kohavi and Lavalee 1967; Murakami et al. 1970; Sheppart and Vranesic 1974; Fujiwara and Kinoshita 1978; Pradhan 1983; Bhattacharyya 1983). The work in Yannakakis and Lee (1994) has shown that it is PSPACE-complete to verify whether or not a DFA has a distinguishing sequence (as there exist machines whose shortest distinguishing sequence is exponential in length). However, they have also shown that one can determine in polynomial time whether a DFA has an "adaptive" distinguishing sequence, and, if that is the case, one can find such a sequence (whose length is argued to be $O(N^2)$) (where $N = |Q|$ is the number of states of the given automaton) in polynomial time. Recent work has focused on extending these ideas to the case where the underlying system may be nondeterministic (see, for example, Kushik et al. 2016 and references therein).

References

Bhattacharyya A (1983) On a novel approach of fault detection in an easily testable sequential machine with extra inputs and extra outputs. IEEE Trans Comput 32(3):323–325

Caines PE, Greiner R, Wang S (1988) Dynamical logic observers for finite automata. In: Proceedings of IEEE conference on decision and control (CDC), pp 226–233

Caines P, Greiner R, Wang S (1991) Classical and logic-based dynamic observers for finite automata. IMA J Math Control Inf 8(1):45–80

Chow TS (1978) Testing software design modeled by finite-state machines. IEEE Trans Softw Eng 4(3):178–187

Fujiwara H, Kinoshita K (1978) On the complexity of system diagnosis. IEEE Trans Comput 27(10):881–885

Gönenc G (1970) A method for the design of fault detection experiments. IEEE Trans Comput 19(6):551–558

Hadjicostis CN (2012) Resolution of initial-state in security applications of DES. In: Proceedings of 20th Mediterranean conference on control and automation (MED), pp 794–799

Hadjicostis CN, Seatzu C (2016) K-Detectability in discrete event systems. In: Proceedings of 55th IEEE conference on decision and control (CDC), pp 420–425
Hennie FC (1964) Fault detecting experiments for sequential circuits. In: Proceedings of the 5th annual symposium on switching circuit theory and logical design, pp 95–110
Hennie FC (1968) Finite state models for logical machines. Wiley, New York
Hsieh EP (1971) Checking experiments for sequential machines. IEEE Trans Comput 20(10):1152–1166
Keroglou C, Hadjicostis CN (2015) Detectability in stochastic discrete event systems. Syst Control Lett 84:21–26
Keroglou C, Hadjicostis CN (2017) Verification of detectability in probabilistic finite automata. Automatica 86:192–198
Kime CR (1966) An organization for checking experiments on sequential circuits. IEEE Trans Electron Comput 15(1):113–115
Kohavi Z, Lavalee P (1967) Design of sequential machines with fault detection capabilities. IEEE Trans Electron Comput 16(4):473–484
Kushik N, El-Fakih K, Yevtushenko N, Cavalli AR (2016) On adaptive experiments for nondeterministic finite state machines. Int J Softw Tools Technol Transf 18(3):251–264
Lee D, Yannakakis M (1996) Principles and methods of testing finite state machines. Proc IEEE 84(8):1090–1123
Masopust T (2018) Complexity of deciding detectability in discrete event systems. Automatica 93:257–261
Masopust T, Yin X (2019) Deciding detectability for labeled Petri nets. Automatica 104:238–241
Murakami SI, Kinoshita K, Ozaki Z (1970) Sequential machines capable of fault diagnosis. IEEE Trans Comput 19(11):1079–1085
Özveren CM, Willsky AS (1990) Observability of discrete event dynamic systems. IEEE Trans Autom Control 35(7):797–806
Pradhan DK (1983) Sequential network design using extra inputs for fault detection. IEEE Trans Comput 32(3):319–323
Sandberg S (2005) Homing and synchronizing sequences. Model-based testing of reactive systems. Springer, Berlin, pp 39–49
Sheppart DA, Vranesic ZG (1974) Fault detection of binary sequential machines. IEEE Trans Comput 23(4):352–358
Shu S, Lin F (2010) Detectability of discrete event systems with dynamic event observation. Syst Control Lett 59(1):9–17
Shu S, Lin F (2012) I-detectability of discrete-event systems. IEEE Trans Autom Sci Eng 10(1):187–196
Shu S, Lin F (2013a) Delayed detectability of discrete event systems. IEEE Trans Autom Control 58(4):862–875
Shu S, Lin F (2013b) Enforcing detectability in controlled discrete event systems. IEEE Trans Autom Control 58(8):2125–2130
Shu S, Lin F, Ying H (2007) Detectability of discrete event systems. IEEE Trans Autom Control 52(12):2356–2359
Shu S, Lin F, Ying H, Chen X (2008) State estimation and detectability of probabilistic discrete event systems. Automatica 44(12):3054–3060
Shu S, Huang Z, Lin F (2013) Online sensor activation for detectability of discrete event systems. IEEE Trans Autom Sci Eng 10(2):457–461
Vasilevskii MP (1973) Failure diagnosis in automata. Kybernetika 9(4):98–108
Yannakakis M, Lee D (1994) Testing finite-state machines: state identification and verification. IEEE Trans Comput 43(3):209–227
Yin X (2017) Initial-state detectability of stochastic discrete-event systems with probabilistic sensor failures. Automatica 80:127–134
Zhang K (2017) The problem of determining the weak (periodic) detectability of discrete event systems is PSPACE-complete. Automatica 81:217–220

Zhang K, Giua A (2018) Weak (approximate) detectability of labeled Petri net systems with inhibitor arcs. IFAC-PapersOnLine 51(7):167–171

Zhao P, Shu S, Lin F, Zhang B (2019) Detectability measure for state estimation of discrete event systems. IEEE Trans Autom Control 64(1):433–439

Chapter 7
Diagnosability

7.1 Introduction and Motivation

Starting with the work in Sampath et al. (1995), many researchers have studied fault diagnosis in systems that can be modeled using finite automata formulations. The most commonly studied fault diagnosis setting deals with a labeled automaton under a natural projection mapping, with some of the events being unobservable. Under these assumptions, fault events are taken to be unobservable; otherwise, they will be trivially detected and identified (as soon as they occur), because they will be associated with a unique label. The assumption of knowledge of the system model is rather harmless and natural in many engineered systems (where the system design and implementation—and thus the system model—are known precisely); as a result, the model-based approach to fault diagnosis that is described in this chapter has found success in a number of applications, such as diagnostics of complex machinery (e.g., copier machines and printers, baggage handling systems), network protocol design, manufacturing systems, and others.

Given a set of fault events, fault detection aims to unambiguously determine that a fault has happened following a *finite* number of events after its occurrence. Similarly, given a set of mutually exclusive fault classes (i.e., classes that do not share fault events), fault classification aims to unambiguously determine whether or not faults from one or more of the given classes have occurred, and again requires that this can be accomplished within a *finite* number of events following the occurrence of the fault(s). Fault identification is a special case of fault classification where each fault class is a singleton set. Thus, fault identification amounts to precise characterization of the fault that has taken place, again following a finite number of events after its occurrence. When multiple fault occurrences are allowed, one might also be interested in determining how many times each fault (or a fault from each class of faults) has occurred, or in knowing which fault (or which fault class) has occurred first. In this chapter, we focus mostly on the problems of fault detection and fault classification, and only briefly discuss solutions to the variants of these latter problems.

Chapter 5 introduced different versions of the state isolation problem and discussed ways to verify them using state estimator constructions (including a current-state estimator, an initial-state estimator, and a D-delayed-state estimator). As

C. N. Hadjicostis, *Estimation and Inference in Discrete Event Systems*, Communications and Control Engineering,
https://doi.org/10.1007/978-3-030-30821-6_7

mentioned in Chap. 5, the problem of fault diagnosis is closely related to the problem of state isolation, and this chapter establishes these connections more explicitly. Apart from describing ways to perform fault diagnosis and event detection/identification using recursive estimators, we will also discuss how one can verify the system property of *diagnosability*, i.e., the ability to detect/classify all faults of interest in a given finite automaton following a *finite* sequence of events after the occurrence of the fault (see Sect. 7.3). We will see that diagnosability is a property that is closely related to state isolation. Thus, it can be verified by constructing (current) state estimators (as described in Chap. 5) for an appropriately modified version of the given finite automaton; as discussed in Sect. 7.3.1, the complexity of these estimators (which in the context of diagnosis are called *diagnosers*) is, in the worst-case, exponential in the number of states of the given finite automaton. In Sect. 7.3.2 of this chapter, we will also see that a different set of techniques (namely the construction of a *verifier*) can be used to check diagnosability, with complexity that is polynomial in the number of states of the given automaton.

7.2 Fault Diagnosis and Event Inference

7.2.1 Problem Formulation: Fault Inference from a Sequence of Observations

As mentioned in the previous section, most formulations of fault diagnosis problems in finite automata typically consider labeled automata under a natural projection mapping; thus, the faults (events) that need to be inferred are taken to be unobservable (otherwise, since each fault will be associated with a unique label, the task of fault detection/identification will become trivial). In this chapter, we consider a more general setup where we are given a deterministic finite automaton (DFA) with outputs and possibly silent transitions; based on our knowledge of the system model (including possibly partial knowledge of the initial state) and the sequence of observations generated by underlying activity in the system, our goal is to diagnose (specifically, detect and/or classify/identify) certain events. Note that we use the terms "diagnosis of faults" and "inference of events" interchangeably to refer to the underlying task of fault/event detection and/or classification/identification (the exact meaning will be made precise if it cannot be inferred from context).

We now describe the setting more formally, sometimes making reference to the more popular labeled automaton case. We are given a DFA with outputs and possibly silent transitions $G = (Q, \Sigma, Y \cup \{\epsilon\}, \delta, \lambda, Q_0)$, with state set Q, input set Σ, output set Y, (possibly partially defined) next-state transition function $\delta : Q \times \Sigma \to Q$, output function $\lambda : Q \times \Sigma \to Y \cup \{\epsilon\}$ (with ϵ representing the empty output), and a set of possible initial states $Q_0 \subseteq Q$ (see Definition 3.15 in Chap. 3). Furthermore, we are given a set of fault events $F \subseteq \Sigma$ that is partitioned into C mutually exclusive fault classes, i.e.,

$$F = F_1 \dot{\cup} F_2 \dot{\cup} \cdots \dot{\cup} F_C ,$$

where $F_{c_1} \cap F_{c_2} = \emptyset$ for $c_1, c_2 \in \{1, 2, \ldots, C\}$, $c_1 \neq c_2$. We use $\mathcal{F} = \{F_1, F_2, \ldots, F_C\}$ to represent the set of fault classes. Note that the restriction to deterministic automata is mostly done to simplify notation and many of the results we present can be generalized to nondeterministic finite automata (NFA) in a rather straightforward manner (but at the cost of heavier notation).

Given a sequence of observations y_0^k (generated by some unknown underlying sequence of inputs applied to the system), fault diagnosis aims to infer the occurrence of fault events in F. [The case, where certain inputs are observable, can be handled by enhancing the set of observable outputs (appending to them the corresponding observable input) as discussed in Chap. 4; thus, the developments here will not discuss such extensions.] We will be interested in the three tasks described below.

Fault Detection: Following the sequence of observations $y_0^k \in Y^{k+1}$, we say that "a fault has been detected with certainty" if we are sure that some fault in F has occurred. In other words, given that the system started from an unknown initial state in Q_0 and given the observed sequence of outputs y_0^k, we are certain that *some* fault $f \in F$ appears in each compatible sequence of inputs σ_0^{k+m}, $m \geq 0$ (note that $\sigma_0^{k+m} \in \Sigma^{k+1}\Sigma^*$), that might have taken place. For this to happen, we need the following condition to hold: for all $q_0 \in Q_0$ and all $\sigma_0^{k+m} \in \Sigma^{k+1}\Sigma^*$ (where $m \geq 0$), we have

$$\left\{E(\lambda_{seq}(q_0, \sigma_0^{k+m})) = y_0^k\right\} \Rightarrow \{\exists i \in \{0, 1, \ldots, k+m\},\ \exists f \in F,\ \sigma[i] = f\} .$$

[Recall that $\sigma_0^{k+m} = \sigma[0], \sigma[1], \ldots, \sigma[k+m]$, and the functions E and λ_{seq} were defined in Chap. 3 (see Eq. (3.9) and pointers there). Here, $E(\lambda_{seq}(q_0, \sigma_0^{k+m}))$ captures the sequence of outputs observed when the system starts at state q_0 and the sequence of inputs $\sigma[0], \sigma[1], \ldots, \sigma[k+m]$ is applied; note that $E(\lambda_{seq}(q_0, \sigma_0^{k+m}))$ could be the empty sequence (due to the presence of silent transitions) and is undefined if the sequence of inputs cannot be applied starting at state q_0.] It is worth pointing out that the fault f does not have to be the same fault for the different sequences of events that match the system model, the initial-state constraints, and the observed sequence of outputs; however, some fault appears in each such sequence and that is the reason we are certain that a fault has taken place. Note that if the same fault f appeared in each sequence that matches the system model, the initial-state constraints, and the observed sequence of outputs, then we would actually be able to identify the fault that has taken place; in general, however, the above condition does not allow us to pinpoint the exact fault that has taken place.

The reverse situation is also of interest. Specifically, following the sequence of observations y_0^k, we say that "a fault has definitely *not* occurred" if the following condition holds: for all $q_0 \in Q_0$ and all $\sigma_0^{k+m} \in \Sigma^{k+1}\Sigma^*$ (where $m \geq 0$), we have

$$\left\{E(\lambda_{seq}(q_0, \sigma_0^{k+m})) = y_0^k\right\} \Rightarrow \{\forall i \in \{0, 1, \ldots, k+m\},\ \forall f \in F,\ \sigma[i] \neq f\}.$$

Finally, there may be uncertainty about the occurrence of a fault if, among all sequences of events that match the system model, the initial-state constraints, and the observed sequence of outputs y_0^k, some contain a fault and some do not contain a fault. For this to occur, both of the following two conditions need to hold:
(i) there exists $\sigma_0^{k+m} \in \Sigma^{k+1}\Sigma^*$ (where $m \geq 0$) such that $E(\lambda_{seq}(q_0, \sigma_0^{k+m})) = y_0^k$ and $\exists i \in \{0, 1, \ldots, k+m\}$, $\exists f \in F$, $\sigma[i] = f$, and
(ii) there exists (a different) $\sigma_0^{k+m} \in \Sigma^{k+1}\Sigma^*$ (where $m \geq 0$) such that $E(\lambda_{seq}(q_0, \sigma_0^{k+m})) = y_0^k$ and $\forall i \in \{0, 1, \ldots, k+m\}$, $\forall f \in F$, $\sigma[i] \neq f$.

Single Fault Classification: Following the sequence of observations $y_0^k \in Y^{k+1}$, we say that "fault class F_c has been identified" if we are certain that some fault in F_c has occurred. In other words, given that the system started from an unknown state in Q_0 and given the observed sequence of outputs y_0^k, any compatible sequence of events σ_0^{k+m}, $m \geq 0$ (note that $\sigma_0^{k+m} \in \Sigma^{k+1}\Sigma^*$) that matches the system model, the initial-state constraints, and the observed sequence of outputs necessarily contains a fault in class F_c for some $c \in \{1, 2, \ldots, C\}$ (i.e., some $f_c \in F_c$ appears in each such σ_0^{k+m}). For this to happen, we need the following condition to hold: for all $q_0 \in Q_0$ and all $\sigma_0^{k+m} \in \Sigma^{k+1}\Sigma^*$ (where $m \geq 0$), we have

$$\left\{E(\lambda_{seq}(q_0, \sigma_0^{k+m})) = y_0^k\right\} \Rightarrow \{\exists i \in \{0, 1, \ldots, k+m\},\ \exists f_c \in F_c,\ \sigma[i] = f_c\}.$$

Note that the fault f_c does not have to be the same for the different sequences of events that match the system model, the initial-state constraints, and the observed sequence of outputs, as long as f_c belongs in the fault class F_c. Clearly, if each fault class forms a singleton set, then fault classification amounts to fault identification. The reason for introducing fault classes is to allow for more flexibility in the fault diagnosis process, as in many cases we might be content with identifying faults with respect to fault classes (in fact, in some cases we may not even have the ability to identify faults exactly).
The reverse situation might also be of interest in the case of fault classification. Specifically, following the sequence of observations y_0^k, we say that "a fault in class F_c has definitely *not* occurred" if the following condition holds: for all $q_0 \in Q_0$ and all $\sigma_0^{k+m} \in \Sigma^{k+1}\Sigma^*$ (where $m \geq 0$), we have

$$\left\{E(\lambda_{seq}(q_0, \sigma_0^{k+m})) = y_0^k\right\} \Rightarrow \{\forall i \in \{0, 1, \ldots, k+m\},\ \forall f_c \in F_c,\ \sigma[i] \neq f_c\}.$$

This would imply that no fault in class F_c has occurred, though it might still be the case that other faults (in one or more other fault classes) have occurred (i.e., this situation does not exclude fault detection).

Finally, there may be uncertainty about the occurrence of a fault in class F_c if, among all sequences of events that match the system model, the initial-state constraints, and the observed sequence of outputs y_0^k, some contain a fault in F_c and some do not contain a fault in F_c. For this to occur, both of the following two conditions need to hold:

(i) there exists $\sigma_0^{k+m} \in \Sigma^{k+1}\Sigma^*$ (where $m \geq 0$) such that $E(\lambda_{seq}(q_0, \sigma_0^{k+m})) = y_0^k$ and $\exists i \in \{0, 1, \ldots, k+m\},\ \exists f_c \in F_c,\ \sigma[i] = f_c$, and

(ii) there exists (a different) $\sigma_0^{k+m} \in \Sigma^{k+1}\Sigma^*$ (where $m \geq 0$) such that $E(\lambda_{seq}(q_0, \sigma_0^{k+m})) = y_0^k$ and $\forall i \in \{0, 1, \ldots, k+m\},\ \forall f_c \in F_c,\ \sigma[i] \neq f_c$.

Note that the above conditions for fault classification resemble the conditions for fault detection if one replaces the set F with the set F_c. However, the situation gets a bit more complicated when one has to deal with the possible occurrence of multiple faults.

Multiple Fault Detection/Classification: When multiple faults occur, the same questions about fault detection and fault classification can be asked (i.e., for fault detection we can still ask whether a fault—from one or more fault classes—has occurred at least once, and for fault classification, we can ask, for each fault class, whether at least one fault in that particular class has occurred). For example, suppose that following the observation of the sequence of outputs y_0^k, we establish the following: for all $q_0 \in Q_0$ and all $\sigma_0^{k+m} \in \Sigma^{k+1}\Sigma^*$ (where $m \geq 0$), we have

$$\left\{E(\lambda_{seq}(q_0, \sigma_0^{k+m})) = y_0^k\right\} \Rightarrow$$
$$\Rightarrow \left\{\exists i_1 \in \{0, 1, \ldots, k+m\},\ \exists f_{c_1} \in F_{c_1},\ \sigma[i_1] = f_{c_1}\right\}$$

and

$$\left\{E(\lambda_{seq}(q_0, \sigma_0^{k+m})) = y_0^k\right\} \Rightarrow$$
$$\Rightarrow \left\{\exists i_2 \in \{0, 1, \ldots, k+m\},\ \exists f_{c_2} \in F_{c_2},\ \sigma[i_2] = f_{c_2}\right\}.$$

Then, we can conclude that at least one fault in class F_{c_1} and at least one fault in class F_{c_2} have occurred. The above is equivalent to the condition: for all $q_0 \in Q_0$ and all $\sigma_0^{k+m} \in \Sigma^{k+1}\Sigma^*$ (where $m \geq 0$), we have

$$\left\{E(\lambda_{seq}(q_0, \sigma_0^{k+m})) = y_0^k\right\} \Rightarrow$$
$$\Rightarrow \left\{\exists i_1, i_2 \in \{0, \ldots, k+m\},\ \exists f_{c_1} \in F_{c_1}, \exists f_{c_2} \in F_{c_2},\ \sigma[i_1] = f_{c_1}, \sigma[i_2] = f_{c_2}\right\}.$$

This condition requires that faults from both fault classes F_{c_1} and F_{c_2} appear in each sequence of events that matches the initial conditions and generates the observed sequence of outputs. Clearly, the above approach generalizes to faults from more than two fault classes.

Note that the above approach does not take into account the order in which faults from different fault classes occur (e.g., whether fault $f_{c_1} \in F_{c_1}$ occurs first and fault $f_{c_2} \in F_{c_2}$ occurs later). Also, the above definitions do not pay attention to the number of occurrences of the same fault or the number of faults from different fault classes. Such generalizations can be made at the cost of more complicated detectability/classification requirements. For simplicity, in this chapter, we focus on multiple fault classification that ignores the temporal order of faults as well as the number of occurrences of each fault class (though we will sometimes comment on how such questions can be addressed). Our primary goal will be to determine whether, given a sequence of observations $y_0^k \in Y^{k+1}$, the following condition holds for one or more fault classes $F_c \in \mathcal{F}$: for all $q_0 \in Q_0$ and all σ_0^{k+m} (where $m \geq 0$, i.e., $\sigma_0^{k+m} \in \Sigma^{k+1}\Sigma^*$), we have

$$\left\{E(\lambda_{seq}(q_0, \sigma_0^{k+m})) = y_0^k\right\} \Rightarrow \{\exists i \in \{0, 1, \ldots, k+m\},\ \exists f_c \in F_c,\ \sigma[i] = f_c\}.$$

Note that this condition will need to be verified separately for each fault class $F_c \in \mathcal{F}$. Nevertheless, it should be evident from the above discussion that (in the absence of a need to determine the temporal order of faults and the number of occurrences of faults from each class) the problem of multiple fault detection and classification essentially reduces to multiple instances of single fault detection/classification.

We summarize the discussion in this section as follows. We have considered a DFA with outputs and possibly silent transitions $G = (Q, \Sigma, Y \cup \{\epsilon\}, \delta, \lambda, Q_0)$, with a set of fault events $F \subseteq \Sigma$ that is partitioned into C (mutually exclusive) fault classes $\mathcal{F} = \{F_1, F_2, \ldots, F_C\}$, where $F_{c_1} \cap F_{c_2} = \emptyset$ for $c_1, c_2 \in \{1, 2, \ldots, C\}$, $c_1 \neq c_2$. Following the observation of the sequence of outputs $y_0^k \in Y^{k+1}$, we are interested in determining whether the fault diagnosis conclusions listed below hold.

- (i) **Fault Detection with Certainty**: A fault has occurred with certainty when: for all $q_0 \in Q_0$ and all σ_0^{k+m} (where $m \geq 0$, i.e., $\sigma_0^{k+m} \in \Sigma^{k+1}\Sigma^*$), we have

$$\left\{E(\lambda_{seq}(q_0, \sigma_0^{k+m})) = y_0^k\right\} \Rightarrow \{\exists i \in \{0, 1, \ldots, k+m\},\ \exists f \in F,\ \sigma[i] = f\}. \tag{7.1}$$

- (ii) **Fault Absence with Certainty**: No fault has occurred when: for all $q_0 \in Q_0$ and all σ_0^{k+m} (where $m \geq 0$, i.e., $\sigma_0^{k+m} \in \Sigma^{k+1}\Sigma^*$), we have

$$\left\{E(\lambda_{seq}(q_0, \sigma_0^{k+m})) = y_0^k\right\} \Rightarrow \{\forall i \in \{0, 1, \ldots, k+m\},\ \forall f \in F,\ \sigma[i] \neq f\}. \tag{7.2}$$

- (iii) **Possible Fault Occurrence**: A fault *may* have occurred if neither Condition (7.1) nor Condition (7.2) above holds.
- (iv) **Fault Class Presence with Certainty**: A fault in class $F_c \in \mathcal{F}$ has occurred with certainty when: for all $q_0 \in Q_0$ and all σ_0^{k+m} (where $m \geq 0$, i.e., $\sigma_0^{k+m} \in \Sigma^{k+1}\Sigma^*$), we have

$$\left\{E(\lambda_{seq}(q_0, \sigma_0^{k+m})) = y_0^k\right\} \Rightarrow \{\exists i \in \{0, 1, \ldots, k+m\},\ \exists f_c \in F_c,\ \sigma[i] = f_c\}. \tag{7.3}$$

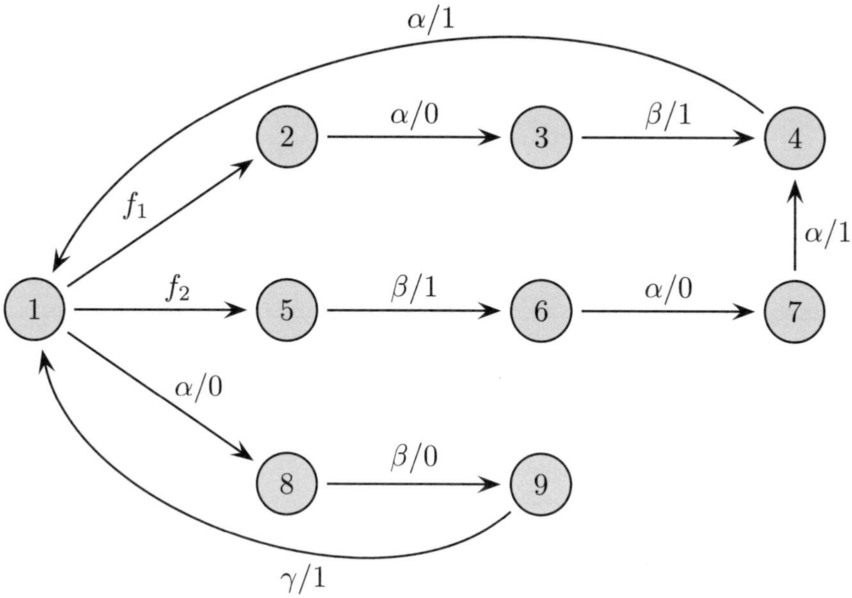

Fig. 7.1 Finite automaton G used to illustrate the conditions for fault identification

This condition has to be verified separately for each fault class $F_c \in \mathcal{F}$. Note that the above condition necessarily implies that Condition (7.1) holds, but not the other way around: Condition (7.1) may hold but Condition (7.3) may not hold for any fault class $F_c \in \mathcal{F}$.

- (v) **Fault Class Absence with Certainty**: No fault in class $F_c \in \mathcal{F}$ has occurred when: for all $q_0 \in Q_0$ and all σ_0^{k+m} (where $m \geq 0$, i.e., $\sigma_0^{k+m} \in \Sigma^{k+1}\Sigma^*$), we have

$$\left\{E(\lambda_{seq}(q_0, \sigma_0^{k+m})) = y_0^k\right\} \Rightarrow \{\forall i \in \{0, 1, \ldots, k+m\},\ \forall f_c \in F_c,\ \sigma[i] \neq f_c\}. \tag{7.4}$$

Again, this condition has to be verified separately for each fault class $F_c \in \mathcal{F}$.

- (vi) **Possible Fault Class Presence**: A fault in class F_c *may* have occurred if neither Condition (7.3) nor Condition (7.4) above holds. Again, this has to be verified separately for each fault class $F_c \in \mathcal{F}$.

Example 7.1 Consider the finite automaton with outputs $G = (Q, \Sigma, Y \cup \{\epsilon\}, \delta, \lambda, Q_0)$ shown in Fig. 7.1 (slightly modified version of the one considered in Fig. 5.1 of Chap. 5), where $Q = \{1, 2, 3, 4, 5, 6, 7, 8, 9\}$, $\Sigma = \{\alpha, \beta, \gamma, f_1, f_2\}$, $Y = \{0, 1\}$, $Q_0 = \{1\}$, and δ and λ are as shown at the top of the figure. We assume two fault classes, $F_1 = \{f_1\}$ and $F_2 = \{f_2\}$, and analyze the problem of fault classification (which in this example amounts to fault identification).

We first consider what might happen when the system starts operation (from state 1). We can easily establish the following:

1. If fault f_1 occurs, one will necessarily observe output 0, followed by output 1 (which will necessarily be followed by output 1). Once the observation sequence $y_0^1 = 01$ is seen, it is not hard to conclude that the only possible explanation is the sequence of events $f_1\alpha\beta$, which allows us to infer that fault f_1 has occurred. Thus, identification of fault f_1 is possible with a delay of two events.
2. Following the occurrence of fault f_2, one will necessarily observe output 1 (which will necessarily be followed by output 0, then output 1, and then output 1). Once the observation sequence $y_0^0 = 1$ is seen, it is not hard to see that the only possible explanation is the sequence of events $f_2\beta$, which allows us to infer that fault f_2 has occurred. In this case, identification of fault f_2 is possible with a delay of one event.
3. If no fault occurs, then we will necessarily observe output 00 which allows to conclude that no fault has occurred. Note that we will only be certain that no fault has occurred only after we observe the second 0. [If a single 0 is observed, the possible explanations involve $f_1\alpha$ and α, thus we will be uncertain about the occurrence of the fault. Notice, however, that eventually (once another event occurs) this issue will be resolved.]

A similar approach can be used to analyze what happens once the system has been in operation. Following a sequence of observations of the form $(001)^*(00)^*$, we know that no fault has occurred. The first fault can be identified as f_1 if we observe $(001)^*01$, or as f_2 if we observe $(001)^*1$.

The above discussion establishes that in this particular example we will always be able to identify the first fault that occurs. The analysis for detecting/identifying a second fault, perhaps of a different type that might occur following the first one, is not difficult for this particular example, but in general the reasoning can become complicated because we will have to consider the possible states that the system can be in (following the observations and the possible fault explanations). The question of how this can be done systematically using a reduction to the state isolation problem is the focus of the next section of this chapter. □

7.2.2 *Reduction of Fault Diagnosis to State Isolation*

Fault detection and classification as defined in the previous section can be easily converted to a state isolation problem as discussed in Chap. 5 (refer to Fig. 5.1 for an illustration of the approach in the case of a simple finite automaton with nine states). In this section we formalize this reduction.

Suppose we are given a DFA with outputs and possibly silent transitions $G = (Q, \Sigma, Y \cup \{\epsilon\}, \delta, \lambda, Q_0)$ with state set $Q = \{q^{(1)}, q^{(2)}, \ldots, q^{(N)}\}$, input set Σ, output set Y, (possibly partially defined) next-state transition function $\delta : Q \times \Sigma \to Q$, output function $\lambda : Q \times \Sigma \to Y \cup \{\epsilon\}$ (with ϵ representing the empty output), and a set of possible initial states $Q_0 \subseteq Q$. Furthermore, we are given a set of fault events $F \subseteq \Sigma$ that is partitioned into C mutually exclusive fault classes, i.e.,

$$F = F_1 \dot{\cup} F_2 \dot{\cup} \cdots \dot{\cup} F_C ,$$

where $F_{c_1} \cap F_{c_2} = \emptyset$ for $c_1, c_2 \in \{1, 2, \ldots, C\}$, $c_1 \neq c_2$. We use $\mathcal{F} = \{F_1, F_2, \ldots, F_C\}$ to represent the different fault classes.

Given G, if we are simply interested in fault detection, we can construct a DFA $G_F = (Q \dot{\cup} Q_F, \Sigma, Y \cup \{\epsilon\}, \delta_F, \lambda_F, Q_0)$ with state set $Q \dot{\cup} Q_F$ where $Q_F = \{q_F^{(1)}, q_F^{(2)}, \ldots, q_F^{(N)}\}$, and δ_F and λ_F defined so that the restriction to states Q_F essentially yields a copy of the original automaton (with state $q^{(i)}$ represented by state $q_F^{(i)}$), whereas the restriction to states Q yields the transition and output functionality of the original automaton with the only difference being that fault events, instead of taking us to states in Q, they take us to the corresponding states in Q_F (i.e., under some fault event $f \in F$, instead of a transition from state $q^{(i)}$ to state $q^{(j)}$ that generates output $y \in Y \cup \{\epsilon\}$, we transition from state $q^{(i)}$ to state $q_F^{(j)}$ while generating the same output y). Formally, this is described as follows:

$$\delta_F(q^{(i)}, \sigma) = \delta(q^{(i)}, \sigma) \ , \ \text{for}\, q^{(i)} \in Q, \sigma \in \Sigma \setminus F \ , \tag{7.5}$$
$$\delta_F(q^{(i)}, \sigma) = q_F^{(j)} \ \text{when}\, q^{(j)} = \delta(q^{(i)}, \sigma) \ , \ \text{for}\, q^{(i)} \in Q, \sigma \in F \ , \tag{7.6}$$
$$\delta_F(q_F^{(i)}, \sigma) = q_F^{(j)} \ \text{when}\, q^{(j)} = \delta(q^{(i)}, \sigma) \ , \ \text{for}\, q_F^{(i)} \in Q_F, \sigma \in \Sigma \ , \tag{7.7}$$

and

$$\lambda_F(q^{(i)}, \sigma) = \lambda(q^{(i)}, \sigma) \ , \ \text{for}\, q^{(i)} \in Q, \sigma \in \Sigma \ , \tag{7.8}$$
$$\lambda_F(q_F^{(i)}, \sigma) = \lambda(q^{(i)}, \sigma) \ , \ \text{for}\, q_F^{(i)} \in Q_F, \sigma \in \Sigma \ . \tag{7.9}$$

[In the above definitions, $\delta_F(q, \sigma)$ (or $\lambda_F(q, \sigma)$) is taken to be undefined if the needed $\delta(q, \sigma)$ (or $\lambda(q, \sigma)$) is undefined.]

Given the above construction, it is not very hard to verify that the following lemma holds.

Lemma 7.1 *Consider a DFA with outputs and possibly silent transitions* $G = (Q, \Sigma, Y \cup \{\epsilon\}, \delta, \lambda, Q_0)$ *with a set of fault events* $F \subseteq \Sigma$. *Suppose that the corresponding finite automaton* G_F *is constructed as described above in* (7.5)–(7.9). *Given a sequence of observations* y_0^k *generated by unknown underlying activity in automaton* G, *a fault is detected with certainty (i.e., Condition* (7.1) *is satisfied) if and only if for automaton* G_F *we have*

$$\hat{q}_{y[k]}(y_0^k) \subseteq Q_F \ ,$$

i.e., the state estimate (for automaton G_F) following observation $y[k]$ is isolated to be within the set Q_F.

The above lemma establishes that the problem of checking Condition (7.1) for an observed sequence of outputs y_0^k is equivalent to state isolation in G_F with respect to the set of states Q_F, given the observed sequence of outputs y_0^k. Similar statements can be made for the absence of a fault (i.e., Condition (7.2) is equivalent to $\hat{q}_{y[k]}(y_0^k) \subseteq Q$), and for uncertainty about the occurrence of a fault (which is equivalent to simultaneously having $\hat{q}_{y[k]}(y_0^k) \cap Q \neq \emptyset$ and $\hat{q}_{y[k]}(y_0^k) \cap Q_F \neq \emptyset$).

Given G, if we are not simply interested in fault detection but we also want to perform fault classification with respect to fault classes $\mathcal{F} = \{F_1, F_2, \ldots, F_C\}$, then we can construct, for each fault class $F_c \in \mathcal{F}$, a DFA $G_{F_c} = (Q \dot{\cup} Q_{F_c}, \Sigma, Y \cup \{\epsilon\}, \delta_{F_c}, \lambda_{F_c}, Q_0)$ with state set $Q \dot{\cup} Q_{F_c}$ where $Q_{F_c} = \{q_{F_c}^{(1)}, q_{F_c}^{(2)}, \ldots, q_{F_c}^{(N)}\}$, and δ_{F_c} and λ_{F_c} defined so that the restriction to states Q_{F_c} essentially yields a copy of the original automaton (with state $q^{(i)}$ represented by state $q_{F_c}^{(i)}$), whereas the restriction to states Q yields the transition and output functionality of the original automaton, with the only difference being that fault events in fault class F_c, instead of taking us to states in Q, they take us to the corresponding states in Q_{F_c} (i.e., under some fault event $f_c \in F_c$, instead of a transition from state $q^{(i)}$ to state $q^{(j)}$ that generates output $y \in Y \cup \{\epsilon\}$, we transition from state $q^{(i)}$ to state $q_{F_c}^{(j)}$ while generating the same output y). In this construction, faults in other fault classes (i.e., faults in the set $F \setminus F_c$) are treated as regular events. Formally, the next-state transition and output functions of automata G_{F_c}, $c = 1, 2, \ldots, C$, are described as follows:

$$\delta_{F_c}(q^{(i)}, \sigma) = \delta(q^{(i)}, \sigma), \text{ for } q^{(i)} \in Q, \sigma \in \Sigma \setminus F_c \,, \tag{7.10}$$

$$\delta_{F_c}(q^{(i)}, \sigma) = q_{F_c}^{(j)} \text{ when } q^{(j)} = \delta(q^{(i)}, \sigma), \text{ for } q^{(i)} \in Q, \sigma \in F_c \,, \tag{7.11}$$

$$\delta_{F_c}(q_{F_c}^{(i)}, \sigma) = q_{F_c}^{(j)} \text{ when } q^{(j)} = \delta(q^{(i)}, \sigma), \text{ for } q_{F_c}^{(i)} \in Q_{F_c}, \sigma \in \Sigma, \tag{7.12}$$

and

$$\lambda_{F_c}(q^{(i)}, \sigma) = \lambda(q^{(i)}, \sigma) \,, \text{ for } q^{(i)} \in Q, \sigma \in \Sigma \,, \tag{7.13}$$

$$\lambda_{F_c}(q_{F_c}^{(i)}, \sigma) = \lambda(q^{(i)}, \sigma) \,, \text{ for } q_{F_c}^{(i)} \in Q_{F_c}, \sigma \in \Sigma \,. \tag{7.14}$$

[Again, $\delta_{F_c}(q, \sigma)$ (or $\lambda_{F_c}(q, \sigma)$) is taken to be undefined if the needed $\delta(q, \sigma)$ (or $\lambda(q, \sigma)$) is undefined.]

Given the above constructions, it is not very hard to verify that the following lemma holds.

Lemma 7.2 *Consider a DFA with outputs and possibly silent transitions $G = (Q, \Sigma, Y \cup \{\epsilon\}, \delta, \lambda, Q_0)$ with a set of fault events $F \subseteq \Sigma$, that is partitioned into C fault classes $\{F_1, F_2, \ldots, F_C\}$ (where $F = F_1 \dot{\cup} F_2 \dot{\cup} \cdots \dot{\cup} F_C$ and $F_{c_1} \cap F_{c_2} = \emptyset$ for $c_1, c_2 \in \{1, 2, \ldots, C\}$, $c_1 \neq c_2$). Suppose that the corresponding finite automata*

G_{F_c}, $c = 1, 2, \ldots, C$, are constructed as described above in (7.10)–(7.14). Given a sequence of observations y_0^k generated by unknown underlying activity in automaton G, a fault in class F_c has certainly occurred (i.e., Condition (7.3) is satisfied) if and only if for automaton G_{F_c} we have

$$\hat{q}_{y[k]}(y_0^k) \subseteq Q_{F_c} ,$$

i.e., the state estimate (for automaton G_{F_c}) following observation $y[k]$ is isolated to be within the set Q_{F_c}.

The above lemma establishes that the problem of checking Condition (7.3) for an observed sequence of outputs y_0^k is equivalent to state isolation in G_{F_c} with respect to the set of states Q_{F_c}, given the observed sequence of outputs y_0^k. Similar statements can be made for the absence of a fault (i.e., Condition (7.4) is equivalent to $\hat{q}_{y[k]}(y_0^k) \subseteq Q$), and for uncertainty about the occurrence of a fault in class F_c (which is equivalent to simultaneously having $\hat{q}_{y[k]}(y_0^k) \cap Q \neq \emptyset$ and $\hat{q}_{y[k]}(y_0^k) \cap Q_{F_c} \neq \emptyset$).

Remark 7.1 The advantage of using state isolation techniques is that state estimation (e.g., in automaton G_F) can be performed recursively in an efficient manner and, if at any point during the recursion, the state can be isolated (e.g., within the set Q_F), then the fault can be detected/classified/identified (in the case of G_F the fault will be detected). The alternative approach, i.e., the construction of all possible explanations for a given sequence of observations, can lead to complexity problems as the number of explanations can increase significantly (potentially exponentially) as the number of observations increases. Naturally, tracking all possible explanations for a given sequence of observations, provides all information necessary to make inferences about faults (as will become evident in our examples later on). Using state isolation techniques in appropriately constructed automata essentially allows us to concisely summarize the information provided by possible explanations in a way that not only reduces the amount of data that we track, but allows us to update it as new observations come in. □

Remark 7.2 As briefly mentioned in Chap. 5, things can become significantly more complex in the case of multiple faults. For example, if one is interested in identifying the *first* class of faults that occurs, one possibility is to replace the C finite automata constructed in Lemma 7.2 with the following (*single*) finite automaton: $G_{\mathcal{F}} = (Q\dot{\cup}Q_{F_1}\dot{\cup}Q_{F_2}\dot{\cup}\cdots\dot{\cup}Q_{F_C}, \Sigma, Y \cup \{\epsilon\}, \delta_{\mathcal{F}}, \lambda_{\mathcal{F}}, Q_0)$ where $Q_{F_c} = \{q_{F_c}^{(1)}, q_{F_c}^{(2)}, \ldots, q_{F_c}^{(N)}\}$ for $c = 1, 2, \ldots, C$, and $\delta_{\mathcal{F}}$ and $\lambda_{\mathcal{F}}$ are defined so that the restriction to states Q_{F_c} essentially yields a copy of the original automaton (with state $q^{(i)}$ represented by state $q_{F_c}^{(i)}$), whereas the restriction to states Q yields the transition and output functionality of the original automaton, with the only difference being that fault events in F_c, instead of taking us to states in Q, take us to the corresponding states in Q_{F_c} (in other words, under some fault event $f_c \in F_c$, instead of a transition from state $q^{(i)}$ to state $q^{(j)}$ that generates output $y \in Y \cup \{\epsilon\}$, we transition from state $q^{(i)}$ to state $q_{F_c}^{(j)}$ while generating output y). It is clear from the above construction that each of the sets of states Q_{F_c}, $c = 1, 2, \ldots, C$, are trapping states and if, following a sequence of observations, we can isolate the current state of the system to be within the set Q_{F_c}, then we are certain that a fault in class F_c was the *first* to occur (however, it is possible that faults from this or other fault classes subsequently occurred). Note that an example of the construction of the automaton $G_{\mathcal{F}}$ for the finite automaton on the top of Fig. 5.1 can be seen in Fig. 5.2 and was discussed in Example 5.2.

Finite automaton $G_{\mathcal{F}}$ allows the use of state estimation/isolation techniques to determine with certainty whether a particular fault class was the first to occur; this should be contrasted with the use of C different finite automata $Q_{F_c}, c = 1, 2, \ldots, C$, in Lemma 7.2, which allows us to perform fault classification. The two requirements are different because no particular fault class might consistently appear first in executions of the system that match the system model, the initial-state constraints, and the observed sequence of outputs (even when a fault class consistently appears in all executions). For example, consider fault classes $F_1 = \{f_{c_1}\}$ and $F_2 = \{f_{c_2}\}$ and a situation where, following the observation of a sequence of outputs, some matching sequences of events have f_{c_1} appearing first and f_{c_2} appearing later, whereas the remaining matching sequences of events have f_{c_2} appearing first and f_{c_1} appearing later. The construction $G_{\mathcal{F}}$ above will be able to determine that $\hat{q}_{y[k]}(y_0^k) \subseteq (Q_{F_1} \cup Q_{F_2})$; obviously, this does not identify any fault class as appearing first (as expected), but it can be used to establish that either fault class F_1, or fault class F_2, or both fault classes F_1 and F_2 have occurred. In contrast, if we use the construction in Lemma 7.2, we will be able to establish that, in this particular case, both fault classes F_1 and F_2 have occurred. Thus, the construction $G_{\mathcal{F}}$ looses some information about the occurrence of multiple faults but retains information about which fault class occurred first (if there is conclusive evidence about that).

More complicated constructions are needed if one wants to track the order in which faults from different fault classes occur (so that it can determine with certainty that, given a sequence of observations y_0^k, a fault in a particular fault class occurred first, followed by a fault in another fault class, etc.). Consider, for instance, a situation where we are given a DFA $G = (Q, \Sigma, Y \cup \{\epsilon\}, \delta, \lambda, Q_0)$ with three fault classes F_1, F_2, and F_3. The problem of keeping track of the first two fault classes that have occurred can be reduced to the problem of state estimation/isolation by using the following finite automaton:

$$G'_{\mathcal{F}} = (Q \ \dot{\cup}_{1 \le i \le 3} \ Q_{F_i} \ \dot{\cup}_{i \neq j, 1 \le i, j \le 3} \ Q_{F_i F_j}, \Sigma, Y \cup \{\epsilon\}, \delta'_{\mathcal{F}}, \lambda'_{\mathcal{F}}, Q_0) \ ,$$

where $Q_* = \{q_*^{(1)}, q_*^{(2)}, \ldots, q_*^{(N)}\}$ are separate copies of the state space Q, and $\delta'_{\mathcal{F}}$ and $\lambda'_{\mathcal{F}}$ are defined as follows:

$$
\begin{aligned}
\delta'_{\mathcal{F}}(q^{(i)}, \sigma) &= \delta(q^{(i)}, \sigma) \ , \ \text{for}\, q^{(i)} \in Q, \sigma \in \Sigma \setminus (F_1 \cup F_2 \cup F_3) \ , \\
\delta'_{\mathcal{F}}(q^{(i)}, \sigma) &= q^{(j)}_{F_1} \ \text{when}\, q^{(j)} = \delta(q^{(i)}, \sigma) \ , \ \text{for}\, q^{(i)} \in Q, \sigma \in F_1 \ , \\
\delta'_{\mathcal{F}}(q^{(i)}, \sigma) &= q^{(j)}_{F_2} \ \text{when}\, q^{(j)} = \delta(q^{(i)}, \sigma) \ , \ \text{for}\, q^{(i)} \in Q, \sigma \in F_2 \ , \\
\delta'_{\mathcal{F}}(q^{(i)}, \sigma) &= q^{(j)}_{F_3} \ \text{when}\, q^{(j)} = \delta(q^{(i)}, \sigma) \ , \ \text{for}\, q^{(i)} \in Q, \sigma \in F_3 \ ,
\end{aligned}
$$

$$
\begin{aligned}
\delta'_{\mathcal{F}}(q^{(i)}_{F_1}, \sigma) &= q^{(j)}_{F_1} \ \text{when}\, q^{(j)} = \delta(q^{(i)}, \sigma) \ , \ \text{for}\, q^{(i)}_{F_1} \in Q_{F_1}, \sigma \in \Sigma \setminus (F_2 \cup F_3), \\
\delta'_{\mathcal{F}}(q^{(i)}_{F_1}, \sigma) &= q^{(j)}_{F_1 F_2} \ \text{when}\, q^{(j)} = \delta(q^{(i)}, \sigma) \ , \ \text{for}\, q^{(i)}_{F_1} \in Q_{F_1}, \sigma \in F_2 \ , \\
\delta'_{\mathcal{F}}(q^{(i)}_{F_1}, \sigma) &= q^{(j)}_{F_1 F_3} \ \text{when}\, q^{(j)} = \delta(q^{(i)}, \sigma) \ , \ \text{for}\, q^{(i)}_{F_1} \in Q_{F_1}, \sigma \in F_3 \ , \\
\delta'_{\mathcal{F}}(q^{(i)}_{F_2}, \sigma) &= q^{(j)}_{F_2} \ \text{when}\, q^{(j)} = \delta(q^{(i)}, \sigma) \ , \ \text{for}\, q^{(i)}_{F_2} \in Q_{F_2}, \sigma \in \Sigma \setminus (F_1 \cup F_3), \\
\delta'_{\mathcal{F}}(q^{(i)}_{F_2}, \sigma) &= q^{(j)}_{F_2 F_1} \ \text{when}\, q^{(j)} = \delta(q^{(i)}, \sigma) \ , \ \text{for}\, q^{(i)}_{F_2} \in Q_{F_2}, \sigma \in F_1 \ , \\
\delta'_{\mathcal{F}}(q^{(i)}_{F_2}, \sigma) &= q^{(j)}_{F_2 F_3} \ \text{when}\, q^{(j)} = \delta(q^{(i)}, \sigma) \ , \ \text{for}\, q^{(i)}_{F_2} \in Q_{F_2}, \sigma \in F_3 \ , \\
\delta'_{\mathcal{F}}(q^{(i)}_{F_3}, \sigma) &= q^{(j)}_{F_3} \ \text{when}\, q^{(j)} = \delta(q^{(i)}, \sigma) \ , \ \text{for}\, q^{(i)}_{F_3} \in Q_{F_3}, \sigma \in \Sigma \setminus (F_1 \cup F_2), \\
\delta'_{\mathcal{F}}(q^{(i)}_{F_3}, \sigma) &= q^{(j)}_{F_3 F_1} \ \text{when}\, q^{(j)} = \delta(q^{(i)}, \sigma) \ , \ \text{for}\, q^{(i)}_{F_3} \in Q_{F_3}, \sigma \in F_1 \ , \\
\delta'_{\mathcal{F}}(q^{(i)}_{F_3}, \sigma) &= q^{(j)}_{F_3 F_2} \ \text{when}\, q^{(j)} = \delta(q^{(i)}, \sigma) \ , \ \text{for}\, q^{(i)}_{F_3} \in Q_{F_3}, \sigma \in F_2 \ ,
\end{aligned}
$$

$$
\begin{aligned}
\delta'_{\mathcal{F}}(q^{(i)}_{F_{c_1} F_{c_2}}, \sigma) = q^{(j)}_{F_{c_1} F_{c_2}} \ \text{when}\, q^{(j)} = \delta(q^{(i)}, \sigma) \ , \ \text{for}\, q^{(i)}_{F_{c_1} F_{c_2}} \in Q_{F_{c_1} F_{c_2}}, \sigma \in \Sigma, \\
\text{for}\, c_1, c_2 \in \{1, 2, 3\}, c_1 \neq c_2 \ ,
\end{aligned}
$$

and

$$
\begin{aligned}
\lambda'_{\mathcal{F}}(q^{(i)}, \sigma) &= \lambda(q^{(i)}, \sigma) \ , \ \text{for}\, q^{(i)} \in Q, \sigma \in \Sigma \ , \\
\lambda'_{\mathcal{F}}(q^{(i)}_{*}, \sigma) &= \lambda(q^{(i)}, \sigma) \ , \ \text{for}\, q^{(i)}_{*} \in Q_{*}, \sigma \in \Sigma \ .
\end{aligned}
$$

[Again, $\delta'_{\mathcal{F}}(q, \sigma)$ (or $\lambda'_{\mathcal{F}}(q, \sigma)$) is taken to be undefined if the needed $\delta(q, \sigma)$ (or $\lambda(q, \sigma)$) is undefined.] It is clear that states in the set Q_{F_1} (or Q_{F_2}, or Q_{F_3}) are reached following sequences of events that involve the occurrence of one or more faults from class F_1 (or F_2, or F_3) and *no* faults from the other two classes F_2 and F_3 (or F_1 and F_3, or F_1 and F_2). Similarly, states in the set $Q_{F_{c_1} F_{c_2}}$, $c_1 \neq c_2$, are reached following sequences of events that, in terms of faults, first involve the occurrence of one or more faults from class F_{c_1}, followed by one or more faults from class F_{c_2}, followed by zero or more faults (from any class). Clearly, state isolation with respect to the set of states of $G_{\mathcal{F}}$ implies certain temporal order information about the occurrence of faults from these different fault classes (for example, if following a sequence of observations y_0^k, we can isolate the state of the system to be within the set $Q_{F_1 F_2}$, then we know that one or more faults in fault class F_1 occurred first and eventually a fault from fault class F_2 also occurred, perhaps followed by additional faults—from any class). □

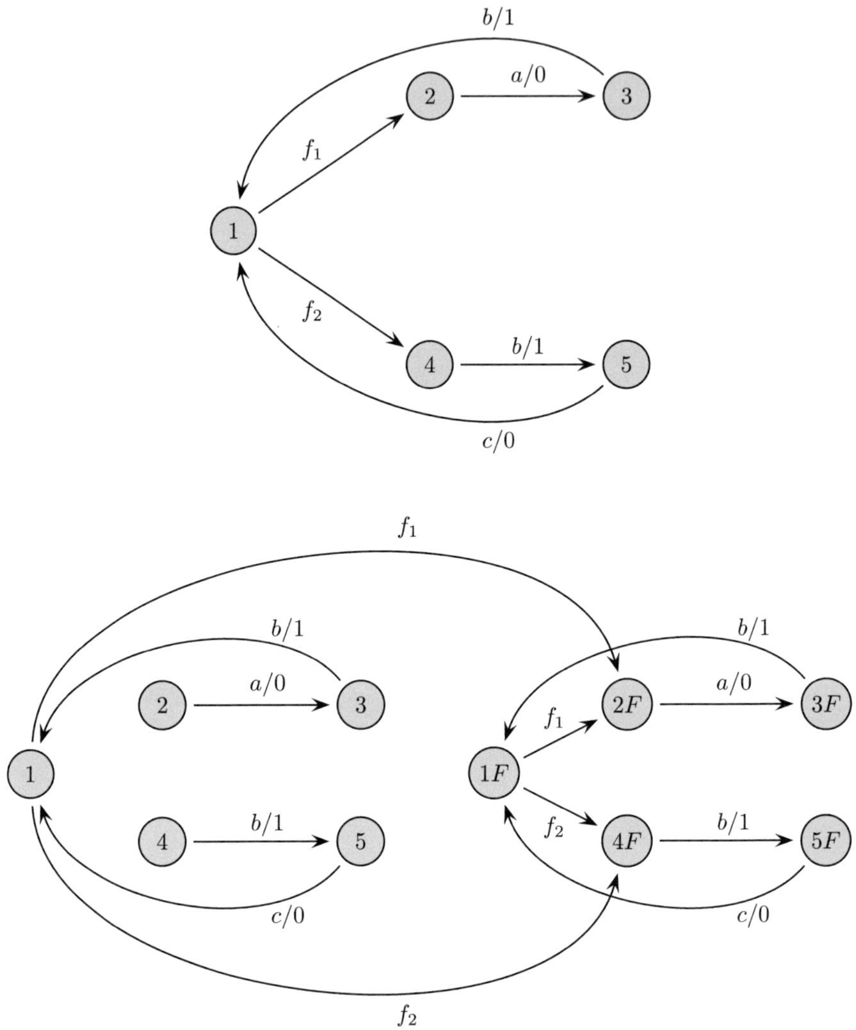

Fig. 7.2 Finite automaton G (top) and the corresponding finite automaton G_F (bottom), illustrating the conversion of the fault detection problem to a state isolation problem

Example 7.2 Consider the finite automaton with outputs $G = (Q, \Sigma, Y \cup \{\epsilon\}, \delta, \lambda, Q_0)$ shown at the top of Fig. 7.2, where

$$Q = \{1, 2, 3, 4, 5\},$$

$\Sigma = \{a, b, c, f_1, f_2\}$, $Y = \{0, 1\}$, $Q_0 = \{1, 2, 3\}$, and δ and λ are as shown at the top of the figure. The fault events are f_1 and f_2, and they belong to different fault classes, i.e., $F_1 = \{f_1\}$, $F_2 = \{f_2\}$, and $F = F_1 \dot{\cup} F_2 = \{f_1, f_2\}$.

Note that if $Q_0 = \{1\}$ then fault identification is obvious: (i) following the observation of 0, one can infer that fault f_1 has occurred, whereas (ii) following the observation of 1, one can infer that fault f_2 has occurred. For our analysis here, we use $Q_0 = \{1, 2, 3\}$ to make fault detection/identification slightly more challenging.

The automaton G_F shown at the bottom of Fig. 7.2 illustrates the reduction of the problem of *fault detection* to a state isolation problem. Using automaton G_F with initial states $Q_0 = \{1, 2, 3\}$, we can state that: following the sequence of observations y_0^k,

1. If the state of automaton G_F can be isolated to be within the set $\{1F, 2F, 3F, 4F, 5F\}$, then at least one fault in the set F has definitely occurred; also
2. If the state of automaton G_F can be isolated to be within the set $\{1, 2, 3, 4, 5\}$, then no fault in the set F has occurred.

Of course, if the state cannot be isolated to either of the two sets described above, a fault may or may not have occurred.

We now illustrate what happens after four different sequences of observations, namely 10, 01, 0101, and 0110.

1. Case 1: $y_0^1 = 10$ (i.e., $y[0] = 1$ and $y[1] = 0$). In this case, we can easily establish that $\hat{q}_{y[1]}(y_0^1) = \{1F, 2F, 3F, 4F\}$ which allows us to conclude that a fault has occurred (because $\hat{q}_{y[1]}(y_0^1) \subseteq Q_F$). We could also reach the same conclusion by considering all explanations of the observation sequence y_0^1, i.e., all event sequences that start from a possible initial state and generate the sequence of outputs y_0^1. Indeed, the possible explanations for $y_0^1 = 10$ are the event sequences f_2bc, f_2bcf_1, f_2bcf_2 (starting from state 1) and also bf_1a (starting from state 3), all of which involve the occurrence of a fault.
2. Case 2: $y_0^1 = 01$ (i.e., $y[0] = 0$ and $y[1] = 1$). In this case, we can easily establish that $\hat{q}_{y[1]}(y_0^1) = \{1, 1F, 2F, 4F\}$ which does not allow us to unambiguously detect a fault (because it is *not* true that $\hat{q}_{y[1]}(y_0^1) \subseteq Q_F$). Note that we would reach the same conclusion if we considered all explanations of the observation sequence: in this case, the possible explanations for $y_0^1 = 01$ include the event sequence ab, abf_1 and abf_2 (starting from state 2), and the event sequences f_1ab, f_1abf_1, f_1abf_2 (starting from state 1); these explanations do not allow us to reach a definite conclusion about the occurrence of a fault.
3. Case 3: $y_0^3 = 0101$ (i.e., $y[0] = 0$, $y[1] = 1$, $y[2] = 0$, and $y[3] = 1$). In this case, we can easily establish that $\hat{q}_{y[3]}(y_0^3) = \{1F, 2F, 4F\}$, which allows us to conclude that a fault has occurred (because we have $\hat{q}_{y[3]}(y_0^3) \subseteq Q_F$). Again, note that the same conclusion is reached if we consider the explanations for y_0^3, all of which involve the occurrence of at least one fault, e.g., the sequences abf_1ab, abf_1abf_1, and abf_1abf_2 (starting from state 2), or the sequences f_1abf_1ab, $f_1abf_1abf_1$, and $f_1abf_1abf_2$ (starting from state 1).
4. Case 4: $y_0^3 = 0110$ (i.e., $y[0] = 0$, $y[1] = 1$, $y[2] = 1$, and $y[3] = 0$). In this case, we can easily establish that $\hat{q}_{y[3]}(y_0^3) = \{1F, 2F, 4F\}$, which allows us to conclude that a fault has occurred (because we have $\hat{q}_{y[3]}(y_0^3) \subseteq Q_F$). Note that the explanations for y_0^3 in this case involve the occurrence of at least one fault, e.g.,

the sequence abf_2bc (starting from state 2) or the sequence f_1abf_2bc (starting from state 1).

Note that Cases 1, 3 and 4 above lead to the same set of state estimates (namely, $\{1F, 2F, 4F\}$) which allows us to determine that a fault has unambiguously occurred. If we were interested in fault classification, however, these three cases would not be equivalent. For example, Case 4 implies either fault f_1 or fault f_2 but it is unclear if we can identify whether one of them (or both) has occurred with certainty. This issue can be addressed via state isolation techniques by constructing the automata G_{F_1} and G_{F_2} shown in Fig. 7.3, which illustrate the reduction of the *fault classification* problem (without regards to order of occurrence of faults) to a state isolation problem. [Note that automaton G_{F_1} is identical to the automaton we would get when reducing the fault detection problem with fault set $F = F_1 = \{f_1\}$ to a state isolation problem, whereas automaton G_{F_2} is identical to the automaton we would get when reducing the fault detection problem with fault set $F = F_2 = \{f_2\}$ to a state isolation problem.] Using automata G_{F_1} and G_{F_2} (each with initial states $Q_0 = \{1, 2, 3\}$), we can state that, following the sequence of observations y_0^k,

1. If the state of automaton G_{F_i}, $i = 1, 2$, can be isolated to be within the set $\{1F_i, 2F_i, 3F_i, 4F_i, 5F_i\}$, then at least one fault in the set F_i has definitely occurred; in addition,
2. If the state of automaton G_{F_i}, $i = 1, 2$, can be isolated to be within the set $\{1, 2, 3, 4, 5\}$, then no fault in the set F_i has occurred.

Note that in the case of multiple faults, if both of the above conditions hold, the above inferences do not provide any information about the number of times each fault appears or the order in which they appear. This is also evident in the analysis we provide below, where we illustrate what happens in each automaton after the four different sequences of observations we considered earlier, namely 10, 01, 0101, and 0110.

1. Case 1: $y_0^1 = 10$ (i.e., $y[0] = 1$ and $y[1] = 0$). In this case, we can easily establish the following.
   In automaton G_{F_1}: we have $\hat{q}_{y[1]}(y_0^1) = \{1, 4, 3F_1\}$, which does not allow us to conclude that fault f_1 has unambiguously occurred (because it is *not* true that $\hat{q}_{y[1]}(y_0^1) \subseteq Q_{F_1}$). The same conclusion follows from the possible explanations for y_0^1, namely f_2bc, f_2bcf_1, and f_2bcf_2 (from state 1), and bf_1a (from state 3).
   In automaton G_{F_2}: we have $\hat{q}_{y[1]}(y_0^1) = \{3, 1F_2, 2F_2, 4F_2\}$, which does not allow us to conclude that fault f_2 has unambiguously occurred (because it is *not* true that $\hat{q}_{y[1]}(y_0^1) \subseteq Q_{F_2}$). The same conclusion follows from the possible explanations for y_0^1 mentioned above.
2. Case 2: $y_0^1 = 01$ (i.e., $y[0] = 0$ and $y[1] = 1$). In this case, we can easily establish the following.
   In automaton G_{F_1}: we have $\hat{q}_{y[1]}(y_0^1) = \{1, 4, 1F_1, 2F_1, 4F_1\}$, which does not allow us to conclude that fault f_1 has unambiguously occurred (because it is *not* true that $\hat{q}_{y[1]}(y_0^1) \subseteq Q_{F_1}$). The same conclusion follows from explanations f_1ab, f_1abf_1, and f_1abf_2 (from state 1), and ab, abf_2, abf_1 (from state 2).

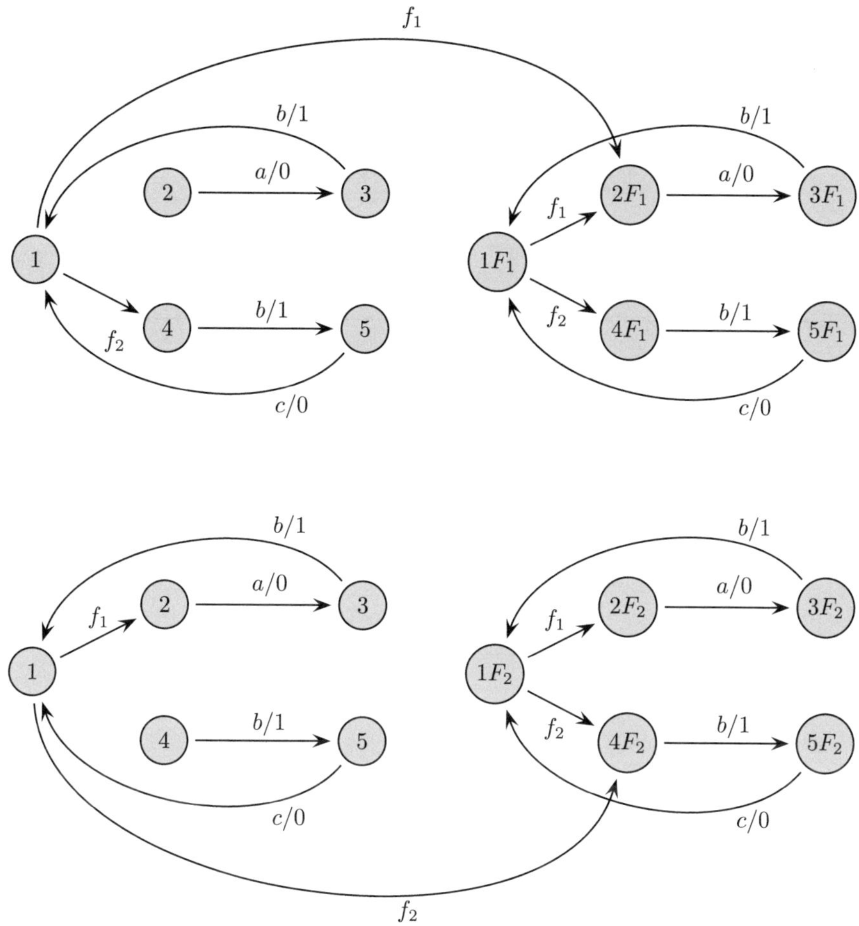

Fig. 7.3 Finite automata G_{F_1} (top) and G_{F_2} (bottom), illustrating the conversion of the fault classification problem (without regards to the ordering of observations) to a state isolation problem

In automaton G_{F_2}: we have $\hat{q}_{y[1]}(y_0^1) = \{1, 2, 4F_2\}$, which does not allow us to conclude that fault f_2 has unambiguously occurred (because it is *not* true that $\hat{q}_{y[1]}(y_0^1) \subseteq Q_{F_2}$). The same conclusion follows from the explanations for y_0^1 mentioned above.

3. Case 3: $y_0^3 = 0101$ (i.e., $y[0] = 0$, $y[1] = 1$, $y[2] = 0$, and $y[3] = 1$). In this case, we can easily establish the following.
   In automaton G_{F_1}: we have $\hat{q}_{y[1]}(y_0^3) = \{1F_1, 2F_1, 4F_1\}$, which allows us to conclude that fault f_1 has unambiguously occurred (because $\hat{q}_{y[1]}(y_0^1) \subseteq Q_{F_1}$). The same conclusion is reached by looking at the explanations for y_0^3, all of which involve the occurrence of fault f_1 at least once, e.g., the sequence abf_1ab (starting from state 2) or the sequence f_1abf_1ab (starting from state 1).

In automaton G_{F_2}: we have $\hat{q}_{y[1]}(y_0^3) = \{1, 2, 4F_2\}$, which does not allow us to conclude that fault f_2 has unambiguously occurred (because it is *not* true that $\hat{q}_{y[1]}(y_0^1) \subseteq Q_{F_2}$). The same conclusion follows from the explanations for y_0^3, which include abf_1ab (starting from state 2) and abf_1abf_2 (starting from state 2).

4. Case 4: $y_0^3 = 0110$ (i.e., $y[0] = 0$, $y[1] = 1$, $y[2] = 1$, and $y[3] = 0$). In this case, we can easily establish the following.
   In automaton G_{F_1}: we have $\hat{q}_{y[1]}(y_0^3) = \{1, 4, 1F_1, 2F_1, 4F_1\}$, which does not allow us to conclude that fault f_1 has occurred (because it is *not* true that $\hat{q}_{y[1]}(y_0^1) \subseteq Q_{F_1}$). The same conclusion is reached by considering the possible explanations for y_0^3, some of which involve the occurrence of fault f_1 (e.g., the sequence f_1abf_2bc starting from state 1), but others do not (e.g., the sequence abf_2bc starting from state 2).
   In automaton G_{F_2}: we have $\hat{q}_{y[1]}(y_0^3) = \{1F_2, 2F_2, 4F_2\}$, which allows us to conclude that fault f_2 has unambiguously occurred (because $\hat{q}_{y[1]}(y_0^1) \subseteq Q_{F_2}$). The same conclusion is reached by considering the explanations for y_0^3, namely abf_2bc, abf_2bcf_1, and abf_2bcf_2 (from state 2), and f_1abf_2bc, $f_1abf_2bcf_1$, and $f_1abf_2bcf_2$ (from state 1), all of which involve the occurrence of f_2.

It is worth pointing out that in Case 1, automaton G_F allowed us to detect a fault (by realizing that $\hat{q}_{y[1]}(y_0^1) \subseteq Q_F$), whereas neither automaton G_{F_1} nor automaton G_{F_2} allowed us to identify fault f_1 and/or fault f_2, because we could neither establish that $\hat{q}_{y[1]}(y_0^1) \subseteq Q_{F_1}$ (using G_{F_1}) nor that $\hat{q}_{y[1]}(y_0^1) \subseteq Q_{F_2}$ (using G_{F_2}). The problem is that all explanations involve a fault, but at least one involves only fault f_1 (namely, bf_1a) and at least one involves only fault f_2 (e.g., f_2bc). This should be contrasted with Case 3, for which automaton G_F allowed us to detect a fault and automaton G_{F_1} identified fault f_1. The difference is that in the latter case all explanations involve fault f_1 (note that not all explanations involve fault f_2, thus we are not able to identify f_2).

Finally, we consider the case where one is interested in also identifying the order in which faults from the two different fault classes, F_1 and F_2, occur. In such case, the problem can be reduced to a state isolation problem in automaton $G_{F_iF_j}$ in Fig. 7.4. Using automaton $G_{F_iF_j}$, we can state that, following the sequence of observations y_0^k,

1. If the states of automaton $G_{F_iF_j}$ can be isolated to be within the set $\{1, 2, 3, 4, 5\}$, then no fault has occurred;
2. If the states of automaton $G_{F_iF_j}$ can be isolated to be within the set $\{1F_1, 2F_1, 3F_1, 4F_1, 5F_1\}$, then fault f_1 has occurred at least once and fault f_2 has not occurred;
3. If the states of automaton $G_{F_iF_j}$ can be isolated to be within the set $\{1F_2, 2F_2, 3F_2, 4F_2, 5F_2\}$, then fault f_2 has occurred at least once and fault f_1 has not occurred;
4. If the states of automaton $G_{F_iF_j}$ can be isolated to be within the set $\{1F_1F_2, 2F_1F_2, 3F_1F_2, 4F_1F_2, 5F_1F_2\}$, then fault f_1 has occurred at least once and fault f_2 has occurred at least once, with fault f_1 occurring first;
5. If the states of automaton $G_{F_iF_j}$ can be isolated to be within the set $\{1F_2F_1, 2F_2F_1, 3F_2F_1, 4F_2F_1, 5F_2F_1\}$, then fault f_1 has occurred at least once and fault f_2 has occurred at least once, with fault f_2 occurring first.

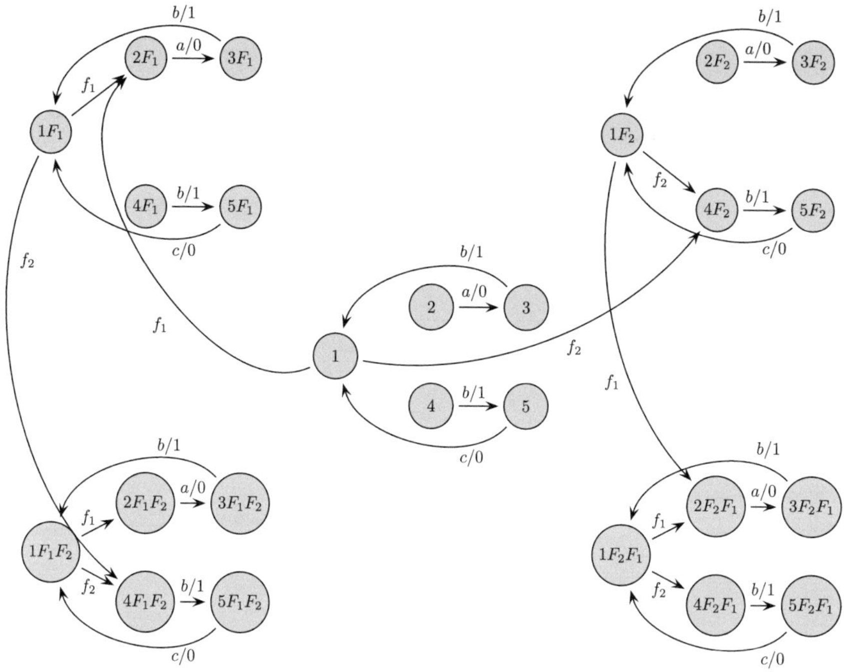

Fig. 7.4 Finite automaton $G_{F_i F_j}$ illustrating the conversion of the fault classification problem to a state isolation problem (in the case where one needs to determine the order in which the different types of faults occur)

We now illustrate state isolation in automaton $G_{F_i} F_j$ following two different sequences of observations, namely 0101, and 101001.

1. Case 1: $y_0^3 = 0101$ (i.e., $y[0] = 0$, $y[1] = 1$, $y[2] = 0$, and $y[3] = 1$). In this case, we can easily establish that $\hat{q}_{y[3]}(y_0^3) = \{1F_1, 2F_1, 4F_1F_2\}$. Interpreting this state estimate, we can conclude that fault f_1 has definitely occurred first and that fault f_2 may or may not have occurred following the occurrence of fault f_1. Of course, we could have reached the same conclusion by looking at the explanations of the observed sequence: state $1F_1$ (respectively, $2F_1, 4F_1F_2$) can be reached from state 1 via strings f_1abf_1ab (respectively, $f_1abf_1abf_1$, $f_1abf_1abf_2$) or from state 2 via strings abf_1ab (respectively abf_1abf_1, abf_1abf_2). In all of these explanations, we see that fault f_1 occurs first, whereas in some (but not all) of them fault f_2 is present.
2. Case 2: $y_0^5 = 101001$ (i.e., $y[0] = 1$, $y[1] = 0$, $y[2] = 1$, $y[3] = 0$, $y[4] = 0$, $y[5] = 1$. In this case, we can easily establish that $\hat{q}_{y[3]}(y_0^3) = \{1F_2F_1, 2F_2F_1, 4F_2F_1\}$. Interpreting this state estimate, we can unambiguously conclude that fault f_2 has occurred first (one or more times) and that fault f_1 also occurred (one or more times). Of course, we could have reached the same conclusion by looking at the explanations of the observed sequence: states $1F_2F_1$, $2F_2F_1$, and

$4F_2F_1$ can be reached from state 1 via strings $f_2bcf_2bcf_1ab$, $f_2bcf_2bcf_1abf_1$, and $f_2bcf_2bcf_1abf_2$, respectively. We see that in all explanations, f_2 occurs first (in fact twice) followed by the occurrence of f_1 (at least once).

Similar constructions, perhaps with more replicas of the original automaton can be used to reduce fault detection to state isolation when additional information is needed (e.g., how many times—up to a bounded number—each fault has occurred). □

7.3 Verification of Diagnosability

Thus far in this chapter, we have considered how to perform fault diagnosis (or, more generally, event inference) given a DFA, with outputs and possibly silent transitions, whose model $G = (Q, \Sigma, Y \cup \{\epsilon\}, \delta, \lambda, Q_0)$ is completely known. More specifically, based on an observed sequence of outputs y_0^k, we focused on the case where one is interested in performing fault detection with respect to a set of fault events $F \subseteq \Sigma$, and fault classification with respect to C fault classes $\{F_1, F_2, \ldots, F_C\}$ (where $F = F_1\dot{\cup}F_2\dot{\cup}\ldots\dot{\cup}F_C$ and $F_{c_1} \cap F_{c_2} = \emptyset$ for $c_1, c_2 \in \{1, 2, \ldots, C\}, c_1 \neq c_2$). The problem (in each case) essentially reduces to a state estimation problem in appropriately constructed finite automata (automaton G_F for fault detection and automata G_{F_c}, $c = 1, 2, \ldots, C$, for fault classification, as explained in the previous section).

Note that the approach described in the previous section is quite straightforward and can be accomplished using any of the techniques for recursive state estimation in Chap. 4. These techniques are rather efficient and can be performed by keeping track the set of states that are possible following each observation (this set has size at most $2N$ in the case of automata G_F and G_{F_c}, $c = 1, 2, \ldots, C$, where $N = |Q|$). Also note that each update requires computational complexity that is polynomial in N (the complexity associated with this update could be $O(N^2)$ due to the presence of unobservable transitions).

In this section, we are interested in the following question. Given that a fault event occurs at some point during the operation of the system, will we always be able to infer with certainty that a fault has occurred, at least within a finite number of events following the fault? Note that this question is independent of the actual sequence of outputs that is observed and, in fact, it has to be checked against all possible sequences of outputs that can be generated by the given system. If the answer to the above question is positive, system G is said to be *diagnosable* with respect to the set of fault events F. Note that a similar question can be asked for each fault class F_c, $c = 1, 2, \ldots, C$ (which would amount to verifying the ability to classify faults with respect to a particular fault class F_c), but for the sake of simplicity, we will concentrate in this section on the question of verifying the ability to detect faults with respect to the set of faults F.

Before we proceed, it is worth pointing out that the majority of existing work on diagnosability considers a labeled (deterministic or nondeterministic) finite automaton G under the natural projection mapping. Our development in this section builds on the more general setting that we developed in the previous chapters. Nevertheless, we will try to make connections with the case of labeled finite automata whenever appropriate. To simplify the development we will make the following (rather common) assumptions for the DFA $G = (Q, \Sigma, Y \cup \{\epsilon\}, \delta, \lambda, Q_0)$:

1. Liveness: Automaton G is live, i.e., for each state $q \in Q$, there exists $\sigma \in \Sigma$ such that $\delta(q, \sigma)$ is defined.
2. Absence of unobservable cycles: There do not exist cycles of events in G that do not generate any observation, i.e., for all $q \in Q$ and all $\sigma_k^{k+m} \in \Sigma^*$, $m \geq 0$, if $\delta(q, \sigma_k^{k+m}) = q$ then
$$E(\lambda_{seq}(q, \sigma_k^{k+m})) \neq \epsilon .$$
[Recall that functions E and λ_{seq} were defined in Chap. 3 (see Eq. (3.9) and pointers there).]

Strictly speaking, the above assumptions are not necessary but they are made to simplify the development and the notation. The motivation behind them should be clear: if, for example, following a fault f we reach a deadlock state, there will be no more system activity and thus there is no hope of any additional observations to aid in identifying the fault; similarly, following a fault f, if we enter a state from which it is possible to execute a cycle of events that produces no outputs, then this cycle could be executed an infinite number of times (without generating any observations). Thus, these two assumptions exclude situations that are generally problematic; however, a given finite automaton may be diagnosable even if these two assumptions are violated.

We are now ready to state the definition of diagnosability for a DFA with outputs and possibly silent transitions G.

Definition 7.1 (*Diagnosability*) Consider a DFA with outputs and possibly silent transitions $G = (Q, \Sigma, Y \cup \{\epsilon\}, \delta, \lambda, Q_0)$ and a set of fault events $F \subseteq \Sigma$. System G is diagnosable with respect to F if the following condition is true: there exists integer $n_0 \geq 0$ such that for all $q_0 \in Q_0$, for all $s \in (\Sigma \setminus F)^*$, for all $f \in F$, for all $s' \in \Sigma^*$ such that $\delta(q_0, sfs')$ is defined and $|s'| \geq n_0$, we have: $\forall q_0' \in Q_0$, $\forall \sigma_0^k \in \Sigma^*$

$$\begin{aligned} &\left\{E(\lambda_{seq}(q_0', \sigma_0^k)) = E(\lambda_{seq}(q_0, sfs'))\right\} \Rightarrow \\ &\quad \Rightarrow \left\{\exists i \in \{0, 1, \ldots, k\},\ \exists f' \in F,\ \sigma[i] = f'\right\} . \end{aligned}$$

In words, diagnosability requires that for all sequences of events sfs' that can be executed in the system starting from some initial state $q_0 \in Q_0$ and in which the first instance of a fault f is followed by at least n_0 events, the observed sequence of outputs $E(\lambda_{seq}(q_0, sfs'))$ is such that it allows us to determine with certainty that a fault has occurred. This requires that any sequence of events that (can be executed in the system starting from some initial state $q_0' \in Q_0$ and) generates the same sequence of observations necessarily contains a fault event.

Example 7.3 Consider again the finite automaton with outputs $G = (Q, \Sigma, Y \cup \{\epsilon\}, \delta, \lambda, Q_0)$ shown on the top of Fig. 7.1, where

$$Q = \{1, 2, 3, 4, 5, 6, 7, 8, 9\} ,$$

$\Sigma = \{\alpha, \beta, \gamma, f_1, f_2\}$, $Y = \{0, 1\}$, $Q_0 = \{1\}$, and δ and λ are as shown at the top of the figure.

Consider the following three different scenarios that might occur from the initial state 1:

1. Fault f_1: This will necessarily be followed by input sequence $\alpha\beta\alpha$ and will generate observation sequence 011, eventually leading back to state 1.
2. Fault f_2: This will necessarily be followed by input sequence $\beta\alpha\alpha\alpha$ and will generate observation sequence 1011, eventually leading back to state 1.
3. No fault: This will be necessarily be followed by input sequence $\alpha\beta\gamma$ and will generate observation sequence 001, eventually leading back to state 1.

Since all observation sequences above are unique, all three possibilities can be identified if one waits long enough. In fact, if the first observation is 1, one knows that fault f_2 has occurred. However, if the first observation is 0, then one has to wait for one more observation: if the second observation is 1, then fault f_1 has occurred; whereas if the second observation is 0, then no fault has occurred.

Clearly, the system is diagnosable. In fact, since the observer knows when the system returns back to state 1, this process can be repeated for multiple faults. The question of systematically verifying diagnosability is addressed in the remainder of this section. □

7.3.1 Diagnoser Construction

The most intuitive way for verifying diagnosability for a given DFA with outputs and possibly silent transitions $G = (Q, \Sigma, Y \cup \{\epsilon\}, \delta, \lambda, Q_0)$ is to construct its diagnoser G_D. In terms of the terminology we have developed in Chap. 5, the diagnoser is essentially an observer (current-state estimator) for the automaton G_F and what we need to verify is whether, following the occurrence of a fault, we can isolate the current state of G within the set Q_F after a bounded number of observations. Note that a bounded number of observations is equivalent to a bounded number of events due to the assumptions that (i) the system is live and (ii) there do not exist any unobservable cycles. In the case of fault classification (i.e., when F consists of C fault classes $\{F_1, F_2, \ldots, F_C\}$, where $F = F_1 \dot{\cup} F_2 \dot{\cup} \cdots \dot{\cup} F_C$ and $F_{c_1} \cap F_{c_2} = \emptyset$ for $c_1, c_2 \in \{1, 2, \ldots, C\}$, $c_1 \neq c_2$), we would have to build separate diagnosers (observers, current-state estimators) for automata G_{F_c}, $c \in \{1, 2, \ldots, C\}$, and use them to separately verify diagnosability for that particular fault class.

Once the diagnoser is build, we have a structure that we can use to track the possible states of automaton G, along with the corresponding fault condition, for any given sequence of observations. Diagnoser states that involve estimates of the system states with label F and without it (sometimes the absence of the label F is denoted with the label N, where N stands for "normal") are states in which we cannot tell—based on the observation sequence seen so far—whether a fault has occurred or not. We call such diagnoser states *indeterminate*. A cycle of solely indeterminate diagnoser states is called an *indeterminate cycle*, and typically (but not always—see discussion and example below) signifies that the system is not diagnosable. The reason is that an indeterminate cycle implies (typically) that there exists a sequence of events that generates a sequence of observations which keeps us confused about whether a fault has occurred (states with label F) or not (states with label N). In other words, an indeterminate cycle corresponds to the situation where there are two traces, of arbitrary long length, that have the same observable projection, and where one trace contains a failure event of a certain type and the other trace does not. Below we formalize the concepts of indeterminate diagnoser state and indeterminate cycle.

Definition 7.2 (*Indeterminate Diagnoser Cycle*) Consider a DFA with outputs and possibly silent transitions $G = (Q, \Sigma, Y \cup \{\epsilon\}, \delta, \lambda, Q_0)$. Let $G_D = AC(2^{Q \dot{\cup} Q_F}, Y, \delta_D, Q_{0D}) =: (Q_D, Y, \delta_D, Q_{0D})$ be its diagnoser, constructed as an observer for DFA G_F (with respect to a set of fault events $F \subseteq \Sigma$). State $v_D \in Q_D$ (recall that $Q_D \subseteq 2^{(Q \dot{\cup} Q_F)}$) is called indeterminate if $v_D \cap Q \neq \emptyset$ *and* $v_D \cap Q_F \neq \emptyset$. A cycle of m ($m > 0$) consecutive states in the diagnoser G_D, denoted by $v_D^{(1)} \rightarrow v_D^{(2)} \rightarrow \cdots \rightarrow v_D^{(m)} = v_D^{(1)}$ (where all states but the last one are distinct) is called an indeterminate cycle if each state $v_F^{(i)}$, $i = 1, 2, \ldots, m-1$ is indeterminate. [The notation $v_D^{(i)} \rightarrow v_D^{(j)}$ means that there exists $y \in Y$ such that $\delta_D(v_D^{(i)}, y) = v_D^{(j)}$.]

Remark 7.3 Since the diagnoser is an observer on system G_F and states in Q_F are absorbing states (i.e., all transitions from a state in Q_F take us to states in Q_F), we can arrive at the following important properties for diagnoser states: (i) all transitions out of a diagnoser state that includes system states with label F have to be to diagnoser states that necessarily include system states with label F; in fact, (ii) all transitions out of a diagnoser state that only includes system states with label F have to be to diagnoser states that only include system states with label F. We refer to this property as the *absorbing property for label* F.

The proof of the following theorem is straightforward and is omitted.

Theorem 7.1 *Consider a DFA with outputs and possibly silent transitions* $G = (Q, \Sigma, Y \cup \{\epsilon\}, \delta, \lambda, Q_0)$. *Let* G_D *be its diagnoser constructed as an observer for DFA* G_F *(with respect to a set of fault events* $F \subseteq \Sigma$*). If* G_D *has no indeterminate cycles, then* G *is diagnosable with respect to* F.

Checking for indeterminate cycles in the diagnoser G_D is a problem that can be solved with complexity that is linear in the number of states of the diagnoser. Essentially, we can treat the diagnoser as a directed graph, whose nodes are the diagnoser states and whose edges are the possible transitions under observable events (note that the exact observable event that causes the transition is not really important in this case). We can mark all indeterminate states and search for cycles involving indeterminate (marked states) using standard graph algorithms (Sampath et al. 1995). The complexity of these algorithms is linear in the number of states of G_D, but one should keep in mind that the number of states of G_D is potentially exponential in the number of states of G_F. In other words, the complexity for checking for indeterminate cycles in G_D is generally of $O(2^{2N}) = O(4^N)$ where $N = |Q|$ is the number of states of G.

Example 7.4 Consider the finite automaton $G = (Q, \Sigma, Y \cup \{\epsilon\}, \delta, \lambda, Q_0)$ on the top of Fig. 7.5, which describes a labeled nondeterministic finite automaton (LNFA) with silent transitions under the natural projection mapping.[1] Specifically, we have $Q = \{1, 2, 3, 4, 5\}$, $\Sigma = \{a, b, c, f\}$, $Q_0 = \{1\}$, and δ as shown in the figure. The set of fault events that need to be detected is given by $F = \{f\}$. We consider two cases for the set of observable events Y and the output mapping λ.

- Case 1: The set of observable events is $\Sigma_{obs} = \{a, b, c\} = Y$ and the set of unobservable events is $\Sigma_{uo} = \{f\}$ (the output mapping λ simply implements the natural projection $P_{\Sigma_{obs}}$, i.e., $\lambda(a) = a$, $\lambda(b) = b$, $\lambda(c) = c$, $\lambda(f) = \epsilon$). The diagnoser G_D in this case is shown in the middle of Fig. 7.5. We can see that the only indeterminate diagnoser state is the state associated with estimates $\{2, 3F, 5\}$ but it is not involved in any cycle; therefore, the system is diagnosable. Specifically, if abc^* is observed, we can conclude with certainty that a fault has occurred, whereas if acc^* is observed, we can conclude with certainty that no fault has occurred. [Note that for diagnosability, we only require that, *if the fault occurs*, then we are able to determine its occurrence after a finite number of observations; the other direction (i.e., if the fault does not occur) is not necessary.]
- Case 2: The set of observable events is $\Sigma_{obs} = \{a, c\} = Y$ and the set of unobservable events is $\Sigma_{uo} = \{b, f\}$ (in this case, the output mapping λ satisfies $\lambda(a) = a$, $\lambda(b) = \epsilon$, $\lambda(c) = c$, $\lambda(f) = \epsilon$). The corresponding diagnoser G_D is shown at the bottom of Fig. 7.5. We can see that the indeterminate diagnoser states are the states associated with estimates $\{2, 3F, 4F, 5\}$ and $\{4F, 5\}$. The latter state actually forms an indeterminate cycle on its own, which generally means that the system is not diagnosable (there are a few exceptions, see the discussion in the next example). In particular, if $afbc^*$ occurs, we observe ac^*, and we cannot tell whether a fault has occurred or not (regardless of how long we wait). □

It is worth pointing out that the presence of indeterminate cycles in G_D does not automatically imply that the system is not diagnosable. Though the presence of such

[1] As mentioned earlier, the extension of the results presented in this chapter to LNFA under the natural projection mapping is rather straightforward (though not explicitly discussed here).

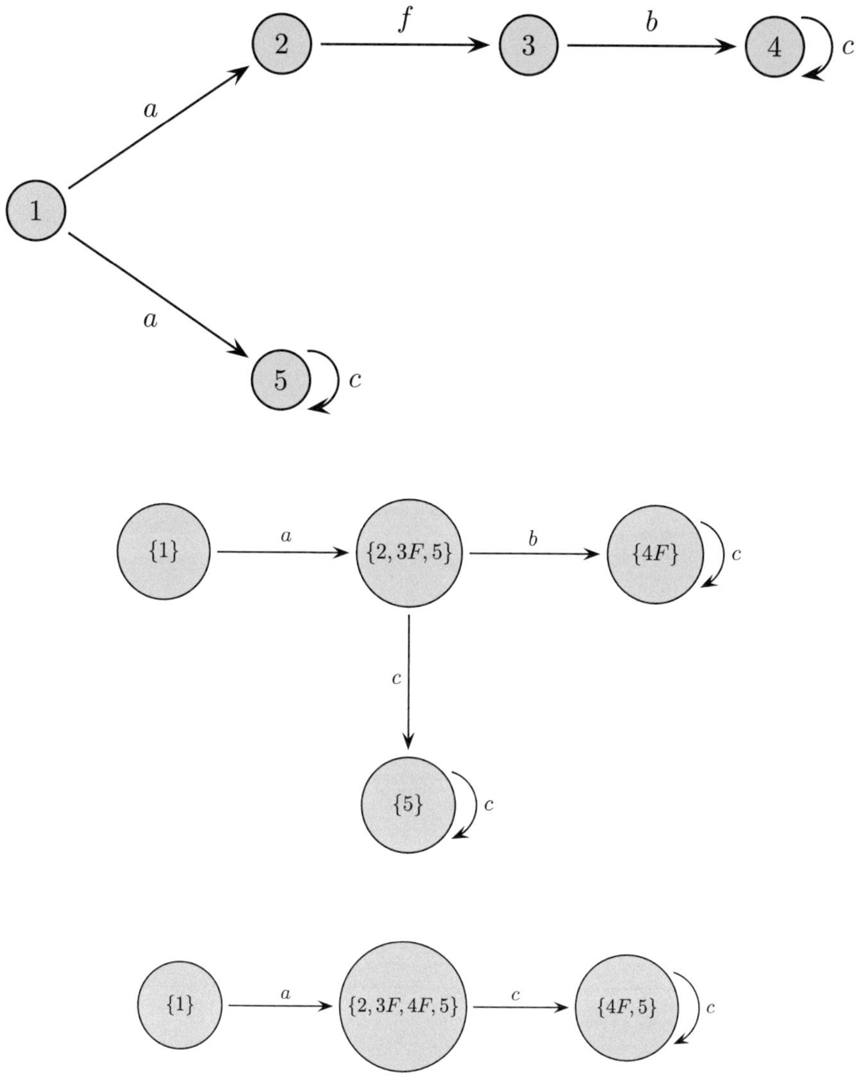

Fig. 7.5 Labeled nondeterministic finite automaton G (top), and its corresponding G_D when $\Sigma_{obs} = \{a, b, c\}$ (middle) and when $\Sigma_{obs} = \{a, c\}$ (bottom)

cycles implies the existence of at least one sequence of events (of arbitrary length) that generates a sequence of observations that keeps the diagnoser in this indeterminate cycle, it is possible that this cycle cannot be executed an arbitrary number of times *after* the occurrence of a fault (which is what we are interested in for diagnosability purposes). Thus, to determine that the system is *not* diagnosable, we have to check that indeterminate cycles can also be executed following the occurrence of a fault. Of course, if there are no indeterminate cycles, we can safely conclude that the system is diagnosable (as stated in Theorem 7.1). An example of a non-problematic indeterminate cycle is illustrated in the example below.

Example 7.5 Consider the finite automaton $G = (Q, \Sigma, Y \cup \{\epsilon\}, \delta, \lambda, Q_0)$ on the top of Fig. 7.6, which describes a labeled deterministic finite automaton (LDFA) with silent transitions under the natural projection mapping. Specifically, we have $Q = \{1, 2, 3, 4\}$, $\Sigma = \{a, b, c, f\}$, $Y = \{a, b, c\}$, $Q_0 = \{1\}$, and δ as shown in the figure. The set of observable events is $\Sigma_{obs} = \{a, b, c\}$ and the set of unobservable events is $\Sigma_{uo} = \{f\}$ (the output mapping λ simply implements the natural projection and is given by $\lambda(a) = a$, $\lambda(b) = b$, $\lambda(c) = c$, $\lambda(f) = \epsilon$). The automaton G_F that reduces the fault diagnosis problem to a state isolation problem is shown in the middle of Fig. 7.6, whereas the diagnoser G_D of G is shown at the bottom of the same figure (one can think of G_D as the observer for automaton G_F).

Looking at the diagnoser G_D, it appears that sequences of the form $(ab)^*$ or $(ab)^*a$ lead us to states $\{1, 3F\}$ or $\{2, 4F\}$ respectively, which form an indeterminate cycle. However, if one examines the system more carefully, one realizes that such sequences of observations can be generated for arbitrary length only *before* the fault occurs. Once the fault actually occurs, c will have to be observed (after one observation). Thus, the cycle of diagnoser states

$$\{1, 3F\} \rightarrow \{2, 4F\} \rightarrow \{1, 3F\}$$

does not really form an indeterminate cycle that is of interest to us in terms of diagnosability. □

7.3.2 Verifier Construction

The discussion in the previous section makes it clear that what is important in determining diagnosability for a given automaton $G = (Q, \Sigma, Y \cup \{\epsilon\}, \delta, \lambda, Q_0)$ is the ability to check whether there exist two (or more) infinite sequences of events s_1 and s_2 that start from valid initial states in Q_0 (not necessarily the same state for both s_1 and s_2) and generate the same sequence of observations, with one (say s_1) containing a fault and the other (say s_2) not containing a fault. Clearly, the system will be diagnosable if and only if no such pair of sequences exist.

The above point of view can actually help us develop a more efficient test for checking diagnosability, with polynomial complexity as opposed to the exponential

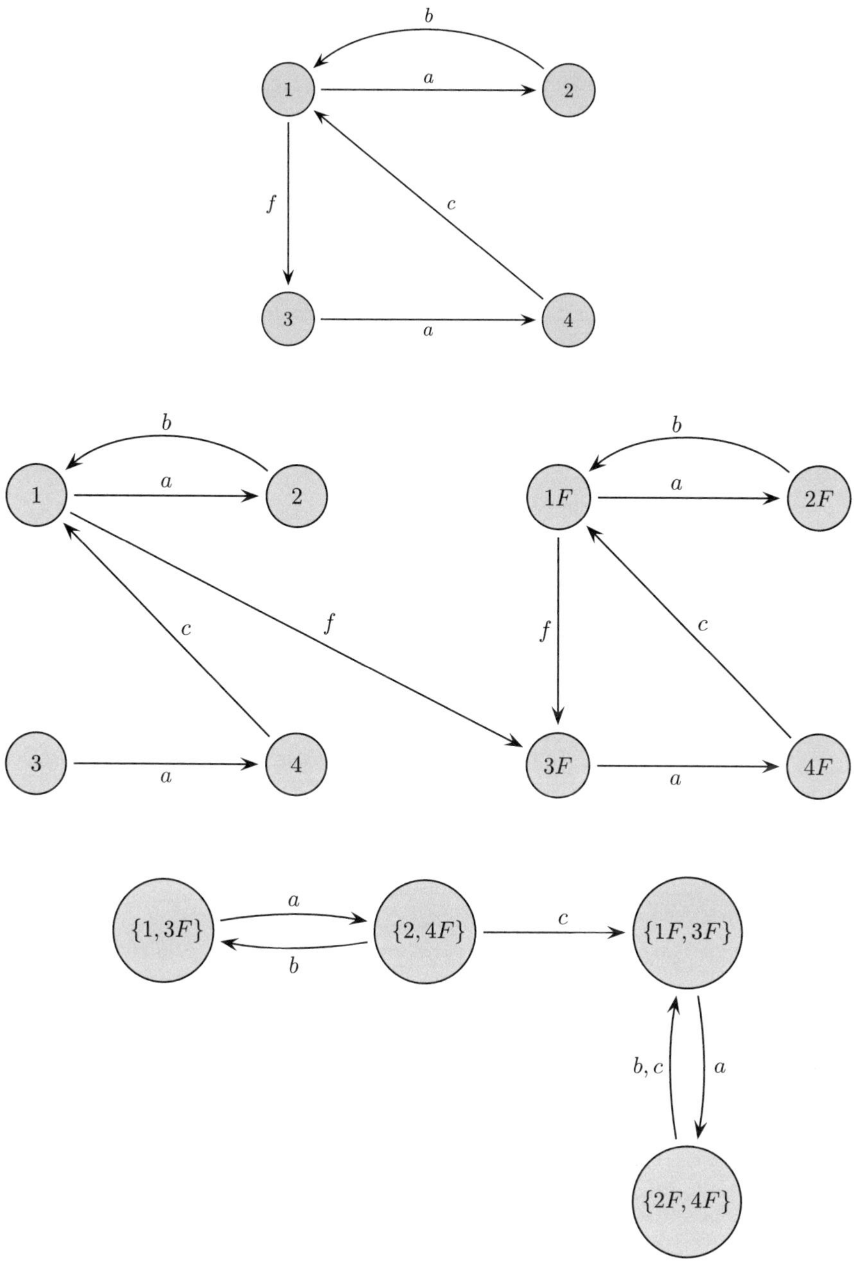

Fig. 7.6 Finite automaton G (top), its corresponding G_F (middle), and the diagnoser G_D (bottom) to illustrate the existence of non-problematic indeterminate cycles

complexity associated with a diagnoser. Below, we describe an approach that is applicable to deterministic systems; this approach generalizes the approach in Yoo and Lafortune (2002), which deals with LDFA under a natural projection mapping, to the DFA with outputs and possibly silent transitions dealt with in this book. More general approaches that can also deal with nondeterministic systems and with more complex decision tasks (e.g., decisions that involve the relative ordering of multiple faults) can be handled in a similar manner (see, for example, Jiang et al. 2001). In what follows we describe verification of diagnosability using the construction of an F-Verifier (introduced in Yoo and Lafortune 2002), appropriately adjusted so that it applies to the observation model we adopt in this book.

Consider a DFA with outputs and possibly silent transitions $G = (Q, \Sigma, Y \cup \{\epsilon\}, \delta, \lambda, Q_0)$. For simplicity, we discuss fault detection of a single fault class $F \subseteq \Sigma$, but multiple fault classes can also be handled by constructing multiple verifiers. The F-Verifier is a nondeterministic finite automaton (NFA) driven by the outputs of G (including the empty output ϵ); it has at most $4 \times N^2$ states (where $N = |Q|$) and is denoted by

$$V_F = AC(Q_{V_F}, Y \cup \{\epsilon\}, \delta_{V_F}, Q_{0V}),$$

where $Q_{V_F} = Q \times \{N, F\} \times Q \times \{N, F\}$, $Q_{0V} = Q_0 \times N \times Q_0 \times N$, and δ_{V_F} is described in detail below. Label N indicates that a fault has not occurred yet (normal behavior), whereas label F indicates that a fault has occurred (faulty behavior). Initially, the verifier starts from any pair of possible initial states, each associated with the normal label N, i.e., $Q_{0V} = \{(q_{01}, N, q_{02}, N) \mid q_{01}, q_{02} \in Q_0\}$.

The next-state transition functionality of the verifier is defined for all possible events in Σ and takes into account the output that they generate as well as the possibility that no output is generated (in case silent transitions are present). The pairs of next states visited need to produce matching outputs from the corresponding starting states, including possible pairings of silent transitions with no transitions (both of which do not produce any output). In addition, the verifier construction ensures that it propagates the label information (N or F) correctly (the label remains unchanged, unless a fault event occurs in which case the label becomes F; once label F is reached, it remains F for subsequent transitions). Specifically, the next-state transition function δ_{V_F} is defined as follows:

1. For each $y \in Y$, and each $q_V \equiv (q_1, L_1, q_2, L_2) \in Q_{V_F}$ we have

$$\delta_{V_F}(q_V, y) = S_n(q_V, y) \cup S_{f_1}(q_V, y) \cup S_{f_2}(q_V, y) \cup S_{f_1 f_2}(q_V, y) ,$$

   where the sets $S_n(q_V, y), S_{f_1}(q_V, y), S_{f_2}(q_V, y), S_{f_1 f_2}(q_V, y) \subseteq Q_{V_F}$ satisfy

$$\begin{aligned} S_n(q_V, y) = \{(q_1', L_1, q_2', L_2) \mid \exists \sigma_1, \sigma_2 \in \Sigma \setminus F, \text{ such that } \\ q_1' = \delta(q_1, \sigma_1),\ q_2' = \delta(q_2, \sigma_2) \text{ and } \lambda(q_1, \sigma_1) = \lambda(q_2, \sigma_2) = y\}, \\ S_{f_1}(q_V, y) = \{(q_1', F, q_2', L_2) \mid \exists f_1 \in F, \sigma_2 \in \Sigma \setminus F, \text{ such that } \\ q_1' = \delta(q_1, f_1),\ q_2' = \delta(q_2, \sigma_2) \text{ and } \lambda(q_1, f_1) = \lambda(q_2, \sigma_2) = y\}, \end{aligned}$$

$$
\begin{aligned}
S_{f_2}(q_V, y) &= \{(q_1', L_1, q_2', F) \mid \exists \sigma_1 \in \Sigma \setminus F,\ f_2 \in F, \text{ such that} \\
&\qquad q_1' = \delta(q_1, \sigma_1),\ q_2' = \delta(q_2, f_2) \text{ and } \lambda(q_1, \sigma_1) = \lambda(q_2, f_2) = y\}, \\
S_{f_1 f_2}(q_V, y) &= \{(q_1', F, q_2', F) \mid \exists f_1, f_2 \in F, \text{ such that} \\
&\qquad q_1' = \delta(q_1, f_1),\ q_2' = \delta(q_2, f_2) \text{ and } \lambda(q_1, f_1) = \lambda(q_2, f_2) = y\}.
\end{aligned}
$$

2. For $y = \epsilon$, and each $q_V \equiv (q_1, L_1, q_2, L_2) \in Q_{V_F}$ we have

$$
\begin{aligned}
\delta_{V_F}(q_V, \epsilon) = {} & S_{n_1}(q_V, \epsilon) \cup S_{n_2}(q_V, \epsilon) \cup S_{n_1 n_2}(q_V, \epsilon) \cup \\
& \cup S_{f_1}(q_V, \epsilon) \cup S_{f_1 n_2}(q_V, \epsilon) \cup S_{f_2}(q_V, \epsilon) \cup \\
& \cup S_{n_1 f_2}(q_V, \epsilon) \cup S_{f_1 f_2}(q_V, \epsilon)\ ,
\end{aligned}
$$

where the sets $S_{n_1}(q_V, \epsilon), S_{n_2}(q_V, \epsilon), S_{n_1 n_2}(q_V, \epsilon), S_{f_1}(q_V, \epsilon), S_{f_1 n_2}(q_V, \epsilon), S_{f_2}(q_V, \epsilon), S_{n_1 f_2}(q_V, \epsilon), S_{f_1 f_2}(q_V, \epsilon) \subseteq Q_{V_F}$ satisfy

$$
\begin{aligned}
S_{n_1}(q_V, \epsilon) &= \{(q_1', L_1, q_2, L_2) \mid \exists \sigma_1 \in \Sigma \setminus F, \\
&\qquad \text{such that } q_1' = \delta(q_1, \sigma_1) \text{ and } \lambda(q_1, \sigma_1) = \epsilon\}\ , \\
S_{n_2}(q_V, \epsilon) &= \{(q_1, L_1, q_2', L_2) \mid \exists \sigma_2 \in \Sigma \setminus F, \\
&\qquad \text{such that } q_2' = \delta(q_2, \sigma_2) \text{ and } \lambda(q_2, \sigma_2) = \epsilon\}\ , \\
S_{n_1 n_2}(q_V, \epsilon) &= \{(q_1', L_1, q_2', L_2), (q_1', L_1, q_2, L_2), (q_1, L_1, q_2', L_2) \mid \\
&\qquad \exists \sigma_1, \sigma_2 \in \Sigma \setminus F, \text{ such that } q_1' = \delta(q_1, \sigma_1), \\
&\qquad q_2' = \delta(q_2, \sigma_2) \text{ and } \lambda(q_1, \sigma_1) = \lambda(q_2, \sigma_2) = \epsilon\}\ , \\
S_{f_1}(q_V, \epsilon) &= \{(q_1', F, q_2, L_2) \mid \exists f_1 \in F \\
&\qquad \text{such that } q_1' = \delta(q_1, f_1) \text{ and } \lambda(q_1, f_1) = \epsilon\}\ , \\
S_{f_1 n_2}(q_V, \epsilon) &= \{(q_1', F, q_2', L_2), (q_1', F, q_2, L_2), (q_1, L_1, q_2', L_2) \mid \\
&\qquad \exists f_1 \in F, \sigma_2 \in \Sigma \setminus F, \text{ such that } q_1' = \delta(q_1, f_1), \\
&\qquad q_2' = \delta(q_2, \sigma_2) \text{ and } \lambda(q_1, f_1) = \lambda(q_2, \sigma_2) = \epsilon\}\ , \\
S_{f_2}(q_V, \epsilon) &= \{(q_1, L_1, q_2', F) \mid \exists f_2 \in F \\
&\qquad \text{such that } q_2' = \delta(q_2, f_2) \text{ and } \lambda(q_2, f_2) = \epsilon\}\ , \\
S_{n_1 f_2}(q_V, \epsilon) &= \{(q_1', L_1, q_2', F), (q_1', L_1, q_2, L_2), (q_1, L_1, q_2', F) \mid \\
&\qquad \exists \sigma_1 \in \Sigma \setminus F, f_2 \in F, \text{ such that } q_1' = \delta(q_1, \sigma_1), \\
&\qquad q_2' = \delta(q_2, f_2) \text{ and } \lambda(q_1, \sigma_1) = \lambda(q_2, f_2) = \epsilon\}\ , \\
S_{f_1 f_2}(q_V, \epsilon) &= \{(q_1', F, q_2', F), (q_1', F, q_2, L_2), (q_1, L_1, q_2', F_2) \mid \exists f_1, f_2 \in F, \\
&\qquad \text{such that } q_1' = \delta(q_1, f_1),\ q_2' = \delta(q_2, f_2) \\
&\qquad \text{and } \lambda(q_1, f_1) = \lambda(q_2, f_2) = \epsilon\}\ .
\end{aligned}
$$

By construction, δ_{V_F} is defined whenever δ is defined for the original system G. Note that the verifier V_F is generally nondeterministic. The next-state transition function δ_{V_F} is defined so that it allows us to track two strings in $L(G)$ that are identical from an observational point of view (i.e., they generate the same output). When no output is generated, one also has to account for the case when no action takes place, thus the need to consider separately S_{n_1}, S_{n_2}, and $S_{n_1 n_2}$ (for normal events), S_{f_1}, S_{f_2}, and $S_{f_1 f_2}$ (for faulty events), as well as $S_{n_1 f_2}$ and $S_{f_1 n_2}$ (for combinations of normal and faulty events). Labels (L_1 and L_2) act as flags to indicate whether a fault has occurred or not (once the flag is raised, it remains raised).

Using the verifier V_F, one can verify diagnosability as in the case of the diagnoser G_D by checking for the presence of indeterminate cycles. We make this more precise via the following definitions.

Definition 7.3 (*Indeterminate Verifier State*) A verifier state $v_F = (q_1, L_1, q_2, L_2) \in Q_{V_F}$ is called indeterminate if $L_1 = N$ and $L_2 = F$ *or* $L_1 = F$ and $L_2 = N$.

Definition 7.4 (*Indeterminate Verifier Cycle*) A cycle of m ($m > 0$) consecutive states in the verifier V_F, denoted by $v_F^{(1)} \rightarrow v_F^{(2)} \rightarrow \cdots \rightarrow v_F^{(m)} = v_F^{(1)}$ (where all states but the last one are distinct) is called an indeterminate cycle if each state $v_F^{(i)}$, $i = 1, 2, \ldots, m-1$ is indeterminate.

The following theorem summarizes the discussion in this section. Its proof is straightforward and is omitted.

Theorem 7.2 *Consider a DFA with outputs and possibly silent transitions $G = (Q, \Sigma, Y \cup \{\epsilon\}, \delta, \lambda, Q_0)$. Let V_F be its verifier constructed with respect to a set of fault events $F \subseteq \Sigma$. Then G is diagnosable with respect to F iff V_F has no indeterminate cycles.*

Example 7.6 Consider again the finite automaton with outputs $G = (Q, \Sigma, Y \cup \{\epsilon\}, \delta, \lambda, Q_0)$ shown at the top of Fig. 7.2 (and discussed in Example 7.2), where

$$Q = \{1, 2, 3, 4, 5\},$$

$\Sigma = \{a, b, c, f_1, f_2\}$, $Y = \{0, 1\}$, $Q_0 = \{1\}$, and δ and λ are as shown at the top of the figure. The fault events are f_1 and f_2, and they belong to different fault classes, i.e., $F_1 = \{f_1\}$, $F_2 = \{f_2\}$, and $F = F_1 \dot{\cup} F_2 = \{f_1, f_2\}$. In this example, we are interested in verifying diagnosability properties of these automaton using a verifier. [Note that the set of initial states $Q_0 = \{1\}$ is different from the set of initial states considered in Example 7.2.]

We first consider the case where we are in interested in determining diagnosability with respect to the set of faults $F = \{f_1, f_2\}$. The verifier V_F with respect to F is the NFA shown in Fig. 7.7: it is driven by the outputs of G, including the empty output ϵ, and has states of the form (q, L, q', L') (written as $(qL, q'L')$ in the figure) where $q, q' \in Q$ and $L, L' \in \{N, F\}$. Since there are no indeterminate cycles, we conclude that system G is diagnosable with respect to faults in the set F.

We also consider the case where we are interested in identifying fault f_1. Thus, we need to construct the V_{F_1}-verifier with respect to the fault set $F_1 = \{f_1\}$, treating f_2 as a normal event (not a fault). The verifier V_{F_1} is the NFA shown in Fig. 7.8:

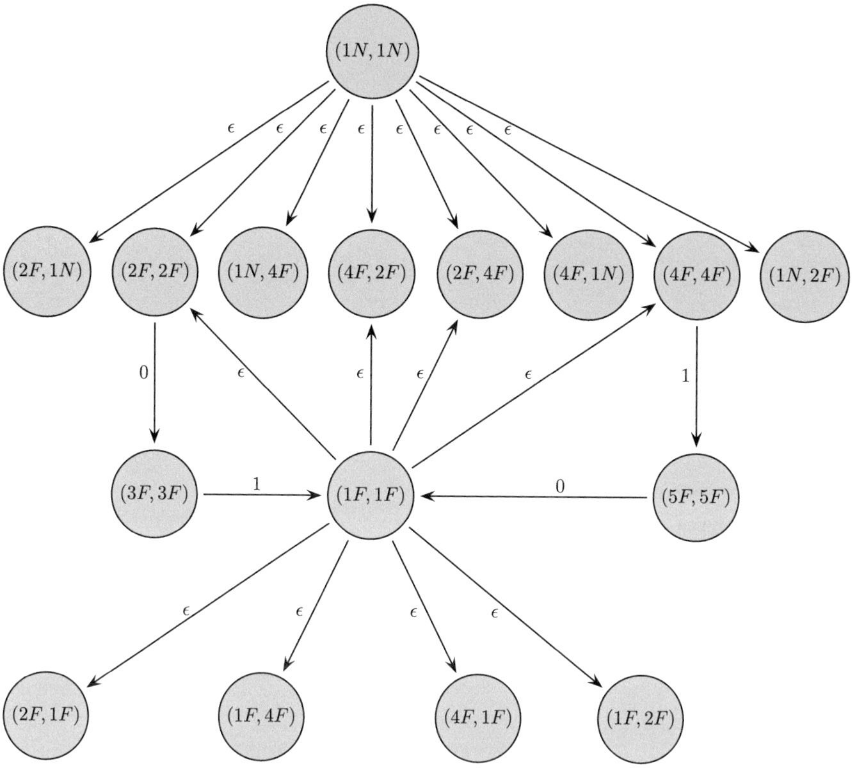

Fig. 7.7 Verifier V_F for determining diagnosability with respect to $F = \{f_1, f_2\}$ for the finite automaton G on the top of Fig. 7.2

it is driven by the outputs of G, including the empty output ϵ and has states of the form (q, L, q', L') (written as $(qL, q'L')$ in the figure) where $q, q' \in Q$ and $L, L' \in \{N, F_1\}$. Again, there are no indeterminate cycles in V_{F_1} and we conclude that the system is diagnosable with respect to F_1. □

It should be clear from the previous example that the particular labeling of the transitions of the verifier automaton (V_F or V_{F_1} in the previous example) is not relevant. Indeed, what is important is the presence of one or more indeterminate cycles (which only depends on the existence of a cycle but not on the particular labeling). This can have implications in the way we construct the verifiers. For example, in the common case of an LDFA under a natural projection mapping, the construction of the verifier is slightly different: given an LDFA $G = (Q, \Sigma, Y \cup \{\epsilon\}, \delta, \lambda, Q_0)$ with $Y = \Sigma_{obs}$ for some set of observable events $\Sigma_{obs} \subseteq \Sigma$ and some set of faults $F \subseteq \Sigma_{uo}$ where $\Sigma_{uo} = \Sigma \setminus \Sigma_{obs}$ (λ in this case is simply the natural projection mapping $P_{\Sigma_{obs}}$), the verifier in V_F in Yoo and Lafortune (2002) has the same set of states as the verifier presented here but it is driven by the set of inputs (rather than the set of outputs). More specifically, the verifier takes the form

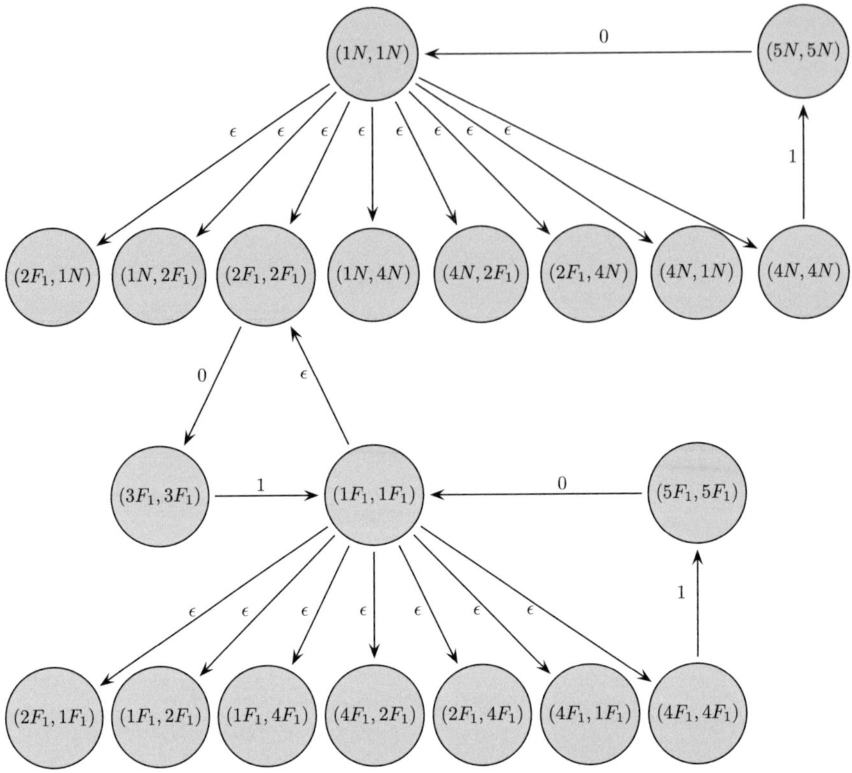

Fig. 7.8 Verifier V_{F_1} for determining diagnosability with respect to $F_1 = \{f_1\}$ for the finite automaton G on the top of Fig. 7.2

$$V_F^L = AC(Q_{V_F}, \Sigma, \delta_{V_F}^L, Q_{0V}),$$

with Q_{V_F} and Q_{0V} as defined earlier, and with $\delta_{V_F}^L$ defined as follows:

1. For each $\sigma \in \Sigma_{obs}$ and each $q_V \equiv (q_1, L_1, q_2, L_2) \in Q_{V_F}$

$$\delta_{V_F}^L(q_V, \sigma) = \{(\delta(q_1, \sigma), L_1, \delta(q_2, \sigma), L_2)\}.$$

2. For each $\sigma \in \Sigma_{uo} \setminus F$ and each $q_V \equiv (q_1, L_1, q_2, L_2) \in Q_{V_F}$

$$\delta_{V_F}^L(q_V, \sigma) = \{(q_1, L_1, \delta(q_2, \sigma), L_2), (\delta(q_1, \sigma), L_1, q_2, L_2), (\delta(q_1, \sigma), L_1, \delta(q_2, \sigma), L_2)\}.$$

3. For each $\sigma \in F$ and each $q_V \equiv (q_1, L_1, q_2, L_2) \in Q_{V_F}$

$$\delta_{V_F}^L(q_V, \sigma) = \{(q_1, L_1, \delta(q_2, \sigma), F), (\delta(q_1, \sigma), F, q_2, L_2), (\delta(q_1, \sigma), F, \delta(q_2, \sigma), F)\}.$$

The definition of $\delta_{V_F}^L$ appears to be slightly different from the definition of δ_{V_F} we provided earlier, but in the case of an LDFA under the natural projection mapping, they can be shown to be equivalent. For example, for each $y \in Y$ ($Y = \Sigma_{obs}$ in this case), and each $q_V \equiv (q_1, L_1, q_2, L_2) \in Q_{V_F}$ we have

$$\delta_{V_F}(q_V, y) = S_n(q_V, y) \cup S_{f_1}(q_V, y) \cup S_{f_2}(q_V, y) \cup S_{f_1 f_2}(q_V, y) ,$$

where the sets $S_n(q_V, y)$, $S_{f_1}(q_V, y)$, $S_{f_2}(q_V, y)$, $S_{f_1 f_2}(q_V, y) \subseteq Q_{V_F}$ were defined earlier. In this case, we have $y = \sigma \in \Sigma_{obs}$, and the set

$$\begin{aligned} S_n(q_V, \sigma)) = \{(q_1', L_1, q_2', L_2) \mid \exists \sigma_1, \sigma_2 \in \Sigma \setminus F, \text{ such that} \\ q_1' = \delta(q_1, \sigma_1),\ q_2' = \delta(q_2, \sigma_2) \text{ and } \lambda(q_1, \sigma_1) = \lambda(q_2, \sigma_2) = \sigma\} , \\ = \{(q_1', L_1, q_2', L_2) \mid \exists \sigma \in \Sigma, \text{ such that} \\ q_1' = \delta(q_1, \sigma),\ q_2' = \delta(q_2, \sigma) \text{ and } \lambda(q_1, \sigma) = \lambda(q_2, \sigma) = \sigma\} , \end{aligned}$$

which is equivalent to

$$\delta_{V_F}^L(q_V, \sigma) = \{(\delta(q_1, \sigma), L_1, \delta(q_2, \sigma), L_2)\}$$

as described above. The only minor difference arises in the way unobservable events are treated. Specifically, if multiple unobservable events are possible from particular states of automaton G, the verifier V_F considers all possible combinations of resulting states (because it is driven by their common output ϵ), whereas the verifier V_F^L considers the uncertainty introduced by one event each time; however, since an unobservable event is tracked in the verifier by also considering the case of no event, this difference in the two constructions is easily seen to have no implication in the presence of indeterminate cycles.

Example 7.7 Consider again the LDFA $G = (Q, \Sigma, Y \cup \{\epsilon\}, \delta, \lambda, Q_0)$, shown on the top of Fig. 7.6 and discussed in Example 7.5. Specifically, we have $Q = \{1, 2, 3, 4\}$, $\Sigma = \{a, b, c, f\}$, $Y = \{a, b, c\}$, $Q_0 = \{1\}$, and δ as shown in the figure. The set of observable events is $\Sigma_{obs} = \{a, b, c\}$ and the set of unobservable events is $\Sigma_{uo} = \{f\}$ (so that the output mapping λ simply implements the natural projection $P_{\Sigma_{obs}}$, i.e., $\lambda(a) = a$, $\lambda(b) = b$, $\lambda(c) = c$, $\lambda(f) = \epsilon$). The verifier V_F^L for this automaton is shown in Fig. 7.9; one can easily verify that there are no indeterminate cycles in the verifier, thus system G is diagnosable. Note that the problem with indeterminate cycles in the diagnoser G_D that cannot be executed after the occurrence of a fault (highlighted in Example 7.5) is not an issue when using the verifier to determine diagnosability. In this particular example, the verifier V_F is identical in structure with the verifier V_F^L (the only difference is that transitions with label f in the verifier V_F^L of Fig. 7.9 are labeled with ϵ in V_F). □

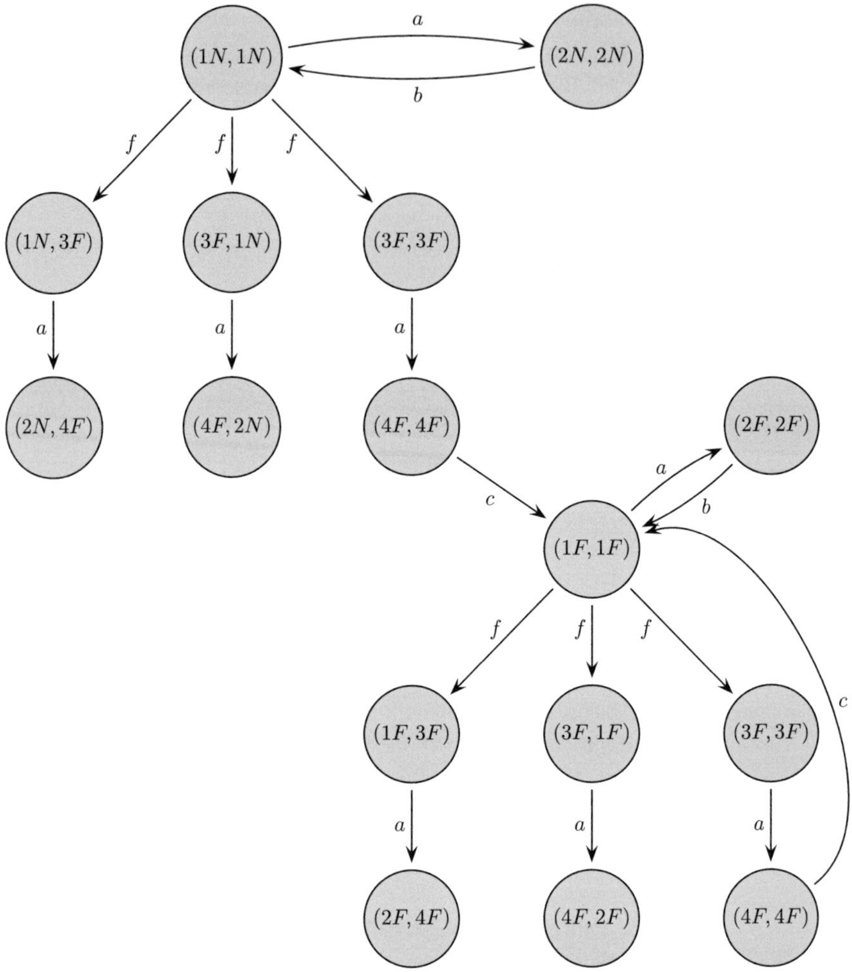

Fig. 7.9 Verifier V_F^L for the finite automaton G shown at the top of Fig. 7.6

7.4 Comments and Further Reading

Failure detection and identification is an important aspect of control system design and operation (Frank 1990; Gertler 2013; Blanke et al. 2006). In the context of DES, fault diagnosis has been extensively studied, with popular applications ranging from transportation systems and heating/ventilation systems, to communication networks and medical diagnostics. In the paragraphs below, we provide some key related literature.

The works in Lin (1994), Bavishi and Chong (1994), Sampath et al. (1995) were among the first to study fault detection and isolation in DES. These works focused on systems that can be modeled as NFA, and aimed at designing a single entity, referred to as *diagnoser*, to perform fault detection and isolation. In particular, Sampath et al. (1995) introduced the notion of *diagnosability* as the system property that captures the fact that an external observer can diagnose *all* faults, at least after a finite number of observations following their occurrence.

The work in Sampath et al. (1995) inspired many researchers to further investigate the verification of diagnosability, which was shown to be of polynomial complexity in Jiang et al. (2001), Yoo and Lafortune (2002). The approach in Yoo and Lafortune (2002) assumes a deterministic system and constructs a verifier for each fault type, whereas the approach in Jiang et al. (2001) assumes a nondeterministic system and constructs a verifier automaton with language specified by the natural projection mapping. In contrast to Yoo and Lafortune (2002), the verifier in Jiang et al. (2001) also tracks labels of multiple faults. Both approaches have the same computational complexity in terms of the number of events of the system, whereas the approach in Yoo and Lafortune (2002) is of lower complexity in terms of the number of states (at the cost of dealing with deterministic automata and a single failure type at a time).

A number of extensions to the basic setting studied in Sampath et al. (1995) have also been pursued. These include *active fault diagnosis*, where a sensor activation strategy is obtained especially designed to enable diagnosis (Sampath et al. 1998); asynchronous diagnosis via *net unfoldings* (Benveniste et al. 2003); failure diagnosis of DES under *linear-time temporal logic specifications* (Jiang and Kumar 2004); and fault (event) *prediction* and *predictability*, where the objective is to correctly predict the occurrence of a fault in the near future (see, for example, Genc and Lafortune 2009; Kumar and Takai 2009). Significant efforts have also been devoted to *decentralized/distributed* diagnosis, which are discussed in more detail in Chaps. 9 and 10.

Fault/event diagnosis has also been extended to stochastic settings by various researchers (Lunze and Schröder 2001; Thorsley and Teneketzis 2005; Hadjicostis 2005; Athanasopoulou et al. 2010). In particular, the work in Thorsley and Teneketzis (2005) considered fault diagnosis in probabilistic finite automata (PFA) and studied two different notions of stochastic diagnosability, namely A-*diagnosability* and AA-*diagnosability*. Both of these notions are weaker than (logical) diagnosability on the NFA that results from the given PFA when one ignores the probabilities associated with the transitions (as studied in this chapter).

- A-diagnosability considers problematic sequences of events that contain a fault but generate observations that do not allow the external observer to conclude with certainty that a fault has occurred. A-diagnosability requires that the total (prior) probability of such sequences of events goes to zero as the number of events that occur after the fault increases. Effectively, this means that if one builds a (logical) diagnoser (using the NFA that results from the given PFA when one ignores probabilities), the resulting diagnoser can have problematic indeterminate cycles as long as these cycles are *transient*, i.e., there are events (with nonzero probability) that will eventually drive us out of these indeterminate cycles. For example, suppose we are given a logical diagnoser like the one at the bottom of

Fig. 7.6. The indeterminate cycle that involves states $\{1, 3F\}$ and $\{2, 4F\}$ would, in general, be considered a problem for logical diagnosability (though in the particular example it was not because it could not be executed after the fault f). However, for A-diagnosability this cycle would not be a problem because it is transient: eventually, event c would occur (with probability one as we wait longer and longer), and this would force us into the (recurrent) cycle that involves states $\{1F, 3F\}$ and $\{2F, 4F\}$. This would allow us to eventually diagnose the fault.

- AA-diagnosability relaxes A-diagnosability even further by considering the posterior probability on the various states and fault occurrences, *conditioned on the observed sequence*. This means that even though a particular sequence of observations may not allow fault diagnosis in the logical sense (because there is a nonzero posterior probability that the fault has not occurred), one may consider whether the (posterior) probability that the fault has occurred is close enough to unity. More specifically, AA-diagnosability requires that the prior probability of sequences of events that allow the external observer to determine the fault occurrence with probability arbitrarily close to unity goes to zero as the number of events after the fault occurrence increases.

Researchers have also performed extensive studies on fault detection and diagnosis in Petri nets. One difference in Petri nets is that there are generally two types of fault events that one needs to take into account: faults that result in the disruption of the firing of a transition (event) or faults that result in the (partial) corruption on the marking (state) of the Petri net. Early works used such type of fault modeling to capture faults of interest and develop encoded Petri nets, which could be used to detect, diagnose and eventually correct the fault (see, for example, Sifakis 1979). In particular, the work in Prock (1991) tries to exploit structural properties of a given Petri net (specifically, place invariants) in order to detect and identify faults, whereas the works in Hadjicostis and Verghese (1999), Wu and Hadjicostis (2005) construct Petri net embeddings with specific structure, in order to facilitate the task of fault detection and identification.

Following the work on fault diagnosis and diagnosability in partially observed NFA in Sampath et al. (1995, 1998), researchers payed significant attention to fault diagnosis formulations in Petri nets under a variety of assumptions on observability. For instance, Ushio et al. (1998) assumes that all transitions are unobservable but markings are partially observable, Chung (2005) allows some of the transitions to be observable, whereas works employing *interpreted* Petri nets consider observations on both the markings (marking changes) and the firing of transitions (Ramírez-Treviño et al. 2007; Lefebvre and Delherm 2007; Ru and Hadjicostis 2009). The work in Genc and Lafortune (2007) considers modular Petri nets and uses marking information in certain shared observable places in order to distributively perform fault detection and identification.

The most direct translation of the fault diagnosis setting in Sampath et al. (1995, 1998) to Petri nets is a setting where markings are not observable but the firing of each transition produces a label (including the empty label). In general, labels can be shared among transitions, whereas the empty label corresponds to unobservable transitions, some of which could comprise faults that need to be detected/identified. Fault diagnosis in so-called *labeled* Petri nets has been studied by several researchers (Dotoli et al. 2009; Cabasino et al. 2010, 2011, 2013). For bounded Petri nets (i.e., Petri nets with a bounded number of tokens—and thus a bounded number of markings), the translation of the approach in NFAs to Petri nets is rather straightforward, but the challenge is to be able to use the structure of the Petri net to perform diagnosis and verify diagnosability in an efficient manner. Since, however, Petri nets can in general have an unbounded number of markings, the construction of a diagnoser or a verifier (for checking diagnosability) may not be straightforward (Cabasino et al. 2012). This type of methodologies have built on the work in Giua and Seatzu (2005) and obtain *minimal explanations* and *basis markings* that match a given sequence of observations (Jiroveanu et al. 2008; Jiroveanu and Boel 2010; Cabasino et al. 2010, 2011).

Fault diagnosis approaches have also been pursued in time Petri nets (Basile et al. 2015); some initial efforts to characterize the likelihood of faults in probabilistic Petri net settings appear in Aghasaryan et al. (1998), Ru and Hadjicostis (2009), Cabasino et al. (2015).

A very good discussion on fault diagnosis and diagnosability in discrete event systems can be found in Zaytoon and Lafortune (2013). Some discussions can also be found in Cassandras and Lafortune (2007), Lafortune et al. (2018).

References

Aghasaryan A, Fabre E, Benveniste A, Boubour R, Jard C (1998) Fault detection and diagnosis in distributed systems: an approach by partially stochastic Petri nets. Discrete Event Dyn Syst Theory Appl 82(2):203–231

Athanasopoulou E, Li L, Hadjicostis CN (2010) Maximum likelihood failure diagnosis in finite state machines under unreliable observations. IEEE Trans Autom Control 55(3):579–593

Basile F, Cabasino MP, Seatzu C (2015) State estimation and fault diagnosis of labeled time Petri net systems with unobservable transitions. IEEE Trans Autom Control 60(4):997–1009

Bavishi S, Chong E (1994) Automated fault diagnosis using a discrete event systems framework. In: Proceedings of IEEE symposium on intelligent control, pp 213–218

Benveniste A, Fabre E, Haar S, Jard C (2003) Diagnosis of asynchronous discrete-event systems: a net unfolding approach. IEEE Trans Autom Control 48(5):714–727

Blanke M, Kinnaert M, Lunze J, Staroswiecki M, Schröder J (2006) Diagnosis and fault-tolerant control. Springer

Cabasino MP, Giua A, Seatzu C (2010) Fault detection for discrete event systems using Petri nets with unobservable transitions. Automatica 46(9):1531–1539

Cabasino MP, Giua A, Pocci M, Seatzu C (2011) Discrete event diagnosis using labeled Petri nets: an application to manufacturing systems. Control Eng Pract 19(9):989–1001

Cabasino MP, Giua A, Lafortune S, Seatzu C (2012) A new approach for diagnosability analysis of Petri nets using verifier nets. IEEE Trans Autom Control 57(12):3104–3117

Cabasino MP, Giua A, Seatzu C (2013) Diagnosis using labeled Petri nets with silent or undistinguishable fault events. IEEE Trans Syst Man Cybern Syst 43(2):345–355

Cabasino MP, Hadjicostis CN, Seatzu C (2015) Probabilistic marking estimation in labeled Petri nets. IEEE Trans Autom Control 60(2):528–533

Cassandras CG, Lafortune S (2007) Introduction to discrete event systems. Springer

Chung SL (2005) Diagnosing PN-based models with partial observable transitions. Int J Comput Integr Manuf 18(2–3):158–169
Dotoli M, Fanti MP, Mangini AM, Ukovich W (2009) On-line fault detection in discrete event systems by Petri nets and integer linear programming. Automatica 45(11):2665–2672
Frank PM (1990) Fault diagnosis in dynamic systems using analytical and knowledge-based redundancy: a survey and some new results. Automatica 26(3):459–474
Genc S, Lafortune S (2007) Distributed diagnosis of place-bordered Petri nets. IEEE Trans Autom Sci Eng 4(2):206–219
Genc S, Lafortune S (2009) Predictability of event occurrences in partially-observed discrete-event systems. Automatica 45(2):301–311
Gertler J (2013) Fault detection and diagnosis. Springer
Giua A, Seatzu C (2005) Fault detection for discrete event systems using Petri nets with unobservable transitions. In: Proceedings of IEEE conference on decision and control and European control conference (CDC-ECC), pp 6323–6328
Hadjicostis CN (2005) Probabilistic detection of FSM single state-transition faults based on state occupancy measurements. IEEE Trans Autom Control 50(12):2078–2083
Hadjicostis CN, Verghese G (1999) Monitoring discrete event systems using Petri net embeddings. In: Proceedings of 20th international conference on application and theory of Petri nets (ICATPN), pp 689–689
Jiang S, Kumar R (2004) Failure diagnosis of discrete-event systems with linear-time temporal logic specifications. IEEE Trans Autom Control 49(6):934–945
Jiang S, Huang Z, Chandra V, Kumar R (2001) A polynomial algorithm for testing diagnosability of discrete-event systems. IEEE Trans Autom Control 46(8):1318–1321
Jiroveanu G, Boel RK (2010) The diagnosability of Petri net models using minimal explanations. IEEE Trans Autom Control 55(7):1663–1668
Jiroveanu G, Boel RK, Bordbar B (2008) On-line monitoring of large Petri net models under partial observation. Discrete Event Dyn Syst 18(3):323–354
Kumar R, Takai S (2009) Decentralized prognosis of failures in discrete event systems. IEEE Trans Autom Control 55(1):48–59
Lafortune S, Lin F, Hadjicostis CN (2018) On the history of diagnosability and opacity in discrete event systems. Annu Rev Control 45:257–266
Lefebvre D, Delherm C (2007) Diagnosis of DES with Petri net models. IEEE Trans Autom Sci Eng 4(1):114–118
Lin F (1994) Diagnosability of discrete event systems and its applications. Discrete Event Dyn Syst 4(2):197–212
Lunze J, Schröder J (2001) State observation and diagnosis of discrete-event systems described by stochastic automata. Discrete Event Dyn Syst 11(4):319–369
Prock J (1991) A new technique for fault detection using Petri nets. Automatica 27(2):239–245
Ramírez-Treviño A, Ruiz-Beltrán E, Rivera-Rangel I, Lopez-Mellado E (2007) Online fault diagnosis of discrete event systems: a Petri net-based approach. IEEE Trans Autom Sci Eng 4(1):31–39
Ru Y, Hadjicostis CN (2009) Fault diagnosis in discrete event systems modeled by partially observed Petri nets. Discrete Event Dyn Syst Theory Appl 19(4):551–575
Sampath M, Sengupta R, Lafortune S, Sinnamohideen K, Teneketzis D (1995) Diagnosability of discrete-event systems. IEEE Trans Autom Control 40(9):1555–1575
Sampath M, Lafortune S, Teneketzis D (1998) Active diagnosis of discrete-event systems. IEEE Trans Autom Control 43(7):908–929
Sifakis J (1979) Realization of fault tolerant systems by coding Petri nets. J Des Autom Fault Toler Comput 3(2):93–107
Thorsley D, Teneketzis D (2005) Diagnosability of stochastic discrete-event systems. IEEE Trans Autom Control 50(4):476–492
Ushio T, Onishi I, Okuda K (1998) Fault detection based on Petri net models with faulty behaviors. In: Proceedings of IEEE international conference on systems, man, and cybernetics, pp 113–118

Wu Y, Hadjicostis CN (2005) Algebraic approaches for fault identification in discrete-event systems. IEEE Trans Autom Control 50(12):2048–2055

Yoo TS, Lafortune S (2002) Polynomial-time verification of diagnosability of partially observed discrete-event systems. IEEE Trans Autom Control 47(9):1491–1495

Zaytoon J, Lafortune S (2013) Overview of fault diagnosis methods for discrete event systems. Annu Rev Control 37(2):308–320

Chapter 8
Opacity

8.1 Introduction and Motivation

Opacity is a privacy notion that focuses on characterizing the *information flow* from the system to the intruder, which is typically modeled as a passive[1] observer (Focardi and Gorrieri 1994; Bryans et al. 2005a). Specifically, a subset of the system's possible behavior, referred to as *secret behavior* and usually captured by a predicate, is deemed critical in the sense that it may reveal important details about the system operation. Opacity requires that this secret behavior is kept hidden (*opaque*) from the outside intruder/observer: the specific requirement is that the intruder will never be able to establish the truth of the predicate, regardless of the underlying activity that occurs in the system (and the corresponding observations it generates).

Despite the fact that we have not yet formalized any notions of opacity, it should be evident that they will be inversely related to the notions of detectability and diagnosability studied in Chaps. 6 and 7 respectively. While detectability (respectively, diagnosability) requires that, for all possible behavior in the system, the observer (respectively, diagnoser) is able to gather sufficient information to determine the exact state of the system (respectively, distinguish between the presence or absence of one or more faults), opacity requires in some sense the opposite: under all possible behavior in the system, the observer/intruder cannot gather sufficient information to distinguish between secret and non-secret behaviors. We revisit this discussion later in this chapter, once we have the opportunity to formally describe notions of opacity.

In the context of finite automata, opacity manifests itself in terms of two main viewpoints, which are equivalent modulo a transformation of polynomial complexity (Wu and Lafortune 2013).

Language-based opacity: Language-based opacity defines the secret behavior as a sublanguage L_S of the given finite automaton and requires that, for all possible

[1]A passive observer is an observer that cannot issue any commands to influence the operation of the system.

C. N. Hadjicostis, *Estimation and Inference in Discrete Event Systems*,
Communications and Control Engineering,
https://doi.org/10.1007/978-3-030-30821-6_8

activity in the system, the outside observer/intruder is never able to determine that the system's behavior definitely belongs in the secret sublanguage.

State-based opacity: State-based opacity defines the secret behavior with respect to a given subset of secret states S, and requires that, for all possible activity in the system, the outside observer/intruder is never able to determine that the current (or initial, or D-delayed) state of the system belongs exclusively to this secret set S. In terms of the terminology in Chap. 5, the observer/intruder should never be able to isolate the current (or initial, or D-delayed) state of the system within the secret set S.

Opacity has found application in emerging technologies that rely on shared cyber-infrastructures (such as the Internet and wireless networks), including defense, banking, health care, and traffic or resource distribution systems. The notions of opacity that have emerged for finite automata can be classified into three categories.

- *Strong* notions of opacity require that, regardless of the underlying activity that occurs in the system, the corresponding observations can be explained with behavior that is non-secret; this implies that the external observer cannot be *absolutely certain* about the presence of secret behavior.
- *Weak* notions of opacity only require that, for *some* activity that occurs in the system (but not necessarily all activity), the corresponding observations can be explained with behavior that is non-secret; this implies that, when this *subset of activity* occurs, the external observer cannot be absolutely certain about the presence of secret behavior.
- *Probabilistic* notions of opacity, developed for stochastic systems where activity occurs with certain probability, come in many variations (in the context of probabilistic finite automata). They are not discussed explicitly in this chapter, though some pointers are provided at the end of the chapter.

As motivation for studying opacity, we present below some examples of applications (see also Dubreil et al. 2010; Saboori 2010; Saboori and Hadjicostis 2011a).

Example 8.1 (*Non-Exposure of Vehicle Location*) Consider a vehicle capable of moving on a grid, such as the toy 2×2 grid on the left of Fig. 8.1. Each cell is a location in the grid and the moves that are allowed at each location can be captured by a kinematic model, i.e., a finite automaton whose states are associated with the state/cell (position) of the vehicle and whose transitions correspond to the movements of the vehicle that are allowed at each position (up, down, left, right, diagonal, etc., depending on the vehicle, the terrain, and other constraints). An example of a kinematic model for the vehicle that moves in the toy grid on the left of Fig. 8.1 is depicted on the right of the figure (the role of the labels on the transitions is discussed later in this example).

Since we are viewing each cell as the state (position) of the vehicle, then we can think of a state sequence in the kinematic model as a vehicle trajectory. In fact, the starting position of the trajectory will correspond to the initial state of the finite automaton that captures the kinematic model, whereas the ending position of the vehicle will be the current state of this finite automaton.

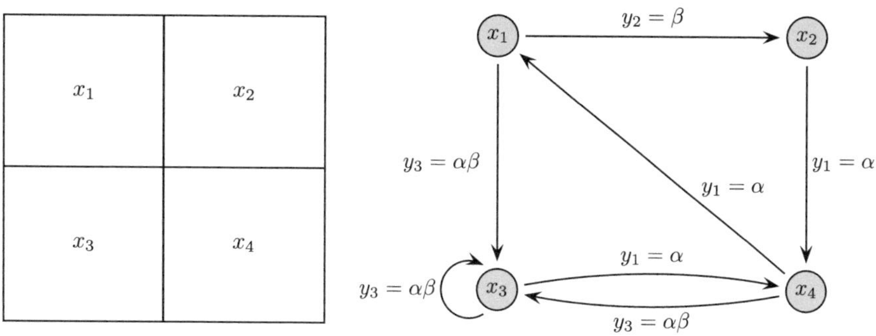

Fig. 8.1 Grid in which a vehicle can move (left) and kinematic model for a vehicle in the grid (right)

Suppose that an external observer/intruder is trying to locate the current (or initial, or D-delayed) position of the vehicle based on some observations that are available. There are multiple ways in which observations could become available to the external observer. For instance, consider a scenario where sensors are dispersed in the grid with each sensor being capable of detecting the presence of the vehicle in a cell or in some aggregation of adjacent cells. Specifically, when the vehicle passes through a cell within the coverage of a sensor, the corresponding sensor emits a signal that indicates this event. An observer with access to this sensory information could attempt to locate the vehicle (i.e., identify its current, initial, or past location) based on the sequence of observations it sees.

In order to analyze this type of problems in the context of opacity for finite automata, we can enhance the kinematic model on the right of Fig. 8.1 by assigning a unique label to all transitions that enter a cell within the coverage area of a particular sensor. Since sensor coverages may overlap, the label of transitions ending in areas that are covered by more than one sensor can be chosen to be a special label that indicates the set of all sensors covering that location. The enhanced kinematic model on the right of Fig. 8.1 corresponds to a (nondeterministic) finite automaton G in which there are sensor readings for two sensors: sensor α covers cells 1, 3, and 4, whereas sensor β covers cells 2 and 3. Note that unobservable transitions (not present in the kinematic model of the example in Fig. 8.1) correspond to locations that are not covered by any sensor. Essentially, the enhanced kinematic model corresponds, in general to a labeled nondeterministic finite automaton (LNFA) under a natural projection mapping (in this case, a labeled deterministic finite automaton (LDFA) with labels α, β, and $\alpha\beta$).

Several security and privacy questions pertaining to the trajectory that the vehicle follows can be formulated in terms of state-based notions of opacity (Saboori 2010). In particular, one of the questions that might arise in the above context is that of understanding whether the sensory information that is available allows us to obtain important information about the present location (current state), the origin (initial state), or previous locations (D-delayed state) of the vehicle. In particular, if we are given a certain subset of locations S that is considered critical (e.g., because the vehicle could be intercepted there), the following question would be directly related to (current, initial, or delayed) state opacity: are we guaranteed that no matter how the vehicle moves, the external observer will never be able to isolate the (current, initial, or delayed) location of the vehicle within the set of critical locations S? Of course, one can also ask the reverse question from the point of view of the vehicle driver: can the vehicle be navigated in such a way so that its (current, initial, or delayed) location cannot be isolated within the set of critical locations S?

A related example that captures privacy notions when offering location-based services can be found in Wu et al. (2014). □

Example 8.2 (*Non-Exposure of Seed of Pseudorandom Number Generator*) In many security applications where encryption is used, one relies on a pseudorandom number generator to come up with "random" numbers needed for encryption. These pseudo-random number generators are typically implemented as autonomous finite automata that cycle through a sequence of states, starting from some initial state known as the *seed* of the generator.

A casual observer will see the pseudorandom number generator cycle through a sequence of states that looks random; however, a sophisticated intruder who knows (or has been able to identify) the structure of the underlying finite automaton might attempt to determine its initial state; since the autonomous finite automaton is deterministic, the combined knowledge of the structure of the finite automaton and the initial state imply complete knowledge of the sequence of states. Thus, the seed of the pseudorandom number generator (or the initial state of the corresponding finite automaton) needs to be kept a secret from an outside observer during the operation of the system. The notion of initial-state opacity, which is discussed in this chapter, is pertinent to the analysis of the risk of such encryption schemes that rely on pseudorandom number generators. □

Remark 8.1 There are also many other examples of privacy/security applications where opacity is a key property. For example, the properties mentioned below can also be expressed in terms of opacity formulations.

Anonymity: Anonymity is the ability to ensure that the identity/data of a certain user/entity of a network can be kept non-identifiable or untrackable from observers/intruders in the system (Reiter and Rubin 1998; Sweeney 2002). The desire to characterize varying levels of anonymity that might be offered by different systems or protocols has resulted in a number of relevant notions (e.g., beyond suspicion, probable innocence, possible innocence, and others Reiter and Rubin 1998).

Noninterference: Noninterference is a notion that tries to quantify the correlation between private strings (that require high security) and observations by an external observer/intruder that records publicly available events (of low security) (Focardi et al. 2000; Hadj-Alouane et al. 2005). More generally, one can have events that are categorized in more than two levels of security, and we can talk about the flow of information from higher levels to lower levels.

Secrecy: Secrecy is a general notion that captures many definitions of security/privacy in the programming language literature and in the verification literature (Alur et al. 2006).

The interested reader is referred to the above references for more details. □

The next sections discuss language-based and state-based notions of opacity in finite automata with outputs and (possibly) silent transitions. Language-based opacity defines the secret behavior of the system with respect to a sublanguage of the system, whereas state-based opacity defines the secret behavior with respect to a subset of states. These sections also discuss methodologies for verifying various notions of opacity, and analyze their complexity. Note that, for notational simplicity, this chapter considers two cases (i) a deterministic finite automaton (DFA) with outputs and (possibly) silent transitions $G = (Q, \Sigma, Y \cup \{\epsilon\}, \delta, \lambda, Q_0)$ (see Definition 3.21), and (ii) a labeled *nondeterministic* finite automaton (LNFA) $G_L = (Q, \Sigma, \Sigma_{obs} \cup \{\epsilon\}, \delta, P_{\Sigma_{obs}}, Q_0)$, under a natural projection mapping $P_{\Sigma_{obs}}$ with respect to the set of observable events Σ_{obs} (see Definition 3.26). The latter case has been the focus of much of the literature on opacity in discrete event systems.

8.2 Language-Based Opacity

Given a DFA with outputs and (possibly) silent transitions, a secret sublanguage is considered opaque if any sequence of outputs that can be generated by some string that belongs to the secret sublanguage can also be generated by at least one string outside this sublanguage (and within the language of the given DFA). More generally, one can talk about strings in a secret sublanguage and strings in a non-secret sublanguage, as in the definition below.

Definition 8.1 (*Language-Based (Strong) Opacity*) Consider a DFA with outputs and (possibly) silent transitions $G = (Q, \Sigma, Y \cup \{\epsilon\}, \delta, \lambda, Q_0)$, a secret language $L_S \subseteq L(G, Q_0) \subseteq \Sigma^*$, and a non-secret language $L_{NS} \subseteq L(G, Q_0) \subseteq \Sigma^*$ (typically, the non-secret language L_{NS} is taken to be $L_{NS} = L(G, Q_0) \setminus L_S$). Automaton G is language-based opaque with respect to L_S and L_{NS} under mapping λ, if for all $t \in L_S$, and all $y_{seq} \in E(\lambda_{seq}(Q_0, t)) := \cup_{q_0 \in Q_0}\{E(\lambda_{seq}(q_0, t))\}$, there exists another string $t' \in L_{NS}$ and initial state $q_0' \in Q_0$ such that

$$E(\lambda_{seq}(q_0', t')) = y_{seq} \ .$$

For the special case of an LNFA under a natural projection mapping $G_L = (Q, \Sigma, \Sigma_{obs} \cup \{\epsilon\}, \delta, P_{\Sigma_{obs}}, Q_0)$, the definition of language-based opacity simplifies to the following.

Definition 8.2 (*Language-Based (Strong) Opacity for LNFA*) Consider an LNFA $G_L = (Q, \Sigma, \Sigma_{obs} \cup \{\epsilon\}, \delta, P_{\Sigma_{obs}}, Q_0)$, under a natural projection mapping $P_{\Sigma_{obs}}$ with respect to the set of observable events Σ_{obs}, $\Sigma_{obs} \subseteq \Sigma$, a secret language $L_S \subseteq L(G_L, Q_0)$, and a non-secret language $L_{NS} \subseteq L(G_L, Q_0)$. Automaton G_L is language-based opaque with respect to L_S and L_{NS} under natural projection $P_{\Sigma_{obs}}$ if for all $t \in L_S$, there exists another string $t' \in L_{NS}$ such that $P_{\Sigma_{obs}}(t) = P_{\Sigma_{obs}}(t')$. Equivalently,[2] $L_S \subseteq P^{-1}_{\Sigma_{obs}}[P_{\Sigma_{obs}}(L_{NS})] \cap L(G_L)$.

Remark 8.2 When languages L_S and L_{NS} are regular languages, one can verify language-based opacity by reducing it (with polynomial complexity Wu and Lafortune 2013) to an initial-state opacity problem, the verification of which is discussed in subsequent sections in this chapter. □

One can also talk about *weak* language-based opacity. The difference from the strong version of opacity is that weak opacity requires the above properties to hold only for some of the strings in the secret sublanguage (and not necessarily all). Formally, the definitions become as follows.

Definition 8.3 (*Weak Language-Based Opacity*) Consider a DFA with outputs and (possibly) silent transitions $G = (Q, \Sigma, Y \cup \{\epsilon\}, \delta, \lambda, Q_0)$, a secret language $L_S \subseteq L(G, Q_0) \subseteq \Sigma^*$, and a non-secret language $L_{NS} \subseteq L(G, Q_0) \subseteq \Sigma^*$ (typically, the non-secret language L_{NS} is taken to be $L_{NS} = L(G, Q_0) \setminus L_S$). Automaton G is weakly language-based opaque with respect to L_S and L_{NS} under mapping λ, if for some $t \in L_S$ we can find another string $t' \in L_{NS}$ and initial states $q_0, q_0' \in Q_0$ such that

$$E(\lambda_{seq}(q_0, t)) = E(\lambda_{seq}(q_0', t'))$$

(and they are defined).

Definition 8.4 (*Weak Language-Based Opacity for LNFA*) Consider an LNFA $G_L = (Q, \Sigma, \Sigma_{obs} \cup \{\epsilon\}, \delta, P_{\Sigma_{obs}}, Q_0)$, under a natural projection mapping $P_{\Sigma_{obs}}$ with respect to the set of observable events Σ_{obs}, $\Sigma_{obs} \subseteq \Sigma$, a secret language $L_S \subseteq L(G_L, Q_0)$, and a non-secret language $L_{NS} \subseteq L(G_L, Q_0)$. Automaton G_L is language-based opaque with respect to L_S and L_{NS} under natural projection $P_{\Sigma_{obs}}$ if for some $t \in L_S$, there exists another string $t' \in L_{NS}$ such that $P_{\Sigma_{obs}}(t) = P_{\Sigma_{obs}}(t')$. Equivalently, $L_S \cap P^{-1}_{\Sigma_{obs}}[P_{\Sigma_{obs}}(L_{NS})] \neq \emptyset$

[2]Recall that in the context of an LNFA $G_L = (Q, \Sigma, \Sigma_{obs} \cup \{\epsilon\}, \delta, P_{\Sigma_{obs}}, Q_0)$, under a natural projection mapping $P_{\Sigma_{obs}}$ with respect to the set of observable events Σ_{obs}, we have $P^{-1}_{\Sigma_{obs}}(L) = \cup_{s \in L} P^{-1}_{\Sigma_{obs}}(s)$, where for a sequence of observable events $s = \sigma_o[0]\sigma_o[1]\ldots\sigma_o[m]\ldots$, $\sigma_o[k] \in \Sigma_{obs}$ for $k = 0, 1, \ldots, m, \ldots$, we have $P^{-1}_{\Sigma_{obs}}(s) = \Sigma^*_{uo}\ \sigma_o[0]\ \Sigma^*_{uo}\ \sigma_o[1]\ \Sigma^*_{uo}\ \cdots\ \Sigma^*_{uo}\ \sigma_o[m]\ \Sigma^*_{uo}\ldots$ and $\Sigma_{uo} = \Sigma \setminus \Sigma_{obs}$ is the set of unobservable events.

8.3 State-Based Opacity

In this section we discuss three different state-based notions of opacity, namely current-state opacity, initial-state opacity, and D-delayed state opacity. As mentioned earlier, throughout this section, we are dealing with either (i) a DFA with outputs and (possibly) silent transitions $G = (Q, \Sigma, Y \cup \{\epsilon\}, \delta, \lambda, Q_0)$ with uncertain initial state in the set Q_0, or with (ii) an LNFA $G_L = (Q, \Sigma, \Sigma_{obs} \cup \{\epsilon\}, \delta, P_{\Sigma_{obs}}, Q_0)$ under a natural projection mapping $P_{\Sigma_{obs}}$ with respect to the set of observable events Σ_{obs} ($\Sigma_{obs} \subseteq \Sigma$). In either case, we are given a subset S of the states ($S \subseteq Q$) that captures the secret behavior of the system. For (strong) state-based opacity, we require that the membership of the system's (current, initial, or D-delayed) state to the secret set S remains opaque (uncertain) to outside observers for all possible behavior that may occur in the system; for weak state-based opacity, we require that the membership of the system's (current, initial, or D-delayed) state to the secret set S remains opaque (uncertain) to outside observers for some possible behavior that may occur in the system.

8.3.1 Current-State Opacity

The following is the formal definition of current-state opacity from Saboori and Hadjicostis (2007, 2011b), Saboori (2010), translated in the context of DFA with outputs and (possibly) silent transitions.

Definition 8.5 *((Strong) Current-State Opacity (CSO))* Consider a DFA with outputs and (possibly) silent transitions $G = (Q, \Sigma, Y \cup \{\epsilon\}, \delta, \lambda, Q_0)$, a set of secret states S, $S \subseteq Q$, and a set of non-secret state NS, $NS \subseteq Q$ (typically, $NS = Q \setminus S$). Automaton G is (strongly) current-state opaque with respect to S and NS, if $\forall t \in \Sigma^*$, $\forall q_0 \in Q_0$, the following holds:

$$\begin{aligned}\delta(q_0, t) \in S \Rightarrow \{&\exists t' \in \Sigma^*, \exists q_0' \in Q_0 \text{ such that}\\ &\{E(\lambda_{seq}(q_0, t)) = E(\lambda_{seq}(q_0', t')) \text{ and } \delta(q_0', t') \in NS\}\}\,.\end{aligned}$$

For the special case of an LNFA $G_L = (Q, \Sigma, \Sigma_{obs} \cup \{\epsilon\}, \delta, P_{\Sigma_{obs}}, Q_0)$, we have the following definition for (strong) current-state opacity.

Definition 8.6 *((Strong) Current-State Opacity (CSO) for LNFA)* Consider an LNFA $G_L = (Q, \Sigma, \Sigma_{obs} \cup \{\epsilon\}, \delta, P_{\Sigma_{obs}}, Q_0)$, under a natural projection mapping $P_{\Sigma_{obs}}$ with respect to the set of observable events Σ_{obs}, $\Sigma_{obs} \subseteq \Sigma$. Let S, $S \subseteq Q$ be the set of secret states and NS, $NS \subseteq Q$, be the set of non-secret states (typically, $NS = Q \setminus S$). Automaton G_L is (strongly) current-state opaque with respect to S, NS, and $P_{\Sigma_{obs}}$, if $\forall q_0 \in Q_0$, $\forall t \in L(G, q_0)$, the following holds:

$$\begin{aligned}\delta(q_0, t) \cap S \neq \emptyset \Rightarrow \{&\exists q_0' \in Q_0, \exists t' \in L(G, q_0') \text{ such that}\\ &\{P_{\Sigma_{obs}}(t) = P_{\Sigma_{obs}}(t') \text{ and } \delta(q_0', t') \cap NS \neq \emptyset\}\}\,.\end{aligned}$$

One can also talk about *weak* current-state opacity. The difference from the strong version of current-state opacity is that the weak version requires the above properties to hold for at least one string that can be generated by the system (and not necessarily all strings). The caveat is that the string should be infinitely extensible,[3] and that the properties should hold for all of its prefixes. Formally, the definition of weak current-state opacity can be stated as follows.

Definition 8.7 (*Weak Current-State Opacity (Weak CSO)*) Consider a DFA with outputs and (possibly) silent transitions $G = (Q, \Sigma, Y \cup \{\epsilon\}, \delta, \lambda, Q_0)$, a set of secret states S, $S \subseteq Q$, and a set of non-secret states NS, $NS \subseteq Q$ (typically, $NS = Q \setminus S$). Automaton G is weakly current-state opaque with respect to S and NS, if there exists an infinite string $t \in L(G)$ and an initial state $q_0 \in Q_0$, such that $\forall s \in \bar{t}$, the following holds:

$$\begin{aligned}\delta(q_0, s) \in S \Rightarrow \{&\exists s' \in \Sigma^*, \exists q_0' \in Q_0 \text{ such that} \\ &\{E(\lambda_{seq}(q_0, s)) = E(\lambda_{seq}(q_0', s')) \text{ and } \delta(q_0', s') \in NS\}\} .\end{aligned}$$

For the special case of an LNFA $G_L = (Q, \Sigma, \Sigma_{obs} \cup \{\epsilon\}, \delta, P_{\Sigma_{obs}}, Q_0)$, we have the following definition for weak current-state opacity.

Definition 8.8 (*Weak Current-State Opacity (Weak CSO) for LNFA*) Consider an LNFA $G_L = (Q, \Sigma, \Sigma_{obs} \cup \{\epsilon\}, \delta, P_{\Sigma_{obs}}, Q_0)$, under a natural projection mapping $P_{\Sigma_{obs}}$ with respect to the set of observable events Σ_{obs}, $\Sigma_{obs} \subseteq \Sigma$. Let S, $S \subseteq Q$, be the set of secret states and let NS, $NS \subseteq Q$, be the set of non-secret states (typically, $NS = Q \setminus S$). Automaton G_L is weakly current-state opaque with respect to S, NS, and $P_{\Sigma_{obs}}$, if there exists an infinite string $t \in \Sigma^*$ and an initial state $q_0 \in Q_0$ such that $\forall s \in \bar{t}$, the following holds:

$$\begin{aligned}\delta(q_0, s) \cap S \neq \emptyset \Rightarrow \{&\exists q_0' \in Q_0, \exists s' \in L(G, q_0') \text{ such that} \\ &\{P_{\Sigma_{obs}}(s) = P_{\Sigma_{obs}}(s') \text{ and } \delta(q_0', s') \cap NS \neq \emptyset\}\} .\end{aligned}$$

One way to verify whether the system G is (weakly) current-state opaque is to construct its observer (or current-state estimator, see Definition 5.1 in Chap. 5) $G_{obs} = AC(2^Q, Y, \delta_{obs}, Q_{0,obs}) =: (Q_{obs}, Y, \delta_{obs}, Q_{0,obs})$ (in the case of the LNFA, the set Y is simply the set of observable events, i.e., Σ_{obs}) and check for states in the observer with certain properties. More specifically, recall that each state q_{obs}, $q_{obs} \in Q_{obs}$, of the observer is associated with a set of system states that correspond to the current-state estimate, i.e., the set of states that are possible in the given system following a sequence of events that generates the sequence of observations $y_{seq} \in Y^*$, which leads us from $Q_{0,obs}$ to this particular state q_{obs} (i.e., $\delta_{obs}(Q_{0,obs}, y_{seq}) = q_{obs}$). Since the observer tracks the set of possible states following any sequence of observations, the presence of an observer state that is associated with states estimates

[3]As in Chaps. 6 and 7, we assume that the system is live and does not possess any unobservable cycles.

that form a (non-empty[4]) subset of S would imply that the sequence of observations that leads to this observer state can be generated by at least one string t that violates the requirement for opacity. This is because the corresponding sequence of observations can only be matched with strings t' that also lead to a state in S. Thus, for the system to be (strongly) current-state opaque, all states in the observer would have to be associated with sets of estimates that are *not* subsets of S.

One can also use the observer G_{obs} to determine whether the given system G is weakly current-state opaque. For weak opacity, one would have to look for the presence of at least one infinitely extensible sequence of observer states that is associated with state estimates that do *not* form a (non-empty) subset of S. Since the sequence of observer states starts from $Q_{0,obs}$, this requirement implies that $Q_{0,obs} \cap NS \neq \emptyset$. Moreover, for the sequence of observer states to be infinitely extensible, the observer needs to have a loop of states that are associated with state estimates that do *not* form a (non-empty) subset of S (and, of course, that this loop can be reached from $Q_{0,obs}$ via a sequence of observer states that are also associated with state estimates that do *not* form a (non-empty) subset of S).

Example 8.3 Consider the finite automaton $G = (Q, \Sigma, \Sigma_{obs} \cup \{\epsilon\}, \delta, P_{\Sigma_{obs}}, Q_0)$ at the top of Fig. 8.2, which was also discussed in Example 5.6 and Fig. 5.4 in Chap. 5 (and it is reproduced here for ease of reference). Automaton G is an LNFA under the natural projection mapping $P_{\Sigma_{obs}}$, with set of observable events $\Sigma_{obs} = \{a, b\}$ and set of unobservable events $\Sigma_{uo} = \{c\}$. We assume that $Q_0 = \{1, 2, 3, 4\}$. The observer (current-state estimator) $G_{obs} = (Q_{obs}, \Sigma_{obs}, \delta_{obs}, Q_{0,obs})$ is also reproduced for convenience at the bottom of Fig. 8.2. [For the details on the construction of the observer refer to Example 5.3; note that the observer state corresponding to the empty set of states $\emptyset$ has not been included for simplicity.]

Suppose that the set of secret states is $S = \{2, 3\}$. It can be seen from the observer at the bottom of Fig. 8.2 that there exists $q_{obs} \in Q_{obs}$ (namely, the state that corresponds to the set of states $\{2, 3\}$), such that $q_{obs} \subseteq S$ (and q_{obs} is non-empty); thus, G is not current-state opaque with respect to S.

If we suppose instead that the set of secret states is $S' = \{3\}$, we realize that we cannot find a state $q_{obs} \in Q_{obs}$ of the observer such that $q_{obs} \subseteq S'$ (and q_{obs} is non-empty); thus, G is current-state opaque with respect to S'.

It should be clear that, once the observer construction is available, the verification of current-state opacity for any secret set S, $S \subseteq Q$, is straightforward (and can be accomplished with linear complexity on the size of the state space of the observer).

□

The following theorem formalizes the above discussions. The theorem is stated when the underlying system is a DFA, but it also applies to an LNFA under a natural projection mapping.

[4]Though the observer construction in Chap. 5 did not include an observer state associated with the empty set of state estimates, many variants of this construction include this so-called *inconsistent state* for convenience. In such case, the inconsistent state is reachable in the observer via sequences of outputs in Y^* that cannot be generated by valid behavior in the system. Thus, the inconsistent state (if present) is not relevant for the verification of opacity.

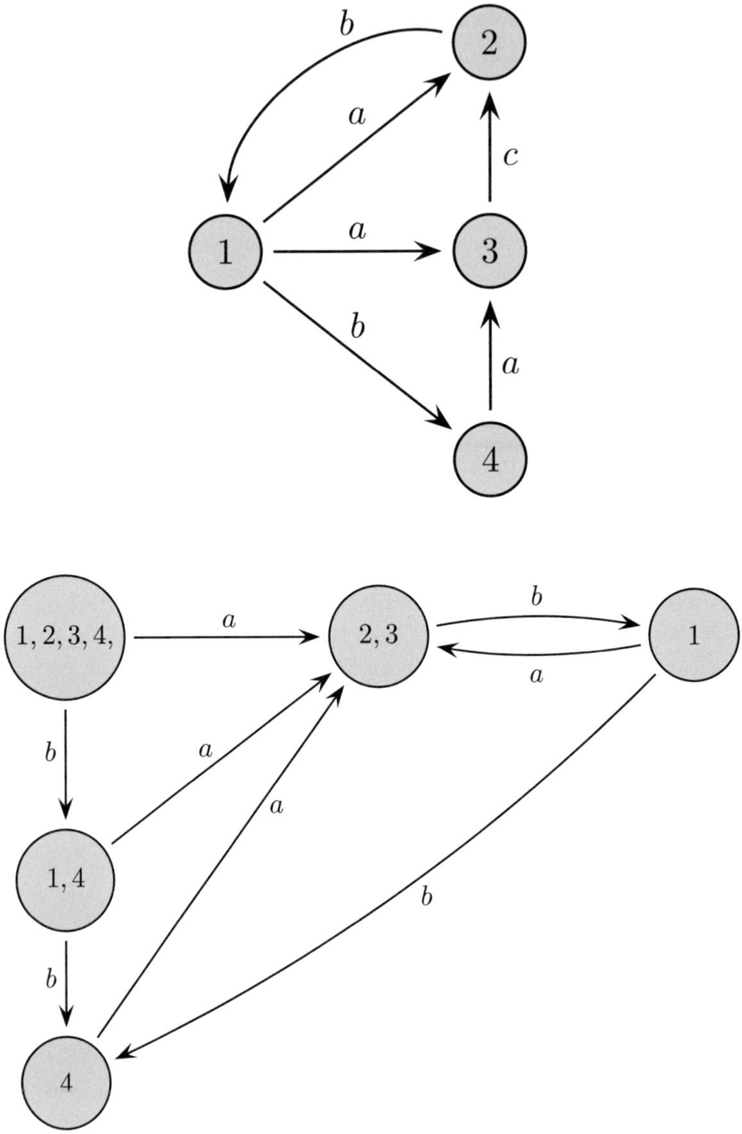

Fig. 8.2 Given labeled nondeterministic finite automaton G (top) and its corresponding observer G_{obs} (bottom)

Theorem 8.1 (Verification of (Weak) Current-State Opacity) *Saboori (2010), Saboori and Hadjicostis (2011b). Consider a DFA with outputs and (possibly) silent transitions* $G = (Q, \Sigma, Y \cup \{\epsilon\}, \delta, \lambda, Q_0)$*, a set of secret states* S*,* $S \subseteq Q$*, and a set of non-secret states* NS*,* $NS \subseteq Q$*. Construct the current-state estimator (observer)* $G_{obs} = AC(2^Q, Y, \delta_{obs}, Q_{0,obs}) =: (Q_{obs}, Y, \delta_{obs}, Q_{0,obs})$ *as in Definition 5.1 in Chap. 5. The following hold true.*

- *Automaton* G *is (strongly) current-state opaque with respect to* S *and* NS*, if*

$$\forall q_{obs} \in Q_{obs} \; \{\{q_{obs} \cap S \neq \emptyset\} \rightarrow \{q_{obs} \cap NS \neq \emptyset\}\} \; . \tag{8.1}$$

- *Automaton* G *is weakly current-state opaque with respect to* S *and* NS*, if there exists an infinitely extensible sequence of observer states*

$$Q_{0,obs} =: q_{obs}[0], q_{obs}[1], \ldots, q_{obs}[k], \ldots ,$$

which (is generated via some infinitely extensible sequence of outputs $y[k] \in Y$*,* $k = 0, 1, 2, \ldots$*, so that* $q_{obs}[k+1] = \delta_{obs}(q_{obs}[k], y[k])$ *and) satisfies*

$$\{q_{obs}[k] \cap S \neq \emptyset\} \rightarrow \{q_{obs}[k] \cap NS \neq \emptyset\} \; , k = 0, 1, 2, \ldots \tag{8.2}$$

Remark 8.3 When $NS = Q \setminus S$, the condition for a DFA with outputs and (possibly) silent transitions $G = (Q, \Sigma, Y \cup \{\epsilon\}, \delta, \lambda, Q_0)$ to be current-state opaque with respect to a set of secret states S, $S \subseteq Q$, simplifies to

$$\forall q_{obs} \in Q_{obs} \{q_{obs} \cap NS \neq \emptyset \} \; .$$

Similarly, for the case of weak current-state opacity, the condition reduces to

$$q_{obs}[k] \cap NS \neq \emptyset \; , k = 0, 1, 2, \ldots$$

for all $q_{obs}[k]$ in the sequence of states. □

Remark 8.4 Since the observer has a finite number of states, the infinitely extensible sequence of states in the definition of weak current-state opacity necessarily has to satisfy $q_{obs}[k_0 + p] = q_{obs}[k_0]$ for some integers $k_0 \geq 0$ and $p > 0$. This means that one can always satisfy the requirements in the definition of weak current-state opacity by picking a sequence of observer states that enters the loop $q_{obs}[k_0], q_{obs}[k_0 + 1], \ldots, q_{obs}[k_0 + p - 1]$, i.e., one can pick the infinite sequence of observer states $q'_{obs}[0], q'_{obs}[1], \ldots, q'_{obs}[k], \ldots$ that satisfies

$$q'_{obs}[k] = \begin{cases} q_{obs}[k] \; , & \text{for } k \leq k_0 + p - 1 \; , \\ q'_{obs}[k-p] \; , & \text{for } k > k_0 + p - 1 \; . \end{cases}$$

i.e., from a certain point onwards, it becomes periodic with period p. □

Example 8.4 Consider again the LNFA $G = (Q, \Sigma, \Sigma_{obs} \cup \{\epsilon\}, \delta, P_{\Sigma_{obs}}, Q_0)$ at the top of Fig. 8.2, which was discussed in Example 8.3, where the set of observable events is $\Sigma_{obs} = \{a, b\}$, and the set of unobservable events is $\Sigma_{uo} = \{c\}$. The observer $G_{obs} = (Q_{obs}, \Sigma_{obs}, \delta_{obs}, Q_{0,obs})$ is shown at the bottom of Fig. 8.2.

Suppose that the set of secret states is $S = \{4\}$. Clearly, the system is not current-state opaque (due to the existence of state $\{4\}$ in the observer). However, the system is weakly current-state opaque because the sequence of observations $a(ba)^*$ (which can be generated by the sequence of events $a(ba)^*$) results in a sequence of states in the observer that does not violate current-state opacity with respect to $S = \{4\}$. □

Current-State Opacity Versus Diagnosability

At this point, it is interesting to make a comparison between the notions of diagnosability and opacity. The basic version of diagnosability discussed in Chap. 7 is concerned with fault detection, i.e., whether, under all possible system behavior that includes at least one fault event, the external observer can determine (perhaps after some finite number of observations) that the fault has occurred. We can compare opacity with diagnosability if we take the system's secret behavior to be the behavior that includes the occurrence of one or more faults. In such case, opacity holds in the given system if under all system behavior (including behavior where the fault occurs), the external observer can never be certain that the fault has occurred. Clearly, the system cannot be opaque if it is diagnosable; however, a system that is not opaque can be diagnosable or non-diagnosable.

In the remainder of this section, we try to make a more explicit connection between diagnosability and weak current-state opacity. Recall that in Chap. 7, we obtained a conversion of a diagnosis problem (detection and/or classification) to a state isolation problem in which states have labels. The simplest case was the case of fault detection, in which the conversion resulted in an automaton with states that had one of two labels, either "F" (for fault) or "N" (for normal). Furthermore, by construction, this converted automaton had a certain "absorption" (trapping) property associated with the label "F," in the sense that transitions out of states with label "F" could only lead to states with label "F."

Consider a (converted) DFA with outputs and (possibly) silent transitions $G_F = (Q \dot{\cup} Q_F, \Sigma, Y \cup \{\epsilon\}, \delta_F, \lambda_F, Q_0)$. Let the set of states Q_F denote the states with label "F," and let Q denote the set of states with label "N." The system is *non-diagnosable* iff its observer/diagnoser $G_{obs} = (Q_{obs}, Y, \delta_{obs}, Q_{0,obs})$ has at least one reachable cycle of observer states with at least one observer state associated with state estimates that involve both label "F" and label "N." In fact, due to the absorption property of states in F, all observer states in this reachable cycle will be associated with state estimates that involve both label "F" and label "N;" such a cycle was called indeterminate cycle in Chap. 7. Also note that the sequence of observer states that takes us to this reachable indeterminate cycle can comprise observer states that are associated with system state estimates that necessarily involve the label "N." The reason is again the absorption property of fault labels: if an observer state on this path was associated with system state estimates that exclusively involved label "F," then all subsequent observer states (including all states in the indeterminate cycle) would have to be associated with system state estimates that exclusively involve label "F," which would be a contradiction.

Let us now consider weak current-state opacity for the same (converted) DFA G_F, taking the set of secret states $S = Q_F$ and the set of non-secret states $NS = Q$. Then, the system is weakly current-state opaque iff (refer to Theorem 8.1 and Remark 8.3) in its observer G_{obs} we can find an infinitely extensible sequence of observer states

$$Q_{0,obs} =: q_{obs}[0], q_{obs}[1], \ldots, q_{obs}[k], \ldots,$$

which satisfies

$$\{q_{obs}[k] \cap NS \neq \emptyset\}, k = 0, 1, 2, \ldots$$

Clearly, if the converted DFA is non-diagnosable with respect to F and N, it will be weakly current-state opaque with respect to $S = Q_F$ and $NS = Q$ (the same reachable indeterminate cycle that causes the system to be non-diagnosable can be used to argue that the system is weakly current-state opaque).

The reverse direction (i.e., that a system that is weakly current-state opaque with respect to $S = Q_F$ and $NS = Q$ will be non-diagnosable) is not necessarily true. The problem is that a system is considered weakly current-state opaque even when its observer involves cycles of observer states that are associated with state estimates involving purely non-secret (normal) states. Since diagnosis is concerned with the detection of a fault only if the fault occurs, the system in question should actually be considered diagnosable.

Of course, the above example is somewhat academic in the sense that, for weak current-state opacity to be interesting, one would rather have secret behavior that is being protected, i.e., in the infinitely extensible sequence of states, we should require that $q_{obs}[k_0] \cap S \neq \emptyset$ *and* $q_{obs}[k_0] \cap NS \neq \emptyset$, at least for some k_0. In such case, if we take into account the absorption property of secret states in the set S (since $S = Q_F$), we can also establish the reverse direction: the key observation is that we have

$$\{q_{obs}[k] \cap S \neq \emptyset\}, k \geq k_0,$$

and, of course, as we already established,

$$\{q_{obs}[k] \cap NS \neq \emptyset\}, k \geq k_0;$$

thus, we conclude that if the (converted) system is weakly current-state opaque, then it will be non-diagnosable.

Before closing this section, we point out that when the DFA G_F is not a converted one (so that a state that is reachable from a state with label "F" does not necessarily have label "F"), then the connection between diagnosability and weak current-state opacity becomes blurred. Note that Lin (2011) established a connection between weak language-based opacity and non-diagnosability.

8.3.2 Initial-State Opacity

The following is the formal definition of initial-state opacity from Saboori and Hadjicostis (2013), translated in the context of DFA with outputs and (possibly) silent transitions. In this case, we are interested in the opacity properties of a set of secret *initial* states S_0 and a given set of non-secret *initial* states NS_0. The sets S_0 and NS_0 can be taken (without loss of generality) to be subsets of the set of possible initial states for the given finite automaton. Typically, NS_0 is taken to be the initial states that do not belong to S_0.

Definition 8.9 (*(Strong) Initial-State Opacity (ISO)*) Consider a DFA with outputs and (possibly) silent transitions $G = (Q, \Sigma, Y \cup \{\epsilon\}, \delta, \lambda, Q_0)$, a set of secret initial states S_0, $S_0 \subseteq Q_0$, and a set of non-secret initial states NS_0, $NS_0 \subseteq Q_0$ (typically, $NS_0 = Q_0 \setminus S_0$). Automaton G is (strongly) initial-state opaque with respect to S_0 and NS_0, if for all $q_0 \in S_0$ and all $t \in L(G, q_0)$, we have

$$\{\exists t' \in \Sigma^*, \exists q_0' \in NS_0 \text{ such that } \{E(\lambda_{seq}(q_0, t)) = E(\lambda_{seq}(q_0', t'))\}\}$$

(which, of course, implies $\delta(q_0, t)$ and $\delta(q_0', t')$ are defined).

For the special case of an LNFA $G_L = (Q, \Sigma, \Sigma_{obs} \cup \{\epsilon\}, \delta, P_{\Sigma_{obs}}, Q_0)$, we have the following definition for (strong) initial-state opacity.

Definition 8.10 (*(Strong) Initial-State Opacity (ISO) for LNFA*) Consider an LNFA $G_L = (Q, \Sigma, \Sigma_{obs} \cup \{\epsilon\}, \delta, P_{\Sigma_{obs}}, Q_0)$, under a natural projection mapping $P_{\Sigma_{obs}}$ with respect to the set of observable events Σ_{obs}, $\Sigma_{obs} \subseteq \Sigma$. Let S_0, $S_0 \subseteq Q_0$, be the set of secret initial states and let NS_0, $NS_0 \subseteq Q_0$, be the set of non-secret initial states (typically, $NS_0 = Q_0 \setminus S_0$). Automaton G_L is (strongly) initial-state opaque with respect to S_0, NS_0, and $P_{\Sigma_{obs}}$, if for all $t \in L(G_L, S_0)$, we have

$$\{\exists t' \in \Sigma^*, \exists q_0' \in NS_0 \text{ such that } \{P_{\Sigma_{obs}}(t) = P_{\Sigma_{obs}}(t') \text{ and } \delta(q_0', t') \neq \emptyset\}\} .$$

One can also talk about *weak* initial-state opacity. The difference from the strong version of initial-state opacity is that the weak version requires the above properties to hold only for some (at least one) of the strings that can be generated by the system (and not necessarily all). Formally, the definition for weak initial-state opacity can be stated as follows.

Definition 8.11 (*Weak Initial-State Opacity (Weak ISO)*) Consider a DFA with outputs and (possibly) silent transitions $G = (Q, \Sigma, Y \cup \{\epsilon\}, \delta, \lambda, Q_0)$. Let S_0, $S_0 \subseteq Q_0$, be the set of secret initial states and let NS_0, $NS_0 \subseteq Q_0$, be the set of non-secret initial states (typically, $NS_0 = Q_0 \setminus S_0$). Automaton G is weakly initial-state opaque with respect to S_0 and NS_0, if there exists an infinitely extensible string $t \in L(G, S_0)$ such that $\forall s \in \bar{t}$, the following holds:

$$\{\exists q_0 \in S_0, \exists s' \in \Sigma^*, \exists q_0' \in NS_0 \text{ such that } \{E(\lambda_{seq}(q_0, s)) = E(\lambda_{seq}(q_0', s'))\}\}$$

(which, of course, implies $\delta(q_0, s)$ and $\delta(q_0', s')$ are defined).

For the special case of an LNFA $G_L = (Q, \Sigma, \Sigma_{obs} \cup \{\epsilon\}, \delta, P_{\Sigma_{obs}}, Q_0)$, we have the following definition for weak initial-state opacity.

Definition 8.12 (*Weak Initial-State Opacity (Weak ISO) for LNFA*) Consider an LNFA $G_L = (Q, \Sigma, \Sigma_{obs} \cup \{\epsilon\}, \delta, P_{\Sigma_{obs}}, Q_0)$, under a natural projection mapping $P_{\Sigma_{obs}}$ with respect to the set of observable events Σ_{obs}, $\Sigma_{obs} \subseteq \Sigma$. Let S_0, $S_0 \subseteq Q_0$, be the set of secret initial states and let NS_0, $NS_0 \subseteq Q_0$, be the set of non-secret initial states (typically, $NS_0 = Q_0 \setminus S_0$). Automaton G_L is weakly initial-state opaque with respect to S_0, NS_0, and $P_{\Sigma_{obs}}$, if there exists an infinite string $t \in L(G_L, S_0)$ such that $\forall s \in \bar{t}$, the following holds:

$$\{\exists s' \in \Sigma^*, \exists q_0' \in NS_0 \text{ such that } \{P_{\Sigma_{obs}}(s) = P_{\Sigma_{obs}}(s') \text{ and } \delta(q_0', s') \neq \emptyset\}\} .$$

One way to check whether a given DFA with outputs and (possibly) silent transitions $G = (Q, \Sigma, Y \cup \{\epsilon\}, \delta, \lambda, Q_0)$ is (weakly) initial-state opaque is to construct its initial-state estimator $G_{Iobs} = AC(2^{Q^2}, Y, \delta_{Iobs}, Q_{0I}) =: (Q_{Iobs}, Y, \delta_{Iobs}, Q_{0I})$ as in Definition 5.3, and check for the presence of states in the initial-state estimator with certain properties. The procedure is outlined below and formalized in Theorem 8.2.

1. Recall that each state of the initial-state estimator can be seen as a subset of the set $Q_0 \times Q$, i.e., each state q_{Iobs} of the initial-state estimator is associated with a set of pairs of system states of the form (q_i, q_c), where q_i corresponds to a possible initial-state estimate and q_c corresponds to a matching possible current-state estimate. The initial-state estimator G_{Iobs} can be utilized to obtain the set of possible initial states of automaton G as follows: given a sequence of observations $y_0^k \in Y^*$, $k \geq 0$, generated by underlying activity in G, the set $\hat{q}_0(y_0^k)$ of possible initial-state estimates of G is given by
$$\hat{q}_0(y_0^k) = \Pi_0(q_{Iobs}) = \{q_i \in Q \mid (q_i, q_c) \in q_{Iobs}\} ,$$
where $q_{Iobs} = \delta_{Iobs}(Q_{0I}, y_0^k)$.
2. For the system to be (strongly) initial-state opaque, we need each state q_{Iobs} of the initial-state estimator that involves initial-state estimates in S_0 to also involve initial-state estimates in NS_0, i.e., for each $q_{Iobs} \in Q_{Iobs}$ we need
$$\{\Pi_0(q_{Iobs}) \cap S_0 \neq \emptyset\} \rightarrow \{\Pi_0(q_{Iobs}) \cap NS_0 \neq \emptyset\} .$$
3. For the system to be weakly initial-state opaque, we only need to be able to find at least one infinitely extensible sequence of initial-state estimator states, each of which has the above property. The latter implies that the initial-state estimator has at least one loop of states that possesses the above property (notice that, due to the monotonic refinement property of initial-state estimation discussed in Remark 4.6 in Chap. 4, the existence of a loop with the above properties necessarily implies

that the loop is reached via a sequence of initial-state estimator states that are associated with system initial-state estimates that also have the above property).

The following theorem formalizes the above discussions. The theorem is stated when the underlying system is a DFA, but it also applies to an LNFA under a natural projection mapping.

Theorem 8.2 (Verification of (Weak) Initial-State Opacity) *Saboori (2010), Saboori and Hadjicostis (2013). Consider a DFA with outputs and (possibly) silent transitions* $G = (Q, \Sigma, Y \cup \{\epsilon\}, \delta, \lambda, Q_0)$, *a set of secret initial states* S_0, $S_0 \subseteq Q_0$, *and a set of non-secret initial states* NS_0, $NS_0 \subseteq Q_0$ *(typically,* $NS_0 = Q_0 \setminus S_0$*). Construct the initial-state estimator* $G_{Iobs} = AC(2^{Q^2}, Y, \delta_{Iobs}, Q_{0I}) =: (Q_{Iobs}, Y, \delta_{Iobs}, Q_{0I})$ *as in Definition 5.3 in Chap. 5. The following hold true.*

- *Automaton* G *is (strongly) initial-state opaque with respect to* S_0 *and* NS_0 *if for all* $q_{Iobs} \in Q_{Iobs}$ *we have*

$$\{\Pi_0(q_{Iobs}) \cap S_0 \neq \emptyset\} \rightarrow \{\Pi_0(q_{Iobs}) \cap NS_0 \neq \emptyset\} . \tag{8.3}$$

- *Automaton* G *is weakly initial-state opaque with respect to* S_0 *and* NS_0 *if there exists an infinitely extensible sequence of states in the initial-state estimator*

$$Q_{0I} =: q_{Iobs}[0], q_{Iobs}[1], \ldots, q_{Iobs}[k], \ldots,$$

which (is generated via some infinitely extensible sequence of outputs $y[k] \in Y$, $k = 0, 1, 2, \ldots$ *so that* $q_{Iobs}[k+1] = \delta_{Iobs}(q_{Iobs}[k], y[k])$ *and) satisfies*

$$\{\Pi_0(q_{Iobs}) \cap S_0 \neq \emptyset\} \rightarrow \{\Pi_0(q_{Iobs}) \cap NS_0 \neq \emptyset\} , k = 0, 1, 2, \ldots \tag{8.4}$$

Example 8.5 Consider the LNFA $G = (Q, \Sigma, \Sigma_{obs}, \delta, P_{\Sigma_{obs}}, Q_0)$ shown in Fig. 8.3, where $Q = \{1, 2, 3, 4\}$, $\Sigma = \{\alpha, \beta\}$, δ is as defined by the transitions in the figure, and $Q_0 = Q = \{1, 2, 3, 4\}$. Assume that $\Sigma_{obs} = \{\alpha, \beta\}$ and $\Sigma_{uo} = \emptyset$. The induced mappings associated with the two possible observations are

$$M_\alpha^{(2)} = \{(2, 3), (2, 4), (3, 3), (4, 1), (4, 3)\} ,$$
$$M_\beta^{(2)} = \{(1, 2), (1, 3)\} ,$$

and can be used to construct the initial-state estimator as shown at the top of Fig. 8.4. The initial uncertainty is assumed to be equal to the state space and hence $Q_{0I} = m_0 = \{(1, 1), (2, 2), (3, 3), (4, 4)\}$. The state mappings associated with each state m_0, m_1, ..., m_{15} of the initial-state estimator are shown at the bottom of Fig. 8.4 (constructed as in Definition 5.3 in Chap. 5).

Suppose that the set of secret initial states is $S_0 = \{3\}$ and $NS_0 = Q_0 \setminus S_0 = \{1, 2, 4\}$. We can easily check that

$$\Pi_0(m_j) \cap NS_0 \neq \emptyset , j = 0, 1, 2, \ldots, 15 ,$$

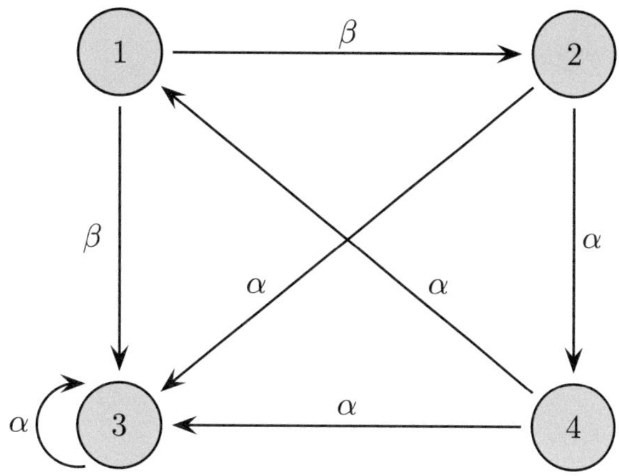

Fig. 8.3 Nondeterministic finite automaton G in Example 8.5

which implies that G_L is (strongly) initial-state opaque with respect to S_0. Of course, G_L is also weakly initial-state opaque with respect to S_0 and NS_0.

Suppose now that the set of secret initial states is $S'_0 = \{3, 4\}$ and the set of non-secret initial states is $NS'_0 = Q_0 \setminus S'_0 = \{1, 2\}$. We can easily check that the system is not (strongly) initial-state opaque with respect to S'_0 and NS'_0 (e.g., $\Pi_0(m_4) \cap NS'_0 = \emptyset$). It turns out, however, that G_L is weakly initial-state opaque with respect to S'_0: for example, if we look at the sequence of states $m_0, m_2, m_5, m_{11}, m_{11}, m_{11}, \ldots$ (obtained via the sequence of outputs $\alpha\alpha\alpha(\alpha)^*$, we see that

$$\Pi_0(m_j) \cap NS'_0 \neq \emptyset\ ,\ j = 0, 2, 5, 11,$$

i.e., the set of secret initial states is not exposed at any point during the observation sequence. □

The above example also serves as a good motivation for the employment of supervisory control strategies to enforce opacity, see for example the works in Saboori and Hadjicostis (2008), Dubreil et al. (2010), Saboori and Hadjicostis (2012). When considering the set of secret initial states to be $S'_0 = \{3, 4\}$ we realize that what needs to be avoided are executions of sequences of events that lead the initial-state estimator to states in the set $\{m_4, m_8, m_9, m_{10}\}$. In other words, if we have a way to ensure that the system does not execute $\alpha\beta$ in the beginning of its operation, we will be able to ensure that initial-state opacity will be preserved (at the cost of restricting the possible behavior in the system). For example, if event β can be disabled, then we will disable it once, after the execution of event α when we start the system (if the first event happens to be β, we do not need to take any action).

Remark 8.5 The verification of initial-state opacity can be shown to be a PSPACE-complete problem by establishing that it is equivalent to the language containment problem. The work in Hadjicostis (2012) takes a different viewpoint and tries to assess the ability of a user (who is dictating the activity in the system and indirectly

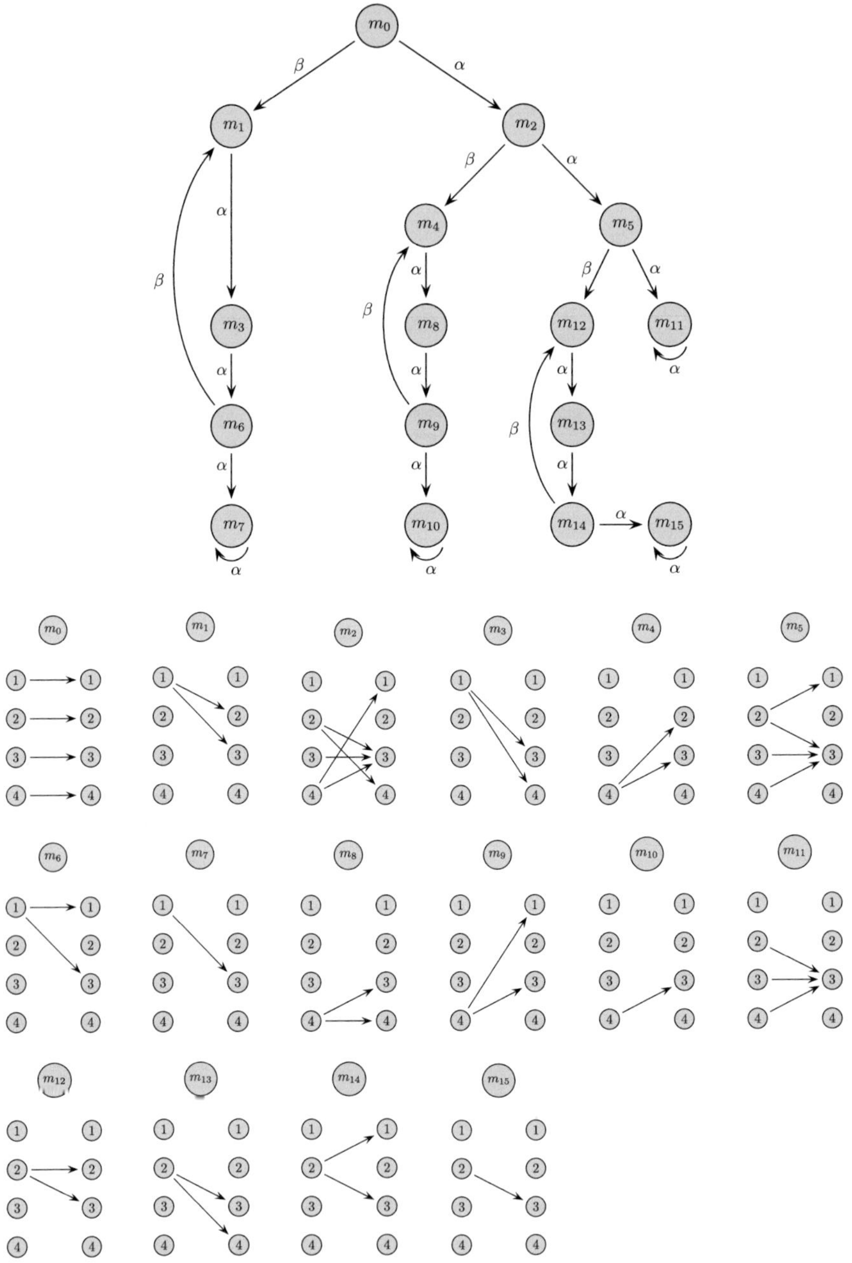

Fig. 8.4 Initial-State Estimator (ISE) G_{Iobs} for the labeled nondeterministic finite automaton in Example 8.5 (top) and corresponding state mappings associated with each state (bottom)

the observations generated) to avoid revealing to the outside observer that the initial state of the system laid within the set of secret initial states S_0. A system that does not allow the user to act indefinitely in such a way is said to posses the property of *resolution of initial state* with respect to S_0. Resolution of initial state is in many ways the absence of weak initial-state opacity; Hadjicostis (2012) established that resolution of initial state can be checked with a polynomial complexity algorithm that resembles the verification of diagnosability using verifiers. Resolution of initial state is important for a variety of security applications where the system designer might be interested in ensuring that no user will be able to indefinitely hide the fact that the initial state of the system laid within the set of secret initial states S_0. □

8.3.3 Delayed-State Opacity

In this section we briefly discuss opacity with respect to delayed states, focusing mostly on (strong) D-delayed-state opacity (see Saboori and Hadjicostis 2011b where this notion of opacity was referred to as K-step opacity with $K = D$). One can also define and verify weak D-delayed-state opacity, along similar lines as in the case of current-state opacity and initial-state opacity. We choose not to elaborate on these variations, since they are directly analogous to the notions we have already seen in previous sections in this chapter. Note that to avoid cluttering this section with definitions, we define D-delayed-state opacity with respect to state estimates following a sequence of observations (and not based on strings and their projections).

Definition 8.13 (*(Strong) Delayed-State Opacity (DSO)*) Consider a DFA with outputs and (possibly) silent transitions $G = (Q, \Sigma, Y \cup \{\epsilon\}, \delta, \lambda, Q_0)$, a set of secret states S, $S \subseteq Q$, and a set of non-secret states NS, $NS \subseteq Q$ (typically, $NS = Q \setminus S$). For a nonnegative integer D, automaton G is (strongly) D-delayed- state opaque with respect to S and NS, if for all $t \in \Sigma^*$, and all $q_0 \in Q_0$, the following holds true:

$$\{\hat{q}_D(y_{t,q_0}) \cap S \neq \emptyset\} \rightarrow \{\hat{q}_D(y_{t,q_0}) \cap NS \neq \emptyset\} ,$$

where $y_{t,q_0} = E(\lambda_{seq}(q_0, t)) = y[0], y[1], \ldots, y[k] =: y_0^k$ is the sequence of outputs generated by t starting from state q_0, and $\hat{q}_D(y_0^k) = \hat{q}_{y[k-D]}(y_0^k)$ (refer to the definitions in Chap. 4) denotes the state estimates at the Dth observation before the last one (taken to be the initial-state estimates if $k < D$).

Remark 8.6 Note that when $D = 0$, D-delayed-state opacity becomes equivalent to current-state opacity. □

Clearly, the verification of (strong) D-delayed-state opacity for a given automaton G can be done using its D-delayed-state estimator $G_{Dobs} = AC(2^{Q^{D+1}}, Y, \delta_{Dobs}, Q_{0D}) =: (Q_{Dobs}, Y, \delta_{Dobs}, Q_{0D})$ (refer to Definition 5.2 in Chap. 5). More specifically, for each state $q_{Dobs} \in Q_{Dobs}$, we need to have

$$\{\Pi_0(q_{Dobs}) \cap S \neq \emptyset\} \rightarrow \{\Pi_0(q_{Dobs}) \cap NS \neq \emptyset\} .$$

When $NS = Q \setminus S$, the above condition reduces to

$$\Pi_0(q_{Dobs}) \cap NS \neq \emptyset \text{ , } \forall q_{Dobs} \in Q_{Dobs}$$

(assuming, of course, that $q_{obs} = \emptyset$ is not included in the construction of the D-delayed observer).

8.4 Complexity Considerations

The complexity of verifying current (or initial or delayed) state opacity using the methodologies outlined in this chapter depends largely on the size of the corresponding estimator. For instance, to verify current-state opacity for a given finite automaton G, one constructs its observer G_{obs}, which has space complexity $O(2^N)$ (where $N = |Q|$ is the number of states of G), and checks whether the conditions in Theorem 8.1 hold. Clearly, for (strong) current-state opacity, the checking of the conditions can be done with complexity that is linear in the number of states of the observer, since each state of the observer needs to be checked against the condition in (8.1).

The situation is a bit different for the verification of weak current-state opacity. In this case, one does not necessarily need to check all states, but ones still needs to find a sequence of states that leads to a loop in the observer, such that all states involved satisfy the condition in (8.2). It turns out the verification of weak current-state opacity can still be answered with linear complexity in the size of observer. To see this, notice that if one ignores labels and views the transition diagram of the observer as a digraph (directed graph), the conditions for weak current-state opacity can be checked by (i) removing nodes in the digraph (states in the observer) that violate the condition in (8.2), and by (ii) finding, in the remaining digraph, an infinite path that starts from the initial state and enters a loop. There are many ways to find such a path with complexity that is linear in the size of the given digraph. For example, one can run a depth-first search (DFS), marking each node that is visited; if a node is visited for the second time, then one has a cycle (which can be obtained by tracing back the DFS tree).

The two strategies described above can also be used to verify, respectively, (strong) initial-state opacity and weak initial-state opacity, as well as (strong) D-delayed-state opacity and weak D-delayed-state opacity (with complexity that is again linear in the size of the corresponding estimator). For example, (strong) initial-state opacity can be verified using the initial-state estimator (ISE), which has space complexity $O(2^{N^2})$ (more precisely, $O(2^{N_0 \times N})$), by checking each state in the ISE against the condition in (8.3). Note that one can reduce the complexity of the verification method for initial-state opacity to $O(4^N)$ via the use of state-status mappings as discussed in Chap. 5—see also Saboori (2010), Saboori and Hadjicostis (2013).

What is not clear from the discussion thus far is whether there are verification methods that are more efficient than the ones proposed in this chapter. The problems of (strong) current-state opacity and (strong) initial-state opacity have been shown to be polynomially reducible into one another (Wu and Lafortune 2013). In fact, they have been shown to be PSPACE-complete problems (Saboori 2010; Cassez et al. 2012) (for example, the verification of initial-state opacity has been shown to be a PSPACE-complete problem for a given LNFA when the number of observable events satisfies $|\Sigma_{obs}| > 1$ Saboori 2010). Thus, the complexity of the algorithms we presented in this chapter for (strong) current-state opacity and (strong) initial-state opacity is expected. *Weak* opacity notions, on the other hand, can generally be verified with polynomial complexity (Lin 2011). Another related problem that can be verified with polynomial complexity is that of *resolution* of initial-state (Hadjicostis 2012), a property that is concerned with whether or not the external of observer will eventually (after a finite number of observations) be able to know that the system started from an initial state in the secret set S_0.

8.5 Comments and Further Reading

Early works that studied opacity in discrete event systems include Bryans et al. (2005a, b), Badouel et al. (2006), Dubreil et al. (2008). The works in Bryans et al. (2005a, b) focus on finite state Petri nets and define opacity with respect to state-based predicates. It can be shown that several privacy/security notions, such as transitive noninterference (Hadj-Alouane et al. 2005) and anonymity (Chaum 1988; Reiter and Rubin 1998), can be formulated as instances of opacity (Bryans et al. 2005a; Lin 2011).

State-based predicates for opacity in finite automata were considered in Saboori and Hadjicostis (2007) and later in Saboori and Hadjicostis (2013). State-based notions of opacity exemplify the use of observers for verifying properties of interest (as discussed in this chapter). A state-based notion of opacity that was not discussed in this chapter is the so-called *infinite-step* opacity introduced in Saboori and Hadjicostis (2009), which requires that any infinite sequence of events generates a sequence of observations that does not allow the intruder to infer that the state of the system, at any point in time (current, past, or initial), is revealed to belong in the given set of secret states. The work in Yin and Lafortune (2017) also studied infinite-step opacity and introduced a reduced complexity verification method that relies on the use of a two-way observer. Other state-based notions of opacity not discussed in this chapter are *initial and final state opacity* (Wu and Lafortune 2013) and *current-state anonymity* (Wu et al. 2014) (the latter was used to assess location privacy when an agent attempts to use location-based services). The work in Wu and Lafortune (2013) established that many of the above state-based notions of opacity, as well as language-based opacity for regular languages, are decision problems that are polynomially reducible from one to another. A characterization of the complexity of various opacity problems can be found in Cassez et al. (2012).

A number of authors have also studied ways to enforce opacity when a given system is not opaque. Apart from the work in Dubreil et al. (2008) mentioned above, other works have also used supervisory control strategies to minimally limit the behavior of the system (by disabling certain events at specific points in time) so as to ensure that the controlled system remains opaque. For example, the authors of Badouel et al. (2006) consider multiple intruders with different observation capabilities and require that no intruder should be able to determine that the actual trajectory of the system belongs to the secret behavior assigned to that intruder. The work in Dubreil et al. (2008) considers a single intruder (that might observe different events than the ones observed/controlled by the supervisor) and establishes that a minimally restrictive supervisor always exists, but might not be regular. Other works that studied maximally permissive supervisory control methods to enforce opacity include Saboori and Hadjicostis (2008, 2012), Dubreil et al. (2010), Saboori (2010). In particular, Tong et al. (2018) considers how supervisory control methodologies need to be designed based on the relationship between three sets of events, namely events observable by the intruder, events observable by the controller, and events controllable by the controller (earlier works typically assume that the set of events observable by the intruder is a subset of the set of events controllable by the controller, which, in turn, is a subset of the set of events observable by the controller).

While supervisory control methods can be successful in enforcing opacity, their success comes at a cost because they rely on limiting the possible behavior of the system. An alternative approach is to *alter* the observations that are emitted by the system, in an effort to confuse the intruder (this class of strategies is also referred to as obfuscation methodologies). The alteration of observations is achieved via the use of insertion functions in Wu and Lafortune (2014, 2016) and it is applied in the context of location-based services in Wu et al. (2014). The challenge, in this case, is to ensure that the insertion of labels is consistent with the knowledge of the intruder (as captured by the observations it has seen so far) and will not raise suspicion to a sophisticated intruder.

Motivated by the absence of likelihood information in most earlier work on opacity, the works in Berard et al. (2010), Saboori and Hadjicostis (2014) extend notions of opacity to probabilistic settings. In particular, state-based notions of opacity have been developed for probabilistic finite automata (PFAs) in Saboori and Hadjicostis (2014) by devising appropriate measures to quantify opacity. Measures of opacity in probabilistic systems can be categorized under two different criteria (and combinations of them).

- One can focus on the prior probability with which a violation of (strong) current-state opacity (or some other notion of opacity) will occur. In other words, one can first identify the sequences of events that lead to violations of current-state opacity (essentially by ignoring probabilities on the given PFA and focusing on the behavior of the resulting NFA), and then assess the prior probability that such problematic sequences of events will occur once the system starts operating. Clearly, this prior probability will be zero if the given system is (strongly) current-state opaque; however, if the system is not current-state opaque, then one can determine the (prior) probability of behavior that leads to current-state opacity violations. The notions of *step-based almost current-state opacity* and *almost current-state opacity* from Saboori and Hadjicostis (2014) fall into this category. For example, *step-based almost current-state opacity* considers the a priori probability of violating current-state opacity, following any sequence of events of length k, and requires that this probability lies below a threshold for all possible lengths ($k = 0, 1, 2, \ldots$). Both step-based almost current-state opacity and almost current-state opacity can be verified with complexity that is exponential in the size of the given PFA (Saboori and Hadjicostis 2014).

- One can also focus on characterizing the confidence of the intruder when making an assessment of a violation of current-state opacity (or some other notion of opacity). More specifically, even though the intruder might not be certain that the current state of the system resides in the set of secret states S, it might be able to determine that the posterior probability that the state of the system does not belong to S is very small. Depending on the application and the value of this probability, such a deduction by the intruder might still be considered alarming (despite the fact that technically the intruder is not certain that the state of the system belongs in S). The notion of *probabilistic current-state opacity* from Saboori and Hadjicostis (2014) requires that, for each possible sequence of observations, the following property holds: the increase in the *conditional* probability that the system current state lies in the set of secret states (conditioned on the given sequence of observations) compared to the prior probability (that the initial state laid in the set of secret states before any observation becomes available) is smaller than a given threshold. Probabilistic current-state opacity is, in general, an undecidable problem (Saboori and Hadjicostis 2014).

Opacity has also been studied in timed models (see, for example, Cassez 2009) and, more recently, in timed stochastic models (Lefebvre and Hadjicostis 2019a, b). The study of opacity in Petri nets models has also received attention recently. For instance, Tong et al. (2017b) uses basis markings and basis reachability graphs to verify current-state opacity in bounded Petri nets in an efficient manner. Decidable and undecidable problems of opacity properties in Petri nets are discussed in Tong et al. (2017a). The work in Bérard et al. (2017) considers opacity in Petri nets, as well as some connections between opacity and diagnosability.

A very good survey of opacity in discrete event systems can be found in Jacob et al. (2015, 2016). A discussion on the history of fault diagnosis and opacity research in discrete event systems can be found in Lafortune et al. (2018).

References

Alur R, Černý P, Zdancewic S (2006) Preserving secrecy under refinement. In: Automata, languages and programming, pp 107–118

Badouel E, Bednarczyk M, Borzyszkowski A, Caillaud B, Darondeau P (2006) Concurrent secrets. In: Proceedings of the 8th international workshop on discrete event systems (WODES), pp 51–57

Berard B, Mullins J, Sassolas M (2010) Quantifying opacity. In: Proceedings of 7th international conference on the quantitative evaluation of systems (QEST), pp 263–272

Bérard B, Haar S, Schmitz S, Schwoon S (2017) The complexity of diagnosability and opacity verification for Petri nets. In: Proceedings of international conference on application and theory of Petri nets and concurrency, pp 200–220

Bryans JW, Koutny M, Mazare L, Ryan PYA (2005a) Opacity generalised to transition systems. In: Proceedings of third international workshop on formal aspects in security and trust, pp 81–95

Bryans JW, Koutny M, Ryan PYA (2005b) Modelling opacity using Petri nets. Electron Notes Theor Comput Sci 121:101–115

Cassez F (2009) The dark side of timed opacity. In: Proceedings of international conference on information security and assurance, pp 21–30

Cassez F, Dubreil J, Marchand H (2012) Synthesis of opaque systems with static and dynamic masks. Form Methods Syst Des 40(1):88–115

Chaum D (1988) The dining cryptographers problem: unconditional sender and recipient untraceability. J Cryptol 1(1):65–75

Dubreil J, Darondeau P, Marchand H (2008) Opacity enforcing control synthesis. In: Proceedings of the 9th international workshop on discrete event systems (WODES), pp 28–35

Dubreil J, Darondeau P, Marchand H (2010) Supervisory control for opacity. IEEE Trans Autom Control 55(5):1089–1100

Focardi R, Gorrieri R (1994) A taxonomy of trace–based security properties for CCS. In: Proceedings of the 7th workshop on computer security foundations, pp 126–136

Focardi R, Gorrieri R, Martinelli F (2000) Non interference for the analysis of cryptographic protocols. In: Proceedings of international colloquium on automata, languages, and programming, pp 354–372

Hadj-Alouane NB, Lafrance S, Feng L, Mullins J, Yeddes MM (2005) On the verification of intransitive noninterference in multilevel security. IEEE Trans Syst Man Cybern Part B (Cybern) 35(5):948–958

Hadjicostis CN (2012) Resolution of initial-state in security applications of DES. In: Proceedings of 20th mediterranean conference on control and automation (MED), pp 794–799

Jacob R, Lesage JJ, Faure JM (2015) Opacity of discrete event systems: models, validation and quantification. IFAC-PapersOnLine 48(7):174–181

Jacob R, Lesage JJ, Faure JM (2016) Overview of discrete event systems opacity: models, validation, and quantification. Annu Rev Control 41:135–146

Lafortune S, Lin F, Hadjicostis CN (2018) On the history of diagnosability and opacity in discrete event systems. Annu Rev Control 45:257–266

Lefebvre D, Hadjicostis CN (2019a) Exposure time as a measure of opacity in timed discrete event systems. In: Proceedings of European control conference (ECC)

Lefebvre D, Hadjicostis CN (2019b) Trajectory-observers of discrete event systems: applications to safety and security issues. In: Proceedings of 6th international conference on decision, information and control technologies (CoDIT)

Lin F (2011) Opacity of discrete event systems and its applications. Automatica 47(3):496–503

Reiter MK, Rubin AD (1998) Crowds: anonymity for web transactions. ACM Trans Inf Syst Secur 1(1):66–92

Saboori A (2010) Verification and enforcement of state-based notions of opacity in discrete event systems. PhD thesis, University of Illinois, Urbana, IL

Saboori A, Hadjicostis CN (2007) Notions of security and opacity in discrete event systems. In: Proceedings of 46th IEEE conference on decision and control (CDC), pp 5056–5061

Saboori A, Hadjicostis CN (2008) Opacity-enforcing supervisory strategies for secure discrete event systems. In: Proceedings of 47th IEEE conference on decision and control (CDC), pp 889–894
Saboori A, Hadjicostis CN (2009) Verification of infinite-step opacity and analysis of its complexity. In: Proceedings of dependable control of discrete systems (DCDS), vol 2, pp 46–51
Saboori A, Hadjicostis CN (2011a) Coverage analysis of mobile agent trajectory via state-based opacity formulations. Control Eng Pract 19(9):967–977
Saboori A, Hadjicostis CN (2011b) Verification of K-step opacity and analysis of its complexity. IEEE Trans Autom Sci Eng 8(3):549–559
Saboori A, Hadjicostis CN (2012) Opacity-enforcing supervisory strategies via state estimator constructions. IEEE Trans Autom Control 57(5):1155–1165
Saboori A, Hadjicostis CN (2013) Verification of initial-state opacity in security applications of discrete event systems. Inf Sci 246:115–132
Saboori A, Hadjicostis CN (2014) Current-state opacity formulations in probabilistic finite automata. IEEE Trans Autom Control 59(1):120–133
Sweeney L (2002) K-anonymity: a model for protecting privacy. Int J Uncertain Fuzziness Knowl Based Syst 10(05):557–570
Tong Y, Li Z, Seatzu C, Giua A (2017a) Decidability of opacity verification problems in labeled Petri net systems. Automatica 80:48–53
Tong Y, Li Z, Seatzu C, Giua A (2017b) Verification of state-based opacity using Petri nets. IEEE Trans Autom Control 62(6):2823–2837
Tong Y, Li Z, Seatzu C, Giua A (2018) Current-state opacity enforcement in discrete event systems under incomparable observations. Discrete Event Dyn Syst 28(2):161–182
Wu YC, Lafortune S (2013) Comparative analysis of related notions of opacity in centralized and coordinated architectures. Discrete Event Dyn Syst 23(3):307–339
Wu YC, Lafortune S (2014) Synthesis of insertion functions for enforcement of opacity security properties. Automatica 50(5):1336–1348
Wu YC, Lafortune S (2016) Synthesis of optimal insertion functions for opacity enforcement. IEEE Trans Autom Control 61(3):571–584
Wu YC, Sankararaman KA, Lafortune S (2014) Ensuring privacy in location-based services: an approach based on opacity enforcement. IFAC Proc Vol 47(2):33–38
Yin X, Lafortune S (2017) A new approach for the verification of infinite-step and K-step opacity using two-way observers. Automatica 80:162–171

Chapter 9
Decentralized State Estimation

9.1 Introduction and Motivation

Over the last decade, digital system and networking technologies have revolutionized many aspects of the scientific and commercial world, and have greatly affected daily life. Emerging control systems are obtained by interconnecting elementary subsystems (sensors, actuators, computing platforms, storage devices, etc.) into feedback loops that involve shared cyber infrastructures (such as wireless networks or the Internet), and they include a variety of systems, with purely discrete dynamics and interactions (referred to as cyber DES) or hybrid dynamics (referred to as cyber-physical systems). Emerging complex networked systems comprise several (possibly semiautonomous) interacting subsystems that may cooperate to reach a common goal or compete to reach individual goals. Examples include automated manufacturing systems, building automation systems, computer grids, traffic networks, and smart grids.

A key task in properly controlling these emerging complex systems is the ability to analyze information from sensors, so as to accurately estimate their state and/or reliably perform event inference tasks (e.g., fault diagnosis). This chapter focuses on two important special cases, namely, state estimation and event inference in a given (monolithic) system that is modeled as a labeled nondeterministic finite automaton (LNFA) and is observed at multiple observation sites. Each of these observation sites is capable of observing a subset of the events; thus, the occurrence of a sequence of events in the system generally causes different observation sequences at each of the observation sites. Assuming these sites have knowledge of the system model (and some basic processing capability), they can form their individual state estimates (or infer the occurrence of an event), essentially operating in isolation based on the techniques described in Chaps. 4 and 5. More generally, however, they may be able to communicate information to a coordinator (or exchange information among themselves) in order to obtain/infer a finer set of estimates/events.

C. N. Hadjicostis, *Estimation and Inference in Discrete Event Systems*,
Communications and Control Engineering,
https://doi.org/10.1007/978-3-030-30821-6_9

The decentralized observation setting adopted in this chapter assumes that each observation site operates in isolation and receives no explicit or implicit information from other sites. The key challenge is to determine how to properly communicate information to a coordinator, i.e., decide what information to send and when to send it. In general, such questions are difficult and can even become undecidable in certain settings (Puri et al. 2002; Witsenhausen 1968) (see the discussion in Sect. 9.9 later in this chapter). However, we will see that, for certain strategies for information exchange, the corresponding decentralized protocols are intuitive and easy to describe, and some of the resulting properties of interest (such as *synchronization-based decentralized* diagnosability) can be verified with low complexity (polynomial in the number of states of the given system). In Chap. 10, we discuss extensions to distributed settings, i.e., settings that allow the observation sites to also receive information from other observation sites (or from the coordinator, if one is present) and incorporate it into their state estimation or event inference processes.

9.2 System Modeling and Observation Architecture

For notational simplicity, we describe the developments in terms of an underlying (monolithic) LNFA $G = (Q, \Sigma, \delta, Q_0)$ (see Definition 3.20); however, the techniques we describe in this chapter and the next chapter (the latter chapter deals with *distributed* state estimation and event inference) can be extended to modular systems, i.e., systems that consist of compositions of nondeterministic finite automata, in a relatively straightforward manner.

We assume that the given system G is monitored by m observation sites O_i, $i = 1, 2, \ldots, m$, each of which is able to (partially) observe activity in the given system. More specifically, observation site O_i observes a subset of events Σ_{o_i}, $\Sigma_{o_i} \subseteq \Sigma$; the remaining events $\Sigma_{uo_i} := \Sigma \setminus \Sigma_{o_i}$ are unobservable at observation site O_i. Thus, the natural projection $P_{\Sigma_{o_i}} : \Sigma^* \rightarrow \Sigma_{o_i}^*$ can be used to map any trace $s \in \Sigma^*$ executed in the system ($s \in L(G)$) to the sequence of observations generated by it at observation site O_i. As described in earlier chapters, this natural projection is defined recursively as $P_{\Sigma_{o_i}}(\sigma s') = P_{\Sigma_{o_i}}(\sigma) P_{\Sigma_{o_i}}(s')$, $s' \in \Sigma^*, \sigma \in \Sigma$, with

$$P_{\Sigma_{o_i}}(\sigma) = \begin{cases} \sigma, & \text{if } \sigma \in \Sigma_{o_i}, \\ \epsilon, & \text{if } \sigma \in \Sigma_{uo_i} \cup \{\epsilon\}, \end{cases}$$

where ϵ represents the empty trace. In the sequel, Σ_{o_i} and $P_{\Sigma_{o_i}}$ will also be denoted, respectively, by Σ_i and P_i when it is clear from context.

We denote the set of events that are observable by at least one site as

$$\Sigma_o = \cup_{i=1}^m \Sigma_{o_i} \ ,$$

and loosely refer to them as the set of observable events. Similarly, we denote the set of events that are not observable at any site by

$$\Sigma_{uo} = \cap_{i=1}^{m} \Sigma_{uo_i} = \Sigma \setminus \Sigma_o ,$$

and loosely refer to them as the set of unobservable events. For comparison purposes, we can think of Σ_o as the set of events that could be observable by a fictitious observer, which we refer to as the *monolithic* or *centralized* observer and denote by O_c. We can then compare the performance of any decentralized state estimation or event inference scheme against the ultimate[1] performance of a centralized state estimation or event inference scheme at O_c.

For comparison purposes, we also make reference to fictitious observers that are able to observe a subset of the events observed by a *collection* of the given observation sites. If we let $\mathbb{M} = \{1, 2, \ldots, m\}$ denote the set of indices for the observation sites, we can talk about any subset of observation sites M, $M \subseteq \mathbb{M}$. The fictitious observer corresponding to M has as set of observable events the set $\Sigma_{oM} := \cup_{i \in M} \Sigma_{o_i}$ (also denoted by Σ_M), i.e., it gets to observe the combined set of observable events at observation sites indexed by M; we also use $P_{\Sigma_{oM}}$ (or P_M) to denote the natural projection with respect to this set of events. In terms of the above notation, the fictitious centralized observation site mentioned above (that gets to observe *all* events seen at any observation site) would correspond to $\Sigma_{o\mathbb{M}} = \Sigma_o$ (also denoted by $\Sigma_{\mathbb{M}}$) and the natural projection mapping associated with it would correspond to $P_{\Sigma_{o\mathbb{M}}}$ (also denoted by $P_{\mathbb{M}}$).

We assume that each observation site has knowledge of the system model (namely, $G = (Q, \Sigma, \delta, Q_0)$ and Σ_{o_i}, $i = 1, 2, \ldots, m$), and some basic processing and storage capability. Thus, if desirable, an observation site can locally process the sequence of events it observes, in order to, depending on the task, determine possible system states and/or infer events. Notice that in the decentralized observation setting studied in this chapter, each observation site O_i is in charge of analyzing its own sequence of observations without receiving any information from any other sites. Thus, given the absence of any additional information from other observation sites, the task at observation site O_i essentially amounts to the centralized state estimation task that was studied in Chap. 4, applied to the LNFA G, with set of observable events Σ_{o_i}; this will result, at each site, in a set of states or a set of fault/event conditions, or combinations of them (pairs of a state and an associated fault/event label) that are *locally consistent*, i.e., consistent with what has been observed at the observation site locally. In particular, we will use the following notation to represent, from the perspective of observation site O_i, the unobservable reach $UR_i(Q')$ from the set of states Q', $Q' \subseteq Q$, and the reachable set of states $R_i(Q', \omega_i)$ from the set of states Q' following the observation of sequence $\omega_i \in \Sigma_{o_i}^*$ at observation site O_i:

[1]Clearly, no decentralized protocol for state estimation or event inference can exceed the performance of the centralized observation site. In fact, due to loss of information (e.g., absence of timing information between events or the limited information communicated from the observation sites), a decentralized/distributed scheme will likely have inferior performance compared to the performance of centralized state estimation or event inference at the fictitious O_c.

$$UR_i(Q') = \{q \in Q \mid \exists q' \in Q', \exists s \in \Sigma^* \text{ such that } \{q \in \delta(q', s) \text{ and } P_i(s) = \epsilon\}\}, \quad (9.1)$$

$$R_i(Q', \omega_i) = \{q \in Q \mid \exists q' \in Q', \exists s \in \Sigma^* \text{ such that } \{q \in \delta(q', s) \text{ and } P_i(s) = \omega_i\}\}. \quad (9.2)$$

Since the functionality and knowledge of each observation site is well-understood, the main challenge in decentralized settings is to determine what information to send to an external coordinator, when to send information to the coordinator, and how to fuse the information received at the coordinator. For simplicity, in the remainder of this chapter, our discussion focuses primarily on current-state estimation, though extensions to event inference and delayed/initial-state estimation can also be made. Clearly, as illustrated by the reductions in Chap. 5, a variety of detectability, fault diagnosis, and opacity properties can be formulated in terms of state isolation problems (possibly for an extended version of the given finite automaton); we will also discuss the verification of such properties for specific protocols in the decentralized observation setting of this chapter. For the verification of properties, we adopt the usual assumptions that (i) G is live and (ii) G possesses no unobservable cycles with respect to the set Σ_o of events that are observable in at least one site (see, for example, the discussions in Chap. 6).

Remark 9.1 In the *distributed* observation setting studied in Chap. 10, observation sites may be able to communicate among themselves and/or a coordinator (if one is present) via some existing communication links. The important difference in Chap. 10 is that each observation site may receive information (from the coordinator and/or the other observation sites), which can be incorporated into subsequent processing at this observation site. □

9.3 Decentralized Information Processing

Consider a decentralized observation architecture with a coordinator or fusion center, as shown in Fig. 9.1. We are given a monolithic system, modeled as an LNFA $G = (Q, \Sigma, \delta, Q_0)$ and m observation sites, such that each site O_i, $i = 1, 2, \ldots, m$, observes a subset of events Σ_{o_i}, $\Sigma_{o_i} \subseteq \Sigma$, under a natural projection mapping $P_{\Sigma_{o_i}}$. In particular, following an unknown sequence of events $s \in \Sigma^*$ that occurs in the system (i.e., $s \in L(G)$), observation site O_i observes the sequence of observations $\omega_i = P_{\Sigma_{o_i}}(s) = P_i(s)$. Each observation site needs to determine what information to report to the coordinator, perhaps after some local processing of the sequence of observations via the computational unit C_i. The coordinator is then tasked with fusing the information it receives, in order to estimate the possible current state of the system, or infer the occurrence of an event of interest or reach some other decision. The task is complex and depends on many parameters (notably what information is sent to the coordinator, and when it is sent). The key feature of the decentralized

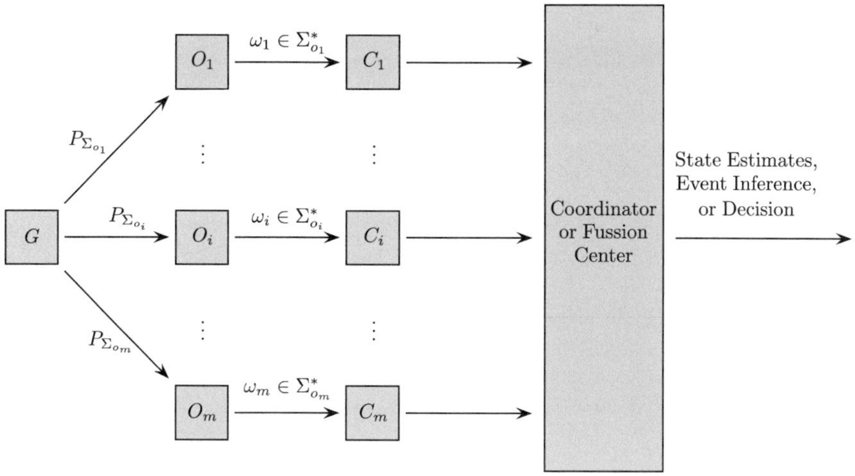

Fig. 9.1 Coordinator or fusion center receiving locally processed information from computational units C_i, $i = 1, 2, \ldots, m$; each computational unit C_i bases its processing on the sequence of observations ω_i seen at each observation site O_i and its (possibly partial) knowledge of the system model

observation architecture is that no information is sent back to the observation sites from the coordinator. This means that the functionality at each observation site cannot be influenced by the coordinator or the other observation sites.

We will assume (for now) that the coordinator (who may or may not have knowledge of system G) is responsible for requesting information from observation sites. When the coordinator issues such a request, all observation sites send information and we have a *synchronization* step (we refer to this as the coordinator initiating a *sync* operation). We leave aside, for now, the question of *how/when* a synchronization step is initiated (we revisit this issue in Sect. 9.9) and focus on the type of information that is sent to the coordinator. We consider three separate cases:

Case I: This is the case when the information sent by each observation site O_i is the sequence of observations that has been observed at O_i since the last synchronization step (in Fig. 9.1 this would correspond to the case when the processing element C_i simply stores the sequence of observations seen at observation site O_i). Notice that in this case, observation sites do not need to have knowledge of the system model G; also, they do not need to process any information but they need the capability to store and transmit sequences of observations. However, to properly process the received information, the coordinator needs to have knowledge of the system model.

Case II: This is the case when the information sent by each observation site O_i is the *local* set of state estimates $\hat{q}_i$, $\hat{q}_i \subseteq Q$, that is locally consistent based on the observations seen at observation site O_i thus far (as shown in Fig. 9.2). In this case, observation sites need to have knowledge of the system model G and be

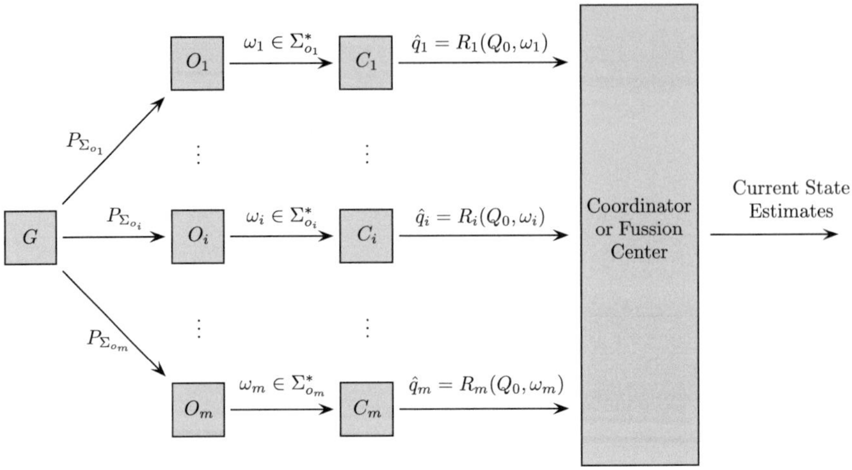

Fig. 9.2 Coordinator or fusion center receiving locally consistent current-state estimates; the processing element C_i is simply an observer at observation site O_i, so that the estimate at the ith site is given by $\hat{q}_i = R_i(Q_0, \omega_i)$

able to process their observations to obtain a set of possible state estimates (i.e., the processing element C_i in Fig. 9.2 is simply a local observer that relies on Σ_{o_i} as the set of observable events). On the other hand, we will see that in Case II the coordinator does not need to have knowledge of the system model G.

Case III: This is the case when the information sent by each observation site O_i is the local decision (e.g., "fault," "no fault," or "uncertain," in the case of fault diagnosis). The requirements for this case are the same as in Case II above.

We now describe the operation of the different protocols in more detail. In terms of notation, we can always break the sequence of events s that occurs in the system ($s \in L(G)$) into subsequences of events $s^{(1)}$, $s^{(2)}$, ..., $s^{(k)}$, so that

$$s = s^{(1)} s^{(2)} \dots s^{(k)}$$

and the ith synchronization step (denoted by $sync^{(i)}$) occurs immediately after the occurrence of the sequence of events $s^{(i)}$. We will also denote this sequence of events/synchronizations by

$$s^{(1)} sync^{(1)} s^{(2)} sync^{(2)} \dots sync^{(k-1)} s^{(k)} sync^{(k)} .$$

Notice that the above sequence results in the following sequence of observations at observation site O_i:

$$\omega_i^{(1)} sync^{(1)} \omega_i^{(2)} sync^{(2)} \dots sync^{(k-1)} \omega_i^{(k)} sync^{(k)} ,$$

where $\omega_i^{(\kappa)} = P_i(s^{(\kappa)})$ for $\kappa = 1, 2, \ldots, k$.

The three protocols differ in terms of the information that is provided by each observation site to the coordinator after the κth synchronization.

Case I: In this case, the information sent by each observation site O_i is the sequence of observations that has been observed at O_i since the last synchronization step, i.e., the information sent to the coordinator at the κth synchronization step is

$$\omega_i^{(\kappa)} = P_i(s^{(\kappa)}) \; , i = 1, 2, \ldots, m \; .$$

Notice that this information does not require an observation site to have access to the estimate at the coordinator (or to remember its own observations seen before the last synchronization).

Case II: In this case, the information sent by each observation site O_i is the local set of current-state estimates that is consistent based on the observations seen at observation site O_i thus far, i.e.,

$$\hat{Q}_i^{(\kappa)} = R_i(Q_0, \omega_i^{(1)}\omega_i^{(2)} \ldots \omega_i^{(\kappa)}) \; , i = 1, 2, \ldots, m \; ,$$

where $\omega_i^{(1)}\omega_i^{(2)} \ldots \omega_i^{(\kappa)}$ is the sequence of observations seen at observation site O_i thus far. As in Case I, this information does not require knowledge of the estimate at the coordinator. Unlike Case I, however, this information depends on earlier observations seen at observation site O_i; it can be maintained recursively at each observation site as follows:

$$\hat{Q}_i^{(\kappa)} = R_i(\hat{Q}_i^{(\kappa-1)}, \omega_i^{(\kappa)}) \; , \kappa = 1, 2, \ldots, k \; ,$$

where $\hat{Q}_i^{(0)} = UR_i(Q_0)$ (note that UR_i and R_i were defined in (9.1) and (9.2)).

Case III: In this case, the information sent by each observation site O_i is the local decision (e.g., "fault," or "no fault," or "uncertain" in the case of fault diagnosis). Notice that this information in most cases will be a function of $\hat{Q}_i^{(\kappa)}$ (for example, if one uses the techniques described in Sect. 7.2.2 of Chap. 7 to reduce the fault diagnosis problem to a state isolation problem, then the decision could be "F" to indicate definite presence of a fault, "N" to indicate definite absence of a fault, and "U" to indicate uncertainty).

Remark 9.2 Clearly, Case III provides less information to the coordinator than Case II, which provides less information to the coordinator than Case I. The reason is simple: the sequence of observations at observation site O_i can be used to infer the local set of state estimates, which in turn allows one to infer the local decision, at least if we assume that the coordinator is aware of the system model and Σ_{o_i} $i \in \mathbb{M}$. Thus, Case I provides at least as much information as Case II, which provides at least as much information as Case III. □

9.4 Totally Ordered Versus Partially Ordered Sequences of Observations

We start with an example to illustrate that, even when the observation sites provide the coordinator with their sequences of observations (Case I), which in some sense is the full information of what they have observed, there is a significant challenge that arises due to the absence of total ordering information.

Example 9.1 Suppose that the given LNFA G has as set of events $\Sigma = \{a_1, a_2, a_3, a_4, a_5, b_1, b_2, b_3, c, d_1, d_2\}$ (the remaining parameters that define G are not important for this example). Assume there are two observation sites O_1 and O_2, such that $\Sigma_{o_1} = \{a_1, a_2, a_3, a_4, a_5, c\}$ and $\Sigma_{o_2} = \{b_1, b_2, b_3, c\}$. Suppose that the sequence of events $s^{(1)} = a_1 d_1 b_1 c d_2 a_2 b_2$ occurs before the first synchronization, so that observation site O_1 records the sequence $\omega_1^{(1)} = P_1(s^{(1)}) = a_1 c a_2$, whereas observation site O_2 records the sequence $\omega_2^{(1)} = P_2(s^{(1)}) = b_1 c b_2$. Note that events d_1 and d_2 are unobservable at both observation sites. Depending on whether time stamping is possible or not, we have two distinct cases.

Availability of Time Stamps: If timing information is available, each observation site can time stamp its observations and send to the coordinator the sequence of events along with their corresponding time stamps. The coordinator can use the time stamps to determine the exact (total) order in which observations took place. This means that the coordinator can piece together the sequence of observable events, reconstruct the observation sequence that would have been seen at the fictitious centralized observation site O_c (in this case, $a_1 b_1 c a_2 b_2$), and obtain the matching set of possible current states in the system. Note that the last step requires that the coordinator is aware of the system model G and Σ_{o_M} (recall that $\Sigma_{o_M} = \Sigma_o$, i.e., the set of events that are observable by at least one observation site). It also implies that, immediately following the synchronization step, the coordinator will have an estimate that is identical to that of the fictitious centralized observation site O_c. However, depending on the number of events that occur between consecutive synchronizations, the processing that might be required at the coordinator could be challenging; also, clock synchronization issues between various observation sites could further complicate the task of the coordinator.

Unavailability of Time Stamps: If no time stamps are available, the coordinator needs to do more work as it can only infer partial orders between events. In this example, since event c is common, we can deduce from $\omega_1^{(1)}$ that a_1 occurred before c, which occurred before a_2, whereas based on $\omega_2^{(1)}$, we can deduce that b_1 occurred before c, which occurred before b_2. However, the relative ordering between a_1 and b_1 (or between a_2 and b_2) has been lost. In general, there are multiple possible (totally ordered) sequences of observable events that match the (partial) orders seen at the observation sites, and not all of them may be consistent with the given system. In this particular example, we have four possible sequences of observable events: $a_1 b_1 c a_2 b_2$ (this is the actual observation sequence that would have been seen at the fictitious centralized observation site O_c), or $b_1 a_1 c a_2 b_2$, or $b_1 a_1 c b_2 a_2$, or $a_1 b_1 c b_2 a_2$. Note that it is possible that the system may not actually be able to generate some of the above observation sequences, but in any case, some processing is needed to determine the possible (totally ordered) observation sequences and the possible current states following these possible (totally ordered) observation sequences.

Note that subsequent synchronizations will operate in the same manner. For instance, if we assume that following the first synchronization, the sequence of events $s^{(2)} = a_3 a_4 d_2 b_3 a_5$ occurs, then O_1 records $\omega_1^{(2)} = P_1(s^{(2)}) = a_3 a_4 a_5$, whereas O_2 records $\omega_2^{(2)} = P_2(s^{(2)}) = b_3$. Again, if timing information is available, the coordinator can reconstruct the observation sequence that would have been seen at the fictitious centralized observation site O_c, namely $a_3 a_4 b_3 a_5$; however, if no timing information is available then one has to consider the following four (totally ordered) sequences of observable events: $a_3 a_4 a_5 b_3$, $a_3 a_4 b_3 a_5$ (which is the observation sequence that would have been seen at the centralized observation site O_c), $a_3 b_3 a_4 a_5$, and $b_3 a_3 a_4 a_5$.

It is worth pointing out that by combining the possible sequences after the first and second synchronization, we can obtain all possible (partially ordered) sequences of observations. In this case we get 16 different sequences, by combining the four sequences before the first synchronization with the four sequences before the second synchronization (and after the first synchronization), as follows:

$$
\begin{array}{l}
a_1 b_1 c a_2 b_2 (sync^{(1)}) a_3 a_4 a_5 b_3 (sync^{(2)}) \\
a_1 b_1 c a_2 b_2 (sync^{(1)}) a_3 a_4 b_3 a_5 (sync^{(2)}) \\
a_1 b_1 c a_2 b_2 (sync^{(1)}) a_3 b_3 a_4 a_5 (sync^{(2)}) \\
a_1 b_1 c a_2 b_2 (sync^{(1)}) b_3 a_3 a_4 a_5 (sync^{(2)}) \\
b_1 a_1 c a_2 b_2 (sync^{(1)}) a_3 a_4 a_5 b_3 (sync^{(2)}) \\
b_1 a_1 c a_2 b_2 (sync^{(1)}) a_3 a_4 b_3 a_5 (sync^{(2)}) \\
b_1 a_1 c a_2 b_2 (sync^{(1)}) a_3 b_3 a_4 a_5 (sync^{(2)}) \\
b_1 a_1 c a_2 b_2 (sync^{(1)}) b_3 a_3 a_4 a_5 (sync^{(2)}) \\
b_1 a_1 c b_2 a_2 (sync^{(1)}) a_3 a_4 a_5 b_3 (sync^{(2)}) \\
b_1 a_1 c b_2 a_2 (sync^{(1)}) a_3 a_4 b_3 a_5 (sync^{(2)}) \\
b_1 a_1 c b_2 a_2 (sync^{(1)}) a_3 b_3 a_4 a_5 (sync^{(2)}) \\
b_1 a_1 c b_2 a_2 (sync^{(1)}) b_3 a_3 a_4 a_5 (sync^{(2)}) \\
a_1 b_1 c b_2 a_2 (sync^{(1)}) a_3 a_4 a_5 b_3 (sync^{(2)}) \\
a_1 b_1 c b_2 a_2 (sync^{(1)}) a_3 a_4 b_3 a_5 (sync^{(2)}) \\
a_1 b_1 c b_2 a_2 (sync^{(1)}) a_3 b_3 a_4 a_5 (sync^{(2)}) \\
a_1 b_1 c b_2 a_2 (sync^{(1)}) b_3 a_3 a_4 a_5 (sync^{(2)})
\end{array}
$$

The above example clearly illustrates that, in the absence of timing information, the decentralized observation setting has to consider all consistent totally ordered sequences of observations, thus resulting in a set of state estimates that necessarily includes the set of state estimates that the fictitious centralized observer would reach (but perhaps also includes other state estimates). It should also be clear that when communicating partially ordered sequences of observations (Case I of the decentralized protocols), all processing needs to be done at the coordinator (who receives the partially ordered sequences of observations); no processing (other than storage) is needed at the observation sites. □

Consider now the general decentralized observation setting described in the previous section: given the LNFA $G = (Q, \Sigma, \delta, Q_0)$ and m observation sites O_i, $i = 1, 2, \ldots, m$, with site O_i observing the subset of events Σ_{o_i}, $\Sigma_{o_i} \subseteq \Sigma$, suppose that an unknown sequence of events $s \in \Sigma^*$ occurs in the system (i.e., $s \in L(G)$), so that observation site O_i observes the sequence of observations $\omega_i = P_i(s)$. Consider any subset of observation sites M, $M \subseteq \mathbb{M}$, that report the activity in system G after the last event in s. When only partial ordering of observation sequences is available, we have to consider the following set of state estimates.

Definition 9.1 (*Possible states following partially ordered sequences of observations*) Suppose that LNFA $G = (Q, \Sigma, \delta, Q_0)$ is known to be in a set of possible states Q', $Q' \subseteq Q$; in the absence of any other timing information between event observations, the set of all possible states after subsequently observing $\omega_i \in \Sigma_{o_i}^*$ at observation site O_i, $i \in M$, is

$$R_{po,M}(Q', \{\omega_i, i \in M\}) := \{q \in Q \mid \exists q' \in Q', \exists s \in \Sigma^*, \text{ such that } \{(P_i(s) = \omega_i, \forall i \in M) \text{ and } q \in \delta(q', s)\}\} .$$

We can also define the set of totally ordered sequences that match the sequences ω_i observed at each observation site O_i, $i \in M$, as follows.

Definition 9.2 (*Totally ordered sequences matching partially ordered sequences of observations*) Given $\omega_i \in \Sigma_{o_i}^*$ at observation site O_i, $i \in M$, the set

$$PO_M(\{\omega_i, i \in M\}) := \{\omega_M \in \Sigma_{o_M}^* \mid P_i(\omega_M) = \omega_i, \ \forall i \in M\}$$

contains all totally ordered observation sequences in $\Sigma_{o_M}^*$ that respect the partial ordering of observations at each site O_i, $i \in M$.

Clearly, $R_{po,M}(Q', \{\omega_i, i \in M\})$ can be calculated by performing state estimation for each sequence ω_M in $PO_M(\{\omega_i, i \in M\})$, i.e.,

$$R_{po,M}(Q', \{\omega_i, i \in M\}) = \cup_{\omega_M \in PO_M(\{\omega_i, i \in M\})} R_M(Q', \omega_M) , \tag{9.3}$$

which also suggests a straightforward way for computing $R_{po,M}(Q', \{\omega_i, i \in M\})$ (by first obtaining all totally ordered sequences that respect the given partial orders and then performing state estimation with respect to Σ_{o_M} on each sequence).

If we allow processing to take place at the observation sites (Case II of the decentralized protocols), since no information is explicitly or implicitly sent to an observation site from other observation sites or the coordinator, we realize that the set of possible states perceived by observation site O_i is independent of the time instants at which previous synchronizations took place. Specifically, assuming an *unknown* sequence of events $s \in \Sigma^*$ (where $s \in L(G)$) has occurred in the given LNFA $G = (Q, \Sigma, \delta, Q_0)$, it will result in sequences of observations $\omega_i = P_i(s)$ at each observation site $O_i, i = 1, 2, \ldots, m$; the set of possible current states perceived at observation site O_i, based on its own local knowledge, is given by $R_i(Q_0, \omega_i)$ where $R_i(Q', \omega_i)$ was defined in (9.2).

Remark 9.3 Note that, according to the notation in Chap. 4, if $\omega_i = \sigma[0]\sigma[1]\ldots\sigma[k]$, where $\sigma[\kappa] \in \Sigma_{o_i}$ for $\kappa = 0, 1, 2, \ldots, k$, then $R_i(Q_0, \omega_i) = \hat{q}_{\sigma[k]}(\omega_i)$. □

Definition 9.3 (*Possible states following a sequence of observations*) Suppose that LNFA $G = (Q, \Sigma, \delta, Q_0)$ is known to be in a set of possible states Q', $Q' \subseteq Q$, and an unknown sequence of events s occurs in G. From the perspective of a fictitious observer that observes events in Σ_{o_M}, the set of all possible current states is given by

$$R_M(Q', \omega_M) = \{q \in Q \mid \exists q' \in Q', \exists s \in \Sigma^*, \text{ such that } \{P_M(s) = \omega_M \text{ and } q \in \delta(q', s)\}\} ,$$

where $\omega_M = P_M(s) \in \Sigma_{o_M}^*$ is the sequence that is observed at the fictitious observer.

The following lemma characterizes information loss due to partial ordering of observation sequences.

Lemma 9.1 *Suppose that LNFA $G = (Q, \Sigma, \delta, Q_0)$, known to be in a set of possible states Q', $Q' \subseteq Q$, is observed at observation sites O_i, $i \in \mathbb{M} = \{1, 2, \ldots, m\}$. Subsequently, an unknown sequence of events s gets executed resulting in observation sequences $\omega_i = P_i(s) \in \Sigma_{o_i}^*$ at each observation site O_i, $i \in \mathbb{M}$. For any $M \subseteq \mathbb{M}$, the following holds true:*

$$R_M(Q', \omega_M) \subseteq R_{po,M}(Q', \{\omega_i, i \in M\}) \subseteq \bigcap_{i \in M} R_i(Q', \omega_i) ,$$

where $\omega_M = P_M(s)$ is the observation sequence that would have been seen at the fictitious observer that gets to observe events in Σ_M.

The above lemma is key to understanding the tradeoffs involved in the three decentralized state estimation strategies that we describe later in this chapter. Case I (with no processing at the observation sites) corresponds to partial-order-based estimation at the coordinator, whereas Case II and Case III (with local processing at the observation sites) correspond to set intersection-based estimation at the coordinator. One should point out, however, that there are different communication and processing requirements for each of the three protocols. We next discuss these three cases, and their underlying assumptions, in more detail.

9.5 Case I: Partial-Order-Based Estimation

Recall that Case I assumes that, at each synchronization, the observation sites send to the coordinator the subsequence of observations they have recorded since the last synchronization. As mentioned earlier, we can always break the sequence of events s that occurs in the system ($s \in L(G)$) into subsequences of events $s^{(1)}, s^{(2)}, \ldots, s^{(k)}$, so that $s = s^{(1)}s^{(2)} \ldots s^{(k)}$ and the κth synchronization step (denoted by $sync^{(\kappa)}$) occurs immediately after the occurrence of subsequence $s^{(\kappa)}$. Recall that this sequence of events/synchronizations is denoted by

$$s^{(1)}sync^{(1)}s^{(2)}sync^{(2)} \ldots sync^{(k-1)}s^{(k)}sync^{(k)} ,$$

and the corresponding sequence of observations at observation site O_i by

$$\omega_i^{(1)}sync^{(1)}\omega_i^{(2)}sync^{(2)} \ldots sync^{(k-1)}\omega_i^{(k)}sync^{(k)} ,$$

where $\omega_i^{(\kappa)} = P_i(s^{(\kappa)})$ for $\kappa = 1, 2, \ldots, k$.

At the first synchronization, the coordinator needs to obtain the set of state estimates

$$\begin{aligned}\hat{Q}_{po}^{(1)} &= \{q \in Q \mid \exists q_0 \in Q_0, \exists t \in \Sigma^*, \text{ such that} \\ &\quad \{(P_i(t) = \omega_i^{(1)}, \ \forall i \in \mathbb{M}) \text{ and } q \in \delta(q_0, t)\}\} \\ &= R_{po,\mathbb{M}}(Q_0, \{\omega_1^{(1)}, \omega_2^{(1)}, \ldots, \omega_m^{(1)}\}) . \end{aligned} \tag{9.4}$$

We can recursively continue this process at subsequent synchronizations. If we use $\hat{Q}_{po}^{(\kappa)}$ to denote the set of state estimates at the coordinator after the κth synchronization, and assume that the sequence of events $s^{(\kappa+1)}$ occurs between the κth and $(\kappa + 1)^{st}$ synchronization, then we have

$$\hat{Q}_{po}^{(\kappa+1)} = R_{po,\mathbb{M}}(\hat{Q}_{po}^{(\kappa)}, \{\omega_1^{(\kappa+1)}, \omega_2^{(\kappa+1)}, \ldots, \omega_m^{(\kappa+1)}\}) \ , \kappa = 0, 1, 2, \ldots, k-1,$$

where $\hat{Q}_{po}^{(0)} = R_{po,\mathbb{M}}(Q_0, \{\epsilon, \epsilon, \ldots, \epsilon\}) = R_{\mathbb{M}}(Q_0, \epsilon)$ and $\omega_i^{(\kappa+1)} = P_i(s^{(\kappa+1)})$.

Remark 9.4 Note that the occurrence of a synchronization event imposes constraints in the possible partial orders. Thus, at the end of sequence s (at the kth synchronization), we have

$$\hat{Q}_{po}^{(k)} \subseteq R_{po,\mathbb{M}}(Q_0, \{P_1(s), P_2(s), \ldots, P_m(s)\}) .$$

The reason is that events in $\omega_i^{(\kappa)} = P_i(s^{(\kappa)})$, $\kappa = 1, 2, 3, \ldots, k$ have to appear later than events in $\omega_j^{(\kappa')} = P_j(s^{(\kappa')})$ for $\kappa' < \kappa$ (and all $j \in \mathbb{M}$), and have to appear earlier than events in $\omega_j^{(\kappa'')} = P_j(s^{(\kappa'')})$ for $\kappa'' > \kappa$ (and all $j \in \mathbb{M}$). If the synchronizations were not there, the above constraints need not necessarily hold (apart from the obvious constraint that events in $\omega_i^{(\kappa)} = P_i(s^{(\kappa)})$, $\kappa = 1, 2, 3, \ldots, k$ have to appear later than events in $\omega_i^{(\kappa')} = P_i(s^{(\kappa')})$ for $\kappa' < \kappa$, and earlier than events in $\omega_i^{(\kappa'')} = P_i(s^{(\kappa'')})$ for $\kappa'' > \kappa$). □

This discussion makes it clear that the computation of possible states following partially ordered sequences of observations is key to our analysis. Next, we describe a recursive way for computing $R_{po,\mathbb{M}}(Q', \{\omega_1, \omega_2, \ldots, \omega_m\})$ (where Q' is the set of possible states that the system was in prior to generating activity that resulted in sequences of observations $\omega_1, \omega_2, \ldots, \omega_m$ at sites $O_1, O_2, \ldots, O_m$, respectively); the computation has complexity that is polynomial in $\Pi_{i=1}^m |\omega_i|$ and can be used at each synchronization step. For simplicity of notation, we first consider a simplified setting in which there are only two observation sites; then, we generalize the results to the case of multiple observation sites.

9.5.1 Simplified Setting: Two Observation Sites

Consider LNFA $G = (Q, \Sigma, \delta, Q_0)$ and suppose that there are two observation sites O_1 and O_2 ($\mathbb{M} = \{1, 2\}$), each capable of observing events in Σ_{o_1} and Σ_{o_2}, respectively. The two observation sites send to a coordinator their sequences of observations since the last time they sent information. Note that our discussion below applies to any strategy the coordinator might follow to determine *when* to request information from the observation sites; we discuss possible strategies for synchronization and their impact on various properties of interest, such as detectability, diagnosability or opacity, later in Sect. 9.9 of this chapter.

From our discussions in previous sections, the key in performing recursive state estimation at the coordinator is the ability to efficiently perform the operation $R_{po,\mathbb{M}}(Q', \{\omega_1, \omega_2\})$. As mentioned earlier, a straightforward way to calculate this set of possible current states would be to enumerate all possible sequences of observations that match the given partial orders; then, for each such sequence, we can obtain the set of matching current-state estimates (using standard techniques), and eventually merge (take the union of) all sets of such current-state estimates (this is the essence of the approach described in (9.3)).

The drawback of the above approach is that it is computationally expensive. For instance, if we denote the length of the observation sequences by $L_1 = |\omega_1|$ and $L_2 = |\omega_2|$, and assume for simplicity that $\Sigma_{o_1} \cap \Sigma_{o_2} = \emptyset$, then we have

$$\binom{L_1 + L_2}{L_1} = \frac{(L_1 + L_2)!}{L_1! L_2!}$$

different sequences of observations that match the given partial orders. To see this, notice that we have a total of $L_1 + L_2$ symbols (total number of observations) to arrange in a total order that matches the partial orders in ω_1 and ω_2. We can think of the problem as follows: there are $L_1 + L_2$ slots available, ordered as slot 1, slot 2, ..., and slot $L_1 + L_2$; we can completely define a totally ordered sequence of observations by picking L_1 (respectively, L_2) slots out of the $L_1 + L_2$ available slots for the observations in ω_1 (respectively, the observations in ω_2); once the slots are selected, observations in each sequence have to be arranged in the partial order defined by the corresponding sequence. Furthermore, each selection of slots corresponds to a different sequence of observations that matches the given partial orders.

The following lemma and theorem can be used to develop a more efficient recursive algorithm for calculating $R_{po,\mathbb{M}}(Q', \{\omega_1, \omega_2\})$. The proofs are straightforward and are omitted.

Lemma 9.2 *Suppose* $\omega_1 = \alpha_1\alpha_2 \ldots \alpha_{L_1}$, *where* $\alpha_\kappa \in \Sigma_{o_1}$ *for* $\kappa = 1, 2, \ldots, L_1$, *and* $\omega_2 = \beta_1\beta_2 \ldots \beta_{L_2}$, *where* $\beta_\kappa \in \Sigma_{o_2}$ *for* $\kappa = 1, 2, \ldots, L_2$. *Assume* $L_1 \geq 1$ *and* $L_2 \geq 1$, *and let* $\omega_1' = \alpha_1\alpha_2 \ldots \alpha_{L_1-1}$ *and* $\omega_2' = \beta_1\beta_2 \ldots \beta_{L_2-1}$, *so that* $\omega_1 = \omega_1'\alpha_{L_1}$ *and* $\omega_2 = \omega_2'\beta_{L_2}$ *(we can take* $\omega_1' = \epsilon$ *if* $L_1 = 1$ *and/or* $\omega_2' = \epsilon$ *if* $L_2 = 1$*). If* $\Sigma_{o_1} \cap \Sigma_{o_2} = \emptyset$, *we have*

$$PO_{\mathbb{M}}(\{\omega_1, \omega_2\}) = \{t\alpha_{L_1} \mid t \in PO_{\mathbb{M}}(\{\omega_1', \omega_2\})\} \cup \{t\beta_{L_2} \mid t \in PO_{\mathbb{M}}(\{\omega_1, \omega_2'\})\} ,$$

where $PO_{\mathbb{M}}(\{\omega_1, \epsilon\}) = \omega_1$ *and* $PO_{\mathbb{M}}(\{\epsilon, \omega_2\}) = \omega_2$.

Remark 9.5 Note that the result in Lemma 9.2 implies that

$$\binom{L_1 + L_2}{L_1} = \binom{L_1 + L_2 - 1}{L_1 - 1} + \binom{L_1 + L_2 - 1}{L_2 - 1} ,$$

which is indeed the well-known Pascal's identity Rosen (2011). □

The above lemma can be used to obtain a recursive way of computing $R_{po,\mathbb{M}}(Q', \{\omega_1, \omega_2\})$ as implied by the following theorem.

Theorem 9.1 *Consider an LNFA* $G = (Q, \Sigma, \delta, Q_0)$ *that is observed at observation sites* O_1 *and* O_2 *(i.e.,* $\mathbb{M} = \{1, 2\}$*) with observable events* Σ_{o_1} *and* Σ_{o_2}*, respectively, where* $\Sigma_{o_1} \subseteq \Sigma$, $\Sigma_{o_2} \subseteq \Sigma$, *and* $\Sigma_{o_1} \cap \Sigma_{o_2} = \emptyset$. *LNFA G is known to be in a set of possible states* Q', $Q' \subseteq Q$ *and, subsequently, an unknown sequence of events s gets executed, resulting in observation sequences* $\omega_1 = P_1(s) \in \Sigma_{o_1}^*$ *and* $\omega_2 = P_2(s) \in \Sigma_{o_2}^*$. *Let* $\omega_1 = \alpha_1\alpha_2 \ldots \alpha_{L_1}$, *where* $\alpha_\kappa \in \Sigma_{o_1}$ *for* $\kappa = 1, 2, \ldots, L_1$, *and* $\omega_2 = \beta_1\beta_2 \ldots \beta_{L_2}$, *where* $\beta_\kappa \in \Sigma_{o_2}$ *for* $\kappa = 1, 2, \ldots, L_2$ *(note that* L_1 *and/or* L_2 *could be zero implying that* ω_1 *and/or* ω_2 *would be the empty sequence). Let* $\omega_1' = \alpha_1\alpha_2 \ldots \alpha_{L_1-1}$ *and* $\omega_2' = \beta_1\beta_2 \ldots \beta_{L_2-1}$, *so that* $\omega_1 = \omega_1'\alpha_{L_1}$ *and* $\omega_2 = \omega_2'\beta_{L_2}$ *(we can take* $\omega_1' = \epsilon$ *if* $L_1 = 1$ *and/or* $\omega_2' = \epsilon$ *if* $L_2 = 1$; *if* $L_1 = 0$, *we can take* $\omega_1' = \epsilon$ *and* $\alpha_{L_1} = \epsilon$, *whereas if* $L_2 = 0$, *we can take* $\omega_1' = \epsilon$ *and* $\beta_{L_2} = \epsilon$*). We have*

$$R_{po,\mathbb{M}}(Q', \{\omega_1, \omega_2\}) = \begin{cases} R_{\mathbb{M}}(R_{po,\mathbb{M}}(Q', \{\omega_1', \omega_2\}), \alpha_{L_1}) \cup \\ \quad \cup\, R_{\mathbb{M}}(R_{po,\mathbb{M}}(Q', \{\omega_1, \omega_2'\}), \beta_{L_2}) \ , \ \textit{if}\ L_1, L_2 \geq 1 \ , \\ \\ R_{\mathbb{M}}(R_{po,\mathbb{M}}(Q', \{\omega_1', \omega_2\}), \alpha_{L_1}) = \\ \quad = R_{\mathbb{M}}(Q', \omega_1) \ , \ \textit{if}\ L_1 \geq 1, L_2 = 0 \ , \\ \\ R_{\mathbb{M}}(R_{po,\mathbb{M}}(Q', \{\omega_1, \omega_2'\}), \beta_{L_2}) = \\ \quad = R_{\mathbb{M}}(Q', \omega_2) \ , \ \textit{if}\ L_1 = 0, L_2 \geq 1 \ , \\ \\ R_{\mathbb{M}}(Q', \epsilon) \ , \ \textit{if}\ L_1 = 0, L_2 = 0 \ . \end{cases} \tag{9.5}$$

Note that $R_{\mathbb{M}}$ in (9.5) is taken with respect to observable events in $\Sigma_{\mathbb{M}} = \Sigma_{o_1} \cup \Sigma_{o_2}$. Also note that the computation can be performed iteratively as follows: if we use $\hat{Q}(k_1, k_2)$ to concisely denote $\hat{Q}(Q_0, \{\alpha_1\alpha_2 \ldots \alpha_{k_1}, \beta_1\beta_2 \ldots \beta_{k_2}\})$, we can proceed as follows:

(0) Set $\hat{Q}(0, 0) = R_{\mathbb{M}}(Q_0, \epsilon) = U R_{\mathbb{M}}(Q_0)$.
(1) Use $\hat{Q}(0, 0)$ to calculate
$\hat{Q}(1, 0) = R_{\mathbb{M}}(\hat{Q}(0, 0), \alpha_1)$ and
$\hat{Q}(0, 1) = R_{\mathbb{M}}(\hat{Q}(0, 0), \beta_1)$.
(2) Use $\hat{Q}(1, 0)$ and $\hat{Q}(0, 1)$ to calculate
$\hat{Q}(2, 0) = R_{\mathbb{M}}(\hat{Q}(1, 0), \alpha_2)$,
$\hat{Q}(0, 2) = R_{\mathbb{M}}(\hat{Q}(0, 1), \beta_2)$, and
$\hat{Q}(1, 1) = R_{\mathbb{M}}(\hat{Q}(0, 1), \alpha_1) \cup R_{\mathbb{M}}(\hat{Q}(1, 0), \beta_1)$.
(3) $\cdots$

$\vdots$

(k) Use $\{\hat{Q}(k_1, k_2) \mid 0 \leq k_1 \leq L_1, 0 \leq k_2 \leq L_2, k_1 + k_2 = k \leq L_1 + L_2 - 1\}$ to calculate, for $0 \leq k_1' \leq L_1, 0 \leq k_2' \leq L_2, k_1' + k_2' = k + 1 \leq L_1 + L_2$,

$$\hat{Q}(k_1', k_2') = R_{\mathbb{M}}(\hat{Q}(k_1' - 1, k_2'), \alpha_{k_1'}) \cup R_{\mathbb{M}}(\hat{Q}(k_1', k_2' - 1), \beta_{k_2'}) \ .$$

$\vdots$

Continue until all $\hat{Q}(k_1, k_2)$ for $0 \leq k_1 \leq L_1$ and $0 \leq k_2 \leq L_2$ have been calculated.

Example 9.2 In Fig. 9.3 we illustrate the order in which the computation takes place for the case when $\omega_1 = \alpha_1\alpha_2\alpha_3$ and $\omega_2 = \beta_1\beta_2$. Arrows in the figure indicate the precedence of various computations. For instance, in order to compute $\hat{Q}(2, 1)$, we need to have $\hat{Q}(1, 1)$ and $\hat{Q}(2, 0)$. The order in which the various sets of state estimates are calculated as described in the algorithm preceding this example is given by

1. $\hat{Q}(0, 0)$;
2. $\hat{Q}(1, 0)$, $\hat{Q}(0, 1)$;

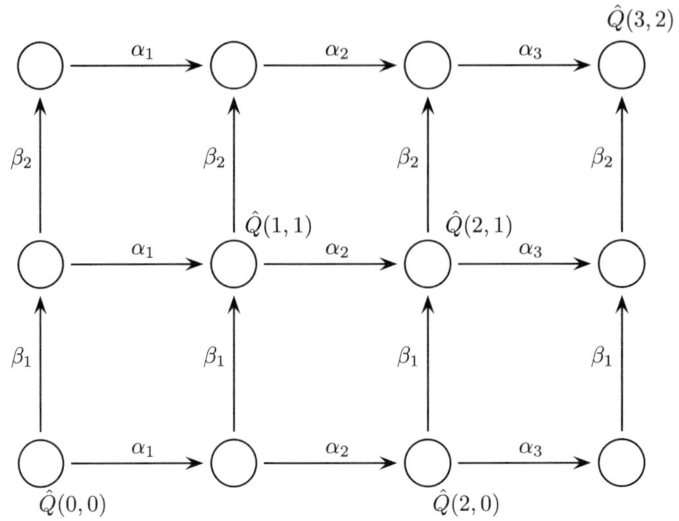

Fig. 9.3 Recursive computation of state sets for computing $R_{\mathbb{M}}(Q', \{\omega_1, \omega_2\})$ for the case when $\omega_1 = \alpha_1\alpha_2\alpha_3$ and $\omega_2 = \beta_1\beta_2$

3. $\hat{Q}(2,0)$, $\hat{Q}(1,1)$, $\hat{Q}(0,2)$;
4. $\hat{Q}(3,0)$, $\hat{Q}(2,1)$, $\hat{Q}(1,2)$;
5. $\hat{Q}(3,1)$, $\hat{Q}(2,2)$;
6. $\hat{Q}(3,2)$.

From Fig. 9.3, it becomes clear that the order of computation does not have to follow the exact order indicated above, as long as it respects the precedence relationships indicated by the arrows in the figure. For example, we can decide to compute first the bottom row of Fig. 9.3: $\hat{Q}(0,0)$, then $\hat{Q}(1,0)$, then $\hat{Q}(2,0)$, and finally $\hat{Q}(3,0)$; after the first row, we can compute the middle row: $\hat{Q}(0,1)$, then $\hat{Q}(1,1)$, then $\hat{Q}(2,1)$, and finally $\hat{Q}(3,1)$; in the end, we can compute the top row in the figure: $\hat{Q}(0,2)$, then $\hat{Q}(1,2)$, then $\hat{Q}(2,2)$, and finally $\hat{Q}(3,2)$. □

It should be clear from the discussions in the above example that to compute $R_{\mathbb{M}}(Q', \{\omega_1, \omega_2\})$ (i.e., in order to compute $\hat{Q}(Q_0, \{\omega_1(L_1), \omega_2(L_2)\})$, we need to perform $L_1 \times L_2$ operations that involve manipulations of state sets (reachability under a particular observation) and set union operations. Thus, the complexity[2] reduces to $O(L_1 L_2 |Q|^2)$, where $|Q|$ is the number of states of the given LNFA.

When $\Sigma_{o_1} \cap \Sigma_{o_2} \neq \emptyset$, the discussion above only has to be modified slightly: if between synchronization $sync^{(k-1)}$ and $sync^{(k)}$, the two observation sites record sequences $\omega_1^{(k)}$ and $\omega_2^{(k)}$ with common symbols, then the number and order of common symbols will necessarily be the same. In fact, the coordinator can treat the points at which these common symbols occur as sub-synchronization points between

[2]We assume that $R_{\mathbb{M}}(Q', \sigma)$ for any $\sigma \in (\Sigma_{o_1} \cup \Sigma_{o_2})$ can be obtained with complexity $|Q|^2$ by precomputing all $R_{\mathbb{M}}(\{q\}, \sigma)$ for each $q \in Q$, and then simply taking unions of sets.

$sync^{(k-1)}$ and $sync^{(k)}$. In the discussion below, we use $sync^{(\kappa.j)}$ to denote the jth sub-synchronization between the $(\kappa-1)^{st}$ and κth synchronizations. The following example discusses this process in more detail.

Example 9.3 Consider Example 9.1 where there are two observation sites, O_1 and O_2, with $\Sigma_{o_1} = \{a_1, a_2, a_3, a_4, a_5, c\}$ and $\Sigma_{o_2} = \{b_1, b_2, b_3, c\}$. Consider again the sequence of events $s^{(1)} = a_1 d_1 b_1 c d_2 a_2 b_2$, which occurs before the first synchronization $sync^{(1)}$. Observation site O_1 records the sequence $\omega_1^{(1)} = P_1(s^{(1)}) = a_1 c a_2$ and observation site O_2 records the sequence $\omega_2^{(1)} = P_2(s^{(1)}) = b_1 c b_2$. Note that events d_1 and d_2 are unobservable at both observation sites. Since c is a shared event, we can obtain the set of possible states

$$\hat{Q}_{po}^{(1)} = R_{po,\mathbb{M}}(Q_0, \{\omega_1^{(1)}, \omega_2^{(1)}\})$$

by modifying the recursive procedure described above. More specifically, we can think that the following sequence of events took place:

- First, O_1 observes $\omega_1^{(1.1)} = a_1$ while O_2 observes $\omega_2^{(1.1)} = b_1$.
- Then, a sub-synchronization happens followed by observation c at both sites. Thus, we can talk about the set of states that are possible before and after the occurrence of event c. More specifically, we have

$$\hat{Q}_{po}^{pre(1.1)} = R_{po,\mathbb{M}}(Q_0, \{a_1, b_1\})$$
$$\hat{Q}_{po}^{(1.1)} = R_{po,\mathbb{M}}(Q_0, \{a_1 c, b_1 c\}) \ .$$

- Finally, O_1 observes $\omega_1^{(1.2)} = a_2$ while O_2 observes $\omega_2^{(1.2)} = b_2$.

The above decomposition also outlines a recursive solution to the problem of state estimation:

(1) First, we obtain $\hat{Q}_{po}^{pre(1.1)} = R_{po,\mathbb{M}}(Q_0, \{a_1, b_1\})$.
(2) We then obtain $\hat{Q}_{po}^{(1.1)} = R_{\mathbb{M}}(\hat{Q}_{po}^{pre(1.1)}, c)$, where the reachability is taken with respect to the set of observable events $\Sigma_o = \Sigma_{o_1} \cup \Sigma_{o_2}$.
(3) Finally, we obtain $\hat{Q}_{po}^{(1)} = R_{po,\mathbb{M}}(\hat{Q}_{po}^{(1.1)}, \{a_2, b_2\})$.

Note that Steps 1 and 3 above can employ the recursive approach to obtain the set of state estimates given partially ordered observation sequences (with no common events). □

When $\Sigma_{o_1} \cap \Sigma_{o_2} \neq \emptyset$, Lemma 9.2 generalizes quite naturally to the following (note that Theorem 9.1 also generalizes in a similar manner, but we do not explicitly state it).

Lemma 9.3 *Suppose* $\omega_1 = \alpha_1 \alpha_2 \ldots \alpha_{L_1}$, *where* $\alpha_\kappa \in \Sigma_{o_1}$ *for* $\kappa = 1, 2, \ldots, L_1$, *and* $\omega_2 = \beta_1 \beta_2 \ldots \beta_{L_2}$, *where* $\beta_\kappa \in \Sigma_{o_2}$ *for* $\kappa = 1, 2, \ldots, L_2$. *Assume that symbols in* $\Sigma_{o_1} \cap \Sigma_{o_2}$ *appear in both sequences, in the same order and multiplicities (which would imply that* $P_2(\omega_1) = P_1(\omega_2)$*).*

Assume $L_1 \geq 1$ *and* $L_2 \geq 1$, *and let* $\omega_1' = \alpha_1\alpha_2 \ldots \alpha_{L_1-1}$ *and* $\omega_2' = \beta_1\beta_2 \ldots \beta_{L_2-1}$, *so that* $\omega_1 = \omega_1'\alpha_{L_1}$ *and* $\omega_2 = \omega_2'\beta_{L_2}$ *(we can take* $\omega_1' = \epsilon$ *if* $L_1 = 1$ *and/or* $\omega_2' = \epsilon$ *if* $L_2 = 1$*). We have the following four cases:*

Case 1. If $\alpha_{L_1} \in \Sigma_{o_1} \setminus \Sigma_{o_2}$ *and* $\beta_{L_2} \in \Sigma_{o_2} \setminus \Sigma_{o_1}$, *then*

$$PO_{\mathbb{M}}(\{\omega_1, \omega_2\}) = \{t\alpha_{L_1} \mid t \in PO_{\mathbb{M}}(\{\omega_1', \omega_2\})\} \cup \{t\beta_{L_2} \mid t \in PO_{\mathbb{M}}(\{\omega_1, \omega_2'\})\}.$$

Case 2. If $\alpha_{L_1} \in \Sigma_{o_1} \cap \Sigma_{o_2}$ *and* $\beta_{L_2} \in \Sigma_{o_2} \setminus \Sigma_{o_1}$, *then*

$$PO_{\mathbb{M}}(\{\omega_1, \omega_2\}) = \{t\beta_{L_2} \mid t \in PO_{\mathbb{M}}(\{\omega_1, \omega_2'\})\} .$$

Case 3. If $\beta_{L_2} \in \Sigma_{o_1} \cap \Sigma_{o_2}$ *and* $\alpha_{L_1} \in \Sigma_{o_1} \setminus \Sigma_{o_2}$, *then*

$$PO_{\mathbb{M}}(\{\omega_1, \omega_2\}) = \{t\alpha_{L_1} \mid t \in PO_{\mathbb{M}}(\{\omega_1', \omega_2\})\} .$$

Case 4. If $\alpha_{L_1} = \beta_{L_2}$ *(i.e., an event in* $\Sigma_{o_1} \cap \Sigma_{o_2}$ *observed by both observation sites), then*[3]

$$PO_{\mathbb{M}}(\{\omega_1, \omega_2\}) = \{t\alpha_{L_1} \mid t \in PO_{\mathbb{M}}(\{\omega_1', \omega_2'\})\} .$$

9.5.2 General Setting: Multiple Observation Sites

When $m > 2$ but there are no shared observable events between observation sites (i.e., $\Sigma_{o_i} \cap \Sigma_{o_j} = \emptyset$ for all $i \neq j, i, j \in \mathbb{M}$), Lemma 9.2 and Theorem 9.1 generalize quite naturally; for example, Theorem 9.1 generalizes to the following.

Theorem 9.2 *Suppose that LNFA* $G = (Q, \Sigma, \delta, Q_0)$ *is observed at* m *observation sites* $O_1, O_2, \ldots,$ *and* O_m, *with observable events* $\Sigma_{o_1}, \Sigma_{o_2}, \ldots,$ *and* Σ_{o_m}, *respectively, such that* $\Sigma_{o_i} \cap \Sigma_{o_j} = \emptyset$ *for all* $i \neq j,\ i, j \in \mathbb{M}$. *Suppose that LNFA* G *is known to be in a set of possible states* Q', $Q' \subseteq Q$, *and subsequently, an unknown sequence of events* s, $s \in \Sigma^*$, *occurs, resulting in observation sequences* $\omega_i = P_i(s) \in \Sigma_{o_i}^*$, $i \in \mathbb{M} = \{1, 2, \ldots, m\}$ *(note that it is possible that* L_i *is zero for one or more sequences, implying that the corresponding* ω_i *is an empty sequence). Let* $\omega_i = \alpha_1^{(i)}\alpha_2^{(i)} \ldots \alpha_{L_i}^{(i)}$, *where* $\alpha_\kappa^{(i)} \in \Sigma_{o_i}$ *for* $\kappa = 1, 2, \ldots, L_i$, *and let* $\omega_i' = \alpha_1^{(i)}\alpha_2^{(i)} \ldots \alpha_{L_i-1}^{(i)}$, *for* $i = 1, 2, \ldots, m$ *(if* $L_i = 1$, *we can take* $\omega_i' = \epsilon$, *whereas, if* $L_i = 0$, *we can take* $\omega_i' = \epsilon$ *and* $\alpha_{L_i}^{(i)} = \epsilon$*). We have*

[3]Note that, under the assumptions in the lemma, it is not possible for $\alpha_{L_1} \in \Sigma_{o_1} \cap \Sigma_{o_2}$ and $\beta_{L_2} \in \Sigma_{o_1} \cap \Sigma_{o_2}$, unless $\alpha_{L_1} = \beta_{L_2}$.

$$
\begin{aligned}
R_{po,\mathbb{M}}(Q', \{\omega_1, \omega_2, \ldots, \omega_m\}) = {} & R_{\mathbb{M}}(R_{po,\mathbb{M}}(Q', \{\omega_1', \omega_2, \omega_3, \ldots, \omega_m\}), \alpha_{L_1}^{(1)}) \cup \\
& \cup R_{\mathbb{M}}(R_{po,\mathbb{M}}(Q', \{\omega_1, \omega_2', \omega_3, \ldots, \omega_m\}), \alpha_{L_2}^{(2)}) \cup \\
& \vdots \\
& \cup R_{\mathbb{M}}(R_{po,\mathbb{M}}(Q', \{\omega_1, \omega_2, \ldots, \omega_{m-1}, \omega_m'\}), \alpha_{L_m}^{(m)}),
\end{aligned}
\tag{9.6}
$$

where the initial condition satisfies $R_{po,\mathbb{M}}(Q', \{\epsilon, \epsilon, \ldots, \epsilon\}) = R_{\mathbb{M}}(Q', \epsilon)$.

When observation sites may share observable events, the above theorem can be generalized quite naturally but notation becomes more cumbersome. The key observation is to realize that the recursion depends on whether the final event $\alpha_{L_i}^{(i)}$ observed at observation site O_i is observable in one or more other observation sites, in which case it has to necessarily be "synchronized" with events observed at that site (or at those sites).

Theorem 9.3 *Suppose that LNFA* $G = (Q, \Sigma, \delta, Q_0)$ *is observed at m observation sites* $O_1, O_2, \ldots$, *and* O_m, *with observable events* $\Sigma_{o_1}, \Sigma_{o_2}, \ldots$, *and* Σ_{o_m}, *respectively. Suppose that LNFA G is known to be in a set of possible states* Q', $Q' \subseteq Q$, *and subsequently, an unknown sequence of events s,* $s \in \Sigma^*$, *occurs, resulting in observation sequences* $\omega_i = P_i(s) \in \Sigma_{o_i}^*$, $i \in \mathbb{M} = \{1, 2, \ldots, m\}$ *(note that it is possible that* L_i *is zero for one or more sequences, implying that the corresponding* ω_i *is an empty sequence). Let* $\omega_i = \alpha_1^{(i)} \alpha_2^{(i)} \ldots \alpha_{L_i}^{(i)}$, *where* $\alpha_\kappa^{(i)} \in \Sigma_{o_i}$ *for* $\kappa = 1, 2, \ldots, L_i$, *and let* $\omega_i' = \alpha_1^{(i)} \alpha_2^{(i)} \ldots \alpha_{L_i-1}^{(i)}$, *for* $i = 1, 2, \ldots, m$ *(if* $L_i = 1$, *we can take* $\omega_i' = \epsilon$, *whereas, if* $L_i = 0$, *we can take* $\omega_i' = \epsilon$ *and* $\alpha_{L_i}^{(i)} = \epsilon$*). We have*

$$
R_{po,\mathbb{M}}(Q', \{\omega_1, \omega_2, \ldots, \omega_m\}) = \cup_{i=1}^{m} \hat{q}_i \ , \tag{9.7}
$$

where $\hat{q}_i$ *depends on whether the final event* $\alpha_{L_i}^{(i)}$ *for observation site* O_i *is shared or not. More specifically, we have the following three cases:*
1. If $\alpha_{L_i}^{(i)} \notin \Sigma_{O_j}$ *for all* $j \in \mathbb{M} \setminus \{i\}$, *then*

$$
\hat{q}_i = R_{\mathbb{M}}(R_{po,\mathbb{M}}(Q', \{\omega_1, \ldots, \omega_{i-1}, \omega_i', \omega_{i+1}, \ldots, \omega_m\}), \alpha_{L_i}^{(i)}) \ .
$$

2. If $\alpha_{L_i}^{(i)} \in \Sigma_{O_j}$ *for* $j \in M$, $M \subseteq \mathbb{M}$, $|M| > 1$, and $\alpha_{L_i}^{(i)} = \alpha_{L_j}^{(j)}$ *for all* $j \in M$ *(i.e.,* $\alpha_{L_i}^{(i)}$ *is shared among all observations sites in M, and the last observation of all of these observation sites matches exactly the observation at observation site* O_i*), then*

$$
\hat{q}_i = R_{\mathbb{M}}(R_{po,\mathbb{M}}(Q', \{\omega_1'', \omega_2'', \ldots, \omega_m''\}), \alpha_{L_i}^{(i)}) \ ,
$$

where ω_j'' *satisfies*

$$
\omega_j'' = \begin{cases} \omega_j' \ , & \text{if } j \in M, \\ \omega_j \ , & \text{otherwise.} \end{cases}
$$

3. If $\alpha_{L_i}^{(i)} \in \Sigma_{O_j}$ for $j \in M$, $M \subseteq \mathbb{M}$, $|M| > 1$, and $\alpha_{L_i}^{(i)} \neq \alpha_{L_j}^{(j)}$ for some $j \in M$ (i.e., $\alpha_{L_i}^{(i)}$ is shared among all observations sites in M, but not all of them match exactly this observation), then

$$\hat{q}_i = \emptyset \ .$$

[Note that the initial condition satisfies $R_{po,\mathbb{M}}(Q', \{\epsilon, \epsilon, \ldots, \epsilon\}) = R_{\mathbb{M}}(Q', \epsilon)$.]

Remark 9.6 Note that some of the sets in the union that appears on the right of (9.7) are identical. For example, if $\alpha_{L_1}^{(1)} = \alpha_{L_2}^{(2)}$ (and this event is not shared by any other observation site), then the first two terms, given by

$$R_{\mathbb{M}}(R_{po,\mathbb{M}}(Q', \{\omega_1', \omega_2', \omega_3, \ldots, \omega_m\}), \alpha_{L_1}^{(1)})$$

and

$$R_{\mathbb{M}}(R_{po,\mathbb{M}}(Q', \{\omega_1', \omega_2', \omega_3, \ldots, \omega_m\}), \alpha_{L_2}^{(2)}) \ ,$$

are identical; however, this makes no difference in the expression in (9.7). □

Example 9.4 Suppose that a given LNFA G has as set of events

$$\Sigma = \{a, b, c_1, c_2, d_{12}, d_{23}, d_{123}\}$$

(the remaining parameters that define G are not important for this example). Assume there are three observation sites O_1, O_2, and O_3, such that

$$\begin{aligned}
\Sigma_{o_1} &= \{a, d_{12}, d_{123}\} \ , \\
\Sigma_{o_2} &= \{b, d_{12}, d_{23}, d_{123}\} \ , \\
\Sigma_{o_3} &= \{c_1, c_2, d_{23}, d_{123}\} \ .
\end{aligned}$$

Suppose that the sequence of events

$$s^{(1)} = ac_1d_{12}d_{123}bc_2d_{23}$$

occurs before the first synchronization, so that the observation sites record

$$\begin{aligned}
\omega_1^{(1)} &= P_1(s^{(1)}) = ad_{12}d_{123} \ , \\
\omega_2^{(1)} &= P_2(s^{(1)}) = d_{12}d_{123}bd_{23} \ , \\
\omega_3^{(1)} &= P_3(s^{(1)}) = c_1d_{123}c_2d_{23} \ .
\end{aligned}$$

When trying to recursively generate the partial orders, we need to consider all events that appear at the end of the above observation sequences. We realize that d_{123} is an event that is shared by all sites, but does not appear as the last event for all observation sequences. Thus, it cannot be considered as the last event. When we consider d_{23} as the last event, we see that it is a common event for O_2 and O_3 and it appears at the end of both $\omega_2^{(1)}$ and $\omega_3^{(1)}$. Thus, we can take d_{23} as the last event and we now have to consider the partial orders of the sequences of observations

$$\begin{gathered} ad_{12}d_{123}\ , \\ d_{12}d_{123}b\ , \\ c_1d_{123}c_2\ . \end{gathered}$$

Again, d_{123} does not appear as the last event for all sequences and cannot be considered. We have to separately consider b and c_2, both of which are events that are exclusive to observation sites O_2 and O_3, respectively.
(i) If we consider b we have to proceed with the partial orders for

$$\begin{gathered} ad_{12}d_{123}\ , \\ d_{12}d_{123}\ , \\ c_1d_{123}c_2\ , \end{gathered}$$

and again we cannot choose d_{123} as the last event. In fact, at the next unwrapping of the recursion, we have to consider c_2.
(ii) If we consider c_2, we have to proceed with the partial orders for

$$\begin{gathered} ad_{12}d_{123}\ , \\ d_{12}d_{123}b\ , \\ c_1d_{123}\ . \end{gathered}$$

Again, we cannot choose d_{123} as the last event and, in this case, in the next unwrapping of the recursion we have to consider b.

In either of the two cases above, we arrive at the partial orders of the following sets of observation sequences:

$$\begin{gathered} ad_{12}d_{123}\ , \\ d_{12}d_{123}\ , \\ c_1d_{123}\ , \end{gathered} \tag{9.8}$$

and d_{123} is the last event to be considered.

Thus, our analysis so far has concluded that the suffix of the underlying sequence has to include d_{123}, followed by bc_2 or c_2b, followed by d_{23}, whereas the prefix needs to contain the partial orders for (9.8) (without event d_{123} in the end). The reason we have identical prefixes for both of the suffixes we have obtained is the fact that d_{123} is shared by all sites and effectively acts as a synchronization event (in general, the prefixes that will be possible for different suffixes of the sequences could be different).

The partial orders for (9.8) (without event d_{123} in the end) can be calculated to be

$$\begin{gathered} ac_1d_{12}\ , \\ c_1ad_{12}\ , \\ ad_{12}c_1\ . \end{gathered}$$

Putting all of the above together, we arrive at the following six possible sequences that match the partially ordered sequences of observations at the three sites:

$$\begin{aligned}&ac_1d_{12}d_{123}bc_2d_{23}\\&ac_1d_{12}d_{123}c_2bd_{23}\\&c_1ad_{12}d_{123}bc_2d_{23}\\&c_1ad_{12}d_{123}c_2bd_{23}\\&ad_{12}c_1d_{123}bc_2d_{23}\\&ad_{12}c_1d_{123}c_2bd_{23}\ .\end{aligned}$$

Notice that the first sequence in the above list is the actual sequence that occurred in the system. □

9.6 Case II: Set Intersection-Based Estimation

In Case II (and Case III in the next section), the bulk of the processing is shifted from the coordinator to the observation sites. In fact, the coordinator no longer needs to be aware of the system model and the observation setting (i.e., it does not need to know $\Sigma_{o_i}, i \in \mathbb{M}$, or even the system model G) because it only needs to perform simple set intersection operations. On the contrary, unlike Case I that required no processing of observations at the observation sites (other than first storing and later transmitting the sequence of observations), in Case II each observation site needs to be aware of the system model G and be able to perform basic state estimation operations (as in Chap. 4). We next describe the protocol in more detail.

In Case II, each observation site O_i communicates to the coordinator the set of possible current-state estimates it has deduced (at the time the synchronization takes place). Since each observation site operates in isolation (with no explicit or implicit information from other sites or the coordinator), the set of current-state estimates at observation site O_i following a sequence of events $s \in L(G)$ that occurs in the system is given by

$$\hat{Q}_i(\omega_i) = R_i(Q_0, \omega_i)\ ,$$

where $\omega_i = P_i(s)$ is the sequence of observations seen at O_i and R_i is taken with respect to the set of events Σ_{o_i} that are observable at observation site O_i, as defined in (9.2). [Note that in decentralized state estimation, since no information flows back from the coordinator to the observation sites, this estimate is independent of the instants at which previous synchronizations took place; however, this will not necessarily be the case in the distributed protocols in Chap. 10.]

When a synchronization is initiated, each observation site sends to the coordinator its set of state estimates thus far (refer to Fig. 9.2). The fusion at the coordinator is a simple set intersection operation, which results in a finer set of current-state estimates at the coordinator (essentially, the coordinator obtains as possible current-state estimates, states that have not been eliminated by any observation site).

To describe the operation of this decentralized state estimation protocol (Case II), consider that the sequence of events s occurs in the given system G. As mentioned in Sect. 9.3, we can always break the sequence of events s that occurs in the system into subsequences of events $s^{(1)}, s^{(2)}, \ldots, s^{(k)}$, so that $s = s^{(1)}s^{(2)} \ldots s^{(k)}$, and we have the following sequence of events and synchronizations:

$$s^{(1)} sync^{(1)} s^{(2)} sync^{(2)} \ldots sync^{(k-1)} s^{(k)} sync^{(k)} ,$$

and the following sequence of observations at observation site O_i:

$$\omega_i^{(1)} sync^{(1)} \omega_i^{(2)} sync^{(2)} \ldots sync^{(k-1)} \omega_i^{(k)} sync^{(k)} ,$$

where $\omega_i^{(\kappa)} = P_i(s^{(\kappa)})$ for $\kappa = 1, 2, \ldots, k$. At observation site O_i, the state estimate at the κth synchronization is given by

$$\hat{Q}_i^{(\kappa)} = R_i(Q_0, \omega_i^{(1)} \omega_i^{(2)} \ldots \omega_i^{(\kappa)}) ,$$

which, as mentioned earlier, is independent of when synchronizations took place. Thus, at the κth synchronization, the state estimate at the coordinator is given by

$$\hat{Q}_{si}^{(\kappa)} = \cap_{i=1}^{m} \hat{Q}_i^{(\kappa)} .$$

Remark 9.7 Notice that the coordinator does not try to incorporate into $\hat{Q}_{si}^{(\kappa)}$ any information it previously obtained (e.g., $\hat{Q}_{si}^{(\kappa-1)}$). A more sophisticated coordinator could attempt to do that instead of simply relying on what the observation sites are reporting; this could potentially result in a finer set of state estimates, but it would require that the coordinator has knowledge of the system model G and significant processing power. It should also be clear from the above discussion that the state estimate after the kth synchronization does not depend on previous synchronizations. More specifically, if the kth synchronization occurs after the occurrence of the sequence of events s, we have

$$\hat{Q}_{si}^{(k)} = \cap_{i=1}^{m} \hat{Q}_i^{(k)} ,$$

where $\hat{Q}_i^{(k)} = R_i(Q_0, P_i(s))$, which is completely independent of previous synchronizations. The reason is that earlier synchronizations do not affect later current-state estimates at the observation sites (nevertheless, earlier synchronizations could be useful in determining certain conditions of interest, such as detecting/diagnosing a fault event). Note that in the *distributed* protocols we discuss in Chap. 10, previous synchronizations will affect later current-state estimates at the observation sites because the coordinator (or other sites) will be allowed to send information to a certain site; this additional flexibility of distributed strategies generally improves the quality of state estimation and event inference. □

9.7 Case III: Processing of Local Decisions

This case is a simplified version of Case II, where each observation site O_i still operates independently from the other observation sites (as in Case II) but, instead of transmitting its state estimates, it communicates to the coordinator a (local) decision. This local decision depends on what the underlying task is; for example, for fault diagnosis (which has been studied quite extensively in this setting), the decision at each observation site is "fault" ("F"), or "no fault" ("N"), or "uncertain" ("U"), depending on whether the local observation site can, respectively, determine that a fault has definitely happened, or determine that a fault has definitely not happened, or cannot reach a definite conclusion on whether a fault has happened or not. This setting and variations of it have appeared under the name of co-diagnosis (see, for example, (Qiu and Kumar 2006; Wang et al. 2007, 2011; Schmidt 2010). Analogously, if the underlying task is to determine the state of the system exactly (detectability), then the decision at each observation site could be "detectable" ("D") or "not detectable" ("ND"), depending on whether the local observation site is able to determine the state of the system exactly or not.

Below, we describe in more detail the setting of co-diagnosis (refer to Fig. 9.4). The operation at the observation sites is identical to Case II, i.e., each observation site maintains a local set of possible states along with a fault label (e.g., F, N, F_1F_2, and so forth, as described in Chap. 7). In the case of a single fault class where the goal is to detect the occurrence of a fault event, we can think of the enhanced system G_F described in Sect. 7.2.2 of Chap. 7 (so as to reduce fault diagnosis to a state isolation problem), in which case the state estimate at the κth synchronization, given by

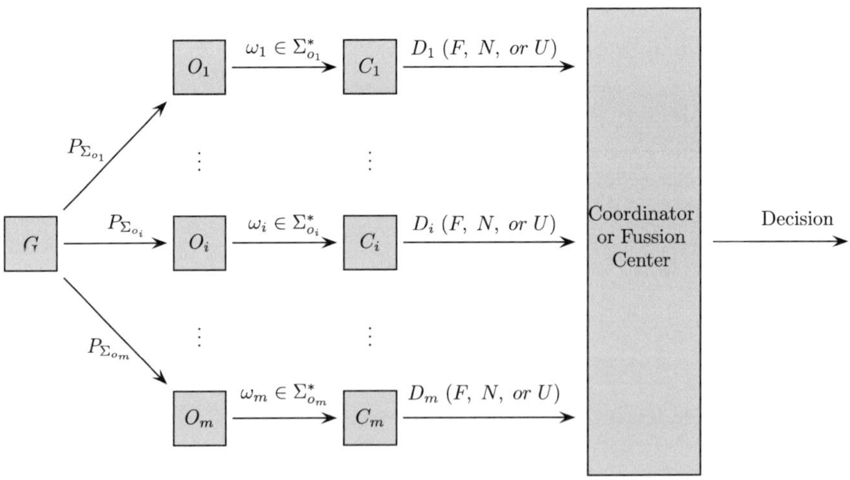

Fig. 9.4 Coordinator or fusion center receiving local decisions $D_1, D_2, \ldots, D_m$, where each decision D_i (F, N, or U) is based on the locally observed sequence of observations ω_i

$$\hat{Q}_i^{(\kappa)} = R_i(Q_0, \omega_i^{(1)}\omega_i^{(2)} \ldots \omega_i^{(\kappa)}) \subseteq Q \times \{F, N\} ,$$

where R_i is taken with respect to system G_F and the observable events Σ_{o_i}. More specifically, $(q, F) \in \hat{Q}_i^{(\kappa)}$ (respectively, $(q, N) \in \hat{Q}_i^{(\kappa)}$) would indicate that, based on the sequence of observations observed at site O_i thus far, state q is reachable via a sequence that includes at least one fault event (respectively, that does not include any fault events). Note that it is possible for both (q, F) and (q, N) to be in $\hat{Q}_i^{(\kappa)}$, indicating that state q can be reached via at least one matching sequence of events that includes at least one fault event, and via at least one matching sequence of events that does not include any fault.

When the κth synchronization is initiated, observation site O_i sends to the coordinator its decision thus far. This decision is rather straightforward:

- If $\hat{Q}_i^{(\kappa)} \subseteq Q \times \{F\}$ then it sends "F" (certainty about the occurrence of a fault).
- If $\hat{Q}_i^{(\kappa)} \subseteq Q \times \{N\}$ then it sends "N" (certainty about the absence of a fault).
- Otherwise, it sends "U" (uncertainty about the occurrence/nonoccurrence of a fault or faults).

The fusion at the coordinator is a simple operation and, as in Case II, it does not require the coordinator to have knowledge of the system model: for example, if one (or more) observation site(s) is (are) certain about the occurrence of a fault (or about the nonoccurrence of any fault), and all other observation sites are uncertain ("U"), then the coordinator can determine that a fault has taken place (or that no fault has occurred). Note that it is not possible for some observation sites to be reporting a fault ("F") and some other observation sites to be reporting normal operation ("N").

9.8 Examples

Example 9.5 Consider LNFA G on the left of Fig. 9.5 with states $Q = \{0, 1, 2, 3\}$, events $\Sigma = \{\alpha, \beta, \gamma\}$, next-state transition function δ as shown in the figure, and $Q_0 = Q$. We assume that there are two observation sites ($\mathbb{M} = \{1, 2\}$), namely O_1 and O_2 with $\Sigma_{o_1} = \{\alpha\}$ and $\Sigma_{o_2} = \{\beta\}$, respectively. Note that event γ is unobservable at both sites. Suppose that the sequence of events $s = \alpha\alpha\beta\beta$ occurs and then a synchronization is initiated.

Following s, the fictitious centralized observer would record $\omega_{\mathbb{M}} = P_{\mathbb{M}}(s) = s$ and would deduce that $\hat{Q}_{\mathbb{M}}(Q_0, \omega_{\mathbb{M}}) = R_{\mathbb{M}}(Q_0, \omega_{\mathbb{M}}) = \{3\}$. This can be seen easily by following the sequence of transitions α, then α, then β, and finally β in the centralized observer shown on the right of Fig. 9.5. [Note that the centralized observer has $\Sigma_{\mathbb{M}} = \Sigma_o = \{\alpha, \beta\}$.]

Let us now consider Case I decentralized state estimation. Following s, we would have the following sequences of observations at the two sites: $\omega_1 = P_1(s) = \alpha\alpha$ and $\omega_2 = P_2(s) = \beta\beta$. There are many possible total orders that match the partial orders reported by O_1 and O_2 (actually, using the discussion above Lemma 9.2, we know

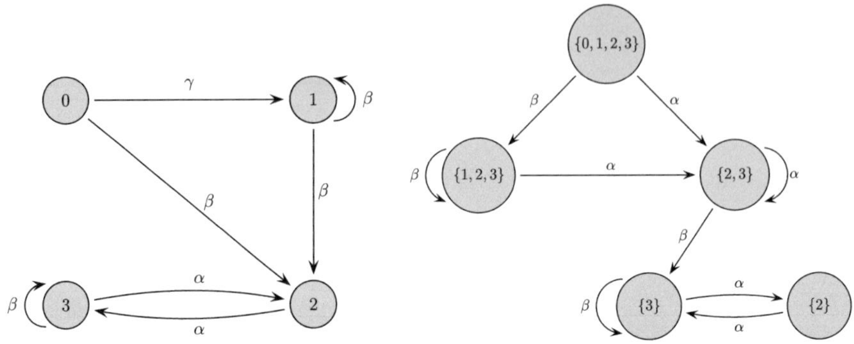

Fig. 9.5 Labeled nondeterministic finite automaton G (left) and its observer (right)

that there are $\frac{4!}{2!2!} = 6$ different totally ordered sequences). We list each one of them below, along with the corresponding sets of current-state estimates that are possible at the end of the sequence (the set of current-state estimates in each case can be obtained simply by traversing in the (centralized) observer the path indicated by the corresponding sequence of observations):

$$\begin{array}{lll} \alpha\alpha\beta\beta \to \{3\}, & \alpha\beta\alpha\beta \to \emptyset, & \alpha\beta\beta\alpha \to \{2\}, \\ \beta\beta\alpha\alpha \to \{2, 3\}, & \beta\alpha\beta\alpha \to \{2\}, & \beta\alpha\alpha\beta \to \{3\}. \end{array}$$

The union of the above sets captures the set of possible current states that we would arrive at using the partial-order constraints imposed by the various observation sites. In this particular case, we have

$$\hat{Q}_{po}^{(1)} = R_{po,\mathbb{M}}(Q_0, \{\omega_1, \omega_2\}) = \{2, 3\} .$$

Note that the above procedure can be performed more efficiently using the recursive approach we described earlier. Using this approach, we would arrive at $\hat{Q}(k_1, k_2)$ as given in the following table (k_1 is indexed column-wise and k_2 is indexed row-wise). As expected, the set of state estimates $R_{po,\mathbb{M}}(Q_0, \{\omega_1, \omega_2\}) = \hat{Q}(2, 2) = \{2, 3\}$ matches the set of possible state estimates reported earlier.

2	$\{1, 2, 3\}$	$\{2, 3\}$	$\{2, 3\}$
1	$\{1, 2, 3\}$	$\{2, 3\}$	$\{2, 3\}$
0	$\{0, 1, 2, 3\}$	$\{2, 3\}$	$\{2, 3\}$
k_2/k_1	0	1	2

To consider Case II decentralized state estimation, notice that, if observation sites O_1 and O_2 separately estimate the possible states of the system, they arrive at $\hat{Q}_1(Q_0, \omega_1) = R_1(Q_0, \omega_1) = \{2, 3\}$ and $\hat{Q}_2(Q_0, \omega_2) = R_2(Q_0, \omega_2) = \{1, 2, 3\}$. [Note that the above computations cannot use the centralized observer on the right of

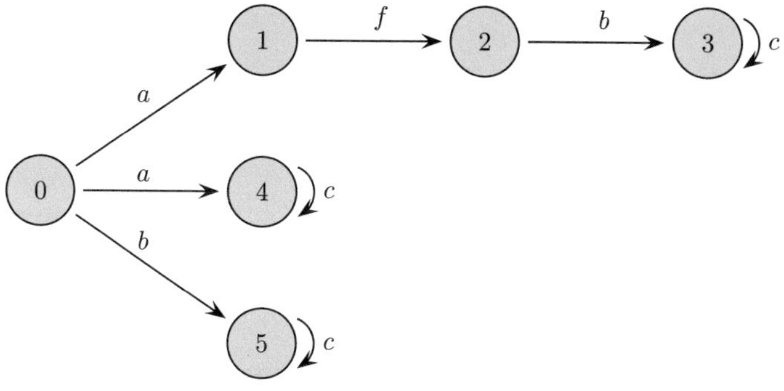

Fig. 9.6 Labeled nondeterministic finite automaton G in Example 9.6

Fig. 9.5 because the sets of observable events are different: for $\hat{Q}_1$, we need to use as unobservable events $\Sigma_{uo_1} = \{\beta, \gamma\}$, whereas for $\hat{Q}_2$, we need to use as unobservable events $\Sigma_{uo_2} = \{\alpha, \gamma\}$.]

When a synchronization is performed immediately after the occurrence of event sequence s, then we obtain the following set of state estimates at the coordinator:

$$\begin{aligned}\hat{Q}^{(1)}_{si} &= \hat{Q}_1(Q_0, \omega_1) \cap \hat{Q}_2(Q_0, \omega_2)\\ &= \{2, 3\} \cap \{1, 2, 3\}\\ &= \{2, 3\}\ ,\end{aligned}$$

which in this case happens to be identical with the state estimate obtained using Case I decentralized state estimation. [Recall that the set of state estimates using Case I is always a subset of the set of state estimates using Case II. An example where the set of state estimates using Case I is a strict subset of the set of state estimates using Case II is given later in this section.] □

Example 9.6 Consider LNFA $G = (Q, \Sigma, \delta, Q_0)$ in Fig. 9.6, with states $Q = \{0, 1, 2, 3, 4, 5\}$, events $\Sigma = \{a, b, c, f\}$, next-state transition function δ as shown in the figure, and $Q_0 = \{0\}$. We assume that there are two observation sites ($\mathbb{M} = \{1, 2\}$), namely O_1 and O_2 with $\Sigma_{o_1} = \{a, c\}$ and $\Sigma_{o_2} = \{b, c\}$, respectively. Event f is a fault event and is unobservable to both observation sites.

If we are interested in diagnosing the fault f, the key sequence of events is $s = afbc^n$, which projects to $\omega_1 = P_1(s) = ac^n$ and $\omega_2 = P_2(s) = bc^n$. In Fig. 9.7, we show the local diagnosers at each observation site (if one ignores the label next to each state, one can think of the constructions in this figure as the local state estimators). Clearly, no observation site can diagnose the occurrence of the fault event f based on its own observations: O_1 will eventually have $\{3F, 4N\}$ as the set of state estimates, whereas O_2 will eventually have $\{3F, 5N\}$ as the set of state estimates.

To analyze the fault diagnosis capability of the various protocols described in this chapter, we need to clarify when a synchronization (synchronizations) takes

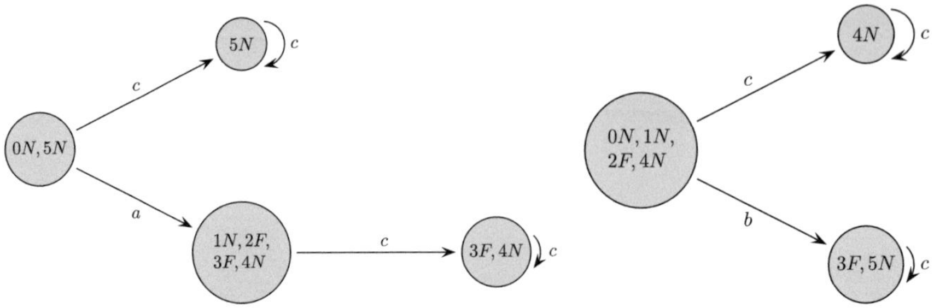

Fig. 9.7 Diagnoser at observation site O_1 (left) and diagnoser at observation site O_2 (right) for the system in Fig. 9.6

(take) place during the sequence of events $s = afbc^n$. Let us consider the case when synchronization is initiated after $afbc$ gets executed ($n = 1$). We describe what happens with the various cases below.

- Case I: In this case, $\omega_1 = ac$ and $\omega_2 = bc$; thus, the set of possible orders to consider are the following two:

$$abc \rightarrow \{3F\} \ ,$$
$$bac \rightarrow \emptyset \ .$$

  After the synchronization, the set of state estimates at the coordinator will be $\hat{Q}^{(1)}_{po} = R_{po,\mathbb{M}}(Q_0, \{\omega_1, \omega_2\}) = \{3F\}$. The discussion in this case does not change significantly if synchronization occurs at some larger n ($n > 1$): since c is a common event, we still have as possible sequences the sequence abc^n, which results in the same set of state estimates $\{3F\}$, and the sequence bac^n, which is not possible.
- Case II: In this case, at the time the first synchronization takes place, we have the following sets of state estimates at the observation sites:

$$\hat{Q}^{(1)}_1 = \{3F, 4N\} \ ,$$
$$\hat{Q}^{(1)}_2 = \{3F, 5N\} \ .$$

  Thus, when the coordinator performs the intersection operation, it obtains

$$\hat{Q}^{(1)}_{si} = \hat{Q}^{(1)}_1 \cap \hat{Q}^{(1)}_2 = \{3F\} \ ,$$

  which allows it to diagnose the fault. Again, the discussion does not change significantly if synchronization occurs for some larger n ($n > 1$).
- Case III: In this case, the decision at each observation site at the time the synchronization takes place is "U" because they are both uncertain; thus, the coordinator will also be uncertain as to the occurrence of the fault. The discussion does not change if synchronization occurs at larger n ($n > 1$).

Based on the above discussion, we would say that system G in Fig. 9.6 is diagnosable using the partial-order-based decentralized protocol (Case I) and the set intersection-based decentralized protocol (Case II), at least when synchronization occurs after the occurrence of the common event c. However, the system is not co-diagnosable (Case III), regardless of when synchronization takes place. □

Notice that the examples in this section are also good illustrations of Lemma 9.1. In Example 9.5, we saw that $R_1(Q_0, \omega_1) \cap R_2(Q_0, \omega_2) = \{2, 3\}$, which happens to be equal to $R_{po,\mathbb{M}}(Q_0, \{\omega_1, \omega_2\}) = \{2, 3\}$; similarly, in Example 9.6, we saw that $\hat{Q}_{si}^{(1)} = \{3F\}$, which happens to be equal to $R_{po,\mathbb{M}}(Q_0, \{\omega_1, \omega_2\}) = \{3F\}$. More generally, however, after the kth synchronization, $\hat{Q}_{po}^{(k)}$ could be a strict subset of $\hat{Q}_{si}^{(k)}$; the following example illustrates this possibility.

Example 9.7 Consider LNFA $G = (Q, \Sigma, \delta, Q_0)$ in Fig. 9.8, with states $Q = \{1, 2, \ldots, 11, 12\}$, events $\Sigma = \{a_1, a_2, b_1, b_2, c\}$, next-state transition function δ as shown in the figure, and $Q_0 = \{1\}$. We assume that there are two observation sites ($\mathbb{M} = \{1, 2\}$), namely O_1 and O_2 with $\Sigma_{o_1} = \{a_1, a_2, c\}$ and $\Sigma_{o_2} = \{b_1, b_2, c\}$, respectively.

Assume that the sequence of events $s = a_1a_2b_1b_2c$ occurs and that a synchronization occurs at the end of this sequence. At synchronization, we have $\omega_1 = P_1(s) = a_1a_2c$ and $\omega_2 = P_2(s) = b_1b_2c$, and it is not hard to see that the partial-order-based estimator (Case I) would obtain

$$\hat{Q}_{po}^{(1)} = R_{po}(Q_0, \{\omega_1, \omega_2\}) = \{5\} ,$$

which also happens to be the state estimate of the centralized observer. However, the set intersection-based estimator would be

$$\hat{Q}_{si}^{(1)} = \hat{Q}_1 \cap \hat{Q}_2 = \{5, 9\} ,$$

where the local estimates are given by

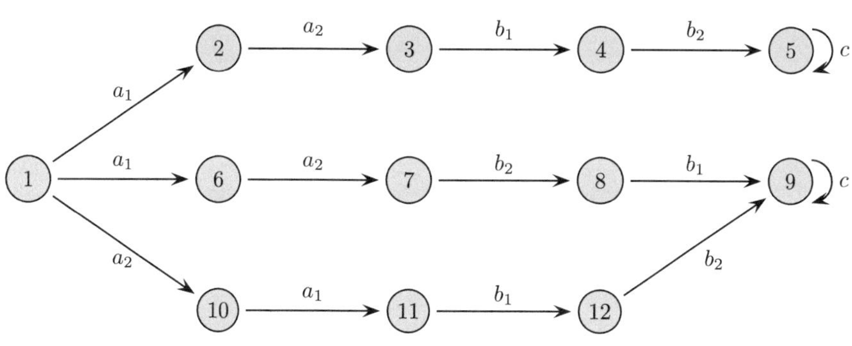

Fig. 9.8 Labeled nondeterministic finite automaton G in Example 9.7

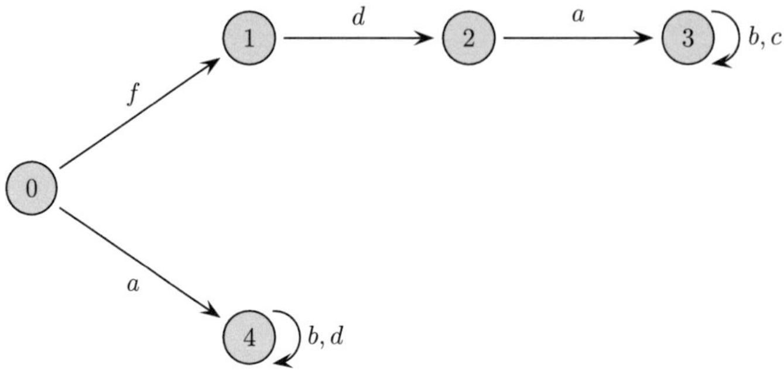

Fig. 9.9 Labeled nondeterministic finite automaton G in Example 9.8

$$\hat{Q}_1 = R_1(Q_0, \omega_1) = \{5, 9\} ,$$
$$\hat{Q}_2 = R_2(Q_0, \omega_2) = \{5, 9\} .$$

Clearly, this is a case where we have $\hat{Q}_{po}^{(1)} \subset \hat{Q}_{si}^{(1)}$. □

Example 9.8 Consider LNFA $G = (Q, \Sigma, \delta, Q_0)$ in Fig. 9.9, with states $Q = \{0, 1, 2, 3, 4\}$, events $\Sigma = \{a, b, c, d, f\}$, next-state transition function δ as shown in the figure, and $Q_0 = \{0\}$. We assume that there are three observation sites ($\mathbb{M} = \{1, 2, 3\}$), namely O_1, O_2, and O_3 with $\Sigma_{o_1} = \{a, b\}$, $\Sigma_{o_2} = \{b, c\}$, and $\Sigma_{o_3} = \{b, d\}$, respectively. Event f is a fault event and is unobservable to all observation sites. We are interested in determining whether fault f is diagnosable using either Case II or Case III decentralized fault diagnosis.

First, let us consider the case where each observation site operates in isolation (Case III decentralized diagnosis). In such case, it is not hard to realize that no observation site will be able to diagnose a fault. We can see this by considering the sequence $s = fdab^n$ for $n > 0$ (one of the sequences that can be generated by the system and contains a fault) and track the local information at each observation site. In Fig. 9.10, we see the local diagnosers at each observation site; in particular, following sequence s, we observe the following:

- Observation site O_1 observes $P_1(s) = ab^n$ and its local diagnoser (on the top of the figure) ends up in state $\{3F, 4N\}$ (which is indeterminate and does not allow it to determine the fault).
- Observation site O_2 observes $P_2(s) = b^n$ and its local diagnoser (in the middle of the figure) ends up in state $\{3F, 4N\}$ (which is indeterminate and does not allow it to determine the fault).
- Observation site O_3 observes $P_3(s) = db^n$ and its local diagnoser (at the bottom of the figure) ends up in state $\{3F, 4N\}$ (which is indeterminate and does not allow it to determine the fault).

The above discussion implies that the fault f cannot be diagnosed using Case III decentralized state estimation (since there exists a string, namely s, which contains

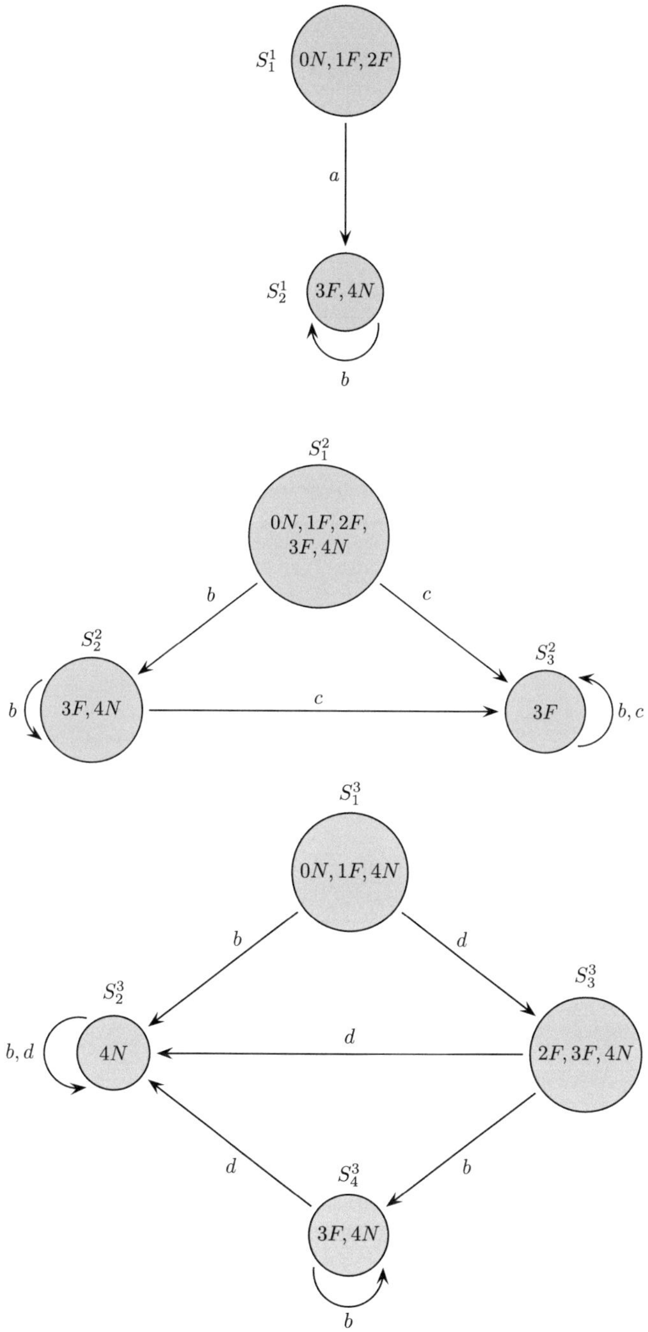

Fig. 9.10 Local diagnosers for observation site O_1 (top), O_2 (center), and O_3 (bottom), in Example 9.8

the fault and can be extended indefinitely without allowing the coordinator to detect the fault).

Let us now consider Case II decentralized state estimation, where each observation site O_i, $i \in \mathbb{M}$, initiates a synchronization (by informing the coordinator) each time it observes an event. Below we track the execution of the protocol for Case II decentralized state estimation for the same string $s = fdab^n$ for $n > 0$ as before.

- When f occurs all local sites do not observe anything and their state estimate is captured by $\hat{Q}_i^{(0)} = UR_i(Q_0)$, i.e.,

$$\hat{Q}_1^{(0)} = \{0N, 1F, 2F\}$$
$$\hat{Q}_2^{(0)} = \{0N, 1F, 2F, 3F, 4N\}$$
$$\hat{Q}_3^{(0)} = \{0N, 1F, 4N\} .$$

- When d occurs, this event is observed by observation site O_3, which updates its state estimate and initiates a synchronization; the state estimates that are sent to the coordinator are:

$$\hat{Q}_1^{(1)} = \{0N, 1F, 2F\}$$
$$\hat{Q}_2^{(1)} = \{0N, 1F, 2F, 3F, 4N\}$$
$$\hat{Q}_3^{(1)} = \{2F, 3F, 4N\} ,$$

  at which point the coordinator takes intersection and obtains $\{2F\}$, i.e., the fault is detected.

We also revisit this example later on when we have the opportunity to define more complex synchronization strategies. □

9.9 Synchronization Strategies

As seen in the last example of the previous section, an important aspect of a decentralized state estimation protocol is to specify *when* to initiate the sending of information from the observation sites to the coordinator. We refer to such an event as a *synchronization* because it synchronizes the information maintained at the coordinator with the information at the observation sites. We have assumed that each synchronization involves all observation sites sending information to the coordinator; more generally, however, one could imagine strategies in which synchronizations require different subsets of the observation sites to send information to the coordinator; in fact, in Chap. 10, we will see such examples in the context of distributed state estimation.

It is rather straightforward to think that a synchronization is initiated by the coordinator (i.e., the coordinator requests from all observation sites information). The difficult question is *how* (i.e., based on what kind of information) the coordinator decides when to initiate a synchronization. In fact, since the coordinator does not directly observe any events, it presumably has no knowledge of what activity (if any) has occurred in the system.

The above discussion implies that synchronization points should perhaps be first initiated by one (or more) observation site(s), based on what it has been locally observed. The observation site could signal the coordinator who could subsequently initiate the synchronization by requesting information from *all* observation sites. The simplest approach would be to have each observation site signal the coordinator for a synchronization based on the sequence of events it has observed so far (or it has observed since the last synchronization). For example, we could have each observation site count the number of events it has observed since the last synchronization, and initiate a synchronization when the count on the number of events reaches a certain constant. Next, we describe a very general class of synchronization strategies that relies on local synchronizing finite automata S_i at each observation site O_i. We refer to the resulting strategies for information exchange between the observation sites and the coordinator as *synchronization-based* strategies.

9.9.1 Synchronizing Automata

A very general class of synchronization strategies is defined below.

Definition 9.4 (*Synchronization Strategy*) Suppose that LNFA $G = (Q, \Sigma, \delta, Q_0)$ is observed at m observation sites $O_1, O_2, \ldots$, and O_m, with observable events $\Sigma_{o_1}, \Sigma_{o_2}, \ldots$, and Σ_{o_m}, respectively. Each observation site O_i is associated with a *synchronizing automaton*, i.e., a marked deterministic finite automaton (DFA) $S_i = (Q_{s,i}, \Sigma_{o_i} \cup \{sync\}, \delta_{s,i}, q_{0s,i}, Q_{ms,i})$ that starts from a specific initial state $q_{0s,i} \in Q_{s,i}$ and is driven by events in $\Sigma_{o_i} \cup \{sync\}$, where $sync$ captures the synchronization event. Based on the observations (and $sync$ operations) seen at site O_i, the state of the synchronizing automaton S_i is updated: whenever S_i reaches a state in the set of marked states $Q_{ms,i}$ ($Q_{ms,i} \subset Q_{s,i}$), then O_i signals the coordinator to initiate a synchronization.

Remark 9.8 The only possible event from each state in $Q_{ms,i}$ is the $sync$ event, and we assume that a $sync$ event results in an unmarked state (in the set $Q_{s,i} \setminus Q_{ms,i}$) to ensure that we avoid situations in which observation site O_i continuously signals to the coordinator for a synchronization (without any activity in the given system G, at least without any activity observable to observation site O_i). Another reasonable assumption about the synchronizing automaton S_i is to require that there are no cycles of events that do not involve a $sync$ event (or, equivalently, a marked state); this requirement guarantees that observation site O_i cannot observe a potentially unbounded number of events before a synchronization is initiated. Finally, we assume that $q_{0s,i} \notin Q_{ms,i}$. □

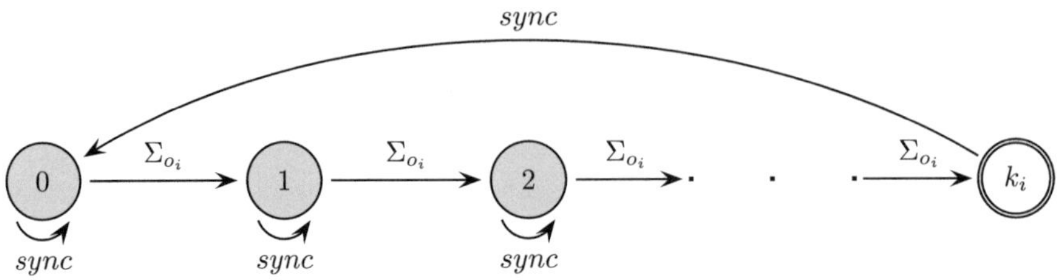

Fig. 9.11 Synchronizing automaton for a periodic synchronization strategy by observation site O_i. The notation Σ_{o_i} on transitions captures the fact that the transition is taken under all events in Σ_{o_i}

Example 9.9 A special case of the above class of synchronization strategies are the periodic strategies that were considered in Keroglou and Hadjicostis (2014), Keroglou and Hadjicostis (2015). In these cases, observation site O_i chooses an integer k_i and asks the coordinator to initiate a synchronization each time it observes k_i events since the last time it signaled the coordinator to initiate a synchronization. This implies that the synchronizing automaton (shown in Fig. 9.11) is given by the DFA $S_i = (Q_{s,i}, \Sigma_{o_i} \cup \{sync\}, \delta_{s,i}, q_{0s,i}, Q_{ms,i})$ with $Q_{s,i} = \{0, 1, 2, \ldots, k_i\}$, initial state $q_{0s,i} = 0$, and state k_i as the only marked state ($Q_{ms,i} = \{k_i\}$). The mapping $\delta_{s,i}$ is defined for all $j \in Q_{s,i}$ and all $\sigma_{o_i} \in \Sigma_{o_i}$ as

$$\delta_{s,i}(j, \sigma_{o_i}) = j + 1, \text{ if } j \neq k_i \ ,$$

whereas $\delta_{s,i}(j, sync) = j$ for all $j \in Q_{s,i} \setminus \{k_i\}$ and $\delta_{s,i}(k_i, sync) = 0$.

Another related strategy would be the strategy where observation site O_i chooses an integer k_i and signals the coordinator to initiate a synchronization each time it observes k_i events since the last time the coordinator initiated a synchronization (not necessarily signaled by this particular observation site). This implies that DFA S_i would have the structure mentioned above with the only difference being that

$$\delta_{s,i}(j, sync) = 0 \ , \forall j \in Q_{s,i} \ ,$$

i.e., each synchronization causes the count on the number of events to be reset to zero. □

Example 9.10 In this example we consider again LNFA $G = (Q, \Sigma, \delta, Q_0)$ in Fig. 9.9, which was considered in Example 9.8. Recall that system G has states $Q = \{0, 1, 2, 3, 4\}$, events $\Sigma = \{a, b, c, d, f\}$, next-state transition function δ as shown in the figure, and $Q_0 = \{0\}$. We assume that there are three observation sites ($\mathbb{M} = \{1, 2, 3\}$), namely O_1, O_2, and O_3 with $\Sigma_{o_1} = \{a, b\}$, $\Sigma_{o_2} = \{b, c\}$, and $\Sigma_{o_3} = \{b, d\}$, respectively. Event f is a fault event and is unobservable to all observation sites.

We are interested in determining whether fault f is diagnosable using Case II decentralized fault diagnosis when a synchronization is initiated by observation site O_i when this site observes k_i events since the last synchronization it initiated (in the analysis in Example 9.8 we had $k_1 = k_2 = k_3 = 1$). It is easy to realize that if $k_3 \geq 2$, the fault f cannot be diagnosed (regardless of how other observation sites initiate synchronizations). This can be seen, for instance, by tracking the execution of the protocol when the sequence $s = fdab^n$ takes place (and $k_3 \geq 2$). In such case, we have the following:

- When f occurs all local sites do not observe anything and their state estimate is captured by $\hat{Q}_i^{(0)} = UR_i(Q_0)$ (as in Example 9.8).
- When d occurs, this event is observed by observation site O_3, which updates its state estimate but does *not* initiate a synchronization (because $k_2 \geq 2$); the state estimates are

$$\hat{Q}_1^{(1)} = \{0N, 1F, 2F\}$$
$$\hat{Q}_2^{(1)} = \{0N, 1F, 2F, 3F, 4N\}$$
$$\hat{Q}_3^{(1)} = \{2F, 3F, 4N\} .$$

- When a occurs, this is visible to O_1 who updates its state estimate and may initiate a synchronization (depending on k_1). In such case, we have

$$\hat{Q}_1^{(1)} = \{3F, 4N\}$$
$$\hat{Q}_2^{(1)} = \{0N, 1F, 2F, 3F, 4N\}$$
$$\hat{Q}_3^{(1)} = \{2F, 3F, 4N\} ,$$

  and, if there is a synchronization, the coordinator will take the intersection and obtain $\{3F, 4N\}$.
- The next observation, b, will be observed at all three observation sites. Thus, each one will update their state estimates to

$$\hat{Q}_1^{(1)} = \{3F, 4N\}$$
$$\hat{Q}_2^{(1)} = \{3F, 4N\}$$
$$\hat{Q}_3^{(1)} = \{3F, 4N\} .$$

  Even if a synchronization occurs (e.g., if $k_3 = 2$), the coordinator will obtain $\{3F, 4N\}$ which does not allow it to diagnose the fault.

Even if more observations (in this case, more b's) are observed, the situation does not change (same local state estimates as above, which implies confusion at the coordinator when a synchronization is initiated). □

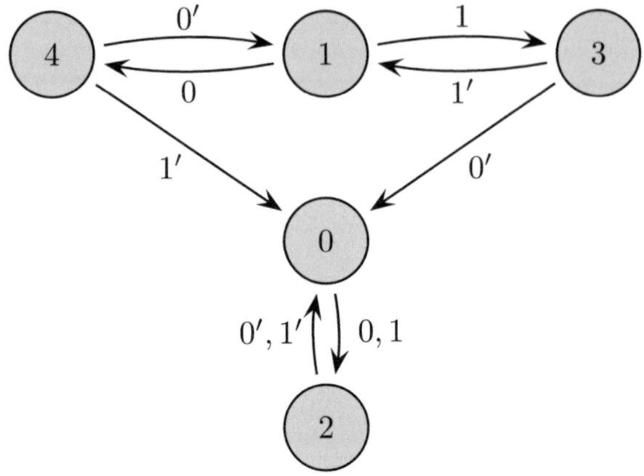

Fig. 9.12 Labeled nondeterministic finite automaton G to illustrate the absence of finite-size local observers

9.9.2 Limitations of Finite Memory Observers

Note that the synchronizing strategies in the previous section are relatively simple and require finite memory at each observation site. One can pursue more complex strategies, aiming for instance at minimizing the number of messages or amount of information that is exchanged. In general, devising such strategies could be a difficult task, as illustrated by the following example taken from Puri et al. (2002).

Example 9.11 Consider the labeled deterministic finite automaton (LDFA) $G = (Q, \Sigma, \delta, Q_0)$ in Fig. 9.12 where $Q = \{0, 1, 2, 3, 4\}$, $\Sigma = \{0, 1, 0', 1'\}$, δ is as defined in the figure, and $Q_0 = \{1\}$. Suppose that observation site O_1 observes $\Sigma_{o_1} = \{0, 1\}$, whereas observation site O_2 observes $\Sigma_{o_2} = \{0', 1'\}$.

The authors of Puri et al. (2002) considered the following decentralized observation problem: suppose that G runs for an unspecified number of events and then stops. Denote the sequence of events s, so that the corresponding sequence of observations at each observation site is $\omega_1 = P_1(s)$ and $\omega_2 = P_2(s)$. It is clear from our discussions in this chapter that in the absence of timing information (and earlier synchronizations) the best state estimate by the coordinator at the end of this process would be $R_{po,\mathbb{M}}(\{1\}, \{\omega_1, \omega_2\})$, which can be calculated recursively.

If observation sites are restricted to have finite memory, we may think of them as being a DFA $O_i = (Q_{o,i}, \Sigma_{o_i}, \delta_{o,i}, q_{0o,i})$, initialized at state $q_{0o,i}$ and driven by the observable events through a *finite* set of states $Q_{o,i}$. When the system stops, we would like to be able to have a joint decision at the coordinator that can be based on a function $\lambda : Q_{o,1} \times Q_{o,2} \rightarrow 2^Q$ that takes as inputs the states of the observation sites and produces as output the set of possible states. Ideally, we would like

$$\lambda(q_{o,1}, q_{o,2}) = R_{po,\mathbb{M}}(\{1\}, \{\omega_1, \omega_2\}) \ ,$$

where $q_{o,1} = \delta_{o,1}(q_{0o,1}, \omega_1)$ is the state reached by the observer DFA at observation site O_1, whereas $q_{o,2} = \delta_{o,2}(q_{0o,2}, \omega_2)$ is the state reached by the observer DFA at observation site O_2.

The authors of Puri et al. (2002) argue that if observation sites are restricted to have finite memory (and can only have one synchronization), then DFA's at the observation sites may not be sufficient. To see this notice that any s of even length will generate ω_1 and ω_2 with identical lengths (i.e., $|\omega_1| = |\omega_2|$) and will cause system G to be either in state 1 or in state 0. In fact, if we have access to ω_1 and ω_2, we can determine exactly the state: the system will be in state 1 if $\omega_1 = \omega_2$, in the sense that wherever string ω_1 has event 0 (respectively, 1), string ω_2 has event $0'$ (respectively, $1'$). Clearly, $R_{po,\mathbb{M}}(\{1\}, \{\omega_1, \omega_2\})$ can make this distinction and determine the exact state of the system; no finite state structures $Q_{o,1}$ or $Q_{o,2}$, however, would be able to capture this: the problem is they need to compare element-wise two strings of potentially unbounded length. □

Remark 9.9 Notice that if we allow multiple synchronizations, then the above problem goes away, at least when we use partial-order-based decentralized estimation in which the synchronization occurs when the two observation sites have observed the same number of events (i.e., a synchronization is initiated each time O_2 observes a certain number of events, such as one event or two events or, more generally, k_2 events). In such case, following a sequence of events s of even length, we would be able to determine whether the system is in state 1 or 0, depending on whether $P_1(s) = P_2(s)$ (where equality is taken in the sense that 0 is equivalent to $0'$, and 1 is equivalent to $1'$). As long as the observation sequences at the two observation sites are identical, the state estimate will be $\{1\}$. Once we know that the system has transitioned to state 0, then every time an even number of events is observed, the state estimate will be $\{0\}$.

Note that if we allow multiple synchronizations but use the set intersection-based decentralized estimation protocol (Case II), we will still have a problem. For example, if $s = 01'$, we will have $\omega_1 = P_1(s) = 0$ and $\omega_2 = P_2(s) = 1'$; the local estimates will be

$$\hat{Q}_1 = R_1(Q_0, \omega_1) = \{0, 1, 4\} ,$$
$$\hat{Q}_2 = R_2(Q_0, \omega_2) = \{0, 1, 2, 3, 4\} ,$$

and

$$\hat{Q}_{si} = \hat{Q}_1 \cap \hat{Q}_2 = \{0, 1, 4\} .$$

In fact, it turns out that $\hat{Q}_2 = R_2(Q_0, P_2(s)) = Q$ for any s of even length, which implies that if synchronization happens at the end of s, we have $\hat{Q}_{si} = \hat{Q}_1 = R_1(Q_0, P_1(s))$ (after a couple of observations, this will be either $\{0, 1, 2, 4\}$ or $\{0, 1, 2, 3\}$). □

Remark 9.10 Early work on co-diagnosis avoided the issue of synchronization by considering the diagnosis of faults and assuming that an observation site communicates information to the coordinator, as soon as it has a definite decision ("F") based on its local observations. Moreover, as soon as the coordinator receives an "F" from one site, then it can reach a decision: effectively, we can think of this as a synchronization step, where the particular observation site (or sites) sent "F", whereas all other (silent) observation sites implicitly sent "U" to the coordinator.

The above strategy applies to Case III of the decentralized protocol, at least for the case of fault diagnosis. It implies that synchronization occurs after a large enough number of events has occurred, so that all local diagnosers have entered a cycle of states that either allows them to determine a fault (F) or not; in the former case, the diagnoser sends decision F; in the latter case, the diagnoser does not send anything to the coordinator. Things do not really change much if one allows for additional synchronizations to occur later (recall that once a local diagnoser determines a fault has occurred, the decision remains unchanged regardless of future or earlier[4] observations). □

9.10 Verification of Properties of Interest

Once we have decided on the type of a decentralized protocol (i.e., Case I, or Case II, or Case III for information exchange) and on a synchronization strategy (i.e., a DFA S_i has been chosen at each observation site O_i, as described in Definition 9.4), then we can ask whether the resulting state estimation or event inference strategy will satisfy a property of interest (e.g., detectability, diagnosability with respect to certain classes of faults, or opacity with respect to a set of secret states). To verify the property, we would have to check whether the property holds for all system behavior under the given choices. In this section, we describe how verification can be systematically accomplished for the types of information exchange and synchronization strategies we have described.

9.10.1 Verification of Diagnosability

In this section we discuss the verification of *decentralized diagnosability*, which is a topic that has attracted recent attention by the research community. We first describe the problem setting, focusing, for simplicity, on the case of detecting a fault from a single fault class. In order to make the problem challenging (and since we are employing a labeled automaton model), we assume that faults cannot be directly observed at any observation site.

Problem Setting (Decentralized Diagnosability): We are given an LNFA $G = (Q, \Sigma, \delta, Q_0)$, which is observed at m observation sites O_1, O_2, ..., and O_m, with

[4]Earlier synchronizations might result in all observation sites reporting "U" (in which case no decision can be made at the coordinator) but eventually, at least if the system is *co-diagnosable*, one diagnoser will report "F" (which will imply that a definite decision "F" can be taken at the coordinator).

observable events Σ_{o_1}, Σ_{o_2}, ..., and Σ_{o_m}, respectively. Each observation site O_i is associated with a synchronizing marked DFA $S_i = (Q_{s,i}, \Sigma_{o_i} \cup \{sync\}, \delta_{s,i}, q_{0s,i}, Q_{ms,i})$ as described in Definition 9.4. Whenever S_i reaches a state in the set of marked states $Q_{ms,i}$ ($Q_{ms,i} \subset Q_{s,i}$), then O_i signals the coordinator to initiate a synchronization and the $sync$ event occurs (involving all observation sites), i.e., the coordinator requests from each observation site their information, namely (i) sequences of events since the last synchronization (Case I), or (ii) pairs of a local state estimate and associated label(s) (Case II), or (iii) local decisions (Case III). We are interested in verifying *decentralized diagnosability*, i.e., verifying whether the coordinator will be able to determine (perhaps after some finite delay) the occurrence of events in the class of fault events F, which are assumed to be unobservable by any observation site, i.e., $F \subseteq \Sigma_{uo}$ where $\Sigma_{uo} = \Sigma \setminus (\cup_{i=1}^{m} \Sigma_{o_i})$. We say that Case I (or Case II or Case III) decentralized diagnosability holds for the given system G (under the given set of observation sites and synchronization strategy) if the occurrence of any fault event $f \in F$ eventually gets determined at the coordinator after a finite number of event occurrences.

9.10.1.1 Verification of Case I Decentralized Diagnosability

We start with the verification of Case I decentralized diagnosability. The basic construction that is used to verify diagnosability is a parallel-like composition that tracks, for each possible sequence of events, the state in which the various components of the overall system are. More specifically, if we imagine that we have a way of storing, at each observation site, the sequence of observations seen at that site since the last synchronization, then we can use a parallel composition to simultaneously track for each possible sequence of events in the system, the following: (i) the state of the given system G, (ii) the sequence of observations recorded since the last synchronization at each observation site O_i (this is captured via the state of a sequence storage DFA called SD_i), (iii) the state of each local synchronizing DFA S_i and whether a $sync$ operation needs to take place, and (iv) the state of the coordinator, which gets updated after each synchronization event based on the partially ordered sequences observed since the last synchronization and reported by the observation sites. Note that the result of this parallel composition is a nondeterministic system as it inherits non-determinism from G.

The above composed system is of finite size and has states that are *indeterminate* at the coordinator (i.e., problematic from the point of view of fault diagnosis) if a fault has occurred and the state of the coordinator does not clarify that. Thus, as in the case of centralized diagnosability, we can verify Case I decentralized diagnosability by checking for the presence of cycles of such indeterminate states (and that these cycles can indeed persist after the occurrence of a fault event).

In more detail, to verify Case I decentralized diagnosability, we perform the following steps:

1. We construct an enhanced version of the system, denoted by G_e. This enhanced system is an LNFA that resembles G_F, as described in Sect. 7.2.2 of Chap. 7 (so as to reduce fault diagnosis to a state isolation problem). Moreover, G_e has one additional event, namely the *sync* event, which acts as a self-transition at each state. In other words, $G_e = (Q\dot{\cup}Q_F, \Sigma \cup \{sync\}, \delta_e, Q_0)$, where $\delta_e(q, sync) = \{q\}$ for all $q \in Q\dot{\cup}Q_F$ and $\delta_e(q, \sigma) = \delta_F(q, \sigma)$ for all $q \in Q\dot{\cup}Q_F$ and $\sigma \in \Sigma$ (note that $\delta_e(q, \sigma)$ is empty/undefined if $\delta_F(q, \sigma)$ is empty/undefined).
2. We construct the local sequence storage at observation site O_i, which is captured by a DFA $SD_i = (Q_{sd,i}, \Sigma_{o_i} \cup \{sync\}, \delta_{sd,i}, q_{0sd,i})$. The states $Q_{sd,i}$ of this automaton are sequences of observations of a maximum length k_i (we elaborate on the choice of k_i next), i.e., $Q_{sd,i} = (\Sigma_{o_i} \cup \{\epsilon\})^{k_i}$. The initial state is $q_{0sd,i} = \epsilon$ and next-state transition function of SD_i, for all $\omega_i \in Q_{sd,i}$ and all $\sigma_{o_i} \in \Sigma_{o_i}$ is given by
$$\delta_{sd,i}(\omega_i, \sigma_{o_i}) = \begin{cases} \omega_i\sigma_{o_i}, & |\omega_i| < k_i \ , \\ \epsilon, & |\omega_i| = k_i \ . \end{cases}$$
Furthermore, we have $\delta_{sd,i}(\omega_i, sync) = \epsilon$ for all $\omega_i \in Q_{sd,i}$.
Note that k_i is chosen so that a sync operation is guaranteed to occur *before* DFA SD_i reaches its maximal storage capacity (thus, the second case of $\delta_{sd,i}(\omega_i, \sigma_{o_i})$, when $|\omega_i| = k_i$ is not supposed to be invoked). This can be ensured, for example, if the synchronizing automaton S_i is chosen to signal a synchronization to the coordinator every k_i observations at local site O_i (see, for instance, the synchronizing automaton in Example 9.9).
3. We construct the product SDS_i of the local sequence storage DFA SD_i and the synchronizing DFA S_i at observation site O_i, i.e., $SDS_i = SD_i \times S_i$, for $i = 1, 2, \ldots, m$.
4. The next step is to construct the parallel composition of the enhanced version of the system and the SDS_i's, given by
$$\begin{aligned} GSDS &= G_e||SDS_1||SDS_2||\ldots||SDS_m \\ &=: (Q_g, \Sigma \cup \{sync\}, \delta_g, Q_{0,g}) \ . \end{aligned}$$
In this parallel composition, we mark all states that have a marked second component in the state of at least one SDS_i component (for some $i = 1, 2, \ldots, m$). These marked states denote states in which a synchronization has been initiated; thus, we remove from them all transitions other than the one associated with the *sync* transition.
5. In the resulting parallel composition $GSDS$, each state is of the form $(q, x_1, x_2, \ldots, x_m) \in (Q\dot{\cup}Q_F) \times (Q_{sd,1} \times Q_{s,1}) \times (Q_{sd,2} \times Q_{s,2}) \times \cdots \times (Q_{sd,m} \times Q_{s,m})$. If we let $x_i = (\omega_i, q_{s_i})$ be the state of SDS_i, then it is easy to track and update the state estimate at the coordinator. To do that, we can annotated the state $(q, x_1, x_2, \ldots, x_m)$ of $GSDS$ with an additional component $\hat{q}$, where $\hat{q} \in 2^{\{Q\dot{\cup}Q_F\}}$ and represents the state estimate at the coordinator. Initially, $\hat{q}_0 = R_{\mathbb{M}}(Q_0, \epsilon)$, so we can annotate the initial state(s) of $GSDS$ with this initial coordinator state. As we track transitions in $GSDS$, the state of the coordinator changes

only when we have a $sync$ operation from a marked state. Specifically, from marked state $(q, x_1, x_2, \ldots, x_m)$, which has been annotated with $\hat{q}$ at the coordinator, we move to the next state (following the $sync$ operation) and update $\hat{q}$ to $R_{po,\mathbb{M}}(\hat{q}, \omega_1, \omega_2, \ldots, \omega_m)$, where ω_i is the first component of x_i; otherwise, we move to the next state with $\hat{q}$ remaining unchanged.

Enhancing $GSDS$ with this additional information regarding the state of the coordinator is rather straightforward (though it might lead in an increase in the number of the states, because it is possible that $(q, x_1, x_2, \ldots, x_m)$ could be associated with different information at the coordinator (e.g., if this state can be reached at $GSDS$ via two different sequences of events, s and s' that cause different synchronizations and result in different information at the coordinator). In any case, however, the construction will be finite. The formal construction of the enhanced version of $GSDS$, which we denote by $GSDS_m$, is described below (notice that this construction also removes any parts of $GSDS$ that are not reachable from its initial state). The algorithm generates the set of states of $GSDS_m$ by sequentially exploring the states of $GSDS$, using a set of states called $UNXES$ to denote states that have not been unexplored yet.

Inputs:
1. $GSDS = (Q_g, \Sigma \cup \{sync\}, \delta_g, Q_{0,g})$
2. $G_e = (Q\dot{\cup}Q_F, \Sigma \cup \{sync\}, \delta_e, Q_0)$
3. $\Sigma_{o_i}, i \in \mathbb{M}$

Output:
$GSDS_m = (Q_m, \Sigma \cup \{sync\}, \delta_m, Q_{0,m})$ where $Q_m \subseteq Q_g \times 2^{Q\dot{\cup}Q_F}$

Set $Q_{0,m} = Q_{0,g} \times \{\hat{q}_0\}$ where $\hat{q}_0 = R_{\mathbb{M}}(Q_0, \epsilon)$
Set $Q_m := \emptyset$; set $UNXES := \{Q_{0,m}\}$
While $UNXES \neq \emptyset$
  For each state $q_m = ((q, x_1, x_2, \ldots, x_m), \hat{q})$ in $UNXES$
    For each $\sigma \in \Sigma$
      For each $(q', x'_1, x'_2, \ldots, x'_m) \in \delta_g((q, x_1, x_2, \ldots, x_m), \sigma)$
        (i.e., the non-deterministic δ_g is non-empty)
        Set $q'_m \in \delta_m(q_m, \sigma)$ where $q'_m = ((q', x'_1, x'_2, \ldots, x'_m), \hat{q})$
          (i.e., no change in $\hat{q}$)
        If $q'_m \notin Q_m$ and $q'_m \notin UNXES$
          $UNXES := UNXES \cup \{q'_m\}$
          END
        END
      END
    For each $(q', x'_1, x'_2, \ldots, x'_m) \in \delta_g((q, x_1, x_2, \ldots, x_m), sync)$
      (i.e., the non-deterministic δ_g is defined for $sync$)
      Set $q'_m \in \delta_m(q_m, \sigma)$ where $q'_m = ((q', x'_1, x'_2, \ldots, x'_m), \hat{q}')$
        with $\hat{q}' = R_{po,\mathbb{M}}(\hat{q}, \omega_1, \omega_2, \ldots, \omega_m)$
        where ω_i is the first component of x_i
      If $q'_m \notin Q_m$ and $q'_m \notin UNXES$

$UNXES := UNXES \cup \{q'_m\}$
END
END

Set $UNXES := UNXES \setminus \{q_m\}$ (i.e., remove q_m from $UNXES$)
Set $Q_m := Q_m \cup \{q_m\}$ (i.e., add q_m to Q_m)
END
END

Notice that the final construction $GSDS_m$ has a finite number of states since each component (and the coordinator) can assume states in a finite set (notice that it will be a nondeterministic construction due to the non-determinism of G). Given any sequence of events s that occurs in G, we can use the above construction to track the information maintained at the various observation sites, the synchronizations that occur, and the resulting state estimates at the coordinator. Furthermore, we can identify in the above construction indeterminate states, i.e., states in which we have uncertainty at the coordinator. Specifically, these will be states in which the estimate $\hat{q}$ at the coordinator involves both normal and faulty states (i.e., $\hat{q} \cap Q \neq \emptyset$ and $\hat{q} \cap Q_F \neq \emptyset$). If we can find "indeterminate cycles" in the final construction (i.e., cycles of states in which we have uncertainty at the coordinator), and these cycles can manifest themselves after the occurrence of a fault, then we conclude that the system is *not* Case I decentralized diagnosable. It should be clear that if we cannot find such cycles (or if such cycles cannot occur after a fault event), then the system *is* Case I decentralized diagnosable.

Remark 9.11 The state complexity of $GSDS_m$ above can be bounded by

$$|Q\dot{\cup}Q_F| \times 2^{|Q\dot{\cup}Q_F|} \times \Pi_{i=1}^{m}((|\Sigma_{o_i}| + 1)^{k_i} \times |Q_{s,i}|) .$$

This expression can be somewhat simplified if we let $k_{\max} = \max_{i \in \mathbb{M}} k_i$ be the maximum sequence length stored in any of the sequence storage SD's, $A_{\max} = \max_{i \in \mathbb{M}} |\Sigma_{o_i}|$ be the maximum number of events observed at an observation site, and $S_{\max} = \max_{i \in \mathbb{M}} |Q_{s,i}|$ be the maximum size of the state space of a synchronizing automaton: in such case, we can bound the size of the state space of the above construction by

$$2|Q| \times 4^{|Q|} \times \left((A_{\max} + 1)^{k_{\max}} \times S_{\max}\right)^m .$$

It might be possible to reduce this complexity, by using verifier-like techniques to track the uncertainty at the coordinator; however, the complexity associated with the storage of the sequences of observations appears to be more difficult to reduce (as it seems inherent to Case I decentralized state estimation). Of course, one easy way to keep this complexity low is to ensure that synchronizations occur often (i.e., the synchronizing automata are such that $k_{\max}$ is a small integer). □

9.10.1.2 Verification of Case II and Case III Decentralized Diagnosability

The basic construction that is used to verify Case II and Case III decentralized diagnosability is based on a parallel-like structure that aims to track, for each possible sequence of events, the state in which the various components of the overall system are. More specifically, if we imagine that we have constructed a local diagnoser D_i at each observation site based on the observable events Σ_{o_i} at that site (see Sect. 7.3.1 in Chap. 7), then we can use a parallel composition to simultaneously track, for each possible sequence of events in the system, the following: (i) the state(s) of the enhanced version of the given system G_e, (ii) the state of each local diagnoser D_i, (iii) the state of each local synchronizing DFA S_i and whether a $sync$ operation needs to take place, and (iv) the state of the coordinator, which is a fusion of the states of the diagnosers after the last synchronization. This composed system has states that are *indeterminate* at the coordinator (i.e., problematic from the point of view of fault diagnosis) if the state of the coordinator does not clarify whether a fault has occurred or not. As in the case of centralized diagnosability, if we can find cycles of such indeterminate states (which can occur after the occurrence of a fault event), then we have a system that is not Case II (or Case III) decentralized diagnosable.

In more detail, to verify Case II or Case III decentralized diagnosability we can do the following:

1. We construct the enhanced version of the system G_e, as described in the previous section (this enhanced system is an LNFA that resembles G_F, as described in Sect. 7.2.2 of Chap. 7, and has $sync$ as an additional event that acts as a self-transition at each state). In other words, $G_e = (Q\dot{\cup}Q_F, \Sigma \cup \{sync\}, \delta_e, Q_0)$, where $\delta_e(q, sync) = \{q\}$ for all $q \in Q\dot{\cup}Q_F$ and $\delta_e(q, \sigma) = \delta_F(q, \sigma)$ for all $q \in Q\dot{\cup}Q_F$ and $\sigma \in \Sigma$ (note that $\delta_e(q, \sigma)$ is empty/undefined if $\delta_F(q, \sigma)$ is empty/undefined).
2. We construct the local diagnoser $D_i = (Q_{d,i}, \Sigma_{o_i}, \delta_{d,i}, q_{0d,i})$ at observation site O_i, as described in Sect. 7.3.1 of Chap. 7, using Σ_{o_i} as the set of observable events. We then construct an enhanced local diagnoser $D_{ei} = (Q_{d,i}, \Sigma_{o_i} \cup \{sync\}, \delta_{ed,i}, q_{0d,i})$, enhanced so that the $sync$ event is a self-transition at each state, i.e., $\delta_{ed,i}(q_{d,i}, sync) = q_{d,i}$ for all $q_{d,i} \in Q_{d,i}$ and $\delta_{ed,i}(q_{d,i}, \sigma_{o_i}) = \delta_{d,i}(q_{d,i}, \sigma_{o_i})$ for all $q_{d,i} \in Q_{d,i}$ and all $\sigma_{o_i} \in \Sigma_{o_i}$ (note that $\delta_{ed,i}(q_{d,i}, \sigma_{o,i})$ is undefined if $\delta_{d,i}(q_{d,i}, \sigma_{o,i})$ is undefined).
3. We construct the product DS_i of the enhanced local diagnoser D_{ei} and the synchronizing DFA S_i at observation site O_i, i.e., $DS_i = D_{ei} \times S_i$, for $i = 1, 2, \ldots, m$.
4. We construct the parallel composition $GDS = G_e||DS_1||DS_2|| \ldots ||DS_m$. We mark states in this parallel composition that have a marked second component in the state of the DS_i component (for some $i = 1, 2, \ldots, m$). These marked states are states in which a synchronization has been initiated; thus, we remove all transitions other than the one associated with the $sync$ transition from marked states in this parallel composition.

5. In the resulting parallel composition GDS, each state is of the form $(q, x_1, x_2, \ldots, x_m)$ and belongs in $(Q\dot{\cup}Q_F) \times (Q_{d,1} \times Q_{s,1}) \times \cdots \times (Q_{d,m} \times Q_{s,m})$. If we let $x_i = (\hat{q}_i, q_{s_i})$ be the state of DS_i, then the state estimate or decision at the coordinator changes each time we have a $sync$ operation from a marked state. In Case II, this state is given by the intersection $\cap_{i=1}^{m} \hat{q}_i$, whereas in Case III it is simply an "F", or "N", or "U" decision based on whether at least one $\hat{q}_i$ has label F (i.e., $\hat{q}_i \subseteq Q_F$), or at least one site has label N (i.e., $\hat{q}_i \subseteq Q$), or all sites have a mixture of F and N labels.
   We can thus enhance the parallel composition GDS with this additional information regarding the state of the coordinator. The coordinator is initialized to $\hat{q}_0 = R_{\mathbb{M}}(Q_0, \epsilon)$ for Case II and to $\hat{q}_0 = F$, or N, or U for Case III (depending on whether $R_{\mathbb{M}}(Q_0, \epsilon) \subseteq Q_F$, $R_{\mathbb{M}}(Q_0, \epsilon) \subseteq Q$, or otherwise). We can follow each transition in GDS and update the state of the coordinator if a synchronization ($sync$) event takes place (otherwise we maintain its state unchanged). Note that it is possible that $(q, x_1, x_2, \ldots, x_m)$ could be associated with different information at the coordinator (e.g., if this state can be reached via two different sequences of events, s and s' that cause different synchronizations and result in different information at the coordinator); this, however, can easily be handled by introducing multiple copies of this particular state, each associated with the corresponding coordinator state.
   The formal construction of the enhanced version of GDS, which we denote by GDS_m, is described below for verifying Case II decentralized diagnosability (notice that this construction also removes any parts of GDS that are not reachable from its initial state). As in the algorithm for $GSDS_m$ in the previous section, the algorithm generates the set of states of GDS_m by sequentially exploring the states of GDS, using a set of states called $UNXES$ to denote states that have been unexplored yet.

Inputs:
1. $GDS = (Q_g, \Sigma \cup \{sync\}, \delta_g, Q_{0,g})$
2. $G_e = (Q\dot{\cup}Q_F, \Sigma \cup \{sync\}, \delta_e, Q_0)$
3. $\Sigma_{o_i}, i \in \mathbb{M}$

Output:
$GDS_m = (Q_m, \Sigma \cup \{sync\}, \delta_m, Q_{0,m})$ where $Q_m \subseteq Q_g \times 2^{Q\dot{\cup}Q_F}$

Set $Q_{0,m} = Q_{0,g} \times \{\hat{q}_0\}$ where $\hat{q}_0 = R_{\mathbb{M}}(Q_0, \epsilon)$
Set $Q_m := \emptyset$; set $UNXES = \{Q_{0,m}\}$
While $UNXES \neq \emptyset$
  For each state $q_m = ((q, x_1, x_2, \ldots, x_m), \hat{q})$ in $UNXES$
    For each $\sigma \in \Sigma$
      for each $(q', x_1', x_2', \ldots, x_m') \in \delta_g((q, x_1, x_2, \ldots, x_m), \sigma)$
        (i.e., the non-deterministic δ_g is non-empty)
        Set $q_m' \in \delta_m(q_m, \sigma)$ where $q_m' = ((q', x_1', x_2', \ldots, x_m'), \hat{q})$
          (i.e., no change in $\hat{q}$)
        If $q_m' \notin Q_m$ and $q_m' \notin UNXES$

$UNXES := UNXES \cup \{q'_m\}$
END
END
END
For each $(q', x'_1, x'_2, \ldots, x'_m) \in \delta_g((q, x_1, x_2, \ldots, x_m), sync)$
(i.e., the non-deterministic δ_g is defined for $sync$)
Set $q'_m \in \delta_m(q_m, \sigma)$ where $q'_m = ((q', x'_1, x'_2, \ldots, x'_m), \hat{q}')$
with $\hat{q}' = \cap_{i=1}^{m} \hat{q}_i$
where $\hat{q}_i$ is the first component of x_i
If $q'_m \notin Q_m$ and $q'_m \notin UNXES$
$UNXES := UNXES \cup \{q'_m\}$
END
END

Set $UNXES := UNXES \setminus \{q_m\}$ (i.e., remove q_m from $UNXES$)
Set $Q_m := Q_m \cup \{q_m\}$ (i.e., add q_m to Q_m)
END
END

Notice that the final construction (obtained in Step 5 of the above process) has a finite number of states since each component (and the coordinator) can assume states in a finite set. Moreover, given any sequence of events s that occurs in G, we can use the above construction to track the state estimates at the various observation sites and the coordinator. In fact, we can identify in the above construction indeterminate states, i.e., states in which we have uncertainty at the coordinator. For Case II decentralized diagnosability, these indeterminate states are associated with a coordinator state $\hat{q}$ that is neither a subset of Q nor a subset of Q_F. If we can find "indeterminate cycles" in the final construction (i.e., cycles of states in which we have uncertainty at the coordinator), and these cycles can manifest themselves after the occurrence of a fault, then we conclude that the system is not decentralized diagnosable under the corresponding protocol.

Remark 9.12 The state complexity of GDS_m above can be bounded by

$$|Q\dot{\cup}Q_F| \times 2^{|Q\dot{\cup}Q_F|} \times \Pi_{i=1}^{m}(2^{|Q\dot{\cup}Q_F|} \times |Q_{s,i}|) \ .$$

If we let $S_{\max} = \max_{i \in \mathbb{M}} |Q_{s,i}|$ be the maximum size of the state space of a synchronizing automaton, then the state complexity of the resulting parallel composition in the above approach could be (in the worst case) $2|Q| \times 4^{|Q|} \times \left(4^{|Q|} \times S_{\max}\right)^m$ for Case II and $2|Q| \times \left(4^{|Q|} \times S_{\max}\right)^m \times 2$ for Case III.

Note that it is possible to reduce this complexity, by using verifier-like techniques to track the uncertainty at each observation site and at the coordinator (Keroglou and Hadjicostis 2018); in such case the complexity reduces to

$$|Q\dot{\cup}Q_F| \times |Q\dot{\cup}Q_F|^2 \times \Pi_{i=1}^{m}(|Q\dot{\cup}Q_F|^2 \times |Q_{s,i}|) \ ,$$

which simplifies to $2|Q| \times \left((2|Q|)^2 \times S_{\max}\right)^m \times (2|Q|)^2$ for Case II decentralized diagnosis and to $2|Q| \times \left((2|Q|)^2 \times S_{\max}\right)^m \times 4$ for Case III decentralized diagnosis. □

Remark 9.13 The construction for Case III decentralized diagnosability can be simplified quite a bit. Notice that, due to the absorption property of the F label (refer to Chap. 7), if a (local) diagnoser can determine "F" at any point in time, then it will be able to determine "F" after each subsequent observation. Thus, since the decision at each site depends only on its own local observations (and not on earlier synchronizations), we can ignore earlier synchronization points and focus on what happens for long enough sequences of events. What would cause problems are infinite sequences of events that keep all local diagnosers confused. Such sequences can be checked without taking into account *sync* operations or the synchronizing automata S_i, $i = 1, 2, \ldots, m$: one simply needs to check whether the parallel composition of $G||D_1||D_2|| \ldots ||D_m$ leads to a cycle of states in which all m local diagnosers D_i, $i = 1, 2, \ldots, m$, are confused as to whether a fault has occurred or not (and, of course, this cycle needs to be executable after the occurrence of a fault). This is the basic premise used in the verification of co-diagnosability in Qiu and Kumar (2006), Wang et al. (2007, 2011), Schmidt (2010), Wang et al. (2011). □

9.10.2 Verification of Detectability

In this section we build on the approach described for the verification of decentralized diagnosability to provide verification algorithms for decentralized detectability. We first describe the problem setting.

Problem Setting (Decentralized Detectability): We are given an LNFA $G = (Q, \Sigma, \delta, Q_0)$, which is observed at m observation sites O_1, O_2, …, and O_m, with observable events Σ_{o_1}, Σ_{o_2}, …, and Σ_{o_m}, respectively. Each observation site O_i is associated with a synchronizing marked DFA $S_i = (Q_{s,i}, \Sigma_{o_i} \cup \{sync\}, \delta_{s,i}, q_{0s,i}, Q_{ms,i})$ as described in Definition 9.4. Whenever S_i reaches a state in the set of marked states $Q_{ms,i}$ ($Q_{ms,i} \subset Q_{s,i}$), then O_i signals the coordinator to initiate a synchronization and the *sync* event occurs (involving all observation sites), i.e., the coordinator requests from each observation site their information, namely (i) sequences of events since the last synchronization (Case I), or (ii) local state estimates (Case II), or (iii) local decisions[5] (Case III). We are interested in verifying *decentralized (strong)*

[5] As mentioned earlier, in Case III decentralized detectability, the decision at each observation site is "detectable" ("D") or "not detectable" ("ND"), depending on whether that local observation site is able, at that particular instant, to determine the state of the system exactly or not. When a synchronization occurs, the coordinator decides "D" if at least one observation site is reporting a "D"; otherwise, the coordinator decides "ND". Note that unlike fault diagnosis (where an "F" decision remains an "F" decision due to the absorbing property of the F label—refer to Remark 7.3 of Chap. 7) the decision about the set of state estimates being a singleton set is not absorbing, i.e., it can change from "D" to "ND" and vice versa. This also implies that at different synchronization

detectability, i.e., verifying whether the coordinator will be able, for long enough sequences of observations, to almost always determine the exact current state of the system. Since between synchronizations the coordinator is not informed of any activity in the system, we only require that the exact state of the system becomes known at the coordinator immediately following the *sync* events. More specifically, for any infinitely extensible[6] sequence of observations y ($y \in Y(G)$ as defined in (6.1)), we let $\kappa_1(y)$, $\kappa_2(y)$, ..., $\kappa_j(y)$, ...denote the points at which a synchronization takes place and define the sets

$$\begin{aligned}
\mathcal{D}_{po,k}(y) &= \{j \in \{1, 2, \ldots, k\} \mid |\hat{q}_{po}^{(j)}(y_0^{\kappa_j})| = 1\} \\
\mathcal{D}_{si,k}(y) &= \{j \in \{1, 2, \ldots, k\} \mid |\hat{q}_{si}^{(j)}(y_0^{\kappa_j})| = 1\} \\
\mathcal{D}_{d,k}(y) &= \{j \in \{1, 2, \ldots, k\} \mid \text{Decision at synchronization } j \text{ is ``}D\text{''}\} \,.
\end{aligned}$$

We would like to determine if G is

1. Case I decentralized detectable, i.e., if
$$\lim_{k \to \infty} \frac{|\mathcal{D}_{po,k}(y)|}{k} = 1$$
for all $y \in Y(G)$.
2. Case II decentralized detectable, i.e., if
$$\lim_{k \to \infty} \frac{|\mathcal{D}_{si,k}(y)|}{k} = 1$$
for all $y \in Y(G)$.
3. Case III decentralized detectable, i.e., if
$$\lim_{k \to \infty} \frac{|\mathcal{D}_{d,k}(y)|}{k} = 1$$
for all $y \in Y(G)$.

Remark 9.14 Note that one can provide analogous definitions for Case I (or Case II or Case III) weak decentralized detectability, strong periodic decentralized detectability, and weak periodic decentralized detectability. □

points, we may have different observation sites that are aware of the exact state of the system; furthermore, in-between synchronization points it is possible that no observation site is aware of the exact state of the system.

[6]Recall that we have adopted the usual assumptions that (i) G is live, and (ii) G possesses no unobservable cycles (see, for example, the discussions in Chap. 6).

9.10.2.1 Verification of Case I Decentralized Detectability

To verify Case I decentralized detectability, we can use the construction of $GSDS_m$ described in Sect. 9.10.1.1. The only difference is that we are only interested in the state estimates in the coordinator and not in the inference of fault events; thus, there is no need to use G_F in the extended version of the system G_e. Moreover, in the case of detectability, problematic states at the coordinator are states that do not allow the coordinator to determine the state of the system exactly, at least not immediately following a *sync* event. Specifically, given a state $q_m = ((q, x_1, x_2, \ldots, x_m), \hat{q})$ of $GSDS_m$, we say that this state *violates detectability* if $|\hat{q}| > 1$ (i.e., if the state estimate at the coordinator involves more than one states. If we can find in $GSDS_m$ a cycle of states in which there is at least one state that violates detectability and this state is reached in the cycle following a *sync* event, then we say that the system is *not* Case I decentralized detectable (in the sense that there is an infinite sequence of events that results in observations and synchronizations that cause to the coordinator uncertainty about the state of the system following the occurrence of a *sync* event). Obviously, the state complexity of $GSDS_m$ (which is primarily responsible for verifying detectability) remains effectively the same, i.e., it is bounded by

$$|Q| \times 2^{|Q|} \times \Pi_{i=1}^{m}((|\Sigma_{o_i}| + 1)^{k_i} \times |Q_{s,i}|) ,$$

where the various sets are as defined in Sect. 9.10.1.1. The only difference in the complexity computations is that we do not have to consider faulty states (there are none).

9.10.2.2 Verification of Case II and Case III Decentralized Detectability

To verify Case II and Case III decentralized detectability, we can use variants of the corresponding structures for Case II and Case III diagnosability, as described in Sect. 9.10.1.2. Again, the main difference is that we do not have faulty states and that we can use observers as opposed to diagnosers. In more detail, to verify Case II or Case III decentralized detectability we can do the following:

1. We construct an enhanced version of the system, denoted by G_e. This enhanced system is an LNFA that resembles G and has one additional event, namely the *sync* event, which acts as a self-transition at each state. In other words, $G_e = (Q, \Sigma \cup \{sync\}, \delta_e, Q_0)$, where $\delta_e(q, sync) = \{q\}$ for all $q \in Q$ and $\delta_e(q, \sigma) = \delta(q, \sigma)$ for all $q \in Q$ and $\sigma \in \Sigma$ (note that $\delta_e(q, \sigma)$ is empty/undefined if $\delta(q, \sigma)$ is empty/undefined).
2. We construct the local observer (state estimator) $SE_i = (Q_{se,i}, \Sigma_{o_i}, \delta_{se,i}, q_{0se,i})$ at observation site O_i, using Σ_{o_i} as the set of observable events. We then construct an enhanced local observer $SE_{ei} = (Q_{se,i}, \Sigma_{o_i} \cup \{sync\}, \delta_{ese,i}, q_{0ese,i})$, enhanced so that the *sync* event is a self-transition at each state, i.e., $\delta_{ese,i}(q_{se,i}, sync) = q_{se,i}$ for all $q_{se,i} \in Q_{se,i}$ and $\delta_{ese,i}(q_{se,i}, \sigma_{o_i}) = \delta_{se,i}(q_{se,i}, \sigma_{o_i})$ for all

$q_{se,i} \in Q_{se,i}$ and all $\sigma_{o_i} \in \Sigma_{o_i}$ (note that $\delta_{ese,i}(q_{se,i}, \sigma_{o_i})$ is undefined if $\delta_{se,i}(q_{se,i}, \sigma_{o_i})$ is undefined).

3. We construct the product SES_i of the enhanced local observer SE_{ei} and the synchronizing DFA S_i at observation site O_i, i.e., $SES_i = SE_{ei} \times S_i$, for $i = 1, 2, \ldots, m$.
4. We construct $GSES = G_e||SES_1||SES_2|| \ldots ||SES_m$ and mark states in this parallel composition that have a marked second component in the state of the SES_i component (for some $i = 1, 2, \ldots, m$). These marked states are states in which a synchronization has been initiated; thus, we remove all transitions other than the one associated with the $sync$ transition from marked states in this parallel composition.
5. In the resulting parallel composition $GSES$, each state is of the form $(q, x_1, x_2, \ldots, x_m)$ and belongs in $(Q \times (Q_{se,1} \times Q_{s,1}) \times \cdots \times (Q_{se,m} \times Q_{s,m}))$. If we let $x_i = (\hat{q}_i, q_{s_i})$ be the state of the ith component, then the state estimate or decision at the coordinator changes each time we have a $sync$ operation from a marked state. In Case II, this state is given by the intersection $\cap_{i=1}^{m} \hat{q}_i$, whereas in Case III it is simply a "D" or an "ND" indicator based on whether at least one observation site, say O_i, is reporting a "D" (when $\hat{q}_i$ is a singleton set) or not.

   We can thus enhance the parallel composition $GSES$ with this additional information regarding the state of the coordinator. The coordinator is initialized to $\hat{q}_0 = R_{\mathbb{M}}(Q_0, \epsilon)$ for Case II and to $\hat{q}_0 = D$ or ND (depending on whether $R_i(Q_0, \epsilon)$ is a singleton set for some $i \in \mathbb{M}$). We can follow each transition in $GSES$ and update the state of the coordinator if a synchronization ($sync$) event takes place (otherwise, we maintain its state unchanged). Note that it is possible that $(q, x_1, x_2, \ldots, x_m)$ could be associated with different information at the coordinator (e.g., if this state can be reached via two different sequences of events, s and s' that cause different synchronizations and result in different information at the coordinator); this, however, can easily be handled by introducing multiple copies of this particular state, each associated with the corresponding coordinator state.

   The formal construction of the enhanced version of $GSES$, which we denote by $GSES_m$, is described below for verifying Case II decentralized detectability (notice that this construction also removes any parts of $GSES$ that are not reachable from its initial state).

Inputs:
1. $GSES = (Q_g, \Sigma \cup \{sync\}, \delta_g, Q_{0,g})$
2. $G_e = (Q, \Sigma \cup \{sync\}, \delta_e, Q_0)$
3. $\Sigma_{o_i}, i \in \mathbb{M}$

Output:
$GSES_m = (Q_m, \Sigma \cup \{sync\}, \delta_m, Q_{0,m})$ where $Q_m \subseteq Q_g \times 2^Q$

Set $Q_{0,m} = Q_{0,g} \times \{\hat{q}_0\}$ where $\hat{q}_0 = R_{\mathbb{M}}(Q_0, \epsilon)$
Set $Q_m := \emptyset$; set $UNXES = Q_{0,m}$
While $UNXES \neq \emptyset$

For each state $q_m = ((q, x_1, x_2, \ldots, x_m), \hat{q})$ in $UNXES$
  For each $\sigma \in \Sigma$
    for each $(q', x'_1, x'_2, \ldots, x'_m) \in \delta_g((q, x_1, x_2, \ldots, x_m), \sigma)$
      (i.e., the non-deterministic δ_g is non-empty)
      Set $q'_m \in \delta_m(q_m, \sigma)$ where $q'_m = ((q', x'_1, x'_2, \ldots, x'_m), \hat{q})$
        (i.e., no change in $\hat{q}$)
      If $q'_m \notin Q_m$ and $q'_m \notin UNXES$
        $UNXES := UNXES \cup \{q'_m\}$
        END
      END
    END
  For each $(q', x'_1, x'_2, \ldots, x'_m) \in \delta_g((q, x_1, x_2, \ldots, x_m), sync)$
    (i.e., the non-deterministic δ_g is defined for $sync$)
    Set $q'_m \in \delta_m(q_m, \sigma)$ where $q'_m = ((q', x'_1, x'_2, \ldots, x'_m), \hat{q}')$
      with $\hat{q}' = \cap_{i=1}^{m} \hat{q}_i$
      where $\hat{q}_i$ is the first component of x_i
    If $q'_m \notin Q_m$ and $q'_m \notin UNXES$
      $UNXES := UNXES \cup \{q'_m\}$
      END
    END

  Set $UNXES := UNXES \setminus \{q_m\}$ (i.e., remove q_m from $UNXES$)
  Set $Q_m := Q_m \cup \{q_m\}$ (i.e., add q_m to Q_m)
  END
END

As in the case of diagnosability, given any sequence of events s that occurs in G, we can use the above construction to track the state estimates at the various observation sites and the coordinator. We can also identify in the above construction problematic states, i.e., states in which we do not know the state of the system exactly. For Case II decentralized detectability, these problematic states are associated with a coordinator state $\hat{q}$ that occurs immediately after a synchronization event and is not a singleton subset of Q. System G is not Case II decentralized detectable (under the given observation setting and the given synchronization strategy) if and only if there exists a cycle of states in $GSES_m$ that involves at least one problematic state.

For Case III decentralized detectability the state of the coordinator would be "D" if at least one observation site knows the state of the system exactly, or "ND" if no observation site knows the state of the system exactly. In such case, system G will not be Case III decentralized detectable (under the given observation setting and the given synchronization strategy) if and only if there exists a cycle of states in $GSES_m$ that involves at least one problematic state (i.e., a state that immediately follows a synchronization event and is marked with "ND" at the coordinator).

Remark 9.15 The complexity of the construction for Case II decentralized detectability is

$$|Q| \times 2^{Q} \times \Pi_{i=1}^{m}(2^{|Q|} \times |Q_{s,i}|) .$$

It may be possible to reduce the complexity of this verification process using detectors as opposed to observers. □

9.11 Comments and Further Reading

Decentralized observation settings for state estimation and event inference of an underlying system that is modeled as an LNFA or an LDFA (labeled deterministic finite automaton) were first considered in Debouk et al. (2000) in the context of fault diagnosis. The authors of Debouk et al. (2000) focused on the case of two observation sites and proposed three protocols for coordinated decentralized diagnosis. Protocols 1 and 2 assume communication from the local diagnosers to a coordinator, whereas Protocol 3 assumes no communication between them or to any coordinator.

- In Protocol 1, the diagnostic information at a local site is generated by an *extended diagnoser*, the states of which consist of pairs of predecessor and successor state estimates of a single observable event, along with its failure label. The information communicated to the coordinator is updated each time an event is observed at a local site and consists of the corresponding states of the extended diagnosers, their unobservable reach, and a status bit depending on the local observability of that event. The decision rule of the coordinator is defined under different set intersection operations applied on the system state estimates and their failure labels, as they are maintained at the two observation sites.
- Protocol 2 uses the (standard) diagnoser to generate the local diagnostic information, and system diagnosis is performed under the same communication and decision rules as Protocol 1. Although the computational complexity of Protocol 2 is reduced compared to Protocol 1, the performance of the former is constrained to traces that exhibit the "ordering problem" of the coordinator, also referred to as *failure-ambiguous* traces in Debouk et al. (2000). Note that Protocol 2 in Debouk et al. (2000) is a special case of Case II decentralized fault diagnosis (as presented in this chapter), with $m = 2$, and synchronizing automata that enforce a synchronization each time an observation is made (i.e., $k_1 = k_2 = 1$ in the synchronizing automata described in Sect. 9.9). On the other hand, the comparison with Protocol 1 is not as clear since it seems to combine features of Case I and Case II decentralized fault diagnosis (still constraining itself to $k_1 = k_2 = 1$).
- Protocol 3 in Debouk et al. (2000) is directly linked to the so-called property of *co-diagnosability*. A system is co-diagnosable (or decentralized diagnosable) if any occurrence of a failure is detected by at least one local diagnoser within a bounded interval of observations. Polynomial complexity algorithms for the verification of co-diagnosability have been proposed in Qiu and Kumar (2006),

Wang et al. (2007, 2011), Basilio and Lafortune (2009), Moreira et al. (2011), Takai and Ushio (2012). In Wang et al. (2007) decentralized diagnosability is studied under a framework of *conditional decisions* issued by the local sites, and polynomial tests are proposed for the verification of equivalent language-based notions of decentralized diagnosability. In Wang et al. (2011) the authors propose polynomial algorithms to transform a problem of co-observability to the problem of co-diagnosability under the assumption of dynamic observations. They also propose a polynomial-time algorithm for testing co-diagnosability of the system based on *cluster automata*. In Basilio and Lafortune (2009) the notion of *robust co-diagnosability* is defined and verified in two ways, one using diagnoser automata and the other using verifier automata. The authors of Takai and Ushio (2012) study co-diagnosability under the condition of state-dependence and non-determinism of partial event observation. The algorithms proposed in Qiu and Kumar (2006), Moreira et al. (2011) are based on constructing a testing automaton that, given a faulty trace, searches for corresponding indistinguishable non-faulty traces at each local site. Lower complexity is achieved in Moreira et al. (2011) by tracking the non-faulty traces in a smaller set than the one assumed in Qiu and Kumar (2006). Several discussions, clarifications, and comparisons of algorithms pertaining to the verification of diagnosability in this setting can also be found in Moreira et al. (2011), Kumar and Takai (2014), Moreira et al. (2016).

Decentralized diagnosability similar to Protocols 1 and 2 of Debouk et al. (2000) has been also studied in Athanasopoulou and Hadjicostis (2006), Panteli and Hadjicostis (2013). In particular, fault diagnosis under the intersection-based distributed diagnosis (IBDD) protocol in Athanasopoulou and Hadjicostis (2006), Panteli and Hadjicostis (2013) is performed locally using a (local, on site) diagnoser and, upon request, diagnostic information (i.e., the sets of state estimates along with matching normal/fault conditions) at each observation site is communicated to the coordinator, who then reaches an overall diagnosis decision by performing set intersection on the diagnostic information provided by the various observation sites. It is important to note that in IBDD, the coordinator does not notify the local diagnosers about the refined diagnostic information that it has obtained (thus, local diagnosers operate exclusively based on locally available observations) and information flows only from the local observation sites to the coordinator (in the form of diagnostic information).

Allowing information to flow from the coordinator back to the local diagnosers, the authors of Keroglou and Hadjicostis (2014, 2015) developed distributed intersection-based strategies that allow the exchange of local state estimates among different observation sites. The verification of this intersection-based scheme (S-IBDD), founded on predetermined synchronization strategies, was solved by constructing a special type of parallel composition of local diagnosers and their product with the system model. These works are discussed in more detail in Chap. 10. Other works that have allowed information to flow from one observation site to another include (Su and Wonham 2005; Fabre et al. 2000), which focus on modular systems (with a local observer for each module) and explore iterative strategies that involve exchange of diagnostic information among neighboring observation sites and set intersection operations for refinement.

References

Athanasopoulou E, Hadjicostis CN (2006) Decentralized failure diagnosis in discrete event systems. In: Proceedings of 2006 American control conference (ACC), pp 14–19

Basilio JC, Lafortune S (2009) Robust codiagnosability of discrete event systems. In: Proceedings of 2009 American control conference (ACC), pp 2202–2209

Debouk R, Lafortune S, Teneketzis D (2000) Coordinated decentralized protocols for failure diagnosis of discrete event systems. Discret Event Dyn Syst: Theory Appl 10(1–2):33–86

Fabre E, Benveniste A, Jard C, Ricker L, Smith M (2000) Distributed state reconstruction for discrete event systems. In: Proceedings of 39th IEEE conference on decision and control (CDC), vol 3, pp 2252–2257

Keroglou C, Hadjicostis CN (2014) Distributed diagnosis using predetermined synchronization strategies. In: Proceedings of 53rd IEEE conference on decision and control (CDC), pp 5955–5960

Keroglou C, Hadjicostis CN (2015) Distributed diagnosis using predetermined synchronization strategies in the presence of communication constraints. In: Proceedings of IEEE conference on automation science and engineering (CASE), pp 831–836

Keroglou C, Hadjicostis CN (2018) Distributed fault diagnosis in discrete event systems via set intersection refinements. IEEE Trans Autom Control 63(10):3601–3607

Kumar R, Takai S (2014) Comments on polynomial time verification of decentralized diagnosability of discrete event systems versus decentralized failure diagnosis of discrete event systems: complexity clarification. IEEE Trans Autom Control 59(5):1391–1392

Moreira MV, Jesus TC, Basilio JC (2011) Polynomial time verification of decentralized diagnosability of discrete event systems. IEEE Trans Autom Control 56(7):1679–1684

Moreira MV, Basilio JC, Cabral FG (2016) Polynomial time verification of decentralized diagnosability of discrete event systems versus decentralized failure diagnosis of discrete event systems: a critical appraisal. IEEE Trans Autom Control 61(1):178–181

Panteli M, Hadjicostis CN (2013) Intersection based decentralized diagnosis: implementation and verification. In: Proceedings of 52nd IEEE conference on decision and control and european control conference (CDC-ECC), pp 6311–6316

Puri A, Tripakis S, Varaiya P (2002) Problems and examples of decentralized observation and control for discrete event systems. In: Synthesis and control of discrete event systems, Springer, Berlin, pp 37–56

Qiu W, Kumar R (2006) Decentralized failure diagnosis of discrete event systems. IEEE Trans Syst Man Cybern Part A: Syst Hum 36(2):384–395

Rosen KH (2011) Discrete mathematics and its applications. McGraw-Hill, New York

Schmidt K (2010) Abstraction-based verification of codiagnosability for discrete event systems. Automatica 46(9):1489–1494

Su R, Wonham WM (2005) Global and local consistencies in distributed fault diagnosis for discrete-event systems. IEEE Trans Autom Control 50(12):1923–1935

Takai S, Ushio T (2012) Verification of codiagnosability for discrete event systems modeled by Mealy automata with nondeterministic output functions. IEEE Trans Autom Control 57(3):798–804

Wang W, Girard AR, Lafortune S, Lin F (2011) On codiagnosability and coobservability with dynamic observations. IEEE Trans Autom Control 56(7):1551–1566

Wang Y, Yoo TS, Lafortune S (2007) Diagnosis of discrete event systems using decentralized architectures. Discret Event Dyn Syst 17(2):233–263

Witsenhausen HS (1968) A counterexample in stochastic optimum control. SIAM J Control 6(1):131–147

Chapter 10
Distributed State Estimation

10.1 Introduction and Motivation

As mentioned in Chap. 9, one of the key tasks for properly controlling emerging networked discrete event systems (cyber DES) is the ability to analyze information from sensors, so as to accurately estimate their state and/or reliably perform event inference tasks (e.g., fault diagnosis). Assuming these sites have knowledge of the system model (and some basic processing capability), they can form their individual state estimates (or infer the occurrence of an event) using the (centralized) techniques described in Chaps. 4 and 5, based exclusively on their own observations. More generally, however, they can exchange information with other observation sites (or with a coordinator if one is present) in order to better resolve the estimation or inference task at hand.

The distributed observation setting adopted in this chapter assumes that each observation site does not necessarily operate in isolation, but can receive explicit or implicit information from other sites. Thus, one of the key challenges is for each observation site to determine how/when to properly send information to its *out-neighboring* observation sites (i.e., the observation sites to whom it can send information) and how to process any information it receives from its *in-neighboring* observation sites (i.e., the observation sites from whom it can receive information). As discussed in Chap. 9, such questions are difficult and can even become undecidable in certain settings (Puri et al. 2002; Witsenhausen 1968) (see the discussion in Sect. 9.9). However, we will see that, for certain strategies for information exchange, the corresponding distributed protocols are intuitive and easy to describe, and some of the resulting properties of interest (such as *synchronization-based distributed diagnosability*) can be verified with complexity that is polynomial in the number of states of the given system and exponential in the number of observation sites.

C. N. Hadjicostis, *Estimation and Inference in Discrete Event Systems*, Communications and Control Engineering,
https://doi.org/10.1007/978-3-030-30821-6_10

10.2 System Modeling and Observation Architecture

The system model and the observation architecture resemble those considered in Chap. 9. The main difference is that a coordinator is not necessarily present, and that each observation site may be able to send/receive information to/from different subsets of the observation sites. Compared to the setting studied in Chap. 9, each observation site may act as a local coordinator with respect to neighboring observation sites. In this section, we describe the network topology that captures the communication capabilities among the various observation sites; for completeness, we also include the description of the system model and the observation capabilities at each observation site, which resemble the ones described in Chap. 9.

We consider an underlying (monolithic) labeled nondeterministic finite automaton (LNFA) $G = (Q, \Sigma, \delta, Q_0)$ (see Definition 3.20); however, the techniques we describe can be extended to modular systems, i.e., systems that consist of compositions of nondeterministic finite automata, in a relatively straightforward manner. We assume that the given system G is monitored by m observation sites O_i, $i = 1, 2, \ldots, m$, each of which is able to (partially) observe activity in the given system. More specifically, observation site O_i observes a subset of events Σ_{o_i}, $\Sigma_{o_i} \subseteq \Sigma$; the remaining events $\Sigma_{uo_i} := \Sigma \backslash \Sigma_{o_i}$ are unobservable at observation site O_i. Thus, the natural projection $P_{\Sigma_{o_i}} : \Sigma^* \rightarrow \Sigma_{o_i}^*$ can be used to map any trace $s \in \Sigma^*$ executed in the system ($s \in L(G)$) to the sequence of observations generated by it at observation site O_i. As described in Chaps. 4 and 5, this natural projection is defined recursively as $P_{\Sigma_{o_i}}(\sigma s') = P_{\Sigma_{o_i}}(\sigma) P_{\Sigma_{o_i}}(s')$, $s' \in \Sigma^*, \sigma \in \Sigma$, with

$$P_{\Sigma_{o_i}}(\sigma) = \begin{cases} \sigma, & \text{if } \sigma \in \Sigma_{o_i}, \\ \epsilon, & \text{if } \sigma \in \Sigma_{uo_i} \cup \{\epsilon\}, \end{cases}$$

where ϵ represents the empty trace. In the sequel, Σ_{o_i} and $P_{\Sigma_{o_i}}$ will also be denoted respectively by Σ_i and P_i when it is clear from context. For notational convenience, we will also use $\mathbb{M} = \{1, 2, \ldots, m\}$ to denote the set of indices associated with the observation sites.

As in Chap. 9, we denote the set of events that are observable by at least one site via

$$\Sigma_o = \cup_{i=1}^{m} \Sigma_{o_i} ,$$

and loosely refer to them as the set of observable events. Similarly, we denote the set of events that are not observable at any site via

$$\Sigma_{uo} = \cap_{i=1}^{m} \Sigma_{uo_i} = \Sigma \backslash \Sigma_o ,$$

and loosely refer to them as the set of unobservable events. For comparison purposes, we can think of Σ_o as the set of events that could be observable by a fictitious observer, which we refer to as the *monolithic* or *centralized* observer and denote by O_c. We can then compare the performance of any distributed state estimation or fault diagnosis scheme against the performance of a centralized state estimation or event inference scheme at O_c. As in previous chapters, for the verification of properties, we adopt the usual assumptions that (i) G is live and (ii) G possesses no unobservable cycles with respect to the set Σ_o of events that are observable in at least one site (see, for example, the discussions in Chap. 6).

As in the decentralized case, we assume that each observation site has knowledge of the system model (namely, G and $\Sigma_{o_i}, i = 1, 2, \ldots, m$), and some basic processing and storage capability. Thus, if desirable, an observation site can locally process the sequence of events it observes, in order to, depending on the task, determine possible system states and/or infer events. Unlike the decentralized observation setting studied in Chap. 9, a coordinator is not necessarily present; however, each observation site O_i cannot only send information, but it can also receive (and process) information from specific subsets of other sites.

We capture the communication capability among observation sites via an interconnection network, also referred to as the *communication topology*. This topology is represented, in general, by a directed graph (digraph) $\mathcal{G}_d = (\mathcal{V}, \mathcal{E})$, where $\mathcal{V} = \{O_1, O_2, \ldots, O_m\}$ is the set of nodes (each representing one of the m observation sites) and $\mathcal{E} \subseteq \mathcal{V} \times \mathcal{V} - \{(O_i, O_i) \mid O_i \in \mathcal{V}\}$ (self-loops excluded) is the set of edges (each representing an existing communication link). Specifically, edge (O_i, O_j) indicates that observation site O_j can send information to observation site O_i. The set of in-neighbors of node O_i is the set of observation sites that can send information to O_i and is denoted by $\mathcal{N}_i^- = \{O_j \in \mathcal{V} \mid (O_i, O_j) \in \mathcal{E}\}$; similarly, the set of out-neighbors of node O_i is the set of observation sites that can receive information from O_i, denoted by $\mathcal{N}_i^+ = \{O_l \in \mathcal{V} \mid (O_l, O_i) \in \mathcal{E}\}$.

Remark 10.1 Note that the decentralized observation architecture with a coordinator studied in Chap. 9 can be seen as a special case of a distributed setting with digraph $\mathcal{G}_d = \{\mathcal{V}, \mathcal{E}\}$, where $\mathcal{V} = \{C, O_1, O_2, \ldots, O_m\}$ (with C denoting the coordinator) and $\mathcal{E} = \{(C, O_i) \mid i = 1, 2, \ldots, m\}$. If the coordinator does not observe anything (as assumed in Chap. 9), we can take $\Sigma_C = \emptyset$. □

Remark 10.2 Note that it is possible that the set of in-neighbors and out-neighbors for observation site O_i coincide. One such example is the case of an *undirected* graph $\mathcal{G}_u = (\mathcal{V}, \mathcal{E}_u)$, which is a digraph with the following property: $(O_i, O_j) \in \mathcal{E}_u \Leftrightarrow (O_j, O_i) \in \mathcal{E}_u$. □

Without much loss of generality, we will assume that the digraph $\mathcal{G}_d$ that captures the communication topology is *strongly connected*, i.e., starting from any node $O_i \in \mathcal{V}$ one can reach any other node $O_j \in \mathcal{V}$ by following a sequence of directed edges, i.e., there exists a sequence of indices $i = i_1, i_2, \ldots, i_t = j$ such that $(O_{i_{\tau+1}}, O_{i_\tau}) \in \mathcal{E}$ for $\tau = 1, 2, \ldots, t-1$. [If the given digraph $\mathcal{G}_d$ is not strongly connected, then there exists at least two observation sites, say O_i and O_j, such that O_i can never receive (either directly or indirectly) information from O_j. Such situations can still be handled by the techniques we develop in this chapter, but information flow will be compartmentalized among different subsets of the observation sites (according to the digraph that captures the structure of the given communication topology). This chapter focuses on the case of a strongly connected digraph, in order to keep the discussions simple.

Recall that, in the absence of any additional information from other observation sites, the task at observation site O_i would essentially amount to the centralized state estimation task that was studied in Chap. 4, applied to the LNFA G, with a set of observable events Σ_{o_i}; this will result, at each site, in a set of states or a set of fault/event conditions, or combinations of them (pairs of a state and an associated fault/event label) that are *locally consistent*, i.e., consistent with what has been observed at observation site O_i locally. However, the presence of the communication topology allows observation sites to send/receive information to/from other observation sites; thus, the main task in a distributed setting is to (i) determine what information to send, when to send it and to whom and (ii) fuse the information at the receiving site and continue the execution of the distributed state estimation or event inference protocol. As in Chap. 9, we focus for simplicity on current-state estimation, though extensions to event inference and delayed/initial-state estimation can also be made. In particular, by taking advantage of the fact that detectability, fault diagnosis, and opacity properties can be formulated in terms of state isolation problems (possibly for an extended version of the given finite automaton), we will be able to discuss the verification of such properties for specific protocols in the distributed observation setting of this chapter.

10.3 Distributed Information Processing

Let us start first with the important special case of a distributed observation architecture with a coordinator or fusion center shown in Fig. 10.1. This observation architecture has the following features, many of which resemble the decentralized architecture we studied in Chap. 9.

- We are given a monolithic system, modeled as an LNFA $G = (Q, \Sigma, \delta, Q_0)$ and m observation sites, such that each site O_i, $i = 1, 2, \ldots, m$, observes a subset of events Σ_{o_i}, $\Sigma_{o_i} \subseteq \Sigma$, under a natural projection mapping $P_{\Sigma_{o_i}}$.
- Following an unknown sequence of events $s \in \Sigma^*$ that occurs in the system (i.e., $s \in L(G)$), site O_i observes the sequence of observations $\omega_i = P_{\Sigma_{o_i}}(s)$. Each observation site can report to the coordinator certain information, perhaps after some local processing of the sequence of observations (via the processing element C_i). The coordinator is then tasked with fusing the information in order to estimate the state of the system or make an inference about the occurrence of an event.

The key difference of the distributed architecture with a coordinator in Fig. 10.1, compared to the decentralized architectures we studied in Chap. 9, is that the coordinator sends information back to the observation sites. The task shares some of the complexities we saw in Chap. 9 (notably what information to send to the coordinator, when to send it, and how to process the information at the coordinator). An additional challenge, however, is to determine when/what information is sent back

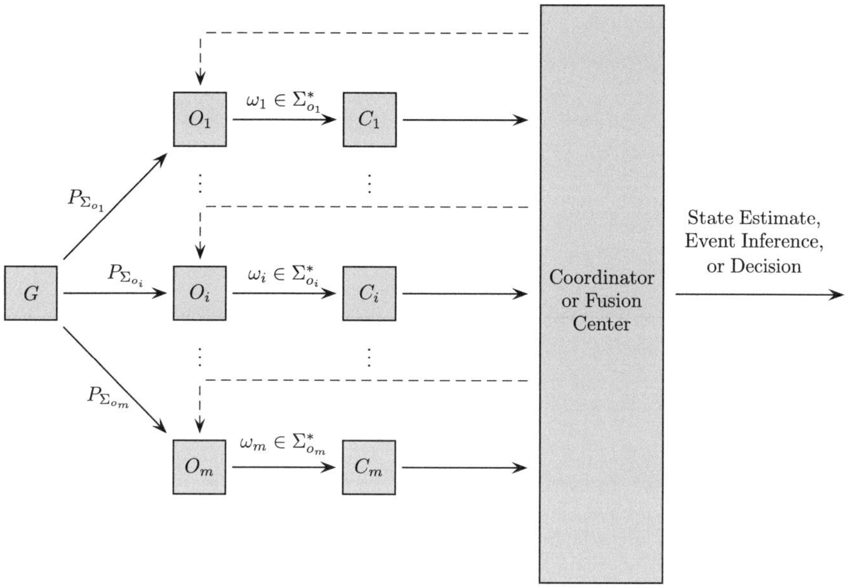

Fig. 10.1 Distributed architecture with a coordinator or fusion center: the coordinator (who is in charge of forming the overall estimate, inference, or decision) receives locally processed information from the observation sites and sends back to them fused information

from the coordinator to the observation sites, and how the observation sites utilize this information. Once these choices are made, we could assume that a synchronization occurs when the coordinator requests information from the observation sites, receives and processes this information, and eventually sends back to the observation sites fused information. This approach would be very similar to the decentralized approaches in Chap. 9, but more flexible,[1] because it allows the coordinator to send back to the observation sites additional information after each synchronization. Of course, to fully specify this approach, one needs to determine when/what information to send back from the coordinator to the observation sites, and to define how the observation sites will utilize this information. We discuss such direct extensions of the decentralized strategies in Chap. 9 to the distributed setting with a coordinator in Sect. 10.5 of this chapter.

An example of a more general distributed observation architecture is shown in Fig. 10.2, where for simplicity we have not drawn the underlying monolithic system G and have focused on the digraph that describes the communication topology among the observation sites. In this case, there is no coordinator but each observation site is able to communicate with different subsets of the observation sites. In the particular example in Fig. 10.2, we have four observation sites, O_1, O_2, O_3, and O_4, with different communication capabilities. For example, observation site O_1 can receive

[1] We will see (also via Example 10.2) later in this chapter that this renders the distributed approach more powerful than the decentralized one.

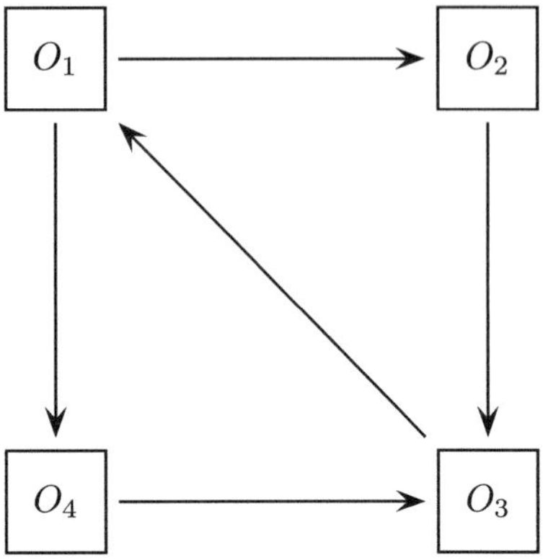

Fig. 10.2 Communication topology of a distributed observation architecture with four observation sites, O_1, O_2, O_3, and O_4

information from O_3 ($\mathcal{N}_1^- = \{O_3\}$) and can send information to O_2 and O_4 ($\mathcal{N}_1^+ = \{O_2, O_4\}$); observation site O_2 receives information from O_1 ($\mathcal{N}_2^- = \{O_1\}$) and sends information to O_3 ($\mathcal{N}_2^+ = \{O_3\}$); and so forth. Note that the interconnection topology is strongly connected, since any observation site can send information, either directly or indirectly, to any other observation site (for instance, information from O_4 can reach O_2 via the path (O_3, O_4), (O_1, O_3), and (O_2, O_1)).

Note that in the case of a distributed observation architecture (as the one in Fig. 10.2) a synchronization becomes more complicated in the sense that it does not necessarily involve all observation sites. More specifically, since a coordinator is not present, each observation site can act as a local coordinator, which means that the synchronization will involve exchange of information among the in-neighbors and out-neighbors associated with the observation site that initiates the synchronization. There are several ways in which this can be done; for instance, if O_1 in Fig. 10.2 initiates a synchronization, then we can imagine that this requires its in-neighbors (i.e., O_3 in this case) to send information to O_1, observation site O_1 to process this information (fuse it with its own information), and then send the fused information to the out-neighbors of O_1 (i.e., O_2 and O_4 in this case). Of course, this is just one way of designing a synchronization step, and many other possibilities exist. Some examples of synchronization strategies are discussed in Sect. 10.4 of this chapter.

In principle, one could imagine numerous ways in which the various decentralized strategies studied in Chap. 9 can be extended to a distributed setting (like the one shown in Fig. 10.2); however, the details and the tracking of different information at different sites can become tedious. For this reason, we focus in Sect. 10.6 on the description of a very intuitive distributed protocol, referred to as DiSIR (Keroglou and Hadjicostis 2018), that relies on set intersection refinements at each synchronization.

10.4 Synchronization Strategies

As in the case of decentralized state estimation and event inference, an important aspect of a distributed protocol is to specify *when* to initiate the sending of information from one or more observation sites to the coordinator (when a coordinator is present) or to one or more other (neighboring) observation sites (when a coordinator is not present); we refer to such an event as a *synchronization* event. A synchronization event can be initiated by an observation site, based on the sequence of events it observes. In this section, we describe the general form that a synchronization event can take, as well as ways for each observation site to determine when to initiate a synchronization. We discuss specific choices for synchronization strategies (in terms of the type of information exchanged and the timing of synchronizations) in later sections.

Presence of a coordinator: When a coordinator is present, a synchronization event in the distributed setting resembles a synchronization event in the decentralized setting. More specifically, once one or more observation sites initiate a synchronization, they inform the coordinator, who subsequently requests and receives information from *all* observation sites; the coordinator then fuses the information received with its own information. The main difference from the decentralized setting is that there is an additional step in the end, in which the coordinator sends the fused information back to all observation sites, who subsequently fuse it with their own information. To summarize, when a coordinator is present, a synchronization initiated by observation site O_i involves the following steps.

Synchronization steps in the presence of a coordinator:

1. Observation site O_i informs the coordinator that a synchronization needs to be initiated.
2. The coordinator requests and receives information from *all* observation sites.
3. The coordinator fuses the information it receives with its own information.
4. The coordinator sends the fused information to *all* observation sites.
5. Each observation site fuses the information it receives from the coordinator with its own information.

Clearly, regardless of which site (or sites) initiate a synchronization, the end result will be identical. Thus, in the presence of a coordinator we will simply use *sync* to denote any synchronization.

Remark 10.3 As in the case of decentralized protocols for state estimation and event inference, we can consider three different types of information sent/received to/from the coordinator: (i) the information sent by each observation site O_i is the sequence of observations that has been observed at O_i since the last synchronization step (Case I); (ii) the information sent by each observation site O_i is the local set of state estimates $\hat{Q}_i$, $\hat{Q}_i \subseteq Q$, that is consistent based on the observations (including *sync* operations and any information sent by the coordinator) seen at observation site O_i thus far (Case II); (iii) the information sent by each observation site O_i is the local decision (Case III), that again is consistent with the observations seen at observation site O_i thus far. □

Absence of a coordinator: When a coordinator is *not* present, different synchronizations will generally involve different subsets of observation sites, depending on which site initiates the synchronization. To distinguish between synchronization events, we will denote by $sync_i$ or $sync_{\{i\}}$ a synchronization initiated by observation site O_i. More generally, we will denote by $sync_M$, where $M \subseteq \mathbb{M}$, synchronizations that are initiated *simultaneously* by observation sites in the set $\{O_i \mid i \in M\}$. There are many ways to execute a synchronization that is initiated by observation site O_i, and we describe some of these choices in the remainder of this section.

Synchronization steps in the absence of a coordinator:
If we assume that O_i has a way of requesting[2] information from its in-neighbors (observation sites in the set $\mathcal{N}_i^-$), then a synchronization initiated by O_i may involve the following steps:

1. Observation site O_i requests and receives information from observation sites in the set $\mathcal{N}_i^-$.
2. Observation site O_i fuses the information it receives from its in-neighbors with its own information.
3. Observation site O_i sends the fused information to observation sites in the set $\mathcal{N}_i^+$.
4. Each observation site in the set $\mathcal{N}_i^+$ fuses the information it receives from O_i (and possibly other observation sites) with its own information.

If, on the other hand, we assume that O_i has *no* way of requesting information from its in-neighbors, then a synchronization initiated by O_i may involve the following steps:

1. Observation site O_i sends its information to observation sites in the set $\mathcal{N}_i^+$.
2. Each observation site in the set $\mathcal{N}_i^+$ fuses the information it receives from O_i (and possibly other observation sites) with its own information.

In both cases above (regardless of whether observation sites are able to request information from their in-neighbors), we assume that the above steps are executed in a synchronized manner (even when multiple observation sites simultaneously initiate a synchronization).

Note that the picture simplifies a bit if we assume that communication is bidirectional (i.e., the communication topology is described by an *undirected* graph). In such case, the sets of in- and out-neighbors for each observation site O_i are identical, i.e., the set of neighbors satisfies $\mathcal{N}_i^- = \mathcal{N}_i^+ =: \mathcal{N}_i$. Under a bidirectional communication topology, a synchronization initiated by O_i may involve the following steps:

[2]In a directed communication topology, a node cannot necessarily communicate with its in-neighbors (not unless they are also out-neighbors); however, there are many applications where a node might be able to send some simple signal to in-neighboring nodes (to request information from them). We also discuss synchronization steps in the case when it is not possible for a node to request information from in-neighboring nodes.

1. Observation site O_i requests and receives information from observation sites in the set $\mathcal{N}_i$.
2. Observation site O_i fuses the information it receives from its neighbors with its own information.
3. Observation site O_i sends the fused information to its neighbors.
4. Each observation site in the set $\mathcal{N}_i$ fuses the information it receives from O_i (and possibly other observation sites) with its own information.

Remark 10.4 As suggested by the discussion above, one should also take into account the possibility that *multiple* observation sites initiate a synchronization *simultaneously*: for example, it is possible that following an event that is observable at multiple observation sites, a subset M, $M \subseteq \mathbb{M}$, of sites initiate a synchronization. We will assume that when observation sites in the set M initiate a synchronization, the steps above are executed in a synchronous manner for all sites (e.g., all sites simultaneously request information from their corresponding in-neighbors; all sites simultaneously receive and fuse information; and so forth). Note that a synchronization that is initiated by all sites in the set M is denoted with $sync_M$. □

Apart from specifying the steps of a synchronization event, we also need to discuss *when* a synchronization event is initiated by each observation site O_i. When a coordinator is present, we can use a synchronizing automaton at each observation site O_i to define synchronization strategy; in this case, the synchronizing automata can be chosen as in the decentralized case (each time the synchronizing automaton of an observation site reaches a marked state, that site informs the coordinator who then initiates a synchronization—see the discussion in Sect. 9.9 of Chap. 9). When a coordinator is *not* present, the synchronizing automata described below can be used to define the synchronization strategy (each synchronization will involve a subset of neighboring observation sites).

Definition 10.1 (*Distributed Synchronization Strategy*) Suppose that LNFA $G = (Q, \Sigma, \delta, Q_0)$ is observed at m observation sites $O_1, O_2, \ldots$, and O_m, with observable events $\Sigma_{o_1}, \Sigma_{o_2}, \ldots$, and Σ_{o_m}, respectively. Each observation site O_i is associated with a *synchronizing automaton*, i.e., a marked deterministic finite automaton (DFA) $S_i = (Q_{s,i}, \Sigma_{o_i} \cup Sync_i, \delta_{s,i}, q_{0s,i}, Q_{ms,i})$ that starts from a specific initial state $q_{0s,i} \in Q_{s,i}$ and is driven by events in $\Sigma_{o_i} \cup Sync_i$, where $Sync_i$ captures the set of synchronization events that involve observation site O_i. Based on the observations (and synchronization operations) seen at site O_i, the state of the synchronizing automaton S_i is updated: whenever S_i reaches a state in the set of marked states $Q_{ms,i}$ ($Q_{ms,i} \subset Q_{s,i}$), then O_i initiates a synchronization. A synchronization initiated by observation site O_i is an event denoted as $sync_i$.

Note that the set of synchronization events that involve observation site O_i, denoted above by $Sync_i$, depends on the steps followed by a synchronization event. For instance, if we assume that each observation site O_i has a way of requesting information from its in-neighbors (observation sites in the set $\mathcal{N}_i^-$), then O_i is involved in synchronization steps that are initiated by one or more observation sites in the set

$$SyncSet_i = \{i\} \cup \{j \mid O_j \in \mathcal{N}_i^-\} \cup \{l \mid O_l \in \mathcal{N}_i^+\} ,$$

and

$$Sync_i = \{sync_M \mid M \subseteq SyncSet_i\} .$$

If we assume instead that each observation site O_i has no way of requesting information from its in-neighbors (observation sites in the set $\mathcal{N}_i^-$), then

$$DirSyncSet_i = \{i\} \cup \{j \mid O_j \in \mathcal{N}_i^-\}$$

(since observation site O_i participates in synchronization steps initiated by observation sites that are its in-neighbors) and

$$Sync_i = \{sync_M \mid M \subseteq DirSyncSet_i\} .$$

Regardless of which type of $Sync_i$ events we consider, we can always partition the $Sync_i$ events into two sets, $ASync_i$ and $ISync_i$, depending on whether or not observation site O_i is directly involved in the initiation of the synchronization step:

$$\begin{aligned} ASync_i &= \{Sync_M \in Sync_i \mid i \in M\} , \\ ISync_i &= \{Sync_M \in Sync_i \mid i \notin M\} . \end{aligned}$$

It is worth pointing out that the set M of observation sites that simultaneously issue a request is highly dependent on the relationships between the various sets of observable events $\Sigma_{o_1}, \Sigma_{o_2}, \ldots, \Sigma_{o_m}$. For example, if $\Sigma_{o_1} \cap \Sigma_{o_2} = \emptyset$, then it is not possible for observation sites O_1 and O_2 to issue a simultaneous request; even if the intersection is non-empty, it is not clear that a common event will cause (under some behavior in the system) simultaneous synchronizations to be initiated by both observation sites.

Analogously to Chap. 9, we require that the only possible events from each state in $Q_{ms,i}$ are events in $ASync_i$, and that each such event results in an unmarked state (in the set $Q_{s,i} \setminus Q_{ms,i}$). Another reasonable assumption about the synchronizing automaton S_i is to require that there are no cycles of locally observable events that do not involve a $sync_i$ event (or, equivalently, a marked state); this requirement guarantees that observation site O_i cannot observe a potentially unbounded number of events before a synchronization is initiated. Finally, we assume that $q_{0s,i} \notin Q_{ms,i}$.

Example 10.1 The simplest approach would be to have each observation site initiate a synchronization based on the number of events it has observed since the last synchronization it was involved in (or since the last synchronization it initiated). For example, we could have each observation site O_i count the number of events it has observed since the last synchronization it initiated, and initiate a synchronization when the count on the number of events reaches a certain constant k_i. In such case, one obtains the synchronizing automaton shown in Fig. 10.3, which resembles some of the features of the synchronizing automaton in Example 9.9 in Chap. 9. The main difference is that we now need to consider multiple different synchronization events.

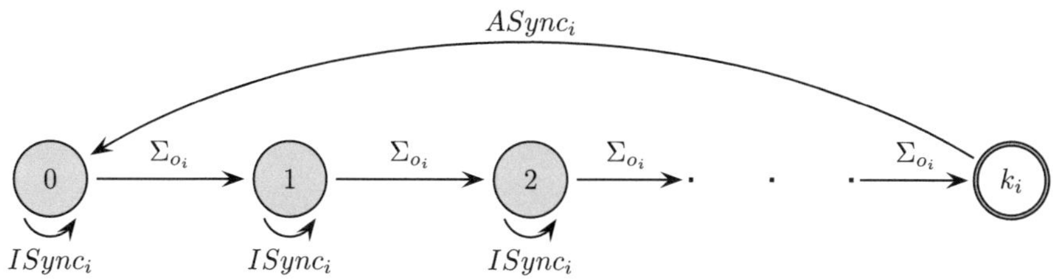

Fig. 10.3 Synchronizing automaton for a periodic synchronization strategy at observation site O_i. The notation Σ_{o_i} (or $ASync_i$ or $ISync_i$) on transitions captures the fact that the transition is taken under all events in Σ_{o_i} (or $ASync_i$ or $ISync_i$)

The DFA $S_i = (Q_{s,i}, \Sigma_{o_i} \cup Sync_i, \delta_{s,i}, q_{0s,i}, Q_{ms,i})$ shown in Fig. 10.3 has set of states $Q_{s,i} = \{0, 1, 2, \ldots, k_i\}$, initial state $q_{0s,i} = 0$, and state k_i as the only marked state ($Q_{ms,i} = \{k_i\}$). The mapping $\delta_{s,i}$ is defined as follows: for all $j \in Q_{s,i} \backslash \{k_i\}$, we have

$$\delta_{s,i}(j, \sigma_{o_i}) = j + 1, \text{ if } j \neq k_i \text{ }, \sigma_{o_i} \in \Sigma_{o_i} \text{ },$$
$$\delta_{s,i}(j, isync) = j, \text{ if } j \neq k_i \text{ }, isync \in ISync_i \text{ }.$$

whereas for the marked state k_i, we have $\delta_{s,i}(k_i, async) = 0$ for all $async \in ASync_i$. □

10.5 Distributed Protocols with a Coordinator

Naturally, all challenges discussed for decentralized observation architectures in Chap. 9 also surface in the case of distributed architectures with a coordinator (as in Fig. 10.1). We can again consider three separate cases that differ based on the type of information sent by each observation site to the coordinator, and the type of information that is sent back from the coordinator to the observation sites, when a synchronization is initiated. Note that in this case, the result of a synchronization that occurs at a particular point in time will be identical, regardless of which observation site (sites) initiates (initiate) the synchronization.

Case I: In this case, the information sent by each observation site O_i is the sequence of observations that have been observed at O_i since the last synchronization step. It should be clear that, in such case, the coordinator obtains the same information as the information it obtained in the decentralized observation setting with a coordinator. The only difference is that the coordinator now sends back information to the observation sites. Regardless of what information is sent back from the coordinator,[3] the information that will be received at the coordinator at subsequent

[3]This information could be, for instance, the state estimates obtained by the coordinator or the totally ordered sequences that match the partially ordered sequences that have been reported at the coordinator by the various observation sites.

synchronization steps remains identical to the information sent in Case I decentralized protocols in Chap. 9 (namely, the sequence of observations seen at each observation site since the last synchronization step). Thus, from the perspective of the coordinator, the two schemes (Case I decentralized observation and Case I distributed observation with a coordinator) will operate in a similar manner (and have identical results and performance). The only difference is that, in the latter case, the observation sites also obtain information about the state estimates maintained at the coordinator (or the matching partially ordered sequences of observations, depending on the type of information sent back from the coordinator).

Case II: In this case, the information sent by each observation site O_i at the κth synchronization is the local set of state estimates $\hat{Q}_i^{(\kappa)}$, $\hat{Q}_i^{(\kappa)} \subseteq Q$, that is consistent based on the observations seen at observation site O_i, as well as any other information received at observation site O_i, up to the κth synchronization. In Case II distributed protocols with a coordinator, the coordinator sends back refined information to the observation sites, namely the set of states $\hat{Q}_c^{(\kappa)} = \cap_{i=1}^m \hat{Q}_i^{(\kappa)}$, which represents the consensus reached by all observation sites at that particular synchronization point. Each observation site O_i subsequently uses this refined set of estimates (note that $\hat{Q}_c^{(\kappa)} \subseteq \hat{Q}_i^{(\kappa)}$) and continues updating its set of state estimates based on the set of events it observes next. Clearly, when the next synchronization occurs, the fact that the coordinator sent back refined information at the end of the previous synchronization may allow one or more observation sites to obtain a more precise set of state estimates; this implies that Case II distributed state estimation with a coordinator will generally result in a refinement of the estimates of Case II decentralized protocol. Case II distributed state estimation is discussed in more detail later in this section.

Case III: In this case, the information sent by each observation site O_i is the local decision, such as "fault," "no fault," or "uncertain," in the case of fault diagnosis. As in Case I, this information received at the observation sites will not influence[4] their subsequent operation (in terms of the information they sent to the coordinator); thus, Case III distributed protocols with a coordinator will perform identically to Case III decentralized protocols.

In the remainder of this section, we focus on the operation of Case II distributed state estimation. We first describe the run-time execution of the protocol and then discuss how to verify that certain properties of interest (such as diagnosability and detectability) hold when using this protocol.

[4]In fault diagnosis applications, we may have situations where one observation site is certain about the "fault" (or "no fault") condition; in such scenarios, a synchronization in Case III distributed state estimation (but not in Case III decentralized state estimation) will enable other observation sites that may be "uncertain" to realize the presence (or absence) of a fault. Clearly, this would influence their subsequent operation (e.g., they only need to retain all estimates associated with the given condition), but this is irrelevant in the sense that the decision ("fault" or "no fault") can already be made.

10.5.1 Run-Time Execution of Case II Distributed Protocol with a Coordinator

In this section we focus, without loss of generality, on the run-time execution of Case II distributed protocol for state estimation. [Recall that the task of event inference can be reduced to a state estimation task via an appropriate transformation of the underlying system. For example, one can use the techniques described in Sect. 7.2.2 of Chap. 7 to reduce the fault diagnosis problem to a state isolation problem.]

As in the case of decentralized protocols, we can break the sequence of events s that occurs in the system ($s \in L(G)$) into subsequences of events $s^{(1)}, s^{(2)}, \ldots, s^{(k)}$, so that

$$s = s^{(1)} s^{(2)} \ldots s^{(k)}$$

and synchronizations (initiated by one or more observation sites, and subsequently by the coordinator) occur immediately after the occurrence of the sequence of events $s^{(\kappa)}$, $\kappa = 1, 2, \ldots, k$. As argued earlier in this chapter, the presence of a coordinator implies that any synchronization, regardless of how it is initiated, involves all observation sites; thus, we do not need to distinguish between synchronizations initiated by different observation sites. If we use $sync^{(\kappa)}$ to denote a synchronization that is initiated after the occurrence of the subsequence of events $s^{(\kappa)}$, we can denote the sequence of events and synchronizations as

$$s^{(1)} sync^{(1)} s^{(2)} sync^{(2)} \ldots sync^{(k-1)} s^{(k)} sync^{(k)} .$$

For Case II distributed protocol, the information sent by each observation site O_i is the local set of current state estimates that is consistent based on the observations seen at observation site O_i thus far (up to the initiation of the κth synchronization), *including* any refined information received from the coordinator. We can recursively track the operation of the protocol as follows:

- At the *initiation* of the first synchronization, we have

  $$\hat{Q}_i^{(1)} = R_i(Q_0, \omega_i^{(1)}) \ , i = 1, 2, \ldots, m \ ,$$

  where $\omega_i^{(1)} = P_i(s^{(1)})$. In other words, each site locally obtains the set of state estimates consistent with what it has observed thus far.
- Subsequently, the coordinator takes the intersection of the state estimates reported by the local sites, sets its own estimate to

  $$\hat{Q}_c^{(1)} = \cap_{i=1}^m \hat{Q}_i^{(1)} \ ,$$

  and reports this refined set of state estimates to the observation sites.
- The observation sites then continue operation from this refined set of state estimates by setting their set of state estimates at the *completion* of the first synchronization) to be

$$\hat{Q}_{si}^{(1)} = UR_i(\hat{Q}_c^{(1)}) = R_i(\hat{Q}_c^{(1)}, \epsilon) \ , i = 1, 2, \ldots, m \ ,$$

where the unobservable reach is taken with respect to the set of events that are not observable at O_i. The important difference from Case II decentralized protocol of Chap. 9 is that the set $\hat{Q}_{si}^{(1)}$ is not necessarily the same as $\hat{Q}_i^{(1)}$ (in fact, it is not hard to argue that $\hat{Q}_{si}^{(1)} \subseteq \hat{Q}_i^{(1)}$, see below).

- At the next synchronization, each observation site O_i observes $\omega_i^{(2)} = P_i(s^{(2)})$ and we have

$$\begin{aligned}
\hat{Q}_i^{(2)} &= R_i(\hat{Q}_{si}^{(1)}, \omega_i^{(2)}) \ , i = 1, 2, \ldots, m \ , \\
\hat{Q}_c^{(2)} &= \cap_{i=1}^m \hat{Q}_i^{(2)} \ , \\
\hat{Q}_{si}^{(2)} &= UR_i(\hat{Q}_c^{(2)}) = R_i(\hat{Q}_c^{(2)}, \epsilon) \ , i = 1, 2, \ldots, m \ .
\end{aligned}$$

- This process continues recursively, so that at synchronization κ, $\kappa = 1, 2, 3, 4, \ldots, k$, we have

$$\begin{aligned}
\hat{Q}_i^{(\kappa)} &= R_i(\hat{Q}_{si}^{(\kappa-1)}, \omega_i^{(\kappa)}) \ , i = 1, 2, \ldots, m \ , \\
\hat{Q}_c^{(\kappa)} &= \cap_{i=1}^m \hat{Q}_i^{(\kappa)} \ , \\
\hat{Q}_{si}^{(\kappa)} &= UR_i(\hat{Q}_c^{(\kappa)}) = R_i(\hat{Q}_c^{(\kappa)}, \epsilon) \ , i = 1, 2, \ldots, m \ ,
\end{aligned}$$

where we have $\omega_i^{(\kappa)} = P_i(s^{(\kappa)})$ and we can take $\hat{Q}_{si}^{(0)} = UR_i(Q_0)$ (without any effect on the protocol we can also take $\hat{Q}_{si}^{(0)} = Q_0$).

The important difference from Case II decentralized protocol in Chap. 9 is that the set $\hat{Q}_{si}^{(\kappa)}$ is not necessarily the same as $\hat{Q}_i^{(\kappa)}$ for $\kappa = 1, 2, 3, \ldots, k$. In fact, it is not hard to establish that

$$\hat{Q}_{si}^{(\kappa)} \subseteq \hat{Q}_i^{(\kappa)} \ , i = 1, 2, \ldots, m \ .$$

To see this, notice that $\hat{Q}_i^{(\kappa)}$ necessarily satisfies $\hat{Q}_i^{(\kappa)} = UR_i(\hat{Q}_i^{(\kappa)})$. Furthermore, $\hat{Q}_c^{(\kappa)}$ necessarily satisfies $\hat{Q}_c^{(\kappa)} \subseteq \hat{Q}_i^{(\kappa)}$. Thus,

$$\hat{Q}_{si}^{(\kappa)} = UR_i(\hat{Q}_c^{(\kappa)}) \subseteq UR_i(\hat{Q}_i^{(\kappa)}) = \hat{Q}_i^{(\kappa)} \ .$$

In other words, at the completion of each synchronization step, the set of state estimates at each observation site may be refined (due to the information sent by the coordinator).

Example 10.2 In this example, we consider LNFA $G = (Q, \Sigma, \delta, Q_0)$ in Fig. 10.4, with states $Q = \{0, 1, 2, 3, 4, 5, 6, 7\}$, events $\Sigma = \{a, b, c, d, e, f\}$, next-state transition function δ as shown in the figure, and $Q_0 = \{0\}$. We assume that there are three observation sites ($\mathbb{M} = \{1, 2, 3\}$), namely O_1, O_2, and O_3 with $\Sigma_{o_1} = \{a, b\}$, $\Sigma_{o_2} = \{b, c\}$, and $\Sigma_{o_3} = \{b, d\}$, respectively. Event f is a fault event and is

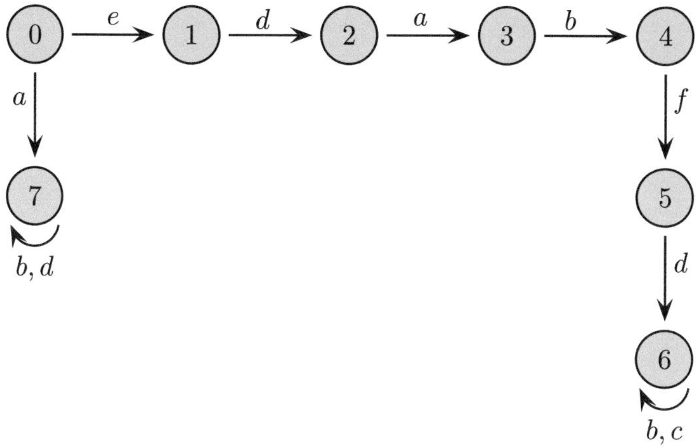

Fig. 10.4 Labeled nondeterministic finite automaton G in Example 10.2

unobservable to all observation sites. Event e is also unobservable to all observation sites, but it is not considered a fault event.

Case III Decentralized Diagnosis. Let us consider first the case where each observation site operates in isolation. In such case, it is not hard to realize that no observation site will be able to diagnose the fault. We can check this by considering the sequence $s = edabfdb^n$ (one of the sequences that contains a fault) and by tracking the local information at each observation site. More specifically, following sequence $s = edabfdb^n$, we can conclude the following:

- Observation site O_1 observes $P_1(s) = abb^n$ and its local set of state estimates are $\{6F, 7N\}$ (at least for $n \geq 1$), which is indeterminate and does not allow it to detect the fault.
- Observation site O_2 observes $P_2(s) = bb^n$ and its local set of state estimates is $\{6F, 7N\}$ (at least for $n \geq 1$), which is indeterminate and does not allow it to determine the fault.
- Observation site O_3 observes $P_3(s) = dbdb^n$ and its local set of state estimates is $\{6F, 7N\}$, which is indeterminate and does not allow it to determine the fault.

Therefore, the fault f cannot be diagnosed using Case III decentralized state estimation.

Case II Decentralized Diagnosis. Before we consider Case II distributed state estimation, let us also consider, for illustration purposes, Case II decentralized state estimation (considered in Chap. 9). Let us assume that each observation site O_i, $i \in \mathbb{M}$, initiates a synchronization (by informing the coordinator) each time it observes an event. Below we track the execution of the Case II decentralized state estimation protocol following the sequence of events $s = edabfdb^n$.

1. When e occurs all local sites do not observe anything and their state estimates remain the same as at initialization, i.e., $\hat{Q}_i^{(0)} = UR_i(Q_0)$:

$$\hat{Q}_1^{(0)} = \{0N, 1N, 2N\}$$
$$\hat{Q}_2^{(0)} = \{0N, 1N, 2N, 3N, 7N\}$$
$$\hat{Q}_3^{(0)} = \{0N, 1N, 7N\} .$$

2. When d occurs, this is observed at O_3, which updates its estimates (the other estimates do not change), and initiates a synchronization, i.e.,

$$\hat{Q}_1^{(1)} = \{0N, 1N, 2N\}$$
$$\hat{Q}_2^{(1)} = \{0N, 1N, 2N, 3N, 7N\}$$
$$\hat{Q}_3^{(1)} = \{2N, 3N, 7N\} .$$

When the coordinator takes the intersection, it obtains $\hat{Q}_c^{(1)} = \{2N\}$.
3. When a occurs, this is observed at O_1, which updates its estimates (the other estimates do not change), and initiates a synchronization, i.e.,

$$\hat{Q}_1^{(2)} = \{3N, 7N\}$$
$$\hat{Q}_2^{(2)} = \{0N, 1N, 2N, 3N, 7N\}$$
$$\hat{Q}_3^{(2)} = \{2N, 3N, 7N\} .$$

When the coordinator takes the intersection, it obtains $\hat{Q}_c^{(2)} = \{3N, 7N\}$.
4. When b occurs, this is observed at all sites, which update their estimates and initiate a synchronization, i.e.,

$$\hat{Q}_1^{(3)} = \{4N, 5F, 6F, 7N\}$$
$$\hat{Q}_2^{(3)} = \{4N, 5F, 6F, 7N\}$$
$$\hat{Q}_3^{(3)} = \{4N, 5F, 7N\} .$$

When the coordinator takes intersection, it obtains $\hat{Q}_c^{(3)} = \{4N, 5F, 7N\}$.
5. When f occurs all local sites do not observe anything and their state estimates do not change.
6. When d occurs, this is observed at O_3, which updates its estimates (the other estimates do not change), and initiates a synchronization, i.e.,

$$\hat{Q}_1^{(4)} = \{4N, 5F, 6F, 7N\}$$
$$\hat{Q}_2^{(4)} = \{4N, 5F, 6F, 7N\}$$
$$\hat{Q}_3^{(4)} = \{6F, 7N\} .$$

When the coordinator takes intersection, it obtains $\hat{Q}_c^{(4)} = \{6F, 7N\}$.
7. When b occurs, this is observed at all sites, which update their estimates and initiate a synchronization, i.e.,

$$\hat{Q}_1^{(5)} = \{6F, 7N\}$$
$$\hat{Q}_2^{(5)} = \{6F, 7N\}$$
$$\hat{Q}_3^{(5)} = \{6F, 7N\} .$$

When the coordinator takes intersection, it obtains $\hat{Q}_c^{(5)} = \{6F, 7N\}$.

8. Subsequent occurrences of b will not change anything. The coordinator will have as its set of state estimates $\{6F, 7N\}$, which does not allow the detection of the fault event.

Therefore, fault f cannot be detected using Case II decentralized fault diagnosis.

Case II Distributed Diagnosis with a Coordinator. Below we track the execution of the Case II distributed state estimation protocol, and argue that it allows the detection of the fault. We again consider the sequence of events $s = edabfdb^n$.

1. When e occurs, all local sites do not observe anything and their state estimates remain the same as at initialization, i.e., $\hat{Q}_{si}^{(0)} = UR_i(Q_0)$:

$$\hat{Q}_{s1}^{(0)} = \{0N, 1N, 2N\}$$
$$\hat{Q}_{s2}^{(0)} = \{0N, 1N, 2N, 3N, 7N\}$$
$$\hat{Q}_{s3}^{(0)} = \{0N, 1N, 7N\} .$$

2. When d occurs, this is observed at O_3, which updates its estimates (the other estimates do not change), i.e.,

$$\hat{Q}_1^{(1)} = \{0N, 1N, 2N\}$$
$$\hat{Q}_2^{(1)} = \{0N, 1N, 2N, 3N, 7N\}$$
$$\hat{Q}_3^{(1)} = \{2N, 3N, 7N\} .$$

Site O_3 then initiates a synchronization; when the coordinator takes the intersection, it obtains $\hat{Q}_c^{(1)} = \{2N\}$; the coordinator then sends this information back to the observation sites so that

$$\hat{Q}_{s1}^{(1)} = UR_1(\hat{Q}_c^{(1)}) = \{2N\}$$
$$\hat{Q}_{s2}^{(1)} = UR_2(\hat{Q}_c^{(1)}) = \{2N, 3N\}$$
$$\hat{Q}_{s3}^{(1)} = UR_3(\hat{Q}_c^{(1)}) = \{2N, 3N\} .$$

3. When a occurs, this is observed at O_1, which updates its estimates (the other estimates do not change), i.e.,

$$\hat{Q}_1^{(2)} = \{3N\}$$
$$\hat{Q}_2^{(2)} = \{2N, 3N\}$$
$$\hat{Q}_3^{(2)} = \{2N, 3N\} \ .$$

Site O_1 then initiates a synchronization; when the coordinator takes the intersection, it obtains $\hat{Q}_c^{(2)} = \{3N\}$; the coordinator then sends this information back to the observation sites so that

$$\hat{Q}_{s1}^{(2)} = UR_1(\hat{Q}_c^{(2)}) = \{3N\}$$
$$\hat{Q}_{s2}^{(2)} = UR_2(\hat{Q}_c^{(2)}) = \{3N\}$$
$$\hat{Q}_{s3}^{(2)} = UR_3(\hat{Q}_c^{(2)}) = \{3N\} \ .$$

4. When b occurs, this is observed at all sites, which update their estimates and initiate a synchronization, i.e.,

$$\hat{Q}_1^{(3)} = \{4N, 5F, 6F\}$$
$$\hat{Q}_2^{(3)} = \{4N, 5F, 6F\}$$
$$\hat{Q}_3^{(3)} = \{4N, 5F\} \ .$$

When the coordinator takes intersection, it obtains $\hat{Q}_c^{(3)} = \{4N, 5F\}$; the coordinator then sends this information back to the observation sites so that

$$\hat{Q}_{s1}^{(3)} = UR_1(\hat{Q}_c^{(3)}) = \{4N, 5F, 6F\}$$
$$\hat{Q}_{s2}^{(3)} = UR_2(\hat{Q}_c^{(3)}) = \{4N, 5F, 6F\}$$
$$\hat{Q}_{s3}^{(3)} = UR_3(\hat{Q}_c^{(3)}) = \{4N, 5F\} \ .$$

5. When f occurs, all local sites do not observe anything and their state estimates do not change.
6. When d occurs, this is observed at O_3, which updates its estimates (the other estimates do not change), i.e.,

$$\hat{Q}_1^{(4)} = \{4N, 5F, 6F\}$$
$$\hat{Q}_2^{(4)} = \{4N, 5F, 6F\}$$
$$\hat{Q}_3^{(4)} = \{6F\} \ .$$

Site O_3 then initiates a synchronization; when the coordinator takes intersection, it obtains $\hat{Q}_c^{(4)} = \{6F\}$ and diagnoses the fault. In fact, when the coordinator sends this information back to the observation sites, all observation sites obtain

$$\hat{Q}_{s1}^{(4)} = UR_1(\hat{Q}_c^{(4)}) = \{6F\}$$
$$\hat{Q}_{s2}^{(4)} = UR_2(\hat{Q}_c^{(4)}) = \{6F\}$$
$$\hat{Q}_{s3}^{(4)} = UR_3(\hat{Q}_c^{(4)}) = \{6F\} ,$$

and also can diagnose the fault.

Therefore, we conclude[5] that fault f can be detected using Case II distributed fault diagnosis. □

10.5.2 *Verification of Case II Distributed Diagnosability with a Coordinator*

The verification of Case II distributed state estimation with a coordinator is very much related to the techniques we described in Chap. 9 for Case II decentralized state estimation with a coordinator. As explained earlier, the main difference is that, following a synchronization, the coordinator not only uses the sets of state estimates provided by the observation sites to obtain a refined set of state estimates, but also sends back to the observation sites this refined set of state estimates. For simplicity, we limit our discussion on the verification of the property of diagnosability, but these discussions can easily be extended toward the verification of detectability or other state isolation properties. We assume that a synchronization at a particular point in time is initiated by one or more observation sites based on the state of their corresponding synchronizing automata (refer to Sect. 10.4 of this chapter). Note that in the presence of a coordinator we do not need to distinguish between synchronizations initiated by different observation sites.

Below, we formally describe the problem setting.

Problem Setting (Case II Distributed Diagnosability with a Coordinator): We are given an LNFA $G = (Q, \Sigma, \delta, Q_0)$, which is observed at m observation sites O_1, $O_2, \ldots$, and O_m, with observable events $\Sigma_{o_1}, \Sigma_{o_2}, \ldots$, and Σ_{o_m}, respectively. Each observation site O_i is associated with a synchronizing marked DFA $S_i = (Q_{s,i}, \Sigma_{o_i} \cup \{sync\}, \delta_{s,i}, q_{0s,i}, Q_{ms,i})$ as described in Definition 10.1. Whenever S_i reaches a state in the set of marked states $Q_{ms,i}$ ($Q_{ms,i} \subset Q_{s,i}$), then O_i signals the coordinator to initiate a synchronization and the $sync$ event occurs (involving all observation sites), i.e., the coordinator requests from each observation site their information, namely pairs of a local state estimate and associated label(s) (Case II). We are interested in verifying *Case II distributed diagnosability*, i.e., verifying whether the coordinator will be able to determine (perhaps after some finite delay) the occurrence of events in the class of fault events F, which are assumed to be unobservable at all observation sites, i.e., $F \subseteq \Sigma_{uo}$ where $\Sigma_{uo} = \Sigma \backslash (\cup_{i=1}^{m} \Sigma_{o_i})$. We say that Case II distributed diagnosability holds for the given system G (under the given set of observation sites

[5]Technically, in order to reach this conclusion, we need to also verify that the fault will be detected for all other executions that contain a fault (i.e., any execution of the form $edabfd(b+c)^*$).

and synchronization strategy) if the occurrence of any fault event $f \in F$ eventually gets determined at the coordinator after a finite number of event occurrences.

The basic construction that is used to verify Case II distributed diagnosability is based on a parallel-like structure that tracks, for each possible sequence of events, the state in which the various components of the overall system are. More specifically, we can use a parallel composition to simultaneously track, for each possible sequence of events in the system, the following: (i) the state(s) of (an enhanced version of) the given system, (ii) the set of state estimates at each observation site as captured by an (enhanced version of an) extended diagnoser, (iii) the state of each local synchronizing DFA S_i and whether a *sync* operation needs to take place, (iv) the state of the coordinator, which is the intersection of the sets of state estimates at each observation site after the last synchronization. In this construction, care will be taken to properly update the state of the coordinator and all observation sites following a synchronization event. Note that the result of this parallel composition is a nondeterministic system as it inherits non-determinism from G.

The main difference from Case II decentralized diagnosis is that when a synchronization takes place, the state estimate at each observation site becomes identical to the state of the coordinator (i.e., observation sites resume operation from this set of state estimates). Note that this new state is not necessarily a state in the local diagnoser D_i (recall that a local diagnoser D_i is obtained for each observation site based on the observable events Σ_{o_i} at that site using the construction in Sect. 7.3.1 in Chap. 7). The reason is that, following a synchronization, the intersection of the sets of state estimates at the various observation sites (i.e., the new set of state estimates that is adopted at each observation site) is not necessarily a set of state estimates that exists in the local diagnoser.[6] Thus, we need an extended local diagnoser D_{ei} that involves all possible sets of state estimates as its states. The extended local diagnoser for a given LNFA G, with respect to a generic set of observable events Σ_o and a single fault class F, is defined below.

Definition 10.2 (*Extended Diagnoser for Distributed Diagnosis*) Consider an LNFA $G = (Q, \Sigma, \delta, Q_0)$ with a set of fault events F ($F \subseteq \Sigma$) and a set of observable events Σ_o ($\Sigma_o \subseteq \Sigma$). Assume, without loss of generality, that $F \cap \Sigma_o = \emptyset$ and let $G_F = (Q \dot{\cup} Q_F, \Sigma, \delta_F, Q_0)$ denote the enhanced version of the system, constructed as described in Sect. 7.2.2 of Chap. 7. The extended diagnoser

$$D_{ed} = (Q_{ed}, \Sigma_o, \delta_{ed}, q_{0ed})$$

is a DFA defined as follows:

- The set of states contains *all* subsets of $Q \dot{\cup} Q_F$ (recall that Q_F contains the states in Q with label F), i.e., $Q_{ed} = 2^{Q \dot{\cup} Q_F}$.
- The set of events is given by the set of observable events Σ_o of system G_F.

[6] In other words, there might not necessarily exist a sequence of events s that generates a projection $P_i(s)$ which drives the local diagnoser D_i to this set of state estimates.

- The next-state transition mapping δ_{ed} is defined for each $q_{ed} \in Q_{ed}$ and each $\sigma_o \in \Sigma_o$ as
$$\delta_{ed}(q_{ed}, \sigma_o) = R_{\Sigma_o}(q_{ed}, \sigma_o) ,$$
where reachability is taken with respect to the set of observable events Σ_o in system G_F.
- The initial state is given by $q_{0ed} = UR_{\Sigma_o}(Q_0)$, i.e., it is the set of states that are reachable in G_F from states in Q_0 without generating any observations in the set Σ_o.

[Note: Typically, the state that corresponds to the empty subset of $Q\dot{\cup}Q_F$ (inconsistent state) is removed (along with all transitions leading to it).]

Remark 10.5 The main difference of the extended diagnoser from the diagnoser in Chap. 7 is that it includes as states all possible subsets of $Q\dot{\cup}Q_F$ (in other words, it has exactly $2^{|Q\dot{\cup}Q_F|}$ states). Many of these states may not be reachable from the initial state q_{0ed}; such states are not present in the standard diagnoser but are needed here in case a local observation site is driven to such state after a synchronization event. □

In more detail, to verify Case II distributed diagnosability we can do the following:

1. We construct the enhanced version of the system G_e, as described in Chap. 9 (this enhanced system is an LNFA that resembles G_F, as described in Sect. 7.2.2 of Chap. 7, and has $sync$ as an additional event that acts as a self-transition at each state). In other words, $G_e = (Q\dot{\cup}Q_F, \Sigma \cup \{sync\}, \delta_e, Q_0)$, where $\delta_e(q, sync) = \{q\}$ for all $q \in Q\dot{\cup}Q_F$ and $\delta_e(q, \sigma) = \delta_F(q, \sigma)$ for all $q \in Q\dot{\cup}Q_F$ and $\sigma \in \Sigma$ (note that $\delta_e(q, \sigma)$ is empty/undefined if $\delta_F(q, \sigma)$ is empty/undefined).
2. We construct the extended local diagnoser $D_{ed,i} = (Q_{ed,i}, \Sigma_{o_i}, \delta_{ed,i}, q_{0ed,i})$ at observation site O_i, as described above in Definition 10.2 using Σ_{o_i} as the set of observable events. We then construct an enhanced extended local diagnoser $D_{ee,i} = (Q_{ed,i}, \Sigma_{o_i} \cup \{sync\}, \delta_{ee,i}, q_{0ed,i})$, enhanced so that the $sync$ event is a self-transition at each state, i.e., $\delta_{ee,i}(q_{ed,i}, sync) = q_{ed,i}$ for all $q_{ed,i} \in Q_{ed,i}$ and $\delta_{ee,i}(q_{ed,i}, \sigma_{o_i}) = \delta_{ed,i}(q_{ed,i}, \sigma_{o_i})$ for all $q_{ed,i} \in Q_{ed,i}$ and all $\sigma_{o_i} \in \Sigma_{o_i}$ (note that $\delta_{ee,i}(q_{ed,i}, \sigma_{o,i})$ is undefined if $\delta_{ed,i}(q_{ed,i}, \sigma_{o,i})$ is undefined).
3. We construct the product DS_i of the enhanced local extended diagnoser $D_{ee,i}$ and the synchronizing DFA S_i at observation site O_i, i.e., $DS_i = D_{ee,i} \times S_i$, for $i = 1, 2, \ldots, m$.
4. We construct the parallel composition $GDS = G_e||DS_1||DS_2|| \cdots ||DS_m$. We mark states in this parallel composition that have a marked second component in the state of the DS_i component (for some $i = 1, 2, \ldots, m$). These marked states are states in which a synchronization has been initiated; thus, we remove all transitions other than the one associated with the $sync$ transition from marked states in this parallel composition.
5. In the resulting parallel composition GDS, each state is of the form $(q, x_1, x_2, \ldots, x_m)$ and belongs in $(Q\dot{\cup}Q_F) \times (Q_{ee,1} \times Q_{s,1}) \times \cdots \times (Q_{ee,m} \times Q_{s,m})$. If we let

$x_i = (\hat{q}_i, q_{s_i})$ be the state of the ith DS component, then the state estimate at the coordinator (and subsequently at all observation sites) changes each time we have a $sync$ operation from a marked state. This state is given by the intersection $\bigcap_{i=1}^{m} \hat{q}_i$.

We can thus enhance the parallel composition GDS with this additional information regarding the $sync$ operations. The coordinator is initialized to $\hat{q}_0 = R_{\mathbb{M}}(Q_0, \epsilon)$. We can follow each transition in GDS and update the state of the coordinator and the observation sites, if a synchronization ($sync$) event takes place. Note that it is possible that $(q, x_1, x_2, \ldots, x_m)$ could be associated with different information at the coordinator (e.g., if this state can be reached via two different sequences of events, s and s' that cause different synchronizations and result in different information at the coordinator); this, however, can easily be handled by introducing multiple copies of this particular state, each associated with the corresponding coordinator state.

The formal construction of the enhanced version of GDS, which we denote by GDS_m, is described below for verifying Case II distributed diagnosability (notice that this construction also removes any parts of GDS that are not reachable from its initial state). As in the algorithm for $GSDS_m$ in Chap. 9, the algorithm generates the set of states of GDS_m by sequentially exploring the states of GDS, using a set of states called $UNXES$ to denote states that have not been unexplored yet.

Inputs:
1. $GDS = (Q_g, \Sigma \cup \{sync\}, \delta_g, Q_{0,g})$
2. $G_e = (Q\dot{\cup}Q_F, \Sigma \cup \{sync\}, \delta_e, Q_0)$
3. $\Sigma_{o_i}, i \in \mathbb{M}$

Output:
$GDS_m = (Q_m, \Sigma \cup \{sync\}, \delta_m, Q_{0,m})$ where $Q_m \subseteq Q_g \times 2^{Q\dot{\cup}Q_F}$

Set $Q_{0,m} := Q_{0,g} \times \{\hat{q}_0\}$ where $\hat{q}_0 = R_{\mathbb{M}}(Q_0, \epsilon)$
Set $Q_m := \emptyset$; set $UNXES := \{Q_{0,m}\}$
While $UNXES \neq \emptyset$
  For each state $q_m = ((q, x_1, x_2, \ldots, x_m), \hat{q})$ in $UNXES$
    For each $\sigma \in \Sigma$
      for each $(q', x_1', x_2', \ldots, x_m') \in \delta_q((q, x_1, x_2, \ldots, x_m), \sigma)$
        (i.e., the nondeterministic δ_g is non-empty)
        Set $q_m' \in \delta_m(q_m, \sigma)$ where $q_m' = ((q', x_1', x_2', \ldots, x_m'), \hat{q})$
          (i.e., no change in $\hat{q}$)
        If $q_m' \notin Q_m$ and $q_m' \notin UNXES$
          $UNXES := UNXES \cup \{q_m'\}$
          END
        END
      END
    For each $(q', x_1', x_2', \ldots, x_m') \in \delta_g((q, x_1, x_2, \ldots, x_m), sync)$
      (i.e., the nondeterministic δ_g is defined for $sync$)
      Set $q_m' \in \delta_m(q_m, \sigma)$ where $q_m' = ((q', x_1', x_2', \ldots, x_m'), \hat{q}')$

$q'_m = ((q', (UR_1(\hat{q}'), s'_1), (UR_2(\hat{q}'), s'_2), \ldots, (UR_m(\hat{q}'), s'_m)), \hat{q}')$
with s'_i being the second component of x'_i and $\hat{q}' = \cap_{i=1}^{m} \hat{q}_i$
where $\hat{q}_i$ is the first component of x_i
If $q'_m \notin Q_m$ and $q'_m \notin UNXES$
$UNXES := UNXES \cup \{q'_m\}$
END
END

Set $UNXES := UNXES \backslash \{q_m\}$ (i.e., remove q_m from $UNXES$)
Set $Q_m := Q_m \cup \{q_m\}$ (i.e., add q_m to Q_m)
END
END

Notice that the final construction (obtained in Step 5 of the above process) has a finite number of states since each component (and the coordinator) can assume states in a finite set (notice that it will be a nondeterministic construction due to the non-determinism of G). Moreover, given any sequence of events s that occurs in G, we can use the above construction to track the state estimates at the various observation sites and the coordinator. In fact, we can identify in the above construction indeterminate states, i.e., states in which we have uncertainty at the coordinator (note that uncertainty in the coordinator necessarily implies uncertainty at all observation sites). For Case II distributed diagnosability, these indeterminate states are associated with a coordinator state $\hat{q}$ that is neither a subset of Q nor a subset of Q_F. If we can find "indeterminate cycles" in the final construction (i.e., cycles of states in which we have uncertainty at the coordinator), and these cycles can manifest themselves after the occurrence of a fault, then we conclude that Case II distributed diagnosability with a coordinator does not hold for the given system and observation setup. On the other hand, if we cannot find such cycles, then we can conclude that Case II distributed diagnosability with a coordinator holds for the given setting.

Remark 10.6 As in the case of decentralized diagnosability, the state complexity of GDS_m above can be bounded by

$$|Q\dot{\cup}Q_F| \times 2^{|Q\dot{\cup}Q_F|} \times \Pi_{i=1}^{m}(2^{|Q\dot{\cup}Q_F|} \times |Q_{s,i}|) .$$

If we let $S_{\max} = \max_{i \in \mathbb{M}} |Q_{s,i}|$ be the maximum size of the state space of a synchronizing automaton, then the state complexity of the resulting parallel composition in the above approach could be (in the worst case) $2|Q| \times 4^{|Q|} \times \left(4^{|Q|} \times S_{\max}\right)^m$.

Note that it is possible to reduce this complexity, by using verifier-like techniques to track the uncertainty at each observation site and at the coordinator (Keroglou and Hadjicostis 2018); in such case the complexity reduces to

$$|Q\dot{\cup}Q_F| \times |Q\dot{\cup}Q_F|^2 \times \Pi_{i=1}^{m}(|Q\dot{\cup}Q_F|^2 \times |Q_{s,i}|) ,$$

which simplifies to $2|Q| \times \left((2|Q|)^2 \times S_{\max}\right)^m \times (2|Q|)^2$. □

10.6 Distributed Protocols Without a Coordinator

All challenges discussed for distributed protocols with a coordinator described in the previous section also surface in the case of distributed protocols without a coordinator. Recall that in the case of distributed protocols without a coordinator, the communication topology between observation sites, denoted by a digraph $\mathcal{G}_d = (\mathcal{V}, \mathcal{E})$ (where $\mathcal{V} = \{O_1, O_2, \ldots, O_m\}$ is the set of observation sites), plays a key role in the execution and properties of the protocol (an example of a digraph can be seen in Fig. 10.2). The reason is that the result of a synchronization depends on which observation site (sites) initiates (initiate) the synchronization, and how this observation site is connected with other observation sites.

We can again consider separate cases that differ based on (i) the type of information sent by each observation site to other (neighboring) observation sites, (ii) the way information is fused, and (iii) the type of information that is sent back (if any), when a synchronization is initiated. For simplicity, we limit our discussion in this section to the case where observation site O_i has *no* way of requesting information from its in-neighbors. More specifically, when a synchronization is initiated by observation site O_i, denoted by $sync_i$, the following two steps are executed:

1. Observation site O_i sends its information to observation sites in the set $\mathcal{N}_i^+$.
2. Each observation site in the set $\mathcal{N}_i^+$ fuses the information it receives from O_i with its own information.

Though not discussed explicitly, many of the techniques we develop extend easily to the cases where (i) observation site O_i can also request information from its in-neighbors when it initiates a synchronization or (ii) the underlying communication topology between observation sites is bidirectional (i.e., $\mathcal{N}_i^+ = \mathcal{N}_i^- =: \mathcal{N}_i$ for each observation site O_i).

One can certainly obtain many different distributed protocols for state estimation and event inference, depending on the type of information that is sent from the observation site that initiates the synchronization and the way this information is fused at the receiving observation site. For brevity, we focus on Case II distributed state estimation, which is one of the most intuitive approaches.

In Case II distributed estimation, when observation site O_i initiates a synchronization, it sends to all of its out-neighbors in $\mathcal{N}_i^+$ the local set of state estimates $\hat{Q}_i$, $\hat{Q}_i \subseteq Q$, that is consistent based on the observations seen at observation site O_i, as well as any other information received at observation site O_i, up to that synchronization. Clearly, this information needs to be incorporated in the set of state estimates maintained by the out-neighbors of node O_i. More specifically, for each $O_l \in \mathcal{N}_i^+$, we have

$$\hat{Q}_{sl} = UR_l(\hat{Q}_l \cap \hat{Q}_i) ,$$

where $\hat{Q}_l$ ($\hat{Q}_i$) is the set of state estimates at observation site O_l (O_i) at the beginning of the synchronization, $\hat{Q}_{sl}$ is the set of state estimates at observation site O_l at the end of the synchronization, and the unobservable reach is taken with respect to the events that are not observable at site O_l.

The situation could become slightly more complicated if multiple observation sites simultaneously initiate a synchronization. In such case, if we use $\mathcal{S}$, $\mathcal{S} \subseteq \mathcal{V}$ (where $\mathcal{V} = \{O_1, O_2, \ldots, O_m\}$) to denote the observation sites that are simultaneously initiating a synchronization, then we have

$$\hat{Q}_{sl} = UR_l(\hat{Q}_l \cap_{i:O_i \in (\mathcal{S} \cap \mathcal{N}_l^-)} \hat{Q}_i) .$$

In fact, we can write the following for each observation site O_i:

$$\hat{Q}_{si} = UR_i(\hat{Q}_i \cap_{j:O_j \in (\mathcal{S} \cap \mathcal{N}_i^-)} \hat{Q}_j) , \tag{10.1}$$

which also covers the case when no synchronization affects observation site O_i (in which situation, we have $\hat{Q}_{si} = UR_i(\hat{Q}_i) = \hat{Q}_i$).

In the remainder of this section, we focus on the operation of Case II distributed state estimation without a coordinator: we first describe the run-time execution of the protocol and then discuss how it can be used to verify properties of interest (such as diagnosability and detectability).

10.6.1 Run-Time Execution of Case II Distributed Protocol Without a Coordinator

As in Case II distributed protocol with a coordinator in the previous section, we focus (without loss of generality) to Case II distributed protocol for state estimation. We can break the sequence of events s that occurs in the system ($s \in L(G)$) into subsequences of events $s^{(1)}, s^{(2)}, \ldots, s^{(k)}$, so that

$$s = s^{(1)}s^{(2)} \ldots s^{(k)}$$

and synchronizations (initiated by one or more observation sites) occur immediately after the occurrence of the sequence of events $s^{(\kappa)}$, $\kappa = 1, 2, \ldots, k$. As argued earlier in this chapter, we need to distinguish between synchronizations initiated by different observation sites or, more generally, different subsets of observation sites. If we assume that the synchronization occurring immediately after $s^{(\kappa)}$ involves observation sites in the set $\mathcal{S}_\kappa$, $\mathcal{S}_\kappa \subseteq \mathcal{V}$ (where $\mathcal{V} = \{O_1, O_2, \ldots, O_m\}$), we can denote the sequence of events and synchronizations as

$$s^{(1)}sync_{\mathcal{S}_1}s^{(2)}sync_{\mathcal{S}_2} \ldots sync_{\mathcal{S}_{k-1}}s^{(k)}sync_{\mathcal{S}_k} .$$

We can recursively track the operation of the protocol as follows:

- At the *initiation* of the first synchronization, we have

$$\hat{Q}_i^{(1)} = R_i(Q_0, \omega_i^{(1)}) \ , i = 1, 2, \ldots, m \ ,$$

where $\omega_i^{(1)} = P_i(s^{(1)})$. In other words, each site locally obtains the set of state estimates consistent with what it has observed thus far.

- Subsequently, the observation sites in $\mathcal{S}_1$ send to their out-neighbors their set of state estimates, and all observation sites update their state estimates. More specifically, observation site O_i sets its set of state estimates at the end of the first synchronization to

$$\hat{Q}_{si}^{(1)} = UR_i(\hat{Q}_i^{(1)} \cap_{j:O_j \in (\mathcal{S}_1 \cap \mathcal{N}_i^-)} \hat{Q}_j^{(1)}) \ ,$$

where $(\mathcal{S}_1 \cap \mathcal{N}_i^-)$ is the set of in-neighbors of observation site O_i that are initiating a synchronization, and the unobservable reach it taken with respect to the events unobservable to observation site O_i.
The important difference from Case II distributed protocol with a coordinator is that the updates of the sets $\hat{Q}_{si}^{(1)}, i = 1, 2, \ldots, m$, are not synchronized and do not rely on the same information.

- At the next synchronization, each observation site O_i observes $\omega_i^{(2)} = P_i(s^{(2)})$ and we have

$$\hat{Q}_i^{(2)} = R_i(\hat{Q}_{si}^{(1)}, \omega_i^{(2)}) \ , i = 1, 2, \ldots, m \ ,$$
$$\hat{Q}_{si}^{(2)} = UR_i(\hat{Q}_i^{(2)} \cap_{j:O_j \in (\mathcal{S}_2 \cap \mathcal{N}_i^-)} \hat{Q}_j^{(2)}) \ , i = 1, 2, \ldots, m \ .$$

- This process continues recursively, so that at synchronization κ, $\kappa = 1, 2, 3, 4, \ldots, k$, we have

$$\hat{Q}_i^{(\kappa)} = R_i(\hat{Q}_{si}^{(\kappa-1)}, \omega_i^{(\kappa)}) \ , i = 1, 2, \ldots, m \ ,$$
$$\hat{Q}_{si}^{(\kappa)} = UR_i(\hat{Q}_i^{(\kappa)} \cap_{j:O_j \in (\mathcal{S}_\kappa \cap \mathcal{N}_i^-)} \hat{Q}_j^{(\kappa)}) \ , i = 1, 2, \ldots, m \ ,$$

where we have $\omega_i^{(\kappa)} = P_i(s^{(\kappa)})$ and we can take $\hat{Q}_{si}^{(0)} = UR_i(Q_0)$ (without any effect on the protocol we can also take $\hat{Q}_{si}^{(0)} = Q_0$).

The important difference from Case II decentralized protocol in Chap. 9 is that the set $\hat{Q}_{si}^{(\kappa)}$ is not necessarily the same as $\hat{Q}_i^{(\kappa)}$ for $\kappa = 1, 2, 3, \ldots, k$. In fact, it is not hard to establish that, for all $\kappa = 1, 2, \ldots, k$, we have

$$\hat{Q}_{si}^{(\kappa)} \subseteq \hat{Q}_i^{(\kappa)} \ , i = 1, 2, \ldots, m \ .$$

To see this, notice that $\hat{Q}_i^{(\kappa)}$ necessarily satisfies $\hat{Q}_i^{(\kappa)} = UR_i(\hat{Q}_i^{(\kappa)})$, whereas $\hat{Q}_{ci}^{(\kappa)} := \hat{Q}_i^{(\kappa)} \cap_{j:O_j \in (\mathcal{S}_\kappa \cap \mathcal{N}_i^-)} \hat{Q}_j^{(\kappa)}$ clearly satisfies $\hat{Q}_{ci}^{(\kappa)} \subseteq \hat{Q}_i^{(\kappa)}$. Thus,

$$\hat{Q}_{si}^{(\kappa)} = UR_i(\hat{Q}_{ci}^{(\kappa)}) \subseteq UR_i(\hat{Q}_i^{(\kappa)}) = \hat{Q}_i^{(\kappa)} .$$

This implies that, at the completion of each synchronization step, the set of state estimates at each observation site may be refined (due to the information sent by the observation sites that are initiating a synchronization). This refinement will likely influence subsequent estimates at each observation site.

Example 10.3 In this example, we consider again LNFA $G = (Q, \Sigma, \delta, Q_0)$ in Fig. 10.4, with states $Q = \{0, 1, 2, 3, 4, 5, 6, 7\}$, events $\Sigma = \{a, b, c, d, e, f\}$, next-state transition function δ as shown in the figure, and $Q_0 = \{0\}$. We consider three observation sites ($\mathbb{M} = \{1, 2, 3\}$), namely O_1, O_2, and O_3 with $\Sigma_{o_1} = \{a, b\}$, $\Sigma_{o_2} = \{b, c\}$, and $\Sigma_{o_3} = \{b, d\}$, respectively. Recall that event f is a fault event and is unobservable to all observation sites; event e is also unobservable to all observation sites, but it is not considered a fault event.

We will consider Case II distributed diagnosis under the two different communication topologies shown in Fig. 10.5, both of which can be thought of as undirected line graphs:

1. Topology G_1 includes edges: (O_1, O_2), (O_2, O_1), (O_3, O_2), (O_2, O_3) (this corresponds to a line graph with O_2 in the middle).
2. Topology G_2 includes edges: (O_1, O_2), (O_2, O_1), (O_3, O_1), (O_1, O_3) (this corresponds to a line graph with O_1 in the middle).

Case II Distributed Diagnosis under Topology G_1. Below we track the execution of the Case II distributed state estimation protocol under topology G_1, assuming that each observation site initiates a synchronization each time it observes an event. We consider the sequence of events $s = edabfdb^n$.

1. When e occurs, all local sites do not observe anything and their state estimates remain the same as at initialization, i.e., $\hat{Q}_{si}^{(0)} = UR_i(Q_0)$:

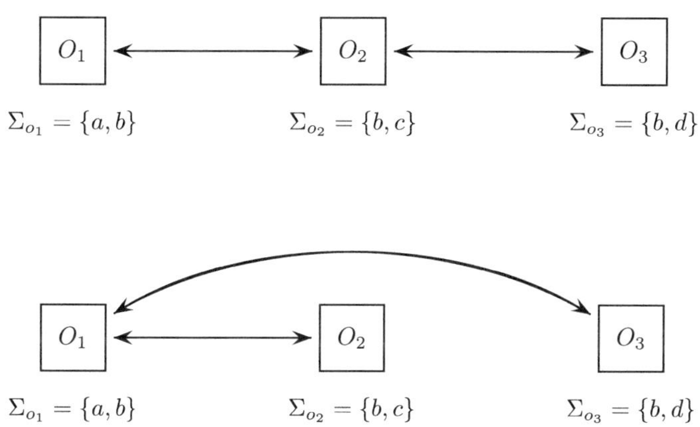

Fig. 10.5 Two communication topologies for a distributed observation architecture with three observation sites, O_1, O_2, and O_3 in Example 10.3: topology G_1 (top) and topology G_2 (bottom)

$$\hat{Q}_{s1}^{(0)} = \{0N, 1N, 2N\}$$
$$\hat{Q}_{s2}^{(0)} = \{0N, 1N, 2N, 3N, 7N\}$$
$$\hat{Q}_{s3}^{(0)} = \{0N, 1N, 7N\} .$$

2. When d occurs, this is observed at O_3, which updates its estimates (the other estimates do not change), i.e.,

$$\hat{Q}_{1}^{(1)} = \{0N, 1N, 2N\}$$
$$\hat{Q}_{2}^{(1)} = \{0N, 1N, 2N, 3N, 7N\}$$
$$\hat{Q}_{3}^{(1)} = \{2N, 3N, 7N\} .$$

Site O_3 then initiates a synchronization and sends its state estimates to its out-neighbors, namely site O_2. Observation site O_2 then updates its set of state estimates (the other observation sites perform no updates), so that

$$\hat{Q}_{s1}^{(1)} = \hat{Q}_{1}^{(1)} = \{0N, 1N, 2N\}$$
$$\hat{Q}_{s2}^{(1)} = UR_2(\hat{Q}_{2}^{(1)} \cap \hat{Q}_{3}^{(1)}) = \{2N, 3N, 7N\}$$
$$\hat{Q}_{s3}^{(1)} = \hat{Q}_{3}^{(1)} = \{2N, 3N, 7N\} .$$

3. When a occurs, this is observed at O_1, which updates its estimates (the other estimates do not change), i.e.,

$$\hat{Q}_{1}^{(2)} = \{3N, 7N\}$$
$$\hat{Q}_{2}^{(2)} = \{2N, 3N, 7N\}$$
$$\hat{Q}_{3}^{(2)} = \{2N, 3N, 7N\} .$$

Site O_1 then initiates a synchronization and sends its state estimates to its out-neighbors, namely site O_2. Observation site O_2 then updates its set of state estimates (the other observation sites perform no updates), so that

$$\hat{Q}_{s1}^{(2)} = \hat{Q}_{1}^{(2)} = \{3N, 7N\}$$
$$\hat{Q}_{s2}^{(2)} = UR_2(Q_{2}^{(2)} \cap Q_{1}^{(2)}) = \{3N, 7N\}$$
$$\hat{Q}_{s3}^{(2)} = \hat{Q}_{3}^{(2)} = \{2N, 3N, 7N\} .$$

4. When b occurs, this is observed at all sites, which update their estimates and initiate a synchronization, i.e.,

$$\hat{Q}_{1}^{(3)} = \{4N, 5F, 6F, 7N\}$$
$$\hat{Q}_{2}^{(3)} = \{4N, 5F, 6F, 7N\}$$
$$\hat{Q}_{3}^{(3)} = \{4N, 5F, 7N\} .$$

Since all observation sites are initiating a synchronization, we have the following updates

$$\begin{aligned}\hat{Q}_{s1}^{(3)} &= UR_1(\hat{Q}_1^{(3)} \cap \hat{Q}_2^{(3)}) = \{4N, 5F, 6F, 7N\} \\ \hat{Q}_{s2}^{(3)} &= UR_2(\hat{Q}_2^{(3)} \cap \hat{Q}_1^{(3)} \cap \hat{Q}_3^{(3)}) = \{4N, 5F, 6F, 7N\} \\ \hat{Q}_{s3}^{(3)} &= UR_3(\hat{Q}_3^{(3)} \cap \hat{Q}_2^{(3)}) = \{4N, 5F, 7N\} .\end{aligned}$$

Note that observation site O_2 takes the intersection of its own set of state estimates with *both* $\hat{Q}_1^{(3)}$ and $\hat{Q}_3^{(3)}$ (since both of them become available to it). Also, observation site O_2 includes $6F$ in its state estimates because this state can be reached from state $5F$ via event d, which is unobservable to O_2.

5. When f occurs, all local sites do not observe anything and their state estimates do not change.
6. When d occurs, this is observed at O_3, which updates its estimates (the other estimates do not change), i.e.,

$$\begin{aligned}\hat{Q}_1^{(4)} &= \{4N, 5F, 6F, 7N\} \\ \hat{Q}_2^{(4)} &= \{4N, 5F, 6F, 7N\} \\ \hat{Q}_3^{(4)} &= \{6F, 7N\} .\end{aligned}$$

Site O_3 then initiates a synchronization and sends its state estimates to its out-neighbors, namely site O_2. Observation site O_2 then updates its set of state estimates (the other observation sites perform no updates), so that

$$\begin{aligned}\hat{Q}_{s1}^{(4)} &= \hat{Q}_1^{(4)} = \{4N, 5F, 6F, 7N\} \\ \hat{Q}_{s2}^{(4)} &= UR_2(\hat{Q}_2^{(4)} \cap \hat{Q}_3^{(4)}) = \{6F, 7N\} \\ \hat{Q}_{s3}^{(4)} &= \hat{Q}_3^{(4)} = \{6F, 7N\} .\end{aligned}$$

7. When b occurs, this is observed at all sites, which update their estimates and initiate a synchronization, i.e.,

$$\begin{aligned}\hat{Q}_1^{(5)} &= \{6F, 7N\} \\ \hat{Q}_2^{(5)} &= \{6F, 7N\} \\ \hat{Q}_3^{(5)} &= \{6F, 7N\} .\end{aligned}$$

Since all observation sites are initiating a synchronization, we have the following updates

$$\begin{aligned}\hat{Q}_{s1}^{(5)} &= UR_1(\hat{Q}_1^{(5)} \cap \hat{Q}_2^{(5)}) = \{6F, 7N\} \\ \hat{Q}_{s2}^{(5)} &= UR_2(\hat{Q}_2^{(5)} \cap \hat{Q}_1^{(5)} \cap \hat{Q}_3^{(5)}) = \{6F, 7N\} \\ \hat{Q}_{s3}^{(5)} &= UR_3(\hat{Q}_3^{(5)} \cap \hat{Q}_2^{(5)}) = \{6F, 7N\} .\end{aligned}$$

Again, notice that observation site O_2 takes the intersection of its own set of state estimates with *both* $\hat{Q}_1^{(5)}$ and $\hat{Q}_3^{(5)}$ (since both of them become available to it).

Subsequent b do not change anything, and we conclude that fault f cannot be detected using Case II distributed fault diagnosis under communication topology G_1 (all observation sites maintain sets of state estimates that are indeterminate).

Case II Distributed Diagnosis under Topology G_2. Below we track the execution of the Case II distributed state estimation protocol under topology G_2, assuming that each observation site initiates a synchronization each time it observes an event. We again consider the sequence of events $s = edabfdb^n$.

1. When e occurs, all local sites do not observe anything and their state estimates remain the same as at initialization, i.e., $\hat{Q}_{si}^{(0)} = UR_i(Q_0)$:

$$\begin{aligned}\hat{Q}_{s1}^{(0)} &= \{0N, 1N, 2N\}\\ \hat{Q}_{s2}^{(0)} &= \{0N, 1N, 2N, 3N, 7N\}\\ \hat{Q}_{s3}^{(0)} &= \{0N, 1N, 7N\}\,.\end{aligned}$$

2. When d occurs, this is observed at O_3, which updates its estimates (the other estimates do not change), i.e.,

$$\begin{aligned}\hat{Q}_{1}^{(1)} &= \{0N, 1N, 2N\}\\ \hat{Q}_{2}^{(1)} &= \{0N, 1N, 2N, 3N, 7N\}\\ \hat{Q}_{3}^{(1)} &= \{2N, 3N, 7N\}\,.\end{aligned}$$

   Site O_3 then initiates a synchronization and sends its state estimates to its out-neighbors, namely site O_1. Observation site O_1 then updates its set of state estimates (the other observation sites perform no updates), so that

$$\begin{aligned}\hat{Q}_{s1}^{(1)} &= UR_1(\hat{Q}_1^{(1)} \cap \hat{Q}_3^{(1)}) = \{2N\}\\ \hat{Q}_{s2}^{(1)} &= \hat{Q}_2^{(1)} = \{0N, 1N, 2N, 3N, 7N\}\\ \hat{Q}_{s3}^{(1)} &= \hat{Q}_3^{(1)} = \{2N, 3N, 7N\}\,.\end{aligned}$$

3. When a occurs, this is observed at O_1, which updates its estimates (the other estimates do not change), i.e.,

$$\begin{aligned}\hat{Q}_{1}^{(2)} &= \{3N\}\\ \hat{Q}_{2}^{(2)} &= \{0N, 1N, 2N, 3N, 7N\}\\ \hat{Q}_{3}^{(2)} &= \{2N, 3N, 7N\}\,.\end{aligned}$$

   Site O_1 then initiates a synchronization and sends its state estimates to its out-neighbors, namely sites O_2 and O_3. Observation sites O_2 and O_3 then update their sets of state estimates, so that

$$\begin{aligned}
\hat{Q}_{s1}^{(2)} &= \hat{Q}_1^{(2)} = \{3N\} \\
\hat{Q}_{s2}^{(2)} &= UR_2(\hat{Q}_2^{(2)} \cap \hat{Q}_1^{(2)}) = \{3N\} \\
\hat{Q}_{s3}^{(2)} &= UR_3(\hat{Q}_3^{(2)} \cap \hat{Q}_1^{(2)}) = \{3N\} .
\end{aligned}$$

4. When b occurs, this is observed at all sites, which update their estimates and initiate a synchronization, i.e.,

$$\begin{aligned}
\hat{Q}_1^{(3)} &= \{4N, 5F, 6F\} \\
\hat{Q}_2^{(3)} &= \{4N, 5F, 6F\} \\
\hat{Q}_3^{(3)} &= \{4N, 5F\} .
\end{aligned}$$

Since all observation sites are initiating a synchronization, we have the following updates

$$\begin{aligned}
\hat{Q}_{s1}^{(3)} &= UR_1(\hat{Q}_1^{(3)} \cap \hat{Q}_2^{(3)} \cap \hat{Q}_3^{(3)}) = \{4N, 5F, 6F\} \\
\hat{Q}_{s2}^{(3)} &= UR_2(\hat{Q}_2^{(3)} \cap \hat{Q}_1^{(3)}) = \{4N, 5F, 6F\} \\
\hat{Q}_{s3}^{(3)} &= UR_3(\hat{Q}_3^{(3)} \cap \hat{Q}_1^{(3)}) = \{4N, 5F\} .
\end{aligned}$$

Note that observation site O_1 takes the intersection of its own set of state estimates with *both* $\hat{Q}_2^{(3)}$ and $\hat{Q}_3^{(3)}$ (since both of them become available to it); however, observation site O_1 includes $6F$ in its state estimates because this state can be reached from state $5F$ via event d, which is unobservable to O_1.

5. When f occurs, all local sites do not observe anything and their state estimates do not change.
6. When d occurs, this is observed at O_3, which updates its estimates (the other estimates do not change), i.e.,

$$\begin{aligned}
\hat{Q}_1^{(4)} &= \{4N, 5F, 6F\} \\
\hat{Q}_2^{(4)} &= \{4N, 5F, 6F\} \\
\hat{Q}_3^{(4)} &= \{6F\} .
\end{aligned}$$

Site O_3 then initiates a synchronization and sends its state estimates to its out-neighbors, namely site O_1. Observation site O_1 then updates its set of state estimates (the other observation sites perform no updates), so that

$$\begin{aligned}
\hat{Q}_{s1}^{(4)} &= UR_1(\hat{Q}_1^{(4)} \cap \hat{Q}_3^{(4)}) = \{6F\} \\
\hat{Q}_{s2}^{(4)} &= \hat{Q}_2^{(4)} = \{4N, 5F, 6F\} \\
\hat{Q}_{s3}^{(4)} &= \hat{Q}_3^{(4)} = \{6F\} .
\end{aligned}$$

7. When b occurs, this is observed at all sites, which update their estimates and initiate a synchronization, i.e.,

$$\hat{Q}_1^{(5)} = \{6F\}$$
$$\hat{Q}_2^{(5)} = \{6F\}$$
$$\hat{Q}_3^{(5)} = \{6F\} .$$

Since all observation sites are initiating a synchronization, we have the following updates

$$\hat{Q}_{s1}^{(5)} = UR_1(\hat{Q}_1^{(5)} \cap \hat{Q}_2^{(5)} \cap \hat{Q}_3^{(5)}) = \{6F\}$$
$$\hat{Q}_{s2}^{(5)} = UR_2(\hat{Q}_2^{(5)} \cap \hat{Q}_1^{(5)}) = \{6F\}$$
$$\hat{Q}_{s3}^{(5)} = UR_3(\hat{Q}_3^{(5)} \cap \hat{Q}_1^{(5)}) = \{6F\} .$$

At this point, the fault has been diagnosed as all observation sites have state estimates that are associated with the label F (note that observation sites O_1 and O_3 were actually aware of the definite presence of the fault one observation earlier). □

10.6.2 Verification of Case II Distributed Diagnosability Without a Coordinator

We now discuss the verification of Case II distributed state estimation without a coordinator. We limit our discussion on the verification of the property of diagnosability, but these discussions can easily be extended toward the verification of detectability or other state isolation properties. We assume that a synchronization by observation site O_i at a particular point in time (perhaps simultaneously with other observation sites) is based on the state of its corresponding synchronizing automaton (refer to Sect. 10.4 of this chapter). When observation site O_i initiates a synchronization, we assume that observation site O_i sends its information to observation sites in the set $\mathcal{N}_i^+$ and, subsequently, each observation site in the set $\mathcal{N}_i^+$ fuses the information it receives from O_i with its own information (i.e., it takes the intersection of the set of pairs of state estimates and labels (that is provided by O_i) with its own corresponding information).

Below, we formally describe the problem setting.

Problem Setting (Case II Distributed Diagnosability without a Coordinator): We are given an LNFA $G = (Q, \Sigma, \delta, Q_0)$, which is observed at m observation sites O_1, $O_2, \ldots,$ and O_m, with observable events $\Sigma_{o_1}, \Sigma_{o_2}, \ldots,$ and Σ_{o_m}, respectively. Each observation site O_i is associated with a synchronizing marked DFA $S_i = (Q_{s,i}, \Sigma_{o_i} \cup Sync_i, \delta_{s,i}, q_{0s,i}, Q_{ms,i})$ as described in Definition 10.1. Whenever S_i reaches a state in the set of marked states $Q_{ms,i}$ ($Q_{ms,i} \subset Q_{s,i}$), then O_i initiates a synchronization by sending to its out-neighbors (observation sites in the set $\mathcal{N}_i^+$) its information, namely pairs of a local state estimate and associated label(s) (Case II); the observation sites in the set $\mathcal{N}_i^+$ receive this set of pairs of a state estimate and associated label(s) and intersect it with their own information. It is possible that multiple observation sites initiate a synchronization simultaneously, in which case the above actions occur in a synchronized manner. We are interested in verifying *distributed diagnosability*, i.e., verifying whether the observation sites will be able to determine (perhaps after some finite delay) the occurrence of events in the class of fault events F, which are assumed to be unobservable at all observation sites, i.e., $F \subseteq \Sigma_{uo}$ where $\Sigma_{uo} = \Sigma \backslash (\cup_{i=1}^{m} \Sigma_{o_i})$. We say that Case II distributed diagnosability holds for the given system G (under the given set of observation sites and synchronization strategy) if the occurrence of any fault event $f \in F$ eventually gets determined by one or more observation sites after a finite number of event occurrences.

The basic construction that is used to verify Case II distributed diagnosability without a coordinator is based on a parallel-like structure that aims to track, for each possible sequence of events, the state in which the various components of the overall system are. More specifically, we can use a parallel composition to simultaneously track, for each possible sequence of events in the system, the following: (i) the state(s) of (an enhanced version of) the given system, (ii) the set of state estimates at each observation site as captured by an (enhanced version of an) extended diagnoser, (iii) the state of each local synchronizing DFA S_i and whether a $sync_i$ operation needs to take place (i.e., whether observation site O_i needs to initiate a synchronization), (iv) the state of the various observation sites after a synchronization takes place (initiated by a single observation site or simultaneously by a set of observation sites). Note that the result of this parallel composition is a nondeterministic system as it inherits non-determinism from G.

The construction below makes use of the extended local diagnoser D_{ei} in Definition 10.2 that involves all possible sets of state estimates as its states. In more detail, to verify Case II distributed diagnosability without a coordinator, we can do the following:

1. We construct the enhanced version of the system G_e, as described in Chap. 9 (this enhanced system is an LNFA that resembles G_F, as described in Sect. 7.2.2 of Chap. 7, and has the set[7] $Sync = \{sync_S \mid S \subseteq \mathcal{V}\}$ as additional events). In other words, $G_e = (Q \dot{\cup} Q_F, \Sigma \cup Sync, \delta_e, Q_0)$, where $\delta_e(q, sync) = \{q\}$ for all $q \in Q \dot{\cup} Q_F$ and all $sync \in Sync$, and $\delta_e(q, \sigma) = \delta_F(q, \sigma)$ for all $q \in Q \dot{\cup} Q_F$ and $\sigma \in \Sigma$ (note that $\delta_e(q, \sigma)$ is empty/undefined if $\delta_F(q, \sigma)$ is empty/undefined).
2. We construct the extended local diagnoser $D_{ed,i} = (Q_{ed,i}, \Sigma_{o_i}, \delta_{ed,i}, q_{0ed,i})$ at observation site O_i, as described in Definition 10.2 using Σ_{o_i} as the set of observable events. We then construct an enhanced extended local diagnoser $D_{ee,i} = (Q_{ed,i}, \Sigma_{o_i} \cup Sync, \delta_{ee,i}, q_{0ed,i})$, enhanced so that any event $sync$ ($sync \in Sync$) is a self-transition at each state, i.e., $\delta_{ee,i}(q_{ed,i}, sync) = q_{ed,i}$ for all $q_{ed,i} \in Q_{ed,i}$ and all $sync \in Sync$, and $\delta_{ee,i}(q_{ed,i}, \sigma_{o_i}) = \delta_{ed,i}(q_{ed,i}, \sigma_{o_i})$ for all $q_{ed,i} \in Q_{ed,i}$ and all $\sigma_{o_i} \in \Sigma_{o_i}$ (note that $\delta_{ee,i}(q_{ed,i}, \sigma_{o,i})$ is undefined if $\delta_{ed,i}(q_{ed,i}, \sigma_{o,i})$ is undefined).
3. We construct the product DS_i of the enhanced local extended diagnoser $D_{ee,i}$ and the synchronizing DFA S_i at observation site O_i, i.e., $DS_i = D_{ee,i} \times S_i$, for $i = 1, 2, \ldots, m$.

[7] In the set $Sync$, event $sync_i$ is the same as event $sync_{\{i\}}$.

4. We construct the parallel composition $GDS = G_e||DS_1||DS_2|| \cdots ||DS_m$, where each state is of the form $(q, x_1, x_2, \ldots, x_m)$ and belongs in $(Q\dot{\cup}Q_F) \times (Q_{ee,1} \times Q_{s,1}) \times \cdots \times (Q_{ee,m} \times Q_{s,m})$. If we let $x_i = (\hat{q}_i, q_{s_i})$ be the state of the ith DS component, then the set
$$\mathcal{S} = \{O_j \mid q_{s_j} \in Q_{ms,j}\}$$
captures the observation sites that have a marked second state component and are therefore initiating a synchronization operation. We mark states in GDS for which $\mathcal{S} \neq \emptyset$ (because they are states in which a synchronization has been initiated by at least one observation site).
5. For each state q_{gds} of the parallel composition GDS of the form $q_{gds} = (q, x_1, x_2, \ldots, x_m)$ (where $x_i = (\hat{q}_i, q_{s_i})$ is the state of DS_i) that has been *marked*, we do the following:
   a. We remove all transitions out of state q_{gds};
   b. We add a transition associated with the $sync_{\mathcal{S}}$ event (where $\mathcal{S} = \{O_j \mid q_{s_j} \in Q_{ms,j}\}$ is the set of observation sites that are initiating a synchronization at q_{gds}), as follows:
   $$\delta_{gds}(q_{gds}, sync_{\mathcal{S}}) = \{q'_{gds}\} ,$$
   where δ_{gds} is the transition function associated with the parallel composition GDS and state q'_{gds} is of the form $(q', x'_1, x'_2, \ldots, x'_m)$ with $x'_i = (\hat{q}'_i, q'_{s_i})$ being the state of the ith DS component. Following a synchronization $sync_{\mathcal{S}}$ that involves the observation sites in the set $\mathcal{S}$, the state q'_{gds} is defined as follows: $q' = q$ and, for each O_i, $i = 1, 2, \ldots, m$, we have
   $$\begin{aligned} \hat{q}'_i &= UR_i(\hat{q}_i \cap_{j:O_j \in (\mathcal{S} \cap \mathcal{N}_i^-)} \hat{q}_j) , \\ q'_{s,i} &= \delta_{s,i}(q_{s,i}, sync_{\mathcal{S}}) . \end{aligned}$$
   In other words, a synchronization event does not change the state the system ($q' = q$), but updates the states of the synchronizing automata at the various sites as well as the state estimates according to (10.1).

Notice that the GDS construction (obtained in Step 5 of the above process) has a finite number of states since the number of states of the system and each component is finite. Moreover, given any sequence of events s that occurs in G, we can use the above construction to track the state estimates at the various observation sites, as they get updated (including refinements due to intersection operations that occur because of the initiation of synchronizations).

Once we have constructed GDS as described above, we can verify distributed diagnosability for the given system and observation setup by checking for the presence of any loop that involves "indeterminate states": for Case II distributed diagnosability, indeterminate states are states that have uncertainty at *all* observation sites, i.e., the component $x_i = (\hat{q}_i, q_{s_i})$ for each observation site O_i, is associated with a set of state estimates $\hat{q}_i$ that involves states both in Q and in Q_F. If we can find "indeterminate cycles" in the final construction (i.e., cycles of indeterminate states), and these cycles can manifest themselves after the occurrence of a fault, then we conclude that Case II distributed diagnosability does not hold for the given system and observation setup. On the other hand, if we cannot find such cycles, then we can conclude that Case II distributed diagnosability holds for the given setting.

Remark 10.7 The state complexity of GDS above can be bounded by

$$|Q\dot{\cup}Q_F| \times \Pi_{i=1}^{m}(2^{|Q\dot{\cup}Q_F|} \times |Q_{s,i}|) \ .$$

If we let $S_{\max} = \max_{i\in\mathbb{M}} |Q_{s,i}|$ be the maximum size of the state space of a synchronizing automaton, then the state complexity of the resulting parallel composition in the above approach could be (in the worst case) $2|Q| \times \left(4^{|Q|} \times S_{\max}\right)^m$.

Note that it is possible to reduce this complexity, by using verifier-like techniques to track the uncertainty at each observation site and at the coordinator (Keroglou and Hadjicostis 2018); in such case the complexity reduces to

$$|Q\dot{\cup}Q_F| \times \Pi_{i=1}^{m}(|Q\dot{\cup}Q_F|^2 \times |Q_{s,i}|) \ ,$$

which simplifies to $2|Q| \times \left((2|Q|)^2 \times S_{\max}\right)^m$. □

The verification of other types of state isolation properties (such as Case II decentralized detectability) can also be performed using the GDS constructed in this section, since this construction captures all possible state estimates (at each observation site) following the synchronization strategy defined by the synchronizing DFA chosen at each observation site. One issue that arises in the distributed setting is that there is no single entity that has access to *all* information available at the various sites. Thus, the definitions of distributed detectability would have to be re-formulated to properly capture that. For example, for distributed detectability one could require that, almost always, the system state becomes known exactly to at least one observation site.

Remark 10.8 Note that in distributed diagnosability state estimates that involve the fault label F can only get updated with state estimates that retain the F label. Therefore, following a sequence of observations, if a certain observation site, say O_i, has all state estimates associated with the F label, then this observation site (O_i) will necessarily have all state estimates associated with the F label for *all* continuations of this sequence of observations. In fact, once a synchronization is initiated by O_i (which will happen after a finite number of steps), then all out-neighbors of O_i will necessarily have all of their state estimates associated with the F label and, after a finite number of steps, this will also be true for the out-neighbors of the out-neighbors of node O_i. Therefore, if the digraph that describes the communication topology between observation sites is strongly connected, all observation sites will have state estimates that are associated with the F label eventually (after a finite number of steps). □

10.7 Comments and Further Reading

In this chapter, we have considered the problem of distributed state estimation and event inference by focusing on an underlying monolithic system, modeled as an LNFA, that is observed by multiple observation sites with different observation capabilities (i.e., each site has its own set of observable events or, equivalently, each site has its own natural projection mapping). Distributed observation architectures allow each observation site to send its observations, estimates, or decisions to other sites (or to a coordinator if one is present), and to also receive and process similar information from other observation sites (or the coordinator if one is present). We discussed a number of different protocols that can be used for distributed state estimation and event inference, described their implementation at run-time, and the verification of various state isolation properties of interest (such as diagnosability, detectability, and opacity). For example, in order to verify diagnosability using the proposed distributed protocol, one can construct appropriate compositions of (extended versions of) local diagnosers or verifiers. These compositions can capture the refinement of information (under the given constraints in the information exchange and the set intersection operations), as well as the influence of the refinement process on immediate or future diagnosis; verification using extended versions of local diagnosers has exponential complexity, whereas verification using extended versions of verifiers has polynomial complexity (in the number of states of the given system).

Though decentralized observation settings for state estimation and event inference have been studied quite extensively in recent years (see the discussions at the end of Chap. 9), very few works have allowed information to flow between different observation sites. Such examples are the works in Panteli and Hadjicostis (2013), Keroglou and Hadjicostis (2014, 2015, 2018). These works studied Case II distributed state estimation or fault diagnosis. The former set of papers developed set intersection-based strategies that allow observation cites to both send local state estimates to a coordinator and also receive refined state estimates from the coordinator, when a synchronization is initiated; this type of protocols were discussed in Sect. 10.5. The latter set of papers considered a set of observations sites that are interconnected via a communication topology that forms an *undirected* graph and developed set intersection-based strategies that allow observation cites to both send local state estimates to their neighboring observation sites and also receive refined state estimates from them, when a synchronization is initiated; a generalization for this type of protocols to the case where the communication topology among observation cites is not necessarily undirected was discussed in Sect. 10.6.

Other works that have allowed information to flow from one observation site to another include Fabre et al. (2000), Su and Wonham (2005), which focus on modular systems (with a local observer for each module) and explore iterative strategies that involve the exchange of diagnostic information among neighboring observation sites and set intersection operations for refinement. It should be noted that the verification methods described in Sects. 10.5 and 10.6 (which allow us to track the result of set intersection operations in an efficient manner) can be adapted to provide polynomial verification for the decentralized protocols in Debouk et al. (2000) (see Chap. 9) and the distributed protocols in Panteli and Hadjicostis (2013), Athanasopoulou and Hadjicostis (2006), Su and Wonham (2005), Fabre et al. (2000), though we did not attempt to make such connections explicit.

The distributed state estimation and event inference protocols that were described in this chapter also relate to the works in Fabre and Benveniste (2007); Benveniste et al. (2003); Li and Hadjicostis (2007). The work in Fabre and Benveniste (2007) considers a distributed/modular system with several modules, each associated with a local observer/supervisor that only has access to the local observations and the model of the local module. Both Benveniste et al. (2003), Li and Hadjicostis (2007) consider Petri nets and multiple observation sites with a partial order model of time. The work in Benveniste et al. (2003) follows a true-concurrency setting (which is not an issue in the monolithic finite automaton setting we considered in this chapter) and relies on Petri net unfoldings. The work in Li and Hadjicostis (2007) considers the problem of reconstructing the possible transition firing sequences in a given Petri net based on asynchronous and partially ordered observations of token changes at different places of the Petri net.

References

Athanasopoulou E, Hadjicostis CN (2006) Decentralized failure diagnosis in discrete event systems. In: Proceedings of 2006 American control conference (ACC), pp 14–19

Benveniste A, Fabre E, Haar S, Jard C (2003) Diagnosis of asynchronous discrete-event systems: a net unfolding approach. IEEE Trans Autom Control 48(5):714–727

Debouk R, Lafortune S, Teneketzis D (2000) Coordinated decentralized protocols for failure diagnosis of discrete event systems. Discret Event Dyn Syst Theory Appl 10(1–2):33–86

Fabre E, Benveniste A (2007) Partial order techniques for distributed discrete event systems: why you cannot avoid using them. Discret Event Dyn Syst 17(3):355–403

Fabre E, Benveniste A, Jard C, Ricker L, Smith M (2000) Distributed state reconstruction for discrete event systems. In: Proceedings of 39th IEEE conference on decision and control (CDC), vol 3, pp 2252–2257

Keroglou C, Hadjicostis CN (2014) Distributed diagnosis using predetermined synchronization strategies. In: Proceedings of 53rd IEEE conference on decision and control (CDC), pp 5955–5960

Keroglou C, Hadjicostis CN (2015) Distributed diagnosis using predetermined synchronization strategies in the presence of communication constraints. In: Proceedings of IEEE conference on automation science and engineering (CASE), pp 831–836

Keroglou C, Hadjicostis CN (2018) Distributed fault diagnosis in discrete event systems via set intersection refinements. IEEE Trans Autom Control 63(10):3601–3607

Li L, Hadjicostis CN (2007) Reconstruction of transition firing sequences based on asynchronous observations of place token changes. In: Proceedings of 46th IEEE conference on decision and control (CDC), pp 1898–1903

Panteli M, Hadjicostis CN (2013) Intersection based decentralized diagnosis: implementation and verification. In: Proceedings of 52nd IEEE conference on decision and control and european control conference (CDC-ECC), pp 6311–6316

Puri A, Tripakis S, Varaiya P (2002) Problems and examples of decentralized observation and control for discrete event systems. In: Synthesis and control of discrete event systems. Springer, Berlin, pp 37–56

Su R, Wonham WM (2005) Global and local consistencies in distributed fault diagnosis for discrete-event systems. IEEE Trans Autom Control 50(12):1923–1935

Witsenhausen HS (1968) A counterexample in stochastic optimum control. SIAM J Control 6(1):131–147

Index

C. N. Hadjicostis, *Estimation and Inference in Discrete Event Systems*, Communications and Control Engineering,
https://doi.org/10.1007/978-3-030-30821-6

MIX
Papier aus verantwortungsvollen Quellen
Paper from responsible sources
FSC® C105338

If you have any concerns about our products,
you can contact us on
ProductSafety@springernature.com

In case Publisher is established outside the EU,
the EU authorized representative is:
Springer Nature Customer Service Center GmbH
Europaplatz 3, 69115 Heidelberg, Germany

Printed by Libri Plureos GmbH
in Hamburg, Germany